S0-CAH-306

Chemical Reactions on Polymers

ACS SYMPOSIUM SERIES 364

Chemical Reactions on Polymers

Judith L. Benham, EDITOR
3M Company

James F. Kinstle, EDITOR
James River Corporation

Developed from a symposium sponsored
by the Division of Polymer Chemistry, Inc.,
and the Division of Polymeric Materials:
Science and Engineering
at the 192nd Meeting
of the American Chemical Society,
Anaheim, California,
September 7-12, 1986

American Chemical Society, Washington, DC 1988

Library of Congress Cataloging-in-Publication Data

Chemical reactions on polymers/editors, Judith L. Benham, James F. Kinstle.

p. cm.—(ACS symposium series; 364)

"Developed from a symposium sponsored by the Division of Polymer Chemistry, Inc., and the Division of Polymeric Materials: Science and Engineering at the 192nd Meeting of the American Chemical Society, Anaheim, California, September 7-12, 1986."

Bibliographies: p.

Includes indexes.

ISBN 0-8412-1448-4

1. Polymers and polymerization—Congresses. 2. Chemical reactions—Congresses.

I. Benham, Judith L., 1947- . II. Kinstle, James F. III. American Chemical Society. Division of Polymer Chemistry, Inc. IV. American Chemical Society. Division of Polymeric Materials: Science and Engineering. V. American Chemical Society. Meeting (192nd: 1986: Anaheim, Calif.) VI. Series.

QD380.C456 1988
668.9—dc19 87-31913
CIP

PRINTED IN THE UNITED STATES OF AMERICA

ACS Symposium Series

M. Joan Comstock, *Series Editor*

Foreword

The ACS SYMPOSIUM SERIES was founded in 1974 to provide a medium for publishing symposia quickly in book form. The format of the Series parallels that of the continuing ADVANCES IN CHEMISTRY SERIES except that, in order to save time, the papers are not typeset but are reproduced as they are submitted by the authors in camera-ready form. Papers are reviewed under the supervision of the Editors with the assistance of the Series Advisory Board and are selected to maintain the integrity of the symposia; however, verbatim reproductions of previously published papers are not accepted. Both reviews and reports of research are acceptable, because symposia may embrace both types of presentation.

Contents

Preface

CHEMICAL REACTIONS ON POLYMERS has emerged as one of the most active fields in polymer science because of its unique ability to produce specialty polymers with desirable chemical and physical properties through modification of readily available polymers. The field is extremely diverse and has grown to encompass many topics. The use of chemical modification to introduce functional or reactive groups into polymers, to alter polymer surfaces, to provide grafts and unusual side-chain substituents, and to aid in analytical characterization of polymers represents a significant portion of the scope of this scientifically interesting and technologically useful field of research. Other topics include radiation chemistry of polymers, use of dopants for improved conductive properties, and polymer modification for biological applications (e.g., drugs and peptide synthesis) and for electronic applications. The rapid growth of these areas, and the continuing identification of new opportunities for specialty polymers, has led to an increasing body of literature reporting research on polymer modification, as well as numerous review articles and monographs, during the last decade.

The symposium from which this book was developed had three goals: to provide a single forum for presentation of research from several diverse areas of polymer modification by chemical reactions; to present to the scientific community the most recent and emerging technical achievements of leading academic, industrial, and government researchers in the field; and to bring together an international group of scientists, specializing in reactions on polymers, for discussion of topics in this rapidly growing and changing field of polymer science. The speakers and subjects were selected with emphasis on six active research areas in chemical reactions on polymers: reactive polymers, new synthesis routes, surface modification of polymers, specialty polymers with polar/ionic groups, chemical modification for analytical characterization, and chemical modification for functionalization and curing. Inclusion of all facets of research within each of the areas was not possible; however, a good balance of topics that had not been covered previously in other symposia or books was sought.

The field of chemical reactions on polymers is so diverse that one book cannot completely cover this continually expanding area of polymer science. Instead, the chapters of this book provide current research results and balanced coverage of six areas of the field. The book is designed to be useful to chemists working with polymers in a variety of application areas, particularly to researchers who are seeking convenient routes for modification of polymers, and to chemists interested in the synthesis and use of polymeric reagents and catalysts. The conclusions presented in this book are those of the authors, whom I thank for their participation in the symposium and for their preparation of manuscripts for publication in this book. I also thank Rebecca C. Nelson for her administrative assistance in preparation of the symposium and this book.

Acknowledgment is made to the donors of The Petroleum Research Fund, administered by the American Chemical Society, for partial support of the symposium. Acknowledgment for financial contributions to the symposium is made to the following companies: Amoco Corporation; ARCO Chemical Company; E. I. du Pont de Nemours and Company; Eastman Kodak Company; General Electric Company; Owens-Corning Fiberglas Corporation; Pfizer Hospital Products Group; Polaroid Corporation; PPG Industries, Inc.; The Procter & Gamble Company; Signal Research Center, Inc.; and Texaco, Inc.

JUDITH L. BENHAM
3M Specialty Chemicals Division
St. Paul, MN 55144

September 22, 1987

REACTIVE POLYMERS

Introduction to Reactive Polymers

In recent years, there has been increasing research activity in the area of reactive polymers. The variety of functionalized polymers has grown, both in the nature of the polymer backbone and in the array of functional groups. The multiplicity of uses for the functionalized polymers has also grown dramatically, to include ion exchange resins, supports for solid phase synthesis, polymeric reagents, polymer bound catalysts, polymeric protective groups, and inert carriers for reactive substrates. The activity in the area has been demonstrated by the numerous research papers, review articles, and monographs which have been published in the past decade.

Functionalized polystyrene has been the support most widely used, first as an ion-exchange resin, and later as a polymer bound reagent or catalyst in a variety of chemical reactions, e.g acylation, alkylation, and oxidation-reduction, and processes, e.g. protection, separation. Polystyrene is well suited to chemical modification since the aromatic rings can be readily functionalized, and the polymer backbone possessses the necessary stability to allow chemical transformations without degradation. Modifications of polystyrene have included electrophilic substitution (e.g. sulfonation) and metallation with subsequent reaction with a nucleophile (e.g. lithiation/phosphonation). Chloromethylated polystyrene is even more suited to chemical modification since the chlorine atom is easily displaced by nucleophilic reagents. A variety of benzylic substituents have been demonstrated, e.g. amines, quaternary ammonium or phosphonium salts, ethers, esters, sulfonamides, alkyl sulfides, silanes, bis-1,3-diketones, and benzophenones. The resulting polymers have found use as reagents, chelating materials, ion-exchange resins, and photoreactive polymers.

Poly(vinyl chloroformate) has been shown to undergo facile modification to provide new functional polymers, since it can react with nucleophilic compounds containing labile hydrogen atoms, e.g. alcohols, phenols, and amines. Since well-defined structures and high molecular weight polymers can be prepared, poly(vinyl chloroformate) is well-suited to kinetic and mechanistic studies, in addition to use as a substrate in preparation of reactive polymers.

Polydienes and other unsaturated polymers have also been widely used as substrates for chemical modification to produce reactive polymers. The reactions have included epoxidization, cyclization (e.g. with carbene), phosphonylation, hydroformylation, hydroxylation, halogenation, and nitration, resulting in new materials for polymer bound reagents, functional polymers for grafting, etc.

Other addition polymers, for example polyolefins, poly(vinyl acetate), poly(vinyl alcohol), and acrylates, have also been the subject of study. Functionalization of polyolefins with mixed halides has produced polymers with improved solubility and physical properties. Poly(vinyl acetate) and poly(vinyl alcohol) continue to be desireable substrates for modification when highly hydrophilic materials are desired. Introduction of reactive groups for biological applications, e.g. immobilizing enzymes, has been a particularly active area. Chemical modification of the acrylic polymers has included hydrolysis of the ester to produce polyacrylic acid, polymethacrylic acid, etc., for hydrophilic applications, and reaction of functional acrylic copolymers, e.g. ring opening of the epoxide ring of glycidyl methacrylate-ethylenedimethacrylate copolymers to produce selective chelating materials. Use of acrylic copolymers and terpolymers derived from functional monomers, e.g. hydroxyethyl acryate, as substrates for further modification is also increasing.

Condensation polymers have received increasing attention recently as substrates for chemical modification, since they can provide improved mechanical properties compared to the addition polymers which have been so widely used. Condensation polymers which have been used historically include polyamides, polyesters, and polyurethanes. More recently, the focus has been on polyphenylene oxides, polysulfones, phenoxy resins, and halogenated polyethers and polysulfides. Electrophilic modification can be used on those polymers which contain activated aromatic groups, including polysulfones, phenoxy resins, and polyphenylene oxides. The electrophilic reactions have included nitration, bromination, chloromethylation, sulfonation, amination, aminomethylation, and acylation. The resulting polymers have shown utility as polymeric supports, polymeric reagents, and permselective materials. Halogenated polyethers and polysulfides are subject to modification by nucleophilic reagents; however, the reactivity of the halogen leaving group is often modest. Recent work has focused on enhancement of the reactivity of these substrates to modification. The condensation polymer area offers many opportunities for new materials.

Reactive polymers and functional polymers continue to evolve as a major area of polymer modification. Many opportunities exist for new modification reactions and new applications of the resulting materials. The papers in this section of the book exemplify many of the recent activities in this area of chemical reactions on polymers.

Chapter 1

Modification of Condensation Polymers

Challenges and Opportunities

William H. Daly, Soo Lee, and C. Rungaroonthaikul

Department of Chemistry, Louisiana State University, Baton Rouge, LA 70803

Electrophilic elaboration including bromination, nitration, chloromethylation, and aminomethylation of poly(2,6-dimethyl-1,4-phenylene oxide), poly-(arylene ether sulfone) and poly(2-cyano-1,3-phenylene arylene ether) is reported. Reaction conditions which minimize degradation of the polymer substrates are defined with the aid of 2,2-bis[4'-(4"-phenylsulfonyl phenoxyl) phenyl]propane, a model for poly (arylene ether sulfone). Further modifications of the functionalized polymers include; reduction of the nitro or cyano substituents to amine or aminomethyl groups, respectively, quaternization of the chloromethyl group, and application of the Gabriel synthesis to the introduction of aminomethyl groups. Applications of the amino- and aminomethyl polymers as substrates for the grafting of γ-benzyl-L-glutamate N-carboxyanhydride to poly(arylene ether sulfone) are reported. Graft copolymers soluble in pyridine, DMSO and halogenated solvents are described.

In recent years, especially following Merrifield's revolutionary invention of solid phase synthesis[1], there has been a steadily increasing interest in the applicability of polymers as participants in chemical reactions. High levels of activity are apparent in the fields of polymer-bound catalysts, polymeric reagents, inert carriers for reactive substrates, and polymeric protective groups. The rapid growth in this area is evident from the many review articles[2] and monographs[3] published during the past decade.

Reactive polymers can be synthesized by either polymerizing or copolymerizing monomers containing the desired functional groups, or performing one or more modifications on a suitable polymer to introduce the essential functionality. Polymers produced directly by polymerization of functionalized monomers have well defined structures, but the physical and mechanical properties of the

0097-6156/88/0364-0004$06.00/0

resultant materials are not always suitable for the applications in mind. Since the purpose for immobilizing the reagent/catalyst is to optimize the environment for the active site, a procedure leading to better defined polymer characteristics is more desirable. Tailored polymeric systems can be produced by modifying preformed polymers such as; polystyrene, polyamides, poly(oxy-2,6-dimethyl-1,4-phenylene), phenoxy resins and poly(arylene ether sulfone).[4] The support most commonly employed, polystyrene, fulfills the major criteria favoring chemical modification, i.e. sufficient chemical stability to allow essential transformations without degrading the backbone, compatibility with organic solvents, and active sites (aromatic rings), which can be functionalized by either electrophilic substitution or metalation followed by treatment with nucleophiles.

Interestingly, only limited work on the changes in the chain extension and flexibility produced by modifying polystyrene resins has been reported.[5] Although polystyrene is a glass at room temperature (Tg=105°), solvent swollen polystyrene derivatives are extremely flexible and extensive crosslinking is necessary to "immobilize" a reagent. The resultant crosslinked resins inhibit the transport of reactants and products to the active sites within the dense polymer matrix. Immobilization of reagents on polymers with relatively rigid chains should restrict segmental motion without inhibiting diffusion. Little data on the influence of restricted chain mobility on the efficacy of a polymer support in enhancing catalyst and/or reagent properties has been reported, so we have focused our attention in this area. Our interest in semipermeable and reactive membranes has lead to a detailed study of the chemical modification of condensation polymers. The excellent mechanical properties exhibited by the modified polymers suggest that they could be employed in speciality applications such as; permselective membranes, conducting films, and catalyst membranes.

Electrophilic Modification of Condensation Polymers.

In contrast to the activating influence of the hydrocarbon backbone attached to aromatic substituents in polystyrene derivatives, the linking units of condensation polymers exhibit varying influences on the reactivity of enchained aromatic backbones. Linking units which promote electrophilic substitution tend to make the corresponding condensation polymers more reactive than polystyrene. The inductively activated aryloxy groups present in the backbone of poly(arylene ether sulfone), 1, poly(2,6-dimethyl-1,4-phenylene oxide), 2, poly(2,6-diphenyl-1,4-phenylene oxide), 3, and phenoxy resin, 4, make them excellent substrates for electrophilic modification at the sites designated in the structures on the next page. Unfortunately, some of the more common commercially available condensation polymers are based upon linking units which deactivate aromatic rings; thus, polyamides, polyesters, and polycarbonates can be modified with electrophilic reagents only under conditions which lead to cleavage of the backbone linkages.

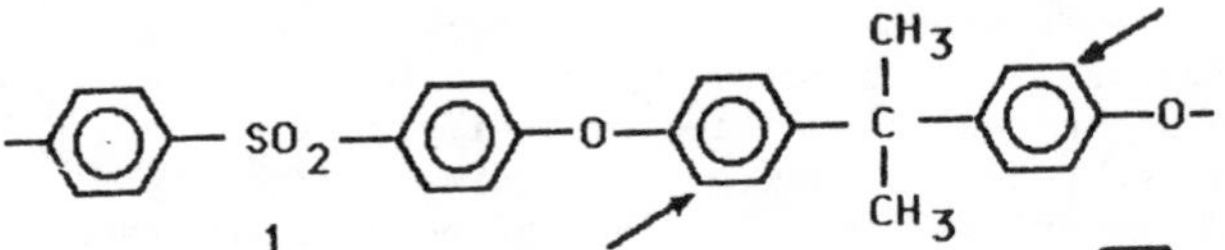

Halogenation. Poly(2,6-dimethyl- and 2,6-diphenyl-1,4-phenylene ether) can be aryl-brominated simply by exposure to a bromine solution; no catalyst is required.[6] In fact, the use of Lewis acid catalysts to promote the chlorination of poly(2,6-dimethyl-1,4-phenylene ether) leads to substantial degradation of the molecular weight of the chlorinated products.[7] Membranes produced from ring brominated PPO (40% wt Br) exhibited enhanced permeability to CH_4 and CO_2 and proved to be more selective in separating CH_4/CO_2 mixtures.[8]

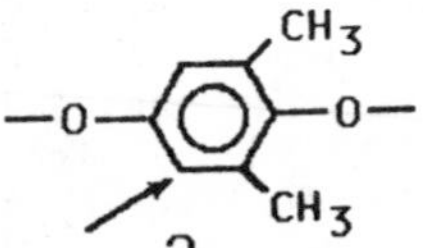

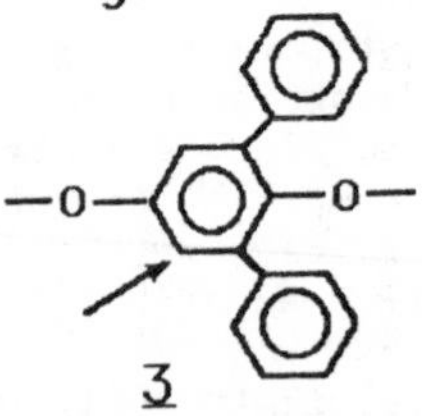

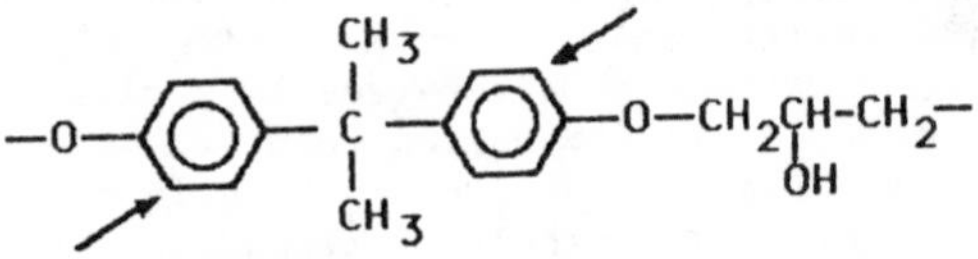

The aryl bromides undergo facile metalation with butyl lithium to produce aryllithium derivatives with the expected organometallic activity.[9] For example, reaction of lithiated PPO with carbon dioxide produces a carboxylated PPO which exhibits unique blending characteristics[10].

In general, if condensation polymers are prepared with methylated aryl repeat units, free radical halogenation can be used to introduce halomethyl active sites and the limitations of electrophilic aromatic substitution can be avoided. The halogenation technique recently described by Ford[11], involving the use of a mixture of hypohalite and phase transfer catalyst to chlorinate poly(vinyl toluene) can be applied to suitably substituted condensation polymers.

Sulfonation. In a classic paper, Noshay and Robeson[12] detailed the conditions for sulfonating poly(arylene ether sulfone) using a sulfur trioxide-triethyl phosphate complex as the sulfonating agent. Chlorosulfonic acid can be used to sulfonate PPO.[13] The inherent instability of the sulfonated PPO[14] could be improved by converting the sulfonate group to a stable sulfone group. Treatment of PPO with arylsulfonyl chloride under Friedel-Crafts conditions produced a series of arylsulfone modified PPO's, which exhibited improved gas permeation characteristics.[15] Since the sulfonated polymers make excellent membranes for reverse osmosis applications, the literature is rather extensive and a complete review is beyond the scope of this paper.

Reactivity of Modified Condensation Polymers

Verdet and Stille[16] employed brominated poly(phenylene oxide) intermediates in an effort to synthesize more stable catalyst supports containing (cyclopentadienyl)metal complexes. Treatment of poly(oxy-2,6-dimethyl-1,4-phenylene) with N-bromosuccinimide under photolytic conditions produced only the bromomethyl derivative if the D.F. did not exceed 0.35. Subsequent treatment of the bromomethylated polymer with sodium cyclopentadienide afforded the cyclopentadienyl functionalized polymer, 5, but the reaction was accompanied by crosslinking and it was not possible to remove the bromomethyl substituents quantitatively.

An alternative synthesis of a thermally stable cyclopentadienyl functionalized polymer involved ring bromination of poly(oxy-2,6-diphenyl-1,4-phenylene), followed by lithiation with butyl lithium to produce an aryllithium polymer. Arylation of 2-norbornen-7-one with the metalated polymer yielded the corresponding 2-norbornen-7-ol derivative. Conversion of the 7-ol to 7-chloro followed by treatment with butyl lithium generated the benzyl anion which undergoes a retro Diels-Alder reaction with the evolution of ethylene to produce the desired aryl cyclopentadiene polymer, 6. The polymer was soluble in a variety of solvents indicating that no crosslinking occurred during functionalization.

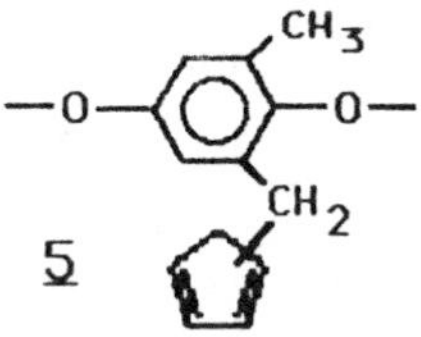

CH_3 —O— —O— CH_2 5

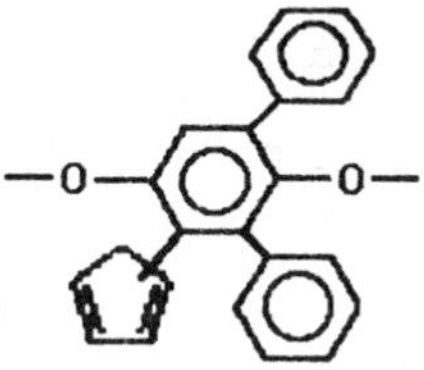

—O— —O— 6

The polymers were converted to supported catalysts corresponding to homogeneous complexes of cobalt, rhodium and titanium. The cobalt catalyst exhibited no reactivity in a Fischer-Tropsch reaction, but was effective in promoting hydroformylation, as was a rhodium analog. A polymer bound titanocene catalyst maintained as much as a 40-fold activity over homogeneous titanocene in hydrogenations. The enhanced activity indicated better site isolation even without crosslinking.

Even in solution the relative rigidity of the polymer support can play a significant role in the reactivity of attached functional groups. Contrasting studies conducted with chloromethylated derivatives of poly(arylene ether sulfone) (Tg= 175°C), phenoxy resin (Tg= 65°C) and polystyrene (Tg= 105°C) allow evaluation of chain rigidity effects. We have shown that the rates of quaternization of chloromethylated poly(arylene ether sulfones) and phenoxy resin deviate from the anticipated second order process at high degrees of conversion; normal kinetics are observed when chloromethylated polystyrene and low molecular weight sulfone model compounds are used as the substrates.[17] Variations in the structure of the tertiary amines used as nucleophiles in this kinetic study changed the initial rates of quaternization, but did not alter the tendency for deviation from second order kinetics. In each case, the process appeared to be second order until the extent of substitution reached 40-50%, beyond that point the rate begins to decrease. The phenomena are not influenced by the addition of electrolytes. Viscosity measurements during the reaction reveal a steady increase in the relative viscosity until the breakpoint is reached and then, the relative viscosity remains constant for the remainder of the reaction. Addition of electrolytes surpresses the chain expansion as the extent of charge along the backbone increases.[18]

The chloromethyl derivatives of 1 and 2 can be converted to the corresponding phosphonium salts by treatment with triphenyl-phosphine.[19] A subsequent phase transfer catalysed Wittig reaction of these salts with formaldehyde introduced pendant vinyl groups. The vinyl substituents could be converted by bromination and dehydrobromination to pendant ethynyl groups.

Our objective in this paper is to illustrate the methods for functionalizing poly(arylene ether sulfone). Particular attention will be paid to bromination, nitration, amination, chloromethyl-ation, and aminomethylation of 1 and its corresponding model compound.

EXPERIMENTAL

Melting points, measured in open capillary tubes using a Thomas-Hoover melting point apparatus, are uncorrected. Elemental analyses were performed by Galbraith Laboratories, Knoxville, Tennesee. ^{1}H and ^{13}C-NMR spectra were generally obtained with an IBM AF-100, if necessary the higher field Bruker WP-200 or AM-400 NMR spectrometer were employed. Chemical shifts are given in parts per million (ppm) on a σ scale downfield from tetramethylsilane (TMS). Infrared spectra were recorded with a Perkin-Elmer 283 spectrophotometer. Low molecular weight solids were dispersed in KBr pellets; polymer films were cast from $CHCl_3$. Intrinsic viscosities were measured by standard procedures using a Cannon Ubbelohde dilution viscometer.

All solvents used for general applications were of reagent grade. For special purposes, purification of solvents was effected using standard procedures. All other reagents were used as supplied commercially except as noted. A solution of chloromethyl methyl ether (6 mmole/mL) in methyl acetate was prepared by adding acetyl chloride (141.2 g, 1.96 mol) to a mixture of dimethoxy methane (180 mL, 2.02 mol) and anhydrous methanol (5.0 mL, 0.12 mol).[20] The solution was diluted with 300 mL of 1,1,2,2-tetrachloroethane and used as a stock solution for the chloromethylation experiments. Poly[oxy-1,4-phenylene-(1'-methylethylidene)-1',4'-phenylene-oxy-(2"-cyano)-1",3"-phenylene], 20, was prepared from bisphenol-A and 2,6-dichlorobenzonitrile according to the precedure of McGrath et al.[21]

Preparation of 2,2-bis[4'-(4"-phenylsulfonylphenoxyl)phenyl] propane,7. Treatment of the disodium salt of bisphenol-A (20 g. 0.088mol) with chlorophenyl phenyl sulfone (44.3 g, 0.175 mol) in a 2:1 v:v toluene:DMSO (100 mL) according the procedure of Johnson et al.[22] afforded 51.2 g of crude adduct. Recrystallization from benzene yielded the pure 7, mp 182-83° C; mol wt.(mass spec) 660.1. ^{1}H NMR ($CDCl_3$): 1.7 (s, 6H, 2-CH_3); 6.7-7.0 (m, 16 H, aromatic H's); 7.8-7.95 (m, 8H, aromatic H's ortho to SO_2). ^{13}C NMR ($CDCl_3$): 30.5 (-CH_3); 42.2 (CH_3-C-CH_3); 117.7, 119.6, 127.2, 128.3, 129.1. 129.7, 132.8, 135.0, 142.0, 147.0, 152.7, 161.9 (aromatic C's). Analysis:Calc'd for $C_{39}H_{32}O_6S_2$: C, 70.91; H, 4.84; O, 14.53; S, 9.72. Found: C, 70.83; H, 4.99, O, 14.54; S, 9.66.

Bromination of 2,2-bis[4'-(4"-phenylsulfonylphenoxyl)phenyl]propane. A solution of 3.0 g (4.70 mmol) of 7 in 25 mL of $CHCl_3$ was stirred in the dark under Ar at room temperature while a solution of 7.51 g (47.0 mmol) of Br_2 in 20 mL of $CHCl_3$ was added. After 5 h of stirring, the excess bromine was removed by purging with Ar for 30 min. The solvent was evaporated under reduced pressure and the residue was washed with 100 mL of methanol. After drying *in vacuo* at 45° for 12 h, 2.95 g (78.9%) of 2,2-bis-[3-bromo-4-(4-phenylsulfonylphenoxy)phenyl]propane, 8, mp 87-91°C was obtained. ^{1}H NMR ($CDCl_3$): 1.68 (s, 6H, 2-CH_3);6.93-7.54 (m, 16 H, aromatic H's); 7.86-7.98 (m, 8H, aromatic H's ortho to SO_2). ^{13}C NMR ($CDCl_3$): 30.6 (-CH_3);42.5 (CH_3-C-CH_3); 115.6, 117.0, 121.9, 127.3, 127.5, 129.1, 129.9, 132.0, 132.9, 135.4, 141.9, 148.5, 149.6, 161.1 (aromatic C's); IR (film), 1050 cm^{-1} (-Br).

Nitration of 2,2-bis[4'-(4"-phenylsulfonylphenoxyl)phenyl]propane. The process reported by Crivello[23] was utilized as follows: a solution of 3.30 g (5.2 mmol) of 7 in 50 mL of $CHCl_3$ was blended with 0.84 g (10.4 mmol) of ammonium nitrate and 5.88 mL (41.6 mmol) of trifluoroacetic anhydride (TFAA) in a 100 mL RBF. The mixture was stirred at room temperature for 11 h. After evaporating the excess TFAA by purging with argon for 30 min, the solution volume was reduced to 5 mL. The product was precipitated by pouring the reaction mixture into 100 mL of methanol, recovered by filtration, washed with 50 mL of water and dried *in vacuo* at 35° for 48 h; 3.85 g (75.3%) of 2,2-bis[3-nitro-4-(4-phenylsulfonyl-phenoxyl)phenyl] propane, 9, mp 104-6° was obtained. ^{1}H NMR ($CDCl_3$): 1.74 (s, 6H, 2-CH_3); 7.19, 7.31, 7.60, 7.68, 8.04 (m, 24 H, aromatic H's); ^{13}C NMR ($CDCl_3$): 30.5 (-CH_3); 42.7 (CH_3-C-CH_3); 117.9, 120.1, 122.9, 124.0, 127.4, 128.3, 129.2, 130.0, 133.1, 136.6, 141.6, 146.4, 146.7, 160.4, (aromatic C's); IR (film); 1535, 1340 cm^{-1} (-NO_2).

Reduction of 2,2-bis[3'-nitro-4'-(4"-phenylsulfonylphenoxyl)phenyl] propane. Compound 9, 200 mg (0.55 meg) was dissolved in a mixture of 30 mL of dichloromethane (DCM) and 30 mL of methanol and 240 mg of 10% palladium on charcoal was added. After purging the solution with argon for 30 min, 520 mg (13.6 mmol) of sodium borohydride was added portionwise over 10 min. The reaction mixture was stirred under argon for 1 hr before addition of 30 mL of DCM. The mixture was filtered, the filtrate evaporated, and the residue extracted with DCM. Evaporation of the extract yielded 140 mg (76.2%) of 2,2-

bis[3'-amino-4'-(4"-phenylsulfonylphenoxyl)phenyl]propane, 10, mp 85-9°. ^{1}H NMR ($CDCl_3$): 1.76 (s, 6H, 2-CH_3); 7.02, 7.11, 7.31, 7.46, 7.52, 7.88, 7.95, 7.98 (m, 24 H, aromatic H's); ^{13}C NMR ($CDCl_3$): 30.8 (-CH_3); 42.4 (CH_3-C-CH_3); 115.6, 116.6, 117.4, 120.6, 127.4. 129.2, 129.9, 132.9, 134.8, 138.2, 139.1, 142.1 148.8,161.8 (aromatic C's); IR (film): 3460, 3380, 1625 cm^{-1} (-NH_2).

Chloromethylation of 2,2-bis[4'-(4"-phenylsulfonylphenoxyl)phenyl] propane. A solution of 7, 40 g (0.06 mol) was blended with 400 mL of 6 M chloromethyl methyl ether solution and 0.7 mL (6 mmol) of $SnCl_4$ in a 500 mL resin kettle, and the mixture was refluxed for 8 h under argon. The solvent was distilled until the volume was reduced to 20 mL and 300 mL of ethanol was added to produce a clear solution. The product, 2,2-bis[3-chloromethyl-4-(4-phenylsulfonyl-phenoxyl)phenyl]propane, 11, crystallized from the ice cooled solution, 40.2 g (87%) mp 182-83°C was isolated by filtration. ^{1}H NMR ($CDCl_3$): 1.7 (s, 6H, 2-CH_3); 4.5 (s, 4H, -CH_2Cl); 6.7-7.0 (m, 16 H, aromatic H's); 7.8-7.95 (m, 8H, aromatic H's ortho to SO_2). ^{13}C NMR ($CDCl_3$): 30.5 (-CH_3); 40.7 (-CH_2Cl); 42.2 (CH_3-C-CH_3); 117.7, 119.9, 127.3, 128.7, 129.1, 129.3, 129.8, 132.8, 135.5, 142.1, 147.2, 150.9, 161.5 (aromatic C's). Analysis: Calc'd for $C_{41}H_{34}O_6S_2Cl_2$: C, 64.98; H, 4.5; Cl, 9.36; S, 8.56; O, 12.66. Found: C, 64.86; H, 4.69; Cl, 9.47, S, 8.62; O, 12.30.

Phthalimidomethylation of 2,2-bis[3'-chloromethyl-4'-(4"-phenyl-sulfonylphenoxyl)phenyl]propane, 11. To a solution of 1.0 g (2.64 meq) of 11 in 20 mL DMF was added 1.20 g (6.49 mmol) of potassium phthalimide. After stirring for 8 h at 100° C, the unreacted potassium phthalimide was removed by filtration and the filtrate evaporated under reduced pressure. The residue was washed sequentially with 30 mL of distilled water and 15 mL of methanol. After drying *in vacuo* at 40° C for 24 h, 1.16 g (90.1%) of the product, 2,2-bis[3-phthalimidomethyl-4'-(4"-phenylsulfonylphenoxyl)-phenyl]propane, 12, was obtained. Spectral data were consistent with quantitative substitution: ^{1}H NMR ($CDCl_3$): 1.69 (s, 6H, 2-CH_3); 4.76 (s, 4H, -CH_2N Phth); 6.8-7.5 (m, 16 H, aromatic H's); 7.61, 7.64 (d, 8H, aromatic H's in phthalimide); 7.8-7.84 (m, 8H, aromatic H's ortho to SO_2). ^{13}C NMR ($CDCl_3$): 30.5 (-CH_3); 37.2 (-CH_2NPhth); 42.2 (CH_3-C-CH_3); 116.9, 117.1, 119.5, 120.4, 123.0, 127.4, 127.6, 127.8, 129.1, 129.7, 131.7, 132.9, 133.9, 134.9, 141.5, 147.4, 150.3, 161.8, 167.5 (aromatic C's); IR (film); 1772, 1716 cm^{-1} (phthalimide C=O).

Hydrazinolysis of 2,2-bis[3'-phthalimidomethyl-4'-(4"-phenylsulfonyl phenoxyl)phenyl]propane, 12. A solution of 1.00 g (2.0 meq) of 12 and 3 mL (34.5 mmol) of hydrazine hydrate in 25 mL of methanol was heated at reflux for 24 h. After evaporation of solvent, the residue was extracted with 30 mL of $CHCl_3$, the extract was evaporated, and the crude product was washed sequentially with 25 mL of 2% aqueous sodium bicarbonate, 100 mL of water, and 50 mL of methanol. Drying *in vacuo* at 40° C for 24 h afforded 0.70 g (97.5%) of 2,2-bis[3-aminomethyl-4'-(4"-phenylsulfonylphenoxyl)-phenyl]propane, 13. ^{1}H NMR ($CDCl_3$): 1.69 (s, 6H, 2-CH_3); 2.07

(b,-NH_2); 3.73 (s, 4H, -CH_2NH_2; 6.79-7.54 (m, 16 H, aromatic H's); 7.84, 7.95 (d, 8H, aromatic H's ortho to SO_2). ^{13}C NMR ($CDCl_3$): 30.9 (-CH_3); 41.7 (-CH_2NH_2); 42.6 (CH_3-C-CH_3); 117.1, 120.2, 127.0, 127.4, 127.8, 129.2, 129.6, 130.0, 133.0, 134.4, 135.1, 142.0, 147.8, 150.4, 162.0, (aromatic C's); IR (film); 3390, 3340 cm^{-1} (-NH_2).

Bromination of Poly(arylene ether sulfone). A solution of 4.42 g (10.0 meq) of 1 in 50 mL of $CHCl_3$ was stirred in the dark under Ar at room temperature while a solution of 3.55 g (22.2 meq) of Br_2 in 20 mL of $CHCl_3$ was added. After 17 h of stirring, the excess bromine was removed by purging with Ar for 30 min. The solution volume was reduced to 40 mL before pouring into 700 mL of methanol. The precipitated product was recovered by filtration, washed with 100 mL of methanol, and dried *in vacuo* at 45° overnight. A total of 4.86g of brominated polymer, 14, was recovered. ^1H NMR ($CDCl_3$): 1.68 (s, 6H, 2-CH_3); 6.93-7.54 (m, 11 H, aromatic H's); 7.86-7.98 (m, 4H, aromatic H's ortho to SO_2). ^{13}C NMR ($CDCl_3$): 30.6 (-CH_3);42.5 (CH_3-C-CH_3); 115.7, 117.1, 117.8, 119.9, 121.9, 127.6, 128.4, 129.8, 131.4, 135.5, 146.3, 148.6, 150.2, 153.2, 161.1 (aromatic C's); IR (film); 1050 cm^{-1} (-Br).

Nitration of Poly(arylene ether sulfone). A solution of 4.00 g (0.05 mol) of ammonium nitrate in 42 mL (0.3 mol) of trifluoroacetic anhydride was added to 22.10 g (0.05 eq) of 1 dissolved in 200 mL of $CHCl_3$. The slurry was stirred at room temperature for 24 h during which time the inorganic salt dissolved. The volume of the reaction mixture was reduced to 100 mL under reduced pressure before precipitating the product by pouring the reaction mixture into 1 L of methanol. The product was separated by filtration and washed sequentially with 200 mL methanol, 100 mL saturated aqueous $NaHCO_3$, 400 mL of water, and 100 mL of methanol. After drying *in vacuo* for 24 h at 45° C, 22.10 g (99.2%) of poly(nitroarylene ether sulfone),15, D.F.= 1.0, [η] = 0.34 dl/g ($CHCl_3$ at 30° C) was obtained. ^1H NMR ($CDCl_3$): 1.73 (s, 6H, 2-CH_3); 6.89-7.40, 7.50, 7.8-7.9 (m, 15 H, aromatic H's); ^{13}C NMR ($CDCl_3$): 30.6 (-CH_3); 42.7 (CH_3-C-CH_3); 117.5, 117.9, 119.8, 123.9, 128.4, 129.9, 135.4, 136.9, 141.8, 145.4, 145.9, 146.9, 148.8, 153.5, 160.8, 161.9, (aromatic C's); IR (film); 1550, 1360 cm^{-1} (-NO_2).

Reduction of Nitrated Poly(arylene ether sulfone), 15. A 250 mL round bottomed flask was charged with 3.00 g (6.16 meq) of 15, D.F.= 1.0, and 30 mL of THF, 0.01 g of tetrabutylammonium chloride, and 5.00 g (22.1 mmol) of stannous chloride. After stirring for 30 min at room temperature, 100 mL of conc. HCl was added over a 30 min interval and stirring was continued for 48 h at 70° C. The THF was evaporated and the residue was neutralized with solid NaOH in an ice bath. The poly(aminoarylene ether sulfone), 16, was extracted from the residue with 50 mL $CHCl_3$, reprecipitated by adding the $CHCl_3$ solution to 500 mL of methanol, and dried *in vacuo* at 50° C for 16 h. The product, 1.20 g (42.6%), [η]= 0.32 dl/g ($CHCl_3$ at 30° C) exhibited the following spectra: ^1H NMR ($CDCl_3$): 1.64 (s, 6H, 2-CH_3); 3.69 (b, 2H, -NH_2); 6.50-7.29, (m, 11 H, aromatic H's); 7.78, 7.86 (d, 4H, aromatic H's ortho to SO_2); ^{13}C NMR ($CDCl_3$): 30.8

(-CH3); 42.4 (CH_3-C-CH_3); 115.6, 116.6, 117.4, 117.7, 119.7, 120.7, 128.4, 129.7, 135.4, 138.3, 139.1, 147.3, 148.6, 152.8, 161.7, 161.9, (aromatic C's); IR (film); 1620, 3360, 3480 cm^{-1} (-NH_2).

Chloromethylation of Poly(arylene ether sulfone). To a 500 mL three-necked resin kettle equipped with condenser, mechanical stirrer, and pressure equalizing dropping funnel, was charged with 75 mL of chloromethyl methyl ether solution and 0.13 mL (1.13 mmol) of $SnCl_4$. After the solution was heated to 100° C, a solution of 1 (5.0 g, 11.25 meq) in 63 mL of tetrachloroethane was added slowly over a period of 15 min. The reaction mixure was stirred and maintained at 100° C for 3 hours before the catalyst was deactivated by injecting 2.0 mL of methanol. The reaction volume was reduced to 70 mL before precipitating the chloromethylated polymer in methanol. The polymer was recovered by filtration, washed with methanol and dried in a vacuum dessicator for 24 h. The crude polymer was purified by dissolution in dioxane, reprecipitation in methanol, followed by successive washings with 40% aqueous dioxane, 40% aqueous dioxane containing 10 % conc. HCl, water and methanol. After drying *in vacuo* at 50° C for 48 h, 5.81 g (96.7%) of, 17, was recovered. ^{1}H NMR ($CDCl_3$): 1.68 (s, 6H, 2-CH_3): 4.53 (s, 4H, -CH_2Cl); 6.8-7.4 (m, 10 H, aromatic H's); 7.9 (d, 4H, aromatic H's ortho to SO_2). ^{13}C NMR ($CDCl_3$): 30.7 (-CH_3); 40.8 (-CH_2Cl); 42.4 (CH_3-C-CH_3); 117.8, 120.1, 128.9, 129.1, 129.5, 129.8, 135.9, 147.4, 151.1, 162.0 (aromatic C's). Elemental Analysis was consistent with the introduction of 1.89 -CH_2Cl groups/repeat unit. Anal. Found: C, 64.97; H, 4.72; Cl, 12.55; S, 6.20; O. 11.91.

Phthalimidomethylation of Poly(arylene ether sulfone). Electrophilic Approach. To a mechanically stirred solution of 22.10 g (0.05 eq) of 1 in 200 mL of DCM was added dropwise a solution of 8.85 g (0.05 mol) of N-hydroxymethylphthalimide and 5.0 g of trichloromethanesulfonic acid in 200 mL of trifluoroacetic acid. The addition required 1 h during which time the mixture darkened to a deep brown. After stirring for 6 h at room temperature, the polymer was precipitated in 1.5 L of methanol and washed sequentially with 400 mL of conc. ammonium hydroxide, 1.5 L of water, and 1.5 L of methanol. The poly(phthalimidomethylarylene ether sulfone),18, D.F.= 0.5, was isolated as a white powder, 19.90 g (76.3%), [η]= 0.40 dl/g ($CHCl_3$ at 30° C).
Nucleophilic Approach. To a mechanically stirred solution of 5.00 g (18.5 meq) of 17, D. F.= 2.0, in 50 mL of DMF was added a solution of 6.00 g (26.9 meq) of potassium phthalimide in 150 mL of DMF. After stirring for 10 h at 100° C, the precipitated salts were removed by filtration, and the product was isolated by pouring the filtrate into 1 L of methanol. Following a water wash and a methanol wash, the precipitate was dried at 45° C for 24 h; 5.50 g, (99.1%) of the product, 18, D. F.= 2.0, [η] = 0.41 dl/g, was obtained. Spectral data were consistent with quantitative substitution.

The spectra obtained from samples prepared by either technique exhibited the following qualitative absorbances: ^{1}H NMR ($CDCl_3$): 1.69 (s, 6H, 2-CH_3); 4.76 (s, 4H, -CH_2NPhth); 6.8-7.4 (m, 10 H, aromatic H's); 7.66 (d, 8H, aromatic H's in phthalimide); 7.7-7.9 (m, 4H, aromatic H's ortho to SO_2).

^{13}C NMR ($CDCl_3$): 30.9 ($-CH_3$); 38.1 ($-CH_2NPhth$); 42.4 ($CH_3-\underline{C}-CH_3$); 117.1, 120.1, 127.5, 127.6, 129.6, 135.4, 147.5, 150.3, 161.9, (aromatic C's); IR (film); 1773, 1717 cm^{-1} (phthalimide C=O).

Hydrazinolysis of Poly(phthalidimidomethylarylene ether sulfone). To a solution of 5.00 g (4.94 meq) of 18, D.F.= 0.5, in 150 mL of THF and 150 mL of ethanol, was added 3.5 mL (51.6 mmol) of hydrazine hydrate at 70° C. After stirring for 18 h at 70° C, the precipitate was recovered by filtration, and extracted with 300 mL of THF. The filtrate was evaporated and the residue extracted with 50 mL of THF. The extracts were combined, the volume reduced to 50 mL, and the product precipitated by pouring the solution into 500 mL of methanol. The polymer was washed sequentially with 50 mL of saturated aqueous $NaHCO_3$, 200 mL of water, and 200 mL of methanol, dried *in vacuo* at 50° C for 16 h, and 3.50 g (79.9%) of poly(aminomethylarylene ether sulfone), 19, D.F.= 0.5, [η]= 0.39 dl/g, was recovered. ^1H NMR ($CDCl_3$): 1.69 (s, 6H, $2-CH_3$); 1.92 (b, 1H, $-NH_2$); 3.73 (s, 1H, $-\underline{CH_2}NH_2$); 6.78-7.29 (m, 11.5 H, aromatic H's); 7.80-7.89 (d, 4H, aromatic H's ortho to SO_2). ^{13}C NMR ($CDCl_3$): 30.9 ($-CH_3$), 41.8 ($-CH_2NH_2$); 42.5 ($CH_3-\underline{C}-CH_3$); 117.0, 117.7, 119.8, 120.2, 125.7, 128.4, 129.7, 133.4, 147.1, 147.7, 149.9, 152.9, 161.9, (aromatic C's), IR (film), 3400 cm^{-1} ($-NH_2$).

Reduction of Poly(2-cyano-1,3-phenylene arylene ether), 20 Twenty-five mL of a 1.0 M solution of lithium aluminum hydride (LAH) in THF was cooled to 0° C before adding a solution of 1.64 g (5.0 meg) of 20 in 120 mL of THF. The resultant slurry was stirred for 24 h at 0° C, refluxed for 1 h, recooled to 5° C, and the excess LAH decomposed with 2 mL of water. The volume of the solution was reduced to 25 mL before pouring the mixture into 500 mL of 5% HCl to dissociate the amine aluminum salt complex and precipitate the polymer. The polymer was recovered by filtration, reslurried in 20 mL of water and the pH adjusted to 9.0 with NaOH. After recovery of the neutralized polymer was recovered, it was dried *in vacuo* redissolved in $CHCl_3$, and reprecipitated using water as the non-solvent. Final drying *in vacuo* for 24 h at 35° C left 1.2 g (72.3%) of poly[oxy-1,4-phenylene-(1-methylethylidene)-1',4'-phenylene-oxy-(2"-aminomethyl)-1",3"-phenylene], 21, [η] ($CHCl_3$) = 0.3 dl/g. ^1H NMR ($CDCl_3$): 1.67-1.85 (b, 8H, 2 CH_3 and NH_2); 3.91 (s, 2H, $-\underline{CH_2}-N$); 6.66-7.23 (m, 11H, aromatic H's). ^{13}C NMR ($CDCl_3$): 30.95 ($-CH_3$); 35.47 ($-CH_2-NH_2$); 42.02 ($CH_3-\underline{C}-CH_3$); 114.0, 117.7, 128.0, 145.4, 155.2, 157.0 (aromatic C's).
Analysis: Calc'd for $C_{22}H_{21}NO_2$: C, 79.73; H, 6.39; O, 9.65; N, 4.23. Found: C, 79.33; H, 6.42; O, 9.9; N, 4.07.

RESULTS AND DISCUSSION

Since poly(oxy-2,6-dimethyl-1,4-phenylene) has exhibited a high tendency to undergo cleavage, rearrangements and to crosslink in the presence of electrophilic reagents,[24] our attention has been focused on modification of poly(arylene ether sulfone), 1, and phenoxy resin, 4. The active sites in these polymers are the 3-positions of the bisphenol-A repeating units. We will report the extent of substitution by a given reagent as the degree of functionalization (D.F.) Normally only one substituent is introduced per activated aromatic ring, i.e., the two active rings per repeating unit in polymers prepared from bisphenol-A lead to a maximum D.F. of 2.

Selection of appropriate conditions to modify polymers is facilitated by preliminary studies with well designed model compounds. The work with model systems is critical when studying condensation polymers because the main chain linkages have proved to be remarkably labile under certain conditions. Condensation of 4-chlorophenyl phenyl sulfone with the disodium salt of bisphenol-A yields 2,2-bis[4'-(4"-phenylsulfonylphenoxyl)phenyl] propane, 7, an excellent model for the poly(arylene ether sulfone) substrate. Conditions for quantitative bromination, nitration, chloromethylation, and aminomethylation of the model compound were established. Comparable conditions were employed to modify the corresponding polymers.

The NMR and infrared spectra of the derivatized model compounds are useful in establishing the structures and the D.F. of the modified polymers. Careful assignment of all peaks in the ^{13}C-NMR spectra for each of structures 7-13 confirms the regioselectivity of the substitution on the oxyphenyl unit and inertness of the phenyl sulfone units. The chemical shifts of the key carbons for the analysis, those of the oxyphenyl rings, are summarized in Table I.

Table I. ^{13}C Characterization of Model Compounds.

[Structure: Ph–SO_2–C_6H_4–O–C_6H_3(X) (ring positions 1–6, X at 3)]$_2$–C(CH_3)$_2$

Chemical Shift in ppm

Compound	X =	C_1 Obs.	C_1 Calc.	C_2 Obs.	C_2 Calc.	C_3 Obs.	C_3 Calc.	C_4 Obs.	C_4 Calc.
7	H	147.0	147.0	128.3	128.3	119.6	119.6	152.7	152.7
8	Br	148.5	149.2	132.0	131.6	115.6	114.2	149.6	156.0
9	NO_2	146.7	147.8	124.0	123.0	133.1	139.2	146.4	147.4
10	NH_2	148.8	148.3	115.6	115.9	138.2	138.8	139.1	140.3
11	CH_2Cl	147.2	147.2	129.1	128.3	129.3	128.7	150.7	152.7
12	CH_2NPh	147.4	147.1	129.1	128.4	127.6	127.5	150.3	151.1
13	CH_2NH_2	147.8	146.8	127.8	126.7	135.1	134.5	150.4	151.1

The calculated chemical shifts are based upon additivity factors for the sulfone, isopropylidene, and oxyphenyl linkages derived from the spectrum of 7. The spectra of derivatives with D.F.=2 establish the chemical shifts of the 3-substituted rings. Studies on model compounds with D.F.=1 confirm that the spectra of partially substituted samples can be calculated by appropriate combination of the spectra of unsubstituted and 3-substituted rings. Quantitative substitution is difficult to achieve with polymeric substrates, but interpretation of spectra of low D.F. derivatives is straight forward.

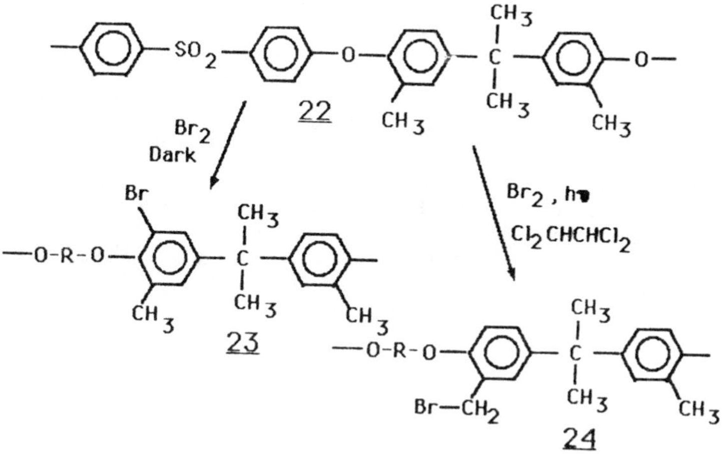

Bromination

Initial studies on bromination of poly(arylene ether sulfone) were directed toward perbromination to produce fire retardants.[25] The conditions used, bromine in the presence of iron catalyst in alkyl bromides at reflux, must have produced substantially degraded products but approximately 6 Br/repeat unit were introduced. We wished to effect bromination under more controlled conditions to produce a range of brominated products to serve as substrates for Heck-type vinylation reactions. We found that aromatic bromination must be conducted at room temperature or below in the presence of a large excess of bromine to assure minimal chain cleavage; reactions run in refluxing $CHCl_3$ or in the presences of thallium chloride, which was used successfully to promote the bromination of polystyrene,[26] lead to significantly degraded products. If poly(3-methylarylene ether sulfone), 22, is employed as the substrate, bromination can be directed either to the aromatic ring, 23, or to the benzyl position, 24, by controlling the bromination conditions. Attack at the benzyl position is promoted by photolysis of the halogen or utilization of N-bromosuccinimide in the presence of free radical initiators.

Nitration and Reduction.

In conjunction with a project involving N-carboxyanhydride (NCA) interaction with aminopropylcellulose to produce peptide grafts, we sought to produce soluble, well characterized polymers with primary amine substituents. The project proved to be one of the challenges associated with modification of condensation polymers. Two substituents were targeted, 3-amino- and 3-aminomethyl, as a primary amine function was needed to initiate NCA polymerization. Polymers with 3-amino substituents were to be produced simply by reduction of the corresponding 3-nitro derivatives.

The usual nitrating agents for aromatic compounds, i.e. fuming nitric acid or nitric-sulfuric acid mixtures tend to decompose condensation polymers. However, Crivello has reported that mixtures of metal nitrates in trifluoroacetic anhydride promote nitration of poly(aryl carbonate)s and poly(phenylene oxide)s, only in the case of the polycarbonates was significant loss of molecular weight during the modification process detected.[23] We utilized the procedure to nitrate poly(arylene ether sulfone) and evaluated the effect of nitration on the hydrolytic stability of the corresponding derivatives.[27] Derivatives with D.F.'s up to two are produced under very mild conditions, but some reduction in molecular weight was observed. The chain cleavage was promoted by the acid medium; surprisingly, in the presence of basic nucleophiles like sodium methoxide, little change in hydrolytic stablity relative to that of the unsubstituted polymer was observed.

15: $-C_6H_4-SO_2-C_6H_4-O-C_6H_3(NO_2)-C(CH_3)_2-C_6H_4-O-$

THF, $SnCl_2$/HCl

16: $-C_6H_4-SO_2-C_6H_4-O-C_6H_3(NH_2)-C(CH_3)_2-C_6H_4-O-$

Bz-O-C(=O)-CH_2CH_2-(NCA ring: C, C=O, O, C=O, HN), THF

25: $-C_6H_4-SO_2-C_6H_4-O-C_6H_3(NH-[C(=O)-CH(CH_2CH_2C(=O)-O-Bz)-NH]_n-H)-C(CH_3)_2-C_6H_4-O-$

Nitrated poly(arylene ether sulfone), 15, proved to be remarkably resistant to reduction. No amino functions could be detected following treatment of the nitrated polymers with sodium dithionite, sodium borohydride, lithium aluminum hydride, palladium and sodium borohydride or low pressure hydrogenation. However, 15 can be reduced with stannous chloride and hydrochloric acid in refluxing THF to the corresponding poly(aromatic amine), 16. The 3-amino derivative dissolved readily in THF and $CHCl_3$ and intrinsic viscosity measurements indicated that little change in the molecular weight occurred during the reduction. We have shown that NCA's can be grafted to 16 to produce soluble graft copolymers, 25, and are currently working on the conditions to control the extent of grafting.

The nitrated model compound, 9, proved even more resistant to reduction than the polymeric analog; the dissolving metal technique used to reduce 15 failed on 9 , but finally the amino model, 10, was produced by treatment of 9 with a 25-fold excess of sodium borohydride. Compound 10 serves as a difunctional initiator for NCA polymerization.

Chloromethylation

We have developed techniques for controlled chloromethylation of condensation polymers containing oxyphenyl repeating units.[16] The chloromethylation can be effected with chloromethyl methyl ether, or 1,4-bis-(chloromethoxy)butane in the presence of stannic chloride. More recently we have been evaluating a mixture of chloromethyl ether and methyl acetate, produced by mixing acetyl chloride with dimethoxymethane in the presence of a catalytic amount of methanol,[20] as a chloromethylating agent. Since the active chloromethyl ether is produced selectively and the reaction mixture can be used directly, the procedure minimizes exposure to potential carcinogens. However, the presence of methyl acetate which is formed as a by-product in the reaction, moderates the activity of the chloromethylating agent significantly (Table II). Note that polystyrene does not react with this reagent under conditions which are effective in introducing chloromethyl groups on more activated aromatic rings.

During the chloromethylation process, gelation occurs at high degrees of conversion, unless an effort is made to maximize polymer dilution and the concentration of chloromethylating reagent. Although soluble derivatives with D.F.'s less than 1.5 can be produced using a chloromethyl ether to substrate ratio of 10:1, a 20:1 ratio is required to eliminate crosslinking as the D.F. approaches 2.0. Selection of the proper catalyst also minimizes the extent of concomitant crosslinking reactions. Evaluation of a series of Lewis Acid catalysts by comparing the extent of active chloride introduction on poly(arylene ether sulfone) within one hour at 110°C established the following order of reactivity: $SbCl_5 > AsCl_3 > SnCl_4 > ZnCl_2 > AlCl_3 > TiCl_4 > SnCl_2 > FeCl_3$. Poly(2,6-dimethylphenylene oxide) exhibits a unique dependence upon the nature of the catalyst because the phenoxy ether backbone tends to form stable inactive catalysts with each Lewis acid described above except stannic chloride. This fact, coupled with the relatively small broadening of the molecular weight distribution observed when stannic chloride was employed as the catalyst in poly(arylene ether sulfone) modifications, prompted us to select stannic chloride as the general catalyst for choromethylation of condensation polymers.

Table II. Chloromethylation of Condensation Polymers

Polymer ([η])	Reagent	Reactant Ratio Reagent: Polymer: $SnCl_4$			Temp(C)	Time(h)	% Subst (DF)	[η]
PPO, 2								
(0.50)	CH_3OCH_2Cl[a]	10	1	0.1[b]	60	1	100(1.0)	0.54
	1,4-bis[d]	2	1	.05[b]	25	2	97(0.97)	
Poly(Arylene ether sulfone), 1								
(0.51)	CH_3OCH_2Cl[a]	20	1	0.1[c]	96	1	60(1.2)	0.51
		20	1	0.1[c]	96	3	95(1.9)	0.51
		10	1	0.1[c]	113	3	75(1.5)	0.65
	1,4-bis[d]	4.2	1	0.5[c]	110	2	96(1.9)	
Sulfone Model, 7								
	CH_3OCH_2Cl[a]	20	1	0.1[c]	85	8	100(2.0)	
Acetylated Phenoxy Resin								
(0.53)	CH_3OCH_2Cl[a]	20	1	0.1[c]	90	3	100(2.0)	0.45
Polystyrene								
	CH_3OCH_2Cl[a]	10	1	0.1[b]	50	24	0	
	1,4-bis[d]	3.6	1	1.3[b]	25	17	100(1.0)	

a) 50:50 mixture of chloromethyl methyl ether and methyl acetate; b) solvent, chloroform; c) solvent, tetrachloroethane; d) 1,4-bis(chloromethoxy)butane.

The substrate catalyst ratio plays an important role in controlling reaction selectivity; a 10:1 substrate:catalyst ratio provided an acceptable balance between the rate of active chloride introduction and the extent of crosslinking. Table II summarizes the conditions required to prepare derivatives of poly(oxy-2,6-dimethyl-1,4-phenylene), polysulfone, and phenoxy resin with varying degrees of active sites. Comparison of the initial viscosities of the polymer substrates with those of the derivatives indicates that little variation in the molecular weight has occurred during the modification. Condensation polymers composed of deactivated aromatic rings, i.e. polycarbonates. polyesters or polyamides undergo chloromethylation only in the presence of stoichiometric amounts of Lewis acid catalysts; the reaction is accompanied by significant reductions in the molecular weight.

The chloromethylated polymers are very reactive substrates for nucleophilic attack; further elaboration can be accomplished under homogeneous conditions in aprotic solvents, or under heterogeneous conditions in the presence of phase transfer catalysts. The following examples are representative of approaches to functionalized condensation polymers via chloromethylated intermediates.

Aminomethylation

Conversion of chloromethylated polymers to our second target system, aminomethylated polymers, was approached in several different ways. Two of the approaches, which were used successfully to convert model compound 11 to the desired aminomethyl products failed when applied to the polymer system. The first of these, the Delépine reaction, appeared to be the most reasonable and economical, but only insoluble, apparently crosslinked products could be isolated.

17

CH_2Cl CH_3 CH_2Cl

2 eq

EtOH/THF
reflux, 5 hr

26

Insoluble Resin

CH_2 Cl^-

1) HCl, anhy
EtOH, reflux
2) NaOH (aq)

Partially Hydrolysed Insoluble Product

The Delépine reaction involves nucleophilic displacement of active halides by hexamethylenetetramine, followed by hydrolysis of intermediate quaternary ammonium salt to release the amine. Normally the reaction is useful for the conversion of alkyl halides to primary amines without concomitant formation of secondary amines.[29] Treatment of polymer 17 with hexamethylenetetramine in a mixture of ethanol/THF afforded an insoluble resin. Using diazabicyclooctane (DABCO), we demonstrated that the reaction could be limited to attack by a single nitrogen in a multifunctional amine, so we did not anticipate crosslinking via bis-quat salt formation. Hydrolysis of 26 with anhydrous HCl in ethanol generated free amino groups as evidenced by a positive ninhydrin test, but quantitative hydrolysis could not be achieved and the product remained insoluble. One would have expected a simple bis-quat to hydrolyse and open the crosslinked structure.

Reduction of azides is a classical approach to primary amine synthesis. Treatment of 17 with sodium azide in DMF or in THF/H_2O mixtures in the presence of phase transfer catalysts effects a quantitative conversion to the corresponding polymeric azide, 27. Recently the reduction of azides to primary amines via hydrolysis of iminophosphoranes produced by interaction of the azide with triethyl phosphite was reported.[30] Application of this technique to the azidomethyl polymer, 27, as shown below, failed to produce a soluble polyamine.

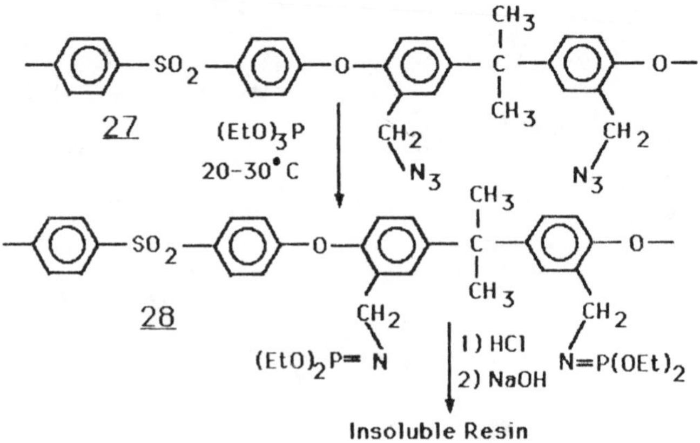

The most successful approach to producing an aminomethyl derivative was the Gabriel synthesis. A phthalimide substituent can be introduced by Sn2 displacement of the chloride on 17 with potassium phthalimide under homogeneous conditions in DMF. The reaction is quantitative in all D.F. ranges and the phthalimidomethyl intermediates, 18, are quite soluble in organic solvents.

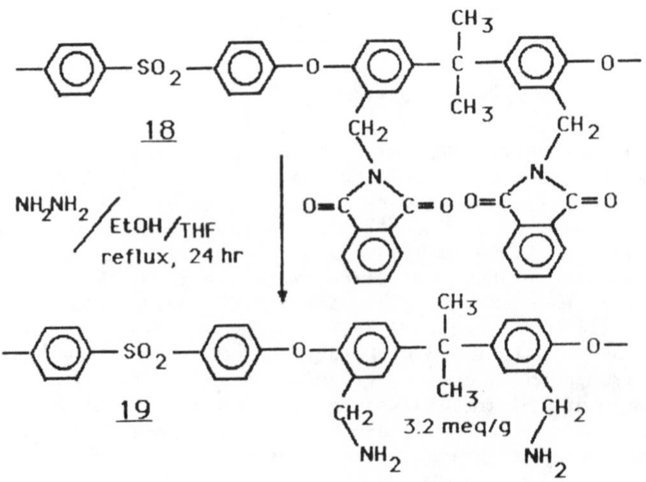

Direct aminomethylation of 1 or 2 can be effected by phthalimidomethylation with N-hydroxymethylphthalimide in trifluoroacetic acid;[31] however, some reduction in the molecular weight of the substituted polymers can be detected by viscosity measurements when the electrophilic phthalimidomethylation procedure is employed. The aminomethyl substituent is released by hydrazinolysis of the phthalimide substituents in mixtures of THF/ethanol. The mixed solvent system was necessary; hydrazinolysis in pure ethanol failed to remove the blocking groups quantitatively. Interestingly, the solubility of the aminomethyl polymer, 19, depends upon the D. F. Polymers, which are soluble in THF, are formed if the D.F. does not exceed 1. Using 19 with a D.F. = 1.0, NCA's were allowed to react at various NCA/amine ratios. All graft copolymers of 19 and the NCA of benzyl glutamate were soluble in pyridine and DMSO. Polymers with short grafts 1-3 peptide units, were also soluble in $CHCl_3$.

The most interesting aminomethyl derivative of condensation polymers that we have prepared to date is derived from direct reduction of poly(2-cyano-1,3-phenylene arylene ether), 20. Enchainment of benzonitrile repeat units is accomplished by coupling 2,6-dichlorobenzonitrile with the potassium salt of bisphenol-A; copolymers with lower nitrile contents can be produced by copolycondensation of bisphenol-A, 2,6-dichlorobenzonitrile and 4,4'-dichlorodiphenyl sulfone.[21] The pendent nitrile function provides an active site for further elaboration.

We have shown that cyanoethylcellulose can be reduced to aminopropylcellulose using borane complexes.[31] However, polymer 20 could not be reduced under similar conditions. The more powerful reducing agent, LAH, was required to effect the reduction of the enchained benzonitrile. Addition of a THF solution of 20 to a solution of LAH in THF produced a homogeneous reaction mixture, but, within one hour, the reduced polymeric complexes began to precipitate. We used high dilution techniques and long reaction times to assure complete reduction. Isolation of the desired poly(2-aminomethyl-1,3-phenylene arylene ether), 21, was complicated by the formation of polyamine-metal complexes and a gelatinous precipitate of hydroxides. The multistep isolation procedure required to free the polyamine of metal ions reduced the actual yields of 21 to about 70%. Both 20 and 21 are soluble in $CHCl_3$; the intrinsic viscosities of the two polymers were 0.59 and 0.30 dl/g, respectively. Thus, the reduction appears to be accompanied by some chain cleavage.

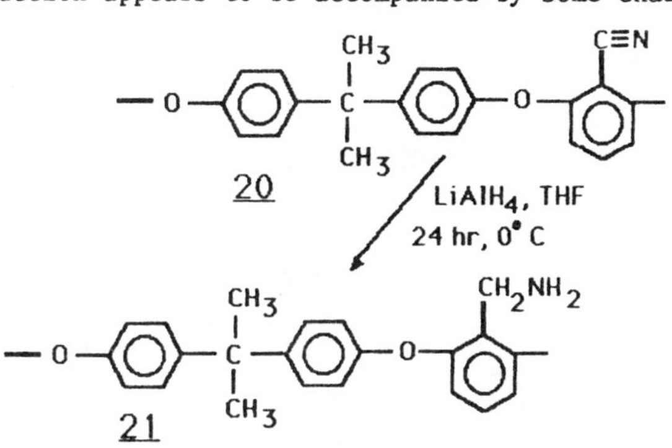

20

$LiAlH_4$, THF
24 hr, 0° C

21

The solubility properties of 21 render it an excellent support for benzyl glutamate grafting. Reaction of various ratios of NCA/amine ratios yielded grafts of the predicted chain length. The graft copolymers were completely soluble in THF and $CHCl_3$ and could be cast into films. The unique properties of these materials are under investigation.

Conclusions

Among the condensation polymers derived from bisphenol-A, poly(arylene ether sulfone) exhibits the best balance between reactivity and backbone stability for subsequent modification. We have shown that electrophilic substitution of the aryleneoxy units, i.e. bromination, nitration and chloromethylation, can be achieved. Introduction of amino and aminomethyl substituents provides a nucleophilic resin with applications in ion-exchange and graft polymerization technology. All of the modified poly(arylene ether sulfone)s can be cast into membranes; further research will focus on the permselectivity imparted by specific functional groups.

Literature Cited

1. Merrifield, R. B. J. Am. Chem. Soc. 1963, 85, 2149.
2. Bailey, D. C.; Langer, S. H. Chem. Rev. 1981, 82, 109; Frechet, J. M. J. Tetrahedron 1981, 37, 663; Akelah A.; Sherrington, D.C. Polymer, 1983, 24, 1369; and references cited therein.
3. Polymer Supported Reactions in Organic Synthesis; Hodge, P.; Sherrington, D. C. Eds. Wiley: New York, 1980; Mathur, N. K.; Narang, C. K.; Williams, R. E. Polymers as Aids in Organic Chemistry; Academic: New York, 1980; Polymeric Reagents and Catalysts; Ford, Warren T., Ed.; ACS Symposium Series 308; American Chemical Society: Washington, DC, 1986.
4. Correct IUPAC nomenclature:poly(oxy-2,6-dimethyl-1,4-pheylene), PPO, poly(oxy-1,4-phenylenesulfonyl-1',4'-phenyleneoxy-1",4" phenylene (1'''-methylethylidene)-1''',4'''-phenylene), poly(arylene ether sulfone), and poly(oxy-(2-hydroxytri-methylene) oxy-1,4-phenylene(1'-methylethylidene)-1',4'-phenylene), phenoxy resin.
5. Ford, Warren T. in Polymeric Reagents and Catalysts; Ford, Warren T., Ed.; ACS Symposium Series 308; American Chemical Society: Washington, DC, 1986; pp 247-285.
6. White, D. M.;Orlando,C.M.in Polyethers; Vandenberg, E.,Ed.; ACS Symposium Series 6; American Chemical Society: Washington, DC, 1975; pp 178-184.
7. Vollmert, B.;Schoene, W. Angew. Makromol. Chem. 1969, 7, 15.
8. Percec, S.; Li, G. Polymer Preprints 1986, 27(2),19.
9. Chalk, A. J.; Hay, A.S. J. Polym. Sci., Polym. Chem. Ed. 1969, 7, 691.
10. Xie, S.; MacKnight, W.J.; Karasz, F.E. J. Appl. Polym. Sci., 1984, 29, 2679.
11. Mohanraj, S.; Ford, Warren T. Macromolecules 1986, 19, 2470.
12. Noshay, A.; Robeson, L. M. J. Appl. Polym. Sci. 1976, 20, 1885.
13. Huang, R. Y.; Kim, J.J. J. Appl. Polym. Sci. 1984, 29, 4017.
14. Ward,III, W.J.; Salemme, R.M. U. S. Patent 3 780 496, 1973.
15. Li, G. U. S. Patent 4 521 224, 1985.
16. Verdet, L.; Stille, J.K. Organometallics 1982, 1, 380.

17. Daly, W. H.; Wu, S. J. in New Monomers and Polymers ed. by Culbertson, Bill M.; Pittman, C.U.; Plenum: New York, 1984, pp 201-222.
18. Daly, W. H. J. Macromol. Sci. Chem. 1985, A22, 713.
19. Percec, V.; Auman, B. C. Makromol. Chem. 1984, 185, 2319.
20. Amoto, J. S.; Karady, S.; Sletzinger, M.; Weinstock, L. M. Synthesis 1979, 970.
21. Mohanty, D. K.; Hedrick, J. L.; Gobetz, K. Johnson, B.C.; Yilgor, I.; Yang, R.; McGrath, J. E. Polymer Preprints 1982, 23(1), 284; Heath, D. R.;Wirth, J. C. U. S. Patent 3 730 946, 1973; Chem. Abstr. 1973, 77, P127232.
22. Johnson, R. N.; Farnham, A. G.; Clendinning, R.A.; Hale, W.F.; Merrian, C.N. J.Polym. Sci. Part A-1, 1967, 5, 2375.
23. Crivello, J. V. J. Org. Chem. 1981, 46, 3056.
24. Wu, S.J. Ph.D. Dissertation, Louisiana State University, December 1983.
25. Ohmae, I; Takeuchi, Y.Japan Kokai 7,550,352, 1975; Chem. Abstr. 1976, 84, P31968k.
26. Farrall, M. J.; Frechet, M. J. J. Org. Chem. 1976, 41, 3877.
27. Rungaroonthaikul, C. Ph.D. Dissertation, Louisiana State University, May 1986.
28. Daly, W. H.; Chotiwana, S.; Nielson, R. A. Polymer Preprints 1979, 20, 835.
29. Blazevic, N.; Kolbah, D.; Belin, B.;Sunjic, V.; Kajfez, F. Synthesis 1979, 161.
30. Koziara, A.; Osowska-Poacewicka, K.; Zawadski, S.; Zwierzak, A. Synthesis 1985, 202.
31. Daly, W. H.; Chotiwana, S.; Liou, Y-C. in Polymeric Amines and Ammonium Salts, ed. by Goethals, E. J., Pergamon: New York 1980, p 37.
32. Daly, W, H.; Munir, A. J. Polym. Sci. Polym. Chem. Ed. 1984, 22, 975.

RECEIVED August 27, 1987

Chapter 2

Dimethylene Spacers in Functionalized Polystyrenes

Graham D. Darling[1] and Jean M. J. Fréchet[2]

Department of Chemistry, University of Ottowa, Ottowa, Ontario K1N 9B4, Canada

Modes of attachment of functional groups to crosslinked polystyrene are discussed (1). Attention is drawn to improved stability and activity of polymer-bound reagents and catalysts incorporating dimethylene spacer between polystyrene aryl and functional group heteroatom, and the simplicity and versatility of their synthesis through high-conversion functional group modifications.

Beginning as ion-exchange resins (2), and blossoming as supports for solid-phase peptide synthesis (3), functionalized polystyrenes have progressively expanded in variety and importance as insoluble catalysts or reagents in the segregation, separation, protection, acylation, alkylation, dehydrohalogenation, oxidation or reduction of small molecules, reactions which are often followed by simple purification and regeneration of the solid phase (4, 5). Currently, most of these reactive polymers are prepared from (chloromethyl) polystyrene, which itself is easily made by controlled functionalization (2, 6) of commercially available styrene-divinylbenzene copolymer. Facile nucleophilic displacement of labile benzylic chloride, frequently under phase-transfer conditions (7), allows functional free molecules bearing nucleophilic heteroatoms to become attached to the macromolecular matrix. However, the new carbon-heteroatom bonds formed in these functional group modifications are themselves relatively fragile, since resonance stabilizes the partial charges developed during their rupture (8, 9) as shown below.

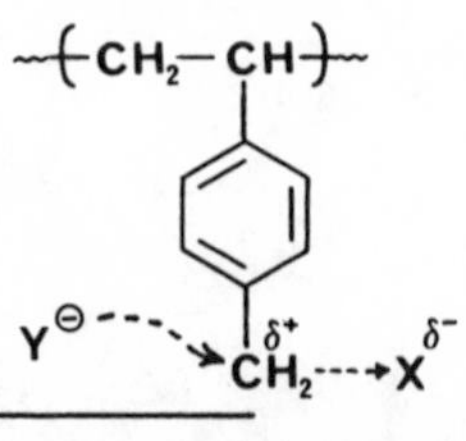

[1]Current address: IBM Almaden Research Laboratory, San Jose, CA 95114

[2]Current address: Department of Chemistry, Cornell University, Ithaca, NY 14853–1301

0097–6156/88/0364–0024$06.00/0

Many examples of such benzylic instability are provided by classical (free-molecule) organic chemistry, as well as a few by the smaller and more specialized field of polymer-supported reagents and catalysts. Hence:

- Benzylic quaternary phosphonium and ammonium salts are dealkylated by mild heating and/or nucleophilic anions, particularly iodide (9) and thiolate (10), but also hydroxide (11). Most N-benzylpyridinium or quaternary aryl ammonium compounds are particularly susceptible (12). Decompositions of this sort have seriously limited the usefulness of solid phase-transfer catalysts derived from (chloromethyl)polystyrene (13, 14).
- Benzyl groups are cleaved from amines heated with inorganic acids (15), carboxylic anhydrides (16) or cyanogen bromide (17). Acids also catalyze the disproportionation of benzylamine to dibenzylamine and ammonia (18).
- N-benzyl-sulfonamides are cleaved under "relatively mild conditions" with potassium alkoxide (19).
- Benzyl ethers are amongst the easiest to cleave by Lewis acids (20). Significant clipping of such bonds, with consequent loss of functionality, resulted during attempted HCl-catalyzed hydrolyses of polystyrene-supported oxazoline intermediates (21, 22) and chiral supports (23, 24).
- Benzyl-type esters hydrolyze relatively easily. This is well-known for polymeric systems (22, 25, 26).
- Acetyl bromide cleaves dibenzyl sulfide, while diethyl or diisopropyl are resistant (27).
- Benzyl-silanes are not immune to hydroxide or other nucleophiles (28, 29).
- Benzyl-nitrogen, sulfur, or -oxygen bonds are somewhat susceptible to hydrogenolysis, either during catalytic hydrogenation (30, 31), or upon treatment with L-selectride (32), or even lithium aluminium hydride (33).

Thus, the chemist or chemical engineer may legitimately fear further attack by undesired nucleophiles on benzylic carbon, as a main or side-reaction in the course of assembly, application or regeneration of many members of this structural class of functional polymer, leading to loss of reactive sites or mobile groups from the solid phase. As the last step of a solid-phase synthesis, such as that of polypeptides by Merrifield, controlled separation of pendant from polystyrene support may even be useful (3, 25) -- but generally such decomposition of a reactive polymer only serves to deactivate it while introducing contaminants to both solid and liquid phases. In addition, oxidation (15, 34) or alkylation (35, 36) may take place at benzylic methylene carbons made particularly acidic by being bound at once to aryl and electron-withdrawing groups or polarizable second-row elements.

Direct connection of pendant heteroatom to polystyrene aryl is a synthetically more difficult, but often still feasible (37), alternative. However, though bonds from phenyl to many common heteroatoms are relatively strong, resonance stabilization of partial positive charge developed on an arylated atom activates it to leave other substituents: alkyl anilium salts (12) and anilines (38), as well as phenolic esters (39), are relatively easy to cleave. Aryl linkages,

resonating with reaction products, make many non-metal atoms easier to oxidize. More electropositive elements, such as silicon (40), are easily displaced from phenyl by protons and other Lewis acids through electrophilic aromatic substitution. Furthermore, an antipodal electron-withdrawing group can activate benzyl hydrogen on the polymer backbone for inopportune ionic or radical side-reactions (41).

On the other hand, inserting even one more methylene group between polymer and pendant would result in a primary bond neither unstable nor destabilizing towards nucleophiles, permitting a chemically more robust working molecule while retaining the attractive physical properties of polystyrene resin. Additionally, extending a reactive site further from polymer backbone improves contact with free substrate in the liquid phase (42, 43), as well as with similar functional neighbours to enhance a cooperativity effect (44), for an overall increase in catalytic or other action. For example, among both polymers and free molecules containing a phenethyl substructure, it has been observed that:

- 2-phenethyl-halopyridinium salts are unusually stable compared to their benzyl, and even methyl or ethyl, counterparts (45, 46);
- N-benzyl-N-phenethyl-N,N-dimethylammonium chloride loses predominantly benzyl when heated with thiophenoxide, giving stable N,N-dimethylphenethylamine (47);
- N,N-dimethylphenethylamine itself is inert to further treatment with hot acetic anhydride (48);
- simple carboxylic acids can be alkylated (26) or photochlorinated (49), cinnamates photocrosslinked (50) and t-butylcarbonylated amino acids deprotected by acidolysis (51), while remaining esterified to polystyrene via dimethylene spacer;
- quaternary phosphonium salts connected through dimethylene spacers are stabler and more active phase-transfer catalysts than benzylic ones for reaction of cyanide or thiolate with alkyl halides; rates further improve with even longer spacers (52, 53);
- phenethyl-supported polystyryl tartarate, functionning as a stable solid-phase chiral auxiliary for the Sharpless epoxidation of allylic alcohols, gives higher chemical and optical yields than its benzylic analogue (54);

Introduction of dimethylene spacers onto polystyrene.
Most of the techniques and reactions described below, though often applicable to soluble polystyrene have been optimized using a 1% divinylbenzene crosslinked gel. To simplify notations, such a polymer having 20% of its aromatic groups substituted with functional group Z would be represented as shown below and be given the molecular formula: $(C_{10}H_{10})_{0.01}\cdot(C_8H_8)_{0.79}\cdot(C_8H_7Z)_{0.20}\cdot$

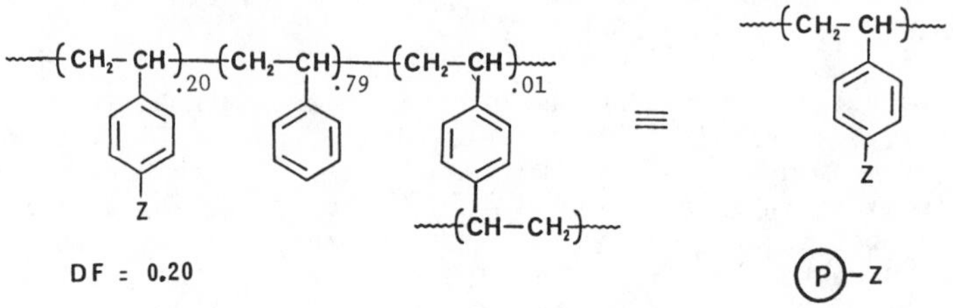

$$-(CH_2-CH)_{.20}-(CH_2-CH)_{.79}-(CH_2-CH)_{.01}- \equiv -(CH_2-CH)-$$

DF = 0.20

Ⓟ-Z

Functionalization through hydroxyethyl group.

A dimethylene spacer, terminated by a versatile hydroxyl "handle", is easily introduced onto commercial crosslinked polystyrene resin, without rearrangement (55) or grafting of ethylene glycol oligomer (56), by treating the cold lithiated polymer with excess ethylene oxide (37). The resulting (hydroxyethyl)polystyrene is virtually identical to product made less simply by cyanation/hydrolysis/ reduction of commercial (chloromethyl)polystyrene (52, 57), or by copolymerization of styrene, divinylbenzene and the protected functional monomer (26, 58). Past efforts to prepare polystyrenes possessing longer functionalized oligomethylene spacers, by Friedel-Crafts alkylation of dihalides, or reaction of dihalides or cyclic halomium ions with metallated polymer, have generally been accompanied by significant additional crosslinking of polymer networks (53, 59, 60). A recent exception is Tomoi et al's (61) alkylation of soluble or crosslinked polystyrene with w-bromoalkenes, catalyzed by 30 mol% trifluoromethanesulfonic acid, a promising polymer functionalization which is reported to proceed controllably and without crosslinking or decomposition of polymer backbone.

(Hydroxyethly)polystyrene has previously been further functionalized without breaking the C-O bond, by transesterification (54), or by carboxylic acid with dicyclohexylcarbodiimide (51), or by acid anhydride (58), or by acyl halides (26, 49) or chlorodialkylphosphites or phosphines (50) and acid acceptor. Derivatizing with 3,5-dinitrobenzoyl chloride in pyridine, followed by nitrogen analysis, is a reliable way of assaying polymer-bound hydroxyl (37, 57). Alkylating and arylating reagents (62) also react with polymeric alcohol in a straightforward manner.

For (hydroxyethyl)polystyrene to succeed as a versatile synthetic intermediate (57), techniques must be evolved to replace oxygen by other atoms which may be more suited to good function of final reagent. Such conversions must be completely quantitative, since unreacted, side-reacted or over-reacted functionalities cannot be removed from the solid product, and may ultimately interfere with its destined activity.

Nucleophilic substitution at 2-phenethyl under SN1 conditions is occasionally accompanied by a curious scrambling of both methylene carbons, attributed by some to the intermediacy of a nonclassical three-membered carbonium ion; however, since the consequences are only detectable in isotopically-labelled molecules, the mechanistic detail is moot (63). The alkene being conjugated, elimination can be a more important side-reaction of phenethyl functionalities than with many others. Fortunately, its occurrence during functional group modification can often be avoided by careful choice of reaction strategy and conditions; for example, by choosing tosylate as a leaving group (64, 65), and by working in polar solvents at low or moderate temperatures with strong nucleophiles that are also weak bases (66). This kind of decomposition is not a problem for most target polymer reagents and catalysts, whose final functionalities are generally poor leaving groups under basic conditions.

Functionalization through tosyloxyethyl group.

(Chloroethyl)polystyrenes and (iodoethyl)polystyrenes are each prepared from the alcohol by common reagents in a single step without complications, but one-pot procedures fail to produce completely pure bromide, which must be prepared from the tosylate by assisted halide exchange (57). The preparation of (toluenesulfonyloxyethyl)polystyrene itself, if performed in ice-cold pyridine as for the free análogue (64, 65), required a week to complete (67) if quaternary ammonium and other side-products (68) are to be avoided. In contrast, with non-nucleophilic diisopropylamine (69) as acid acceptor instead of pyridine, (hydroxyethyl)polystyrene and toluenesulfonyl chloride need only be refluxed in carbon tetrachloride for a few hours to give the desired tosic ester as sole product in quantitative yield (57).

$$\text{(P)}-CH_2CH_2OH \xrightarrow[(i\text{-}Pr)_2NH]{TsCl} \text{(P)}-CH_2CH_2OTs$$

Tosylate is displaced by weak oxyanions with little elimination in aprotic solvents, providing alternative routes to polymer-bound esters and aryl ethers. Alkoxides, unfortunately, give significant functional yields of (vinyl)polystyrene under the same conditions. Phosphines and sulfides can also be prepared from the appropriate anions (57), the latter lipophilic enough for phase-transfer catalysis free from poisonning by released tosylate.

Transformations to polymer-bound amino compounds, which are often useful as ligands for metals ions or other free species (67), employ a wide selection of organic reactions. Quaternary ammonium salts result from heating isolated polymer tosylate with tertiary amine; they may also be prepared in one step from (hydroxyethyl)polystyrene and toluenesulfonyl chloride and a two-fold excess of amine. Secondary amines, such as pyrrolidine, must be alkylated with care: too polar a solvent leads to participation of a second nearby polymer-bound alkylant in the formation of a quaternary ammonium salt, along with the desired immobilized trialkyl amine. The exception, as seen above, is diisopropylamine, which refuses to displace tosylate even in the refluxing pure amine, or in hot dimethylformamide or other polar solvent, while metal diisopropylamide is notorious as a powerful non-nucleophilic base. However, carboxamide is not difficult to form from (carboxymethyl)polystyrene, again using toluenesulfonyl chloride as condensing agent; this can then be reduced to (diisopropyl-ethylaminoethyl)polystyrene, which is of interest as a polymer-bound non-nucleophilic base.

$$\text{(P)}-CH_2COOH \xrightarrow[(i\text{-}Pr)_2NH]{TsCl} \text{(P)}-CH_2C(=O)N(i\text{-}Pr)_2 \xrightarrow{BH_3} \text{(P)}-CH_2CH_2N(i\text{-}Pr)_2$$

An amide anion will prefer substitution if its basicity is sufficiently lowered by resonance, and can be useful where the neutral nitrogen is unreactive or otherwise unsuitable. Quaternization cannot be prevented in the alkylation of even free neutral imidazole, but an imidazolide anion will match with only one of the electrophilic sites terminating dimethylene spacer on polystyrene.

$$\text{(P)}-CH_2CH_2OTs \xrightarrow{\text{imidazolide } N^{\ominus}\ Na^{\oplus}} \text{(P)}-CH_2CH_2N\text{(imidazole)}$$

Phthalimide and N-alkyl-toluenesulfonamide salts are similarly alkylated, and can furthermore be cleaved to polymer-bound secondary and primary amines respectively (57). Potassium pyrrolidonide gives polymer-bound tertiary amide, of interest as a solid cosolvent catalyst;

$$\text{(P)}-CH_2CH_2OTs \xrightarrow{K^{\oplus}\ \text{pyrrolidonide } N^{\ominus}} \text{(P)}-CH_2CH_2N\text{(2-pyrrolidone)}$$

In contrast, the acyclic anions of N-methyl acetamide and N-methyl-o-toluamide prefer protons to carbon, forcing an alternative approach such as acylation of (methylaminoethyl)polystyrene. It is unfortunate too that 4-N-methyl-pyridinamide, whose conjugate acid prefers alkylation on ring rather than side-chain nitrogen, is similarly too basic and/or hindered for good yields of immobilized dialkylaminopyridine acylation catalyst from (toluenesulfonyloxyethyl)polystyrene though not from (bromopropyl)polystyrene (Fréchet, J.M.J.; Deratani, A.; Darling, G.D.; Macromolecules in press), whose synthesis from specialty monomer is however more difficult for the general organic laboratory. Nevertheless, this illustrates one of the few cases where a three carbon spacer shows significant advantages over its two carbon analog.

A number of synthetic approaches were undertaken towards a polymer-bound version of the versatile N-methyl-2-halopyridinium condensation agent developped by Mukaiyama (70), a phenethyl-substituted analogue of which shows exceptional stability and potential for recyclability (71). Polymeric alkylation of 2-chloropyridine or 2-methoxypyridine was incomplete due to hindrance and deactivation of heterocyclic nitrogen, though the impure product still showed some dehydrating ability (72). Alternative formation of 2-pyridone by oxidation (73) of pyridinium chloride with basic ferricyanide was accompanied by small amounts of coloured ring-opening (74) side-products. The same 2-pyridone could be obtained by strongly heating polymeric tosylate in a large excess of 2-methoxypyridine as solvent, followed by hydrolysis. Conditions of formation from the trimethylsilyl ether of the 4-isomer, which ought to display similar behaviour by analogy with the corresponding free molecules (75), are milder and more economical. Treatment of (4-pyridonethyl)polystyrene by thionyl or carbonyl chlorides, concurrently with metathesis by tetrafluoroborate lipophilic counterion, affords a solid-phase reagent capable of transforming a mixture of alcohol and carboxylic acid to the ester. However, the measured activity of the polymer is somewhat less than hoped for, probably due to lack of accessibility of sites deep within a polymer matrix insufficiently swollen by organic solvents. Undoubtedly conditions can still be optimized for this promising reagent.

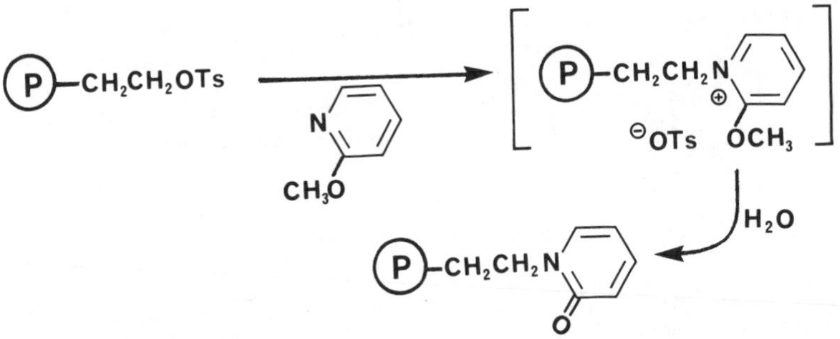

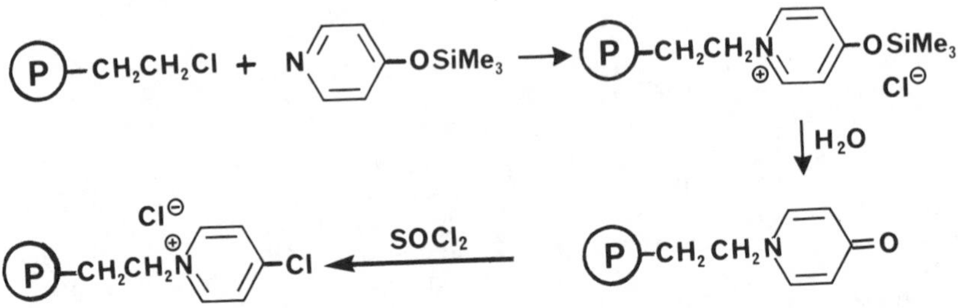

Deliberate production of (vinyl)polystyrene from (toluenesulfoxyethyl)polystyrene or (haloethyl)polystyrenes was best accomplished by quaternization with N,N-dimethylaminoethanol, followed by treatment with base: beta-deprotonation is encouraged in the cyclic zwitterionic intermediate. Reaction was faster and cleaner than with other reagents recommended (64, 76, 77) for eliminations, such as alkoxide, diazabicycloundecene or quaternary ammonium hydroxide; this new and efficient procedure may find application elsewhere. Hydrometallation or other additions to polymer-bound olefin may prove useful steps in future syntheses by polymer modification.

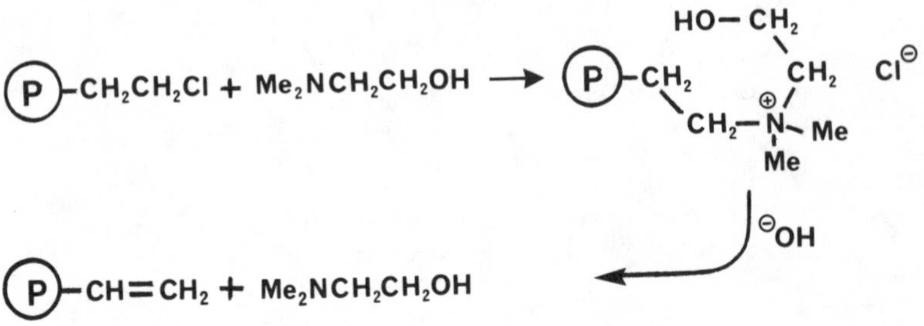

EXPERIMENTAL SECTION

Infrared spectra of KBr pellets were measured on a Nicolet 10DX FTIR. Elemental analyses were performed using Parr peroxide bombs and by MHW Laboratories (Phoenix, AZ).

(Hydroxyethyl)polystyrene. $(C_{10}H_{10})_{0.01}\cdot(C_8H_8)_{0.57}\cdot(C_8H_7Br)_{0.42}$ (Bromo)polystyrene (20.00 g, 60 meq), originally prepared from commercial 1% crosslinked polystyrene resin (37), was suspended in 500 ml dry benzene under nitrogen. Into this pale orange suspension was injected 2.2 M nBuLi/hexane (60 ml, 132 meq), and the whole was stirred at 60°C for 3 hours. The liquid phase was removed by filtration, and the beige residue, still under nitrogen, was suspended in 200 ml dry THF, and cooled to -50°C, before receiving condensed ethylene oxide by injection (9.5 ml, 190 meq), then being further stirred under a CO_2-cooled condenser for 18 hours while gradually warming to room temperature. The pale orange-yellow suspension was then filtered, and the residue washed 1 X THF:H_2O 3:1, 1 X THF:H_2O:conc HCl/H_2O 8:2:1, 3 X H_2O, 1 X THF, 2 X CH_3OH, 1 X Et_2O, then dried under vacuum overnight: IR (KBr) peaks absent at 1487, 1408, 1073, 1010, and 718 cm^{-1} for aryl bromide precursor, and at 1150-1060 cm^{-1} for ether side-products (CH_2-O-CH_2); peaks present at 3400 (br, CH_2O-H) and 1046 (s, CH_2-O) cm^{-1}. Anal. Calcd for $(C_{10}H_{10})_{0.01}\cdot(C_8H_8)_{0.57}\cdot(C_{10}H_{12}O)_{0.42}$: C, 86.58; H, 7.96; O, 5.47; Br, 0. Found: C, 86.51; H, 8.05; O, 5.51; Br, 0.

(3,5-dinitrobenzoyloxyethyl)polystyrene. $(C_{10}H_{10})_{0.01}\cdot(C_8H_8)_{0.57}\cdot(C_{10}H_{12}O)_{0.42}$ (Hydroxyethyl)polystyrene (0.13 g, 0.44 meq) was heated with 3,5-dinitrobenzoyl chloride (0.13 g, 0.55 meq) in 2 ml dry pyridine at 100°C under nitrogen for 2 hours, then the yellow-green suspension was filtered, and the residue washed 3 X H_2O, 1 X MEK, 1 X CH_2Cl_2, 2 X CH_3OH, and dried under vacuum overnight, yielding 0.20 g of a pale yellow-beige powder: IR (KBr) peaks absent at 3400 and 1046 cm^{-1} for alcohol precursor; peaks present at 3098 (w, Ar-H), 1733 (s, C=O), 1547 (s, Ar-NO_2), 1344 (s, Ar-NO_2), 1277 (s, CO-O), 1164 (m, CO-O), 721 (m, Ar-NO_2) cm^{-1} . Anal. Calcd for $(C_{10}H_{10})_{0.01}\cdot(C_8H_8)_{0.57}\cdot(C_{17}H_{14}N_2O_6)_{0.42}$:N, 5.76. Found: N, 5.66.

(2,4-dinitrophenoxyethyl)polystyrene. $(C_{10}H_{10})_{0.01}\cdot(C_8H_8)_{0.63}\cdot(C_{10}H_{12}O)_{0.36}$ (Hydroxyethyl)polystyrene (0.51 g, 1.5 meq) was heated with 2,4-dinitrofluorobenzene (0.30 ml, 2,4 meq) yellow liquid, and triethylamine (0.30 ml, 2.2 meq) in 2 ml dry toluene at 88°C for 40 hours, then the dark red suspension was filtered, and the residue washed 3 X acetone until filtrate was colourless, 3 X H_2O, 1 X MEK, 1 X CH_2Cl_2, 2 X CH_3OH, and dried under vacuum overnight, yielding 0.74 g of bright yellow powder: IR (KBr) peaks absent at 3400 and 1046 cm^{-1} for alcohol precursor; peaks present at 1609 (s, Ar-NO_2), 1536 (s, Ar-NO_2), 1344 (s, Ar-NO_2), 1287 (s, Ar-NO_2), 1152 (m, CH_2-OAr), 1087 (m, CH_2-OAr), 833 (m, Ar-H) cm^{-1} for desired product.

(Toluenesulfonyloxyethyl)polystyrene. $(C_{10}H_{10})_{0.01}\cdot(C_8H_9)_{0.70}\cdot(C_{10}H_{12}O)_{0.29}$ (Hydroxyethyl)polystyrene (10.35 g, 35 meq), toluenesulfonyl chloride (8.4 g, 44 meq), and dry diisopropylamine (10 ml, 71 meq), were stirred together in 70 ml carbon tetrachloride under nitrogen at room temperature, then heated to reflux for 6 hours. The pale yellow suspension was filtered, and the residue washed 1 X acetone, 3 X H_2O, 1 X MEK, 1 X CH_2Cl_2, 2 X Et_2O, and dried under vacuum overnight to yield 14.28 g of a very pale yellow powder: IR (KBr) peaks absent at 3400 and 1046 cm^{-1} for alcohol precursor; peaks present at 1363 (s, SO-OC), 1180 (doublet, s, SO-OC), 1098 (m, C-O), 964 (s, S-O-C), 905 (m, S-O-C), 815 (m, S-O-C), 760 (m, S-O-C), 664 (m, Ar-SO_3) and 555 cm^{-1} (s, Ar-SO_3) cm^{-1}. Anal. Calcd for $(C_{10}H_{10})_{0.01}\cdot(C_8H_8)_{0.70}(C_{17}H_{18}SO_3)_{0.29}$: S, 5,74; N, 0. Found: 9, 5,73; N, 0.

(3-nitrophenoxyethyl)polystyrene. $(C_{10}H_{10})_{0.01}\cdot(C_8H_8)_{0.70}(C_{17}H_{18}SO_3)_{0.29}$ (Toluenesulfonyloxyethyl)polystyrene (0.51 g, 0.91 meq), m-nitrophenol pale yellow crystals (0.25 g, 1.8 meq) and anhydrous potassium carbonate (0.26 g, 2.0 meq) were stirred in 5 ml dry dimethylformamide under nitrogen at 60°C for 16 hours. The red suspension was filtered, and the residue washed 1 X acetome, 3 X H_2O, 1 X MEK, 1 X CH_2Cl_2, 2 X CH_3OH, and dried under vacuum overnight to give a pale yellow powder: IR (KBr) peaks absent at 1363 and 1176 cm^{-1} for tosic ester precursor; peaks present at 1531 (s, Ar-NO_2), 1351 (s, Ar-NO_2), 1245 (m, CH_2-OAr), 1029 (m, CH_2-OAr) and 738 (m, Ar-H) cm^{-1}. Anal. Calcd for $(C_{10}H_{10})_{0.01}\cdot(C_8H_8)_{0.70}\cdot(C_{16}H_{15}NO_3)_{0.29}$: N, 2.67. Found: N, 2.45.

(N-pyridinium)polystyrene tosylate, chloride. $(C_{10}H_{10})_{0.01}\cdot(C_8H_8)_{0.75}(C_{10}H_{12}O)_{0.24}$ (4.99 g, 10.4 meq) and toluensulfonyl chloride (5.0 g, 26 meq) were suspended in dry pyridine (60 ml, 700 meq) chilled in an ice bath to 0°C under nitrogen. The mixture was allowed to warm to room temperature over 4 hours, then refluxed for 2 hours, then cooled and filtered, and the residue washed 3 X H_2O, 1 X MEK, 1 X CH_2Cl_2, 1 X CH_3OH, 1 X Et_2O, and dried under vacuum overnight, yielding 6.72 g of a peach-coloured powder: IR (KBr) peaks absent at 1046 cm^{-1} for alcohol precursor, at 1363 and 1176 cm^{-1} for tosyl ester intermediate, and at 1245 and 657 for alkyl chloride side-product; peaks present at 3500 (s, br, absorbed H_2O), 1633 (m, pyr), 1582 (w, pyr), 1196 (s, SO-O^-), 1123 (s, SO-O^-), 1034 (s, SO-O^-), 1012 (s, SO-O^-), 682 (m, pyr^+) and 569 cm^{-1} (m, $ArSO_3^-$). The product was then washed with concentrated hydrochloric acid, then 1 X HCl/H_2O:THF 1:1, 1 X HCl/H_2O:THF 1:4, 3 X H_2O, 1 X MEK, 1 X CH_2Cl_2, 1 X CH_3OH, 1 X Et_2O, 1 X hexane, and dried under vacuum overnight, yielding 5.99 g of beige powder: IR (KBr) peaks absent at 1196, 1123, 1034, 1012 and 569 cm^{-1} for tosylate anion; peaks still present at 1633, 1582 and 682 cm^{-1} for pyridinium nucleus. Anal. Calcd for $(C_{10}H_{10})_{0.01}\cdot(C_8H_8)_{0.75}(C_{15}H_{16}NCl)_{0.24}$: Cl, 6.15. Found: Cl, 6.19.

(N-pyrrolidinethyl)polystyrene. $(C_{10}H_{10})_{0.01}\cdot(C_8H_8)_{0.70}\cdot(C_{17}H_{18}SO_3)_{0.29}$ (Toluenesulfonyloxyethyl)-polystyrene (0.19 g, 0.34 meq), dry pyrrolidine (0.032 ml, 0.38 meq) and anhydrous potassium carbonate (o.050 g, 0.36 meq) were stirred in 2 ml dry pyridine under nitrogen at room temperature for 2 hours, then slowly warmed to reflux over 5 hours. The red-orange suspension was filtered, and the residue washed 3 X H_2O, 1 X MEK, 1 X CH_2Cl_2, 1 X Et_2O, and dried under vacuum overnight to give a yellow-beige powder: IR (KBr) peaks absent at 1363 and 1176 cm^{-1} for tosic ester precursor, and at 1633, 1582 and 682 cm^{-1} for pyridinium and 1196, 1123, 1034, 1012, 682 and 569 cm^{-1} for tosylate side-products; peak present at 2784 cm^{-1} (CH_2). Anal. Calcd for $(C_{10}H_{10})_{0.01}\cdot(C_8H_8)_{0.70}\cdot(C_{14}H_{19}N)_{0.29}$: N, 3.06. Found: N, 2.98.

(N-pyrrolidinonethyl)polystyrene. $(C_{10}H_{10})_{0.01}\cdot(C_8H_9)_{0.794}\cdot(C_{17}H_{18}SO_3)_{0.196}$ (Toluenesulfonyloxyethyl)-polystyrene (0.31 g, 0.42 meq), dry pyrrolidinone (0.051 ml, 0.66 meq) freshly-crushed potassium hydroxide (0.11 g, 2.0 meq) were rapidly stirred in 4 ml dry dimethyl sulfoxide under nitrogen at room temperature for 3 hours, then at 60°C for 24 hours. The beige-brown suspension was then filtered, and the residue washed 3 X H_2O, 1 X MEK, 1 X CH_2Cl_2, 2 X CH_3OH, and dried under vacuum overnight, yielding 0.26 g of pale yellow-beige powder: IR (KBr) peaks absent at 1363 and 1176 cm^{-1} for tosic ester precursor; peak present at 1691 (s, NC=O) cm^{-1}. Anal. Calcd for $(C_{10}H_{10})_{0.01}\cdot(C_8H_8)_{0.794}\cdot(C_{14}H_{17}NO)_{0.196}$: N, 2.18. Found: N, 2.06.

(3-nitrobenzoyloxyethyl)polystyrene. $(C_{10}H_{10})_{0.01}\cdot(C_8H_8)_{0.794}\cdot(C_{17}H_{18}SO_3)_{0.196}$ (Toluenesulfonyloxyethyl)-polystyrene (0.41 g, 0.55 meq), n-nitrobenzoic acid pale grey powder (0.14 g, 0.83 meq) and anhydrous potassium carbonate (0.17 g, 1.2 meq)were rapidly stirred in a mixture of 4 ml dimethyl sulfoxide and 0.5 ml hexamethylphosphoramide under nitrogen and warmed to 60°C for 22 hours. The pale brown suspension was filtered and washed 1 X acetone, 3 X H_2O (rapidly), 1 X MEK, 1 X CH_2Cl_2, 1 X Et_2O, and dried under vacuum overnight, yielding 0.41 g of yellow-beige powder: IR (KBr) peaks absent at 1363 and 1176 cm^{-1} for tosic ester precursor, and at 1046 cm^{-1} for hydrolysis side-product; peaks present at 1727 (s, C=O), 1535 (s, Ar-NO_2), 1351 (s, NO_2), 1263 (m, CO-O), 1134 (Mm< CO-O) and 720 (m, Ar-H) cm^{-1}. Anal. Calcd for $(C_{10}H_{10})_{0.01}\cdot(C_8H_8)_{0.794}\cdot(C_{17}H_{15}NO_4)_{0.196}$: N, 1.93. Found: N, 1.87.

ACKNOWLEDGMENTS

Financial support of this research by the Natural Sciences and Engineering Research Council of Canada in the form of an operating grant (to JMJF) and a post-graduate scholarship (to GDD) is gratefully acknowledged.

LITERATURE CITED

1. From Darling, G.D.; Ph.D. Dissertation, University of Ottawa, Ottawa, Ontario, Canada, 1987.
2. Pepper, K.W.; Paisley, H.M.; Young, M.A.; J. Chem. Soc., 1953, 4097.
3. Merrifield, R.B.; J. Am. Chem. Soc., 1963, 85, 2149.
4. Akelah, A.; Sherrington, D.C.; Chem. Rev., 1981, 81, 557.
   Fréchet, J.M.J.; Tetrahedron, 1981, 37, 663.
   Hodge, P.; Sherrington, D.C. (Editors); "Polymer-Supported Reactions in Organic Synthesis", J. Wiley & Sons, 1980.
   Mathur, N.K.; Narang, C.K.; Williams, R.E.; "Polymers as Aids in Organic Chemistry", Academic Press, 1980.
5. Ford, W.T. (Editor); "Polymeric Reagents and Catalysts", American Chemical Society, Washington, D.C., 1986.
6. Pinnell, R.P.; Khune, G.D.; Khatri, N.A.; Manatt, S.L.; Tetrahedron Lett., 1984, 25, 3511.
7. Fréchet, J.M.J.; Polym. Sci. Tech., 1984, 24, 1.
   Fréchet, J.M.J.; De Smet, M.D.; Farrall, M.J.; J. Org. Chem., 1979, 44, 1774.
8. Arnett, E.M.; Reich, R.; J. Am. Chem. Soc., 1980, 102, 5892.
9. Westaway, K.C.; Poirier, R.A.; Can. J. Chem., 1975, 53, 3216.
10. Deady, L.W.; Korytsky, O.L.; Tetrahedron Lett., 1979, 451.
11. Dehmlow, E.V.; Slopianka, M.; Heider, J.; Tetrahedron Lett., 1977, 2363.
12. Snyder, H.R.; Speck, J.C.; J. Am. Chem. Soc., 1939, 61, 2895.
13. Tomoi, M.; Ford, W.T.; J. Am. Chem. Soc., 1981, 103, 3821.
14. Cinorium, M.; Colonna, S.; Molinari, H.; Montanari, F.; J. Chem. Soc. (C), 1976, 394.
15. Traynelis, V.J.; Ode, R.H.; J. Org. Chem, 1970, 35, 2207.
16. Mariella, R.P.; Brown, K.H.; Can. J. Chem., 1973, 53, 2177.
17. Snyder, H.R.; Speck, J.C.; J. Am. Chem. Soc., 1939, 61, 668.
18. Vorbruggen, H.; Krolikiewicz, K.; Chem. Ber., 1984, 117, 1523.
19. Searles, S.; Nukina, S.; Chem. Rev., 1959, 59, 1077.
20. Barluenga, J.; Alonso-Cires, L.; Campos, P.J.; Asensio, B.; Synthesis, 1983, 53.
21. Lecavalier, P.; Bald, E.; Jiang, Y.; Fréchet, J.M.J.; Hodge, P.; Reactive Polymers, 1985, 3, 315.
22. Colwell, A.R.; Duckwall, L.R.; Brooks, R.; McManus, S.P.; J. Org. Chem, 1981, 46, 3097.
23. Worster, P.M.; McArthur, C.R.; Leznoff, C.C.; Angew. Chem., Int. Ed. Engl., 1979, 18, 221.
24. Deady, L.W.; Korytsky, D.L.; J. Org. Chem., 1979, 45, 2717.
25. Patchornik, A.; Kraus, M.A.; J. Am. Chem. Soc., 1970, 92, 7587.
26. Camps, F.; Castells, J.; Ferrando, M.J.; Font, J.; Tetrahedron Lett., 1971, 1713.
27. Goldfarb, Y.L.; Ispiryan, R.M.; Belenkii, L.I.; Dokl. Cheml Proc. Acad. Sci. USSR Chem. Sec., 1967, 209.

28. Gilman, H.; Brook, A.G.; Miller, L.S.; J. Am. Chem. Soc., 1953, 75, 4531.
29. Chan, T.H.; Chang, E.; Vinokur, E.; Tetrahedron Lett., 1970, 1137.
30. Hartung, W.H.; Simonoff, R.; Org. React., 1953, 7, 263.
31. Erhardt, P.W.; Synth. Commun., 1983, 13, 103.
32. Newhome, G.R.; Majestic, V.K.; Sauer, J.D.; Org. Prep. Proced. Int., 1980, 12, 345.
33. Tweedie, V.L.; Allabash, J.C.; J. Org. Chem., 1961, 26, 3676.
34. Oda, R.; Hayashi, Y.; Tetrahedron Lett., 1967, 3141.
35. Fréchet, J.M.J.; Eichler, E.; Polym. Bull., 1982, 7, 345.
36. Ford, W.T.; Ref. 5, 155.
37. Farrall, M.J.; Fréchet, J.M.J.; J. Org. Chem., 1976, 41, 3877. Frechet, J.M.J.; Farrall, M.J.; "Chemistry and Properties of Crosslinked Polymers" (Labana, S.S., Editor), Academic Press, 59, 1977.
38. Chambers, R.A.; Pearson, D.E.; J. Org. Chem., 1963, 28, 3144.
39. March. J. "Advanced Organic Chemistry, Third Edition"; Wiley-Interscience, Toronto, 1985.
40. Chan, T.H.; Fleming, I.; Synthesis, 1979, 761.
41. Neckers, D.C.; Ref. 5, 107.
42. Brown, J.M.; Jenkins, J.A.; J. Chem. Soc., Chem Comm., 1976, 458.
43. Molinari. H.; Montanari, F.; Quici, S.; Tundo, P.; J. Am. Chem. Soc., 1979, 101, 3920.
44. Tomoi, M.; Ikeda, M.; Kakiuchi, H.; Tetrahedron Lett., 1978, 3757.
45. Fischer, O.; Chem. Ber., 1899, 32, 1297.
46. M.Kametami, C,; Nomura, Y.; Chem. Pharm. Bull., 1960, 8, 741.
47. Kametami, T.; Kigasawa, K.1 Hiiragi, M.; Wagatsuna, N.; Wakisaka, K.; Tetrahedron Lett., 1969, 635.
48. Tiffeneau, M.; Fuhrer, K.; Bull. Soc. Chim. Fr., 1914, 15, 162; Chem. Abstr. 8, 1567.
49. Bosch, P.; Font, J.; Moral, A.; Sanchez-Ferrando, F.; Tetrahedron, 1978, 34, 947.
50. Lieto, J.; Milstein, D.; Albright, R.L.; Minkiewicz, J.V.; Gates, B.C.; CHEMTECH, 1983, (1), 46.
51. Camps, F.; Castells, J.; Pi, J.; An. Quim., 1974, 70, 848; Chem. Abstr. 83, 10018g.
52. Chiles, M.S.; Jackson, D.D.; Reeves, P.C.; J. Org. Cheml, 1980, 45, 2915.
53. Tomoi, M.; Ogawa, E.; Hosokama, Y.; Kakiuchi, H. J. Polym. Sci., Polym. Chem. Ed., 1982, 20, 3015, 3421.
54. Farrall, M.J.; Alexis, M.; Trecarten, M.; Nouv. J. Chim., 1983, 7, 449.
55. Cristol, S.J.; Douglass, J.R.; Meek, J.S.; J. Am. Chem. Soc., 1951, 73, 816.
56. Lebedev, N.N.; Baranov, Y.I.; Vysokomolekul. Spedin., 1966, 8, 198; Chem. Abstr., 65, 808g.
57. Darling, G.D.; Fréchet, J.M.J.; J. Org. Chem., 1986, 51, 2270.
58. Hirao, A.; Takenada, K.; Yamaguchi, K.; Nakahama, S.; Yamazaki, N.; Polym. Comm., 1983, 24, 339.
59. Tundo, P.; Synthesis, 1978, 315.
60. McManus, S.P.; Olinger, R.D.; J. Org. Chem., 1980, 45, 2717.
61. Tomoi, M.; Kori, N.; Kakiuchi, H.; Reactive Polymers, 1985, 3, 341.

62. Whalley, W.B.; J. Chem. Soc., 1950, 2241.
63. Lee, C.C.; Spinks, J.W.T.; J. Am. Chem. Soc., 1960, 82, 138.
64. Saunders, W.H.; Edison, D.H.; J. Am. Chem. Soc., 1960, 82, 138.
65. Veeravagu, P.; Arnold, R.T.; Eigenmann, E.W.; J. Am. Chem. Soc., 1964, 86, 3072.
66. March, J.; Ref. 39, 895.
67. Cheminat, A.; Benezra, C.; Farrall, M.J.; Fréchet, J.M.J.; Can. J. Chem., 1981, 59, 1405.
68. Tipson, R.S.; J. Org. Chem., 1944, 9, 235.
69. Nagasaka, T.; Ito, H.; Ozawa, N.; Kosugi, Y.; Hamaguchi, F.; Yakugaku Zasshi, 1980, 100, 962.
70. Mukaiyama, T.; Angew. Chem. Int. Ed. Engl, 1979, 18, 707.
71. Paquette, L.A.; Nelson, N.A.; J. Org. Chem., 1962, 27, 1085.
72. Darling, G.D.; Bald, E.; Fréchet, J.M.J.; Polym. Prep., Am. Chem. Soc., Div. Polym. Chem., 1983, 24, 354.
73. Sugasawa, S.; Akahoshi, S.Y.; Suzuki, M.; J. Pharm. Soc. Japan, 1952, 72, 1273; Chem. Abstr., 1953, 47, 10539g.
74. Decker, H.; Chem. Ber., 1903, 36, 2568.
75. Liveris, M.; Miller, J.; Aust. J. Chem., 1958, 11, 297.
76. Nishikubo, T.; Iizawa, T.; Ashi, K.K.; Okawara, M.; Tetrahedron Lett., 1981, 22, 3873.
77. Oediger, H.; Moller, F.; Angew. Chem., Int. Ed. Engl., 1967, 6, 76.

RECEIVED August 27, 1987

Chapter 3

Chemical Modification of Poly(vinyl chloroformate) by Phenol Using Phase-Transfer Catalysis

S. Boivin, P. Hemery, and S. Boileau [1]

College de France, 11 Place Marcelin Berthelot, 75231 Paris Cédex 05, France

The nucleophilic substitution on poly(vinyl chloroformate) with phenol under phase transfer catalysis conditions has been studied. The ^{13}C-NMR spectra of partly modified polymers have been examined in detail in the region of the tertiary carbon atoms of the main chain. The results have shown that the substitution reaction proceeds without degradation of the polymer and selectively with the chloroformate functions belonging to the different triads, isotactic sequences being the most reactive ones.

Functionalized polymers have found increasing use as supports, reagents or catalysts namely in the field of organic synthesis, metal chelation or pharmacology (1). It is well known that chemical modification of polymers containing functional groups is one of the general routes for the synthesis of new polymeric reagents. In the last few years, the study of the reactions of the functional groups of macromolecules has received much attention. Kinetics and mechanisms of polymer-transformation reactions depend on characteristic polymeric effects like neighbouring-groups effects as well as configurational, conformational, electrostatic or supermolecular effects. A theoretical study should take into account all these effects but in practice the consideration of even one of them is very difficult. A mathematical treatment has been proposed by BOUCHER (2) and by PLATE and NOAH (3).

One of the best known examples of polymer-transformation reactions is the quaternization of poly(4-vinyl pyridine) by various alkyl

---

[1]Current address: Laboratoire de Chimie Macromoleculaire associé au Centre National de la Recherche Scientifique: UA24, Paris, France

0097-6156/88/0364-0037$06.00/0

and arylalkyl halides which yields a polyelectrolyte. Kinetic studies have been reported by several authors (4-7). They found retardation of the reaction which has been attributed to a dominant neighbouring-groups effect with the reactivity of a pyridyl group mainly due to the steric hindrance of its modified neighbours.

Similar neighbouring-groups effects as well as electrostatic effects have been observed in the case of nucleophilic substitution of poly(methyl methacrylate) with organolithium reagents and their analogues (8). Another well-known example of polymer-transformation reaction is the hydrolysis of poly(methyl methacrylate). Kinetic and mechanistic studies corresponding to this kind of reaction have shown that the reactivity of ester groups is sensitive to reaction conditions, neighbouring-groups effects and to polymer tacticity (9-13). Under basic conditions it has been found that syndiotactic sequences would be the most reactive.

Such configurational as well as conformational effects have been also reported by MILLAN et al. in the case of nucleophilic substitution of poly(vinyl chloride) with sodium thiophenate (14) and with sodium isooctylthioglycolate or isooctylthiosalicylate (15). The authors have shown that these reactions proceed selectively on the isotactic TT diads which can only exist either in the GTTG* isotactic or in the TTTG heterotactic triads, the former ones being much more reactive than the latter ones.

Nucleophilic substitution of chlorine in poly(vinyl chloroformate) (PVOCCl) has been reported as one of the best procedure for the synthesis of new functional polymers (16). It has been shown that PVOCCl can react with nucleophilic compounds containing labile hydrogen atoms like alcohols, phenols and amines (17). Since it is possible to prepare well-defined high molecular weight PVOCCl in quantitative yields (18,19), convenient conditions have been found for the chemical modification of this polymer with excellent yields of substitution (20). The best results have been obtained by using phase transfer catalysis (21-24). The usefulness of this technique to perform a great deal of reactions is now well established in organic chemistry (25-27).

" In phase transfer catalysis a substrate in an organic phase is reacted chemically with a reagent present in another phase which is usually aqueous or solid. Reaction is achieved by means of the transfer agent; this agent or catalyst is capable of solubilizing or extracting inorganic and organic ions, in the form of ion pairs, into organic media " (E.V. Dehmlow and S.S. Dehmlow (25)). In the case of the chemical modification of PVOCCl, two phases systems have been used : an organic phase (PVOCCl in CH_2Cl_2) and either an aqueous phase (aq. 50% NaOH) or a solid phase (K_2CO_3 , potassium carboxylate) with a catalytic amount of Bu_4NHSO_4 (TBAH), dicyclohexyl-18-crown-6 or cryptand [222]. These phase transfer catalysis conditions are mild enough in order to avoid polymer degradation : they require lower temperatures and shorter reaction times than classical conditions. Furthermore, the addition of a phase transfer catalyst allows to modify PVOCCl with alkaline salts like alkaline carboxylates (22) or with unsaturated heterocyclic amines like indole or carbazole (23,24).

Moreover it has been shown that PVOCCl prepared by free-radical polymerization of vinyl chloroformate (VOCCl) is an atactic polymer having a Bernouillian statistical distribution as expected (19). In order to extend our studies on the chemical modification of PVOCCl, the stereoselective character of the nucleophilic substitution of the chloroformate units with phenol has been examined by the study of the ^{13}C-NMR spectra of partly modified polymers in the region of the aliphatic methine carbon atoms. The results obtained in this field are presented here.

EXPERIMENTAL

Polymer preparation. PVOCCl sample has been prepared by free-radical polymerization of pure VOCCl (acquired from the SNPE, purity 99%) in CH_2Cl_2 at 35°C using dicyclohexyl peroxydicarbonate as initiator. The experimental procedure has been described previously (18,19). The molecular weight $\overline{M}_n$ of this sample is equal to 50,000.

Preparation of modified polymers. Phenol, Bu_4NHSO_4 (TBAH), CH_2Cl_2 are commercial products and they are used without special purification. The method for the chemical modification of PVOCCl with phenol has been described elsewhere (20,21).

The degrees of substitution have been determined by elemental analysis of the remaining Cl and C contents and by ^1H NMR according to the integration of the different peaks observed. The ^1H-NMR spectra have been recorded at 60 MHz in CD_3COCD_3 at 27°C : δ(ppm/TMS) : CH ar. : 7.1 ppm; -CH chain : 5.2 ppm; $-CH_2-$ chain : 2.1 ppm.

^{13}C-NMR spectroscopy. The tacticities for both starting and modified polymers have been determined from the ^{13}C-NMR spectra recorded at 50.3 MHz, at 20°C, with a AM 200 SY Bruker spectrometer. The polymers have been examined as a 10% solution in CD_3COCD_3 (ref : TMS).

RESULTS AND DISCUSSION

In a previous work it has been found that the chemical modification of PVOCCl by phenol in CH_2Cl_2, by using pyridine as HCl scavenger, leads to a soluble polymer containing 88% of vinyl phenyl carbonate units (20). A better yield of substitution has been obtained by using a liquid/liquid two phases system (50% aqueous $NaOH/CH_2Cl_2$) with Bu_4NHSO_4 as phase transfer catalyst (21,24) according to the following reaction scheme :

aq. 50% NaOH

$$ØO^-Na^+ + NBu_4^+X^- \rightleftharpoons ØO^-NBu_4^+ + Na^+X^-$$

$$\sim CH_2-CH\sim(O-C(=O)-O-Ø) + NBu_4^+X^- \leftarrow ØO^-NBu_4^+ + \sim CH_2-CH\sim(O-C(=O)-Cl)$$

CH_2Cl_2 X^- : Cl^- ; HSO_4^-

Under such conditions the substitution reaction of PVOCCl with phenol is quite fast at room temperature as shown by the results of Table I. The degree of substitution reaches 84% after 15 min. (run 1) whereas the modification is quantitative within 45 min. under similar conditions (run 3).

Table I. Chemical modification of PVOCCl by phenol at room temperature (PVOCCl : 5 mmole; [NaOH]/[PVOCCl] = 2; [TBAH]/[PVOCCl] = 0.05)

Run	ØOH (mmole)	CH_2Cl_2 (ml)	time (mn.)	D.S.(%)[a) I[b)]	D.S.(%)[a) II[c)]	D.S.(%)[a) III[d)]
1	5.1	50	15	84	86	-
2	5.0	50	25	91	-	-
3	5.1	50	45	100	-	-
4	5.1	50	120	98	100	100
5	0.7	35	20	14	12	14
6	1.4$_5$	40	35	29	27	29
7	2.3	45	30	46	48	44

a) degree of substitution
b) determined by elemental analysis of the remaining Cl and C.
c) determined by ^1H NMR.
d) calculated from the integration of the methine carbon peaks on ^{13}C-NMR spectra.

Ratios [chloroformate]/[phenol] higher than 1 have been used in order to prepare modified polymers with a degree of substitution lower than 50% (runs 5-7, Table I). Under such conditions it can be noticed that the substitution reactions proceed quantitatively with respect to the initial phenol content.

The evolution of the ^{13}C-NMR spectra of modified polymers with the degree of substitution, in the region of the aliphatic methine carbon atoms, is shown in Figure 1. Spectrum A corresponding to PVOCCl presents three signals located at 78.9, 77.85 and 76.9 ppm which belong to tertiary carbon atoms bearing chloroformate groups whereas the three signals located at 75.3, 73.6 and 72.0 ppm belong to

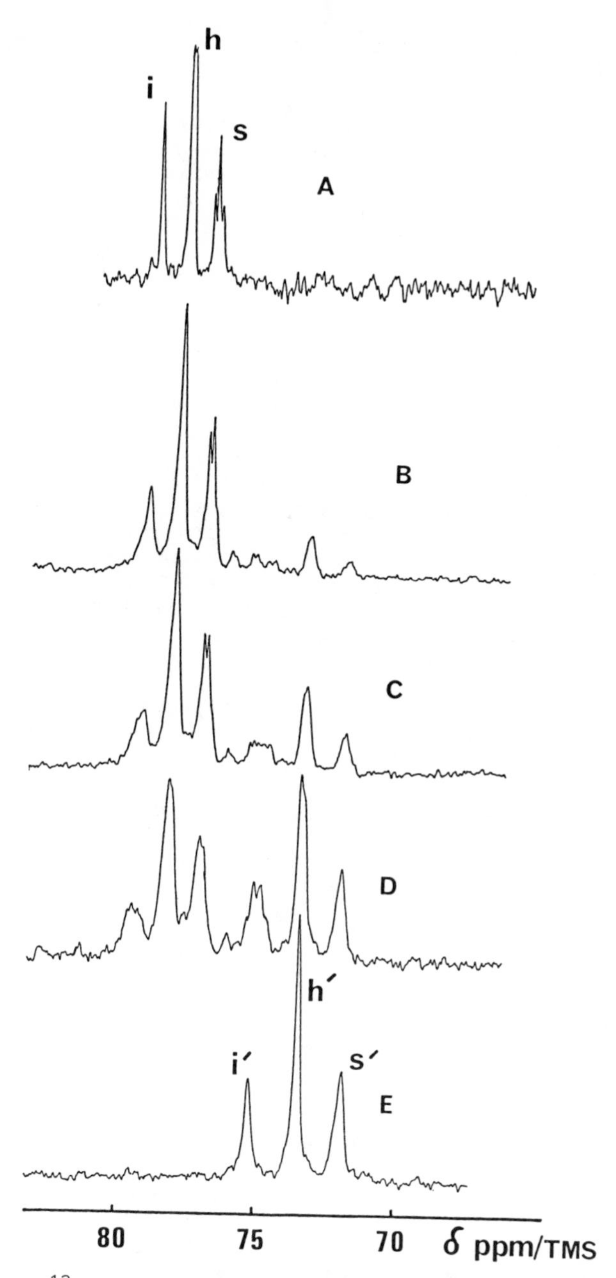

Figure 1. ^{13}C-NMR spectra of modified PVOCCl samples at various conversions : A (0%); B (14%); C (29%); D (46%); E (100%).

tertiary carbon atoms bearing carbonate groups. Spectra B, C and D have been obtained from partly modified polymers with degrees of substitution equal to 14%, 29% and 46% respectively (runs 5-7, Table I).

In the cases of PVOCCl and poly(vinyl phenyl carbonate) (spectra A and E respectively), the tertiary carbons of the main chain are sensitive to triad effects according to a Bernouillian statistical distribution. Moreover, in the case of PVOCCl (spectrum A), the multiplicity of the signals shows that the tertiary carbons of the main chain are likely sensitive to pentad effects. The assignment of the three main signals to iso-, hetero- and syndiotactic triads has been suggested previously for both PVOCCl and poly(vinyl phenyl carbonate) (3). In the present work this assignment has been considered to be true and is indicated in Figure 1.

In the case of partly modified polymers the spectra B, C and D show more complicated structures which can be presumably due to significant neighbouring-groups effects between phenyl carbonate groups and modified or unmodified groups linked to the next aliphatic methine carbon atoms.

As shown in Table I (runs 4-7) the degrees of substitution calculated from the integration of the signals corresponding to the methine carbons of vinyl chloroformate units and of vinyl phenyl carbonate units in the modified polymers are in good agreement with the values obtained by elemental analysis as well as by ^{1}H NMR. Such a result allows to compare the integrations of ^{13}C-NMR spectra of these polymers in the region of the tertiary carbons of the main chain. The percentages of the different triads bearing either a chloroformate or a carbonate function are reported in Table II. They have been calculated from the integration of ^{13}C-NMR spectra shown in Figure 1. The sums of the percentages (i+i'), (h+h') and (s+s') are independent of the degree of substitution and are equal to the values of PVOCCl (i),(h),(s) respectively, within the experimental errors.

From these results it is reasonable to conclude that no inversion in the order of the chemical shifts of the tertiary carbon atoms belonging to the different triads occurs from the starting PVOCCl to the poly(vinyl phenyl carbonate). Moreover the chemical modification of PVOCCl by phenol does not induce any degradation of the polymer.

The degrees of substitution of chloroformate functions belonging to iso-, hetero- and syndiotactic triads respectively have been calculated from the percentages shown in Table II. Their evolution with the total degree of substitution is plotted in Figure 2. It can be concluded that the reactivity of the chloroformate functions decreases in the following order :

$$\text{iso} > \text{hetero} > \text{syndio}$$

in the first steps of substitution (D.S. $\leqslant$ 25%) . From the foregoing it can be suggested that the chloroformate functions belonging to meso diads are more reactive than those belonging to racemic diads. A similar result has been observed previously by MILLAN et al. (14) in the case of the nucleophilic substitution of poly(vinyl chloride) with sodium thiophenate.

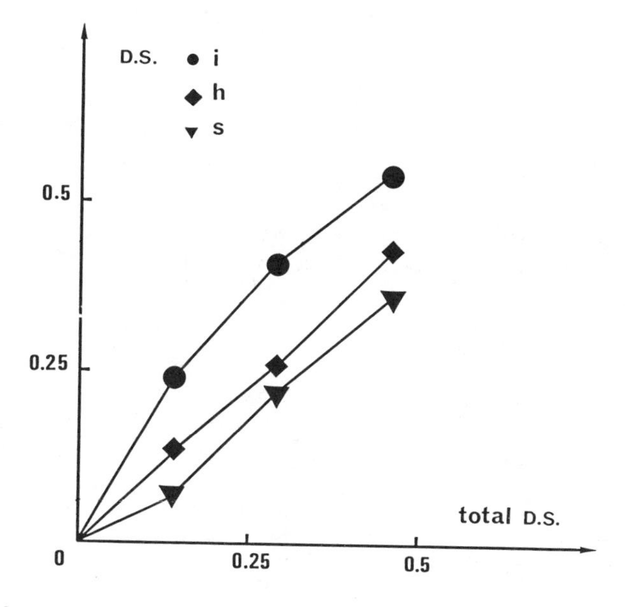

Figure 2. Conversion of chloroformate functions corresponding to different triads (i,h,s) vs the degree of substitution of the polymer.

Table II. Percentages of iso-, hetero- and syndiotactic triads bearing either a chloroformate or a carbonate group in PVOCCl and in the modified polymers[a)]

Run[b)]	spectrum[c)]	δ(ppm)[d)]	H-C-OCOCl 78.9	77.85	76.9	H-C-OCO_2Ø 75.3	73.6	72.0
			i	h	s	i'	h'	s'
PVOCCl	(A)		0.25	0.48	0.27	0	0	0
5	(B)		0.19	0.41	0.25	0.06	0.06	0.02
6	(C)		0.16	0.34	0.21	0.11	0.12	0.06
7	(D)		0.11	0.28	0.18	0.13	0.21	0.10
4	(E)		0	0	0	0.26	0.48	0.26

a) i+h+s+i'+h'+s'=1
b) see Table I.
c) see Fig. 1.
d) chemical shifts of ^{13}C-NMR signals corresponding to the aliphatic methine carbons of the main chain in ppm/TMS.

CONCLUSION

In conclusion the nucleophilic substitution of poly(vinyl chloroformate) with phenol by using a liquid/liquid two phases system (50% aqueous NaOH/CH_2Cl_2) with Bu_4NHSO_4 as phase transfer catalyst proceeds both without degradation of the polymer and selectively with different triads. In order to explain why the chloroformate functions belonging to isotactic sequences are the most reactive, further work is presently undertaken concerning the study of the conformational structure of PVOCCl macromolecules in solution. Along the same line, it would be interesting to examine some substitution reactions of trimeric models having different stereoconfigurations. Moreover it would be worth extending this preliminary study to the examination of the behaviour of PVOCCl samples with different tacticities.

REFERENCES

1. MATHUR,N.K.; NARANG,C.K.; WILLIAMS,R.E. Polymers as Aids in Organic Chemistry; Academic Press : New York, 1980; GECKELER,K.; PILLAI,V.N.R.; MUTTER,M. Adv. Polym. Sci. 1981, 39, 65; AKELAH,A.; SHERRINGTON,D.C. Chem. Rev. 1981, 81, 557; Polymer 1983, 24, 1369; FORD,W.T.; TOMOI,M. Adv. Polym. Sci. 1984, 55, 49.
2. BOUCHER,E.A. Prog. Polym. Sci. 1978, 6, 63.
3. PLATE,N.A.; NOAH,O.V. Adv. Polym. Sci. 1979, 31, 133.

4. TSUCHIDA,E.; IRIE,S. J. Polym. Sci. Polym. Chem. Ed. 1973, 11, 789.
5. NOAH,O.V.; LITMANOVICH,A.D.; PLATE,N.A. J. Polym. Sci. Polym. Phys. Ed. 1974, 12, 1711.
6. MORCELLET-SAUVAGE,J.; LOUCHEUX,C. Makromol. Chem. 1975, 176, 315.
7. BOUCHER,E.A.; MOLLETT,C.C. J. Chem. Soc. Faraday Trans.1 1982, 78, 75; BOUCHER,E.A.; KHOSRAVIBABADI, E. J. Chem. Soc. Faraday Trans. 1 1983, 79, 1951.
8. BOURGUIGNON,J.J.; GALIN,J.J.; BELLISSENT,H.; GALIN,J.C. Polymer 1977, 18, 937; BOURGUIGNON,J.J.; GALIN,J.C. Polymer 1982, 23, 1493.
9. PLATE,N.A. Pure Appl. Chem. 1976, 46, 49.
10. DE LOECKER,W.; SMETS,G. J. Polym. Sci. 1959, 40, 203.
11. BAINES,F.C.; BEVINGTON,J.C. J. Polym. Sci. A-1 1968, 6, 2433.
12. KLESPER,E.; BARTH,V. Polymer 1976, 17, 777, 787.
13. BUGNER, D.E. Am. Chem. Soc. Polym. Prep. 1986, 27(2), 57.
14. MARTINEZ,G.; MIJANGOS,C.; MILLAN,J. J. Macromol. Sci. 1982, A-17, 1129; Polym. Bull. 1985, 13, 151; J.Polym. Sci. Polym. Chem. Ed. 1985, 23, 1077; MIJANGOS,C.; MARTINEZ,G.; MICHEL,A.; MILLAN,J.; GUYOT,A. Eur. Polym. J. 1984, 20, 1.
15. MIJANGOS,C.; MARTINEZ,G.; MILLAN,J. Eur. Polym. J. 1986, 22, 423.
16. MEUNIER,G.; BOIVIN,S.; HEMERY,P.; SENET,J-P.; BOILEAU,S. in Modification of Polymers; CARRAHER,C.E.; MOORE,J.A. Plenum Press : New York, 1983; Vol.21, p.2293.
17. SCHAEFGEN,J.R. Am. Chem. Soc. Polym. Prep. 1967, 8(1), 723; J. Polym. Sci. 1968, C24, 75.
18. BOILEAU,S.; JOURNEAU,S.; MEUNIER,G. Fr. Patent 1980, 2 475 558.
19. MEUNIER,G.; HEMERY,P.; BOILEAU,S.; SENET,J-P.; CHERADAME,H. Polymer 1982, 23, 849.
20. MEUNIER,G.; BOIVIN,S.; HEMERY,P.; BOILEAU,S.; SENET,J-P. Polymer 1982, 23, 861.
21. BOIVIN,S.; CHETTOUF,A.; HEMERY,P.; BOILEAU,S. Polym. Bull. 1983, 9, 114.
22. BOIVIN,S.; HEMERY,P.; SENET,J-P.; BOILEAU,S. Bull. Soc. Chim. France 1984, II 201.
23. KASSIR,F.; BOIVIN,S.; BOILEAU,S.; CHERADAME,H.; WOODEN,G.P.; OLOFSON,R.A. Polymer 1985, 26, 443.
24. BOIVIN,S.; HEMERY,P.; BOILEAU,S. Can. J. Chem. 1985, 63, 1337.
25. WEBER,W.P.; GOKEL,G.W. Phase Transfer Catalysis in Organic Synthesis, Springer-Verlag : New York, 1978; STARKS,C.M.; LIOTTA,C. Phase Transfer Catalysis : Principles and Techniques, Academic Press : New York, 1978; DEHMLOW,E.V.; DEHMLOW,S.S. Phase Transfer Catalysis, Verlag Chemie : Weinheim, 1980.
26. ANTOINE, J.P.; DE AGUIRRE, I.; JANSSENS, F.; THYRION, F. Bull. Soc. Chim. France 1980, 5-6, II-207.
27. MONTANARI, F.; LANDINI, D.; ROLLA, F. Topics in Current Chemistry 1982, 101, 147.

RECEIVED August 27, 1987

Chapter 4

Chemical Modification of Poly(2,6-dimethyl-1,4-phenylene oxide) and Properties of the Resulting Polymers

Simona Percec and George Li

Standard Oil Research and Development, 4440 Warrensville Center Road, Cleveland, OH 44128

The chemical modification of poly (2,6-dimethyl-1,4-phenylene oxide) (PPO) by several polymer analogous reactions is presented. The chemical modification was accomplished by the electrophilic substitution reactions such as: bromination, sulfonylation and acylation. The permeability to gases of the PPO and of the resulting modified polymers is discussed. Very good permeation properties to gases, better than for PPO were obtained for the modified structures. The thermal behavior of the substituted polymers resembled more or less the properties of the parent polymer while their solution behavior exhibited considerable differences.

Although the use of polymers as membrane materials has increased markedly (1-3), the development of structure/permeability relationship has lagged and is an important area of future investigations.

Among the many aromatic polymers which possess high glass transition temperatures, PPO (T_G=212°C) shows the highest permeability to gases. One significant factor in its high permeability stems from large diffusion coefficients of gases in PPO as compared to other glassy polymers with rigid chain backbone (4).

In spite of this, the relatively low selectivity to gases associated with the lack of solubility in conventional membrane forming dipolar aprotic solvents prevented the facile preparation of PPO membrane systems for actual use in gas separations.

Several studies have appeared in the literature (5-9) describing the reactions of various compounds with PPO in order to change the properties of this polymer but none of them have valued the chemical structure in terms of gas permselectivity behavior.

In our research, three chemical modification approaches were investigated: bromination, sulfonylation, and acylation on the aromatic ring. The specific objective of this paper is to present the chemical modification on the PPO backbone by a variety of electrophilic substitution reactions and to examine the features that distinguish modified PPO from unmodified PPO with respect to gas permeation properties, polymer solubility and thermal behavior.

0097-6156/88/0364-0046$06.00/0

Experimental

Materials. The starting PPO was purchased from Aldrich Chemical Co. Two reprecipitations from chloroform into methanol served to purify the polymer. The bromine, chlorosulfonic acid, sulfonyl chlorides, acid chlorides as well as all other reagents and solvents were purchased from Aldrich Chemical Co. and were used without further purification.

Aluminum trichloride was sublimated in a Pyrex glass tube prior to use.

Reactions. The bromination of the aromatic ring was achieved at room temperature using a known procedure (10).

The sulfonylation reaction was performed in a two-step process. The first step involved the reaction of PPO with chlorosulfonic acid according to a literature method (11). The sulfonated PPO was hygroscopic and unstable. We succeeded (12) in converting the sulfonate groups into stable sulfone groups by reacting them with aromatic compounds at elevated temperatures (120°C). The final dark solution was washed with dilute sodium bicarbonate, and the product precipitated in methanol, filtered and dried.

The Friedel-Crafts substitution reactions (13) were carried out in nitrobenzene solution in the presence of $AlCl_3$, under nitrogen and a temperature range of 40-80°C. Sulfonyl chlorides (R_1SO_2Cl) and carboxylic acid chlorides ($R_2C(O)Cl$) respectively, were used as sulfonylating and acylating agents. After the required reaction time, the reaction mixture was washed with water until the pH was neutral. The separated polymer solution was dried over anhydrous $MgSO_4$, filtered and precipitated in methanol. A final purification was carried out by precipitation of the product from chloroform solution into methanol. The polymer was then vacuum dried to constant weight.

The polymer films were prepared by dissolving the polymer in a suitable solvent (chloroform) to form 7 wt% solutions. The solution was then poured over a clean glass plate and spread out evenly to a uniform thickness with the aid of a doctor blade. The films were air dried, removed from the glass plate and further vacuum dried at 60°C for 72 hours.

Measurements. 200 MHz ^1H NMR spectra were recorded from $CDCl_3$ solutions (TMS internal standard) on a Nicolet NT 200 spectrometer. DSC measurements were carried out with a DuPont Differential Scanning Calorimeter (Model 1090) under nitrogen atmosphere. The scanning rate was 10°C/min. in all cases. Indium was used as a standard.

Elemental analyses were performed by Standard Oil Research and Development Analytical Laboratory.

A modified Gilbert cell (14) was used to determine the gas permeation properties of polymer films. The testing area of the film was 45.8 cm^2. The film thickness was in a range of 1.27×10^{-2} mm - 2.81×10^{-2} mm. The test side was exposed to a carbon dioxide : methane : nitrogen mixture in a mole ratio 3.11 : 33.6 : 63.29. The permeant was picked up by a carrier gas, helium, and injected intermittently through a sample valve into a gas chromatograph for analysis. The partial pressure of the test gas was 29.7 psi (0.21 MPa) while the partial pressure of the product gas on the permeant side was held at an insignificant level by purging with 29.7 psi

(0.21 MPa) helium at a flow rate much in excess of the permeation rate.

Results and Discussion

Brominated PPO. The electrophilic aromatic bromination of the PPO was carried out in chloroform solution to yield polymers with different degrees of substitution. For the highly brominated structures the T_G expanded over a considerable temperature range. A maximum value of 273°C was obtained for 100 mol % brominated PPO. The solution properties of the brominated PPO derivatives were similar to those of the parent polymer. The gas permeation properties of PPO and PPO containing between 6.5% to 100% Br groups per repeat unit are summarized in Table I.

The steady state separation factor (α) for the two gases (CO_2 and CH_4) is defined as the ratio of their individual permeabilities. The gas permeability constants (P) are generally expressed by the amount of the gas at standard temperature and pressure normalized for the thickness, membrane area, time and differential pressure of gas as in the following equation:

$$P = \frac{(\text{Amount of the gas})\ (\text{Membrane thickness})}{(\text{Membrane area})\ (\text{Time})\ (\text{Differential pressure of gas})}$$

where P is expressed in units cm^3 (STP) $cm \cdot cm^{-2} \cdot s^{-1} \cdot cmHg$. Permeability is also as reported in Barrier, one Barrier being equal to $1X10^{-10}P$. It can be seen from Table I that no change either in permeability for CO_2 or CO_2/CH_4 selectivity was obtained when the degree of bromination for PPO is low (6.5 mole%). However, at higher levels of bromination, there were increases in both permeability for CO_2 and CO_2/CH_4 selectivity. The maximum effects (an increase of 2.47 times for CO_2 permeability and 1.45 times for CO_2/CH_4 selectivity to that of PPO) were reached at 100% substitution degree of the aromatic ring.

Sulfonylated PPO. The sulfonylation is achieved by sulfonation reaction on PPO, followed by the reaction with aromatic compounds as outlined in Scheme 1.

No comments regarding the thermal and solution behavior of the polymers obtained by this two step procedure are included here since these properties are discussed in the next section for sulfonylated PPO modified under Friedel-Crafts conditions.

The changes of the CO_2 permeability and separation factor for CO_2/CH_4, compared to the typical values of PPO standard are shown in Table II for PPO modified with different sulfonyl groups. The best enhancement was obtained for PPO containing phenyl sulfone groups. The presence of $-SO_2(OH)$ groups reduced the carbon dioxide permeability by a factor of three. This can be explained (15) by the decrease in local segmental mobility of the polymer chains due to the interactions arising from hydrogen bonding. However, the overall transport process for this polymer membrane is more complicated and involves a more pronounced discrimination against methane molecules due to the highly polar nature of the polymer. This leads to a twofold increase in the CO_2/CH_4 separation factor.

Table I. The Permeability to Gases of Brominated PPO

No.	Polymer	Substitution Degree[a] (mol%)	P_{CO_2} (barrer)[b]	α_{CO_2/CH_4}
1.	PPO	0.0	64.00	16.40
2.	BRPPO	6.5	64.00	16.72
3.	BRPPO	21.5	67.20	17.71
4.	BRPPO	46.0	78.08	19.35
5.	BRPPO	61.0	99.84	18.04
6.	BRPPO	100.0	158.08	23.73

a) Determined by elemental analysis

b) 1 Barrer = 10^{-10} [cm^3(STP) cm]/[cm^2•sec•cmHg]

α = Separation factor

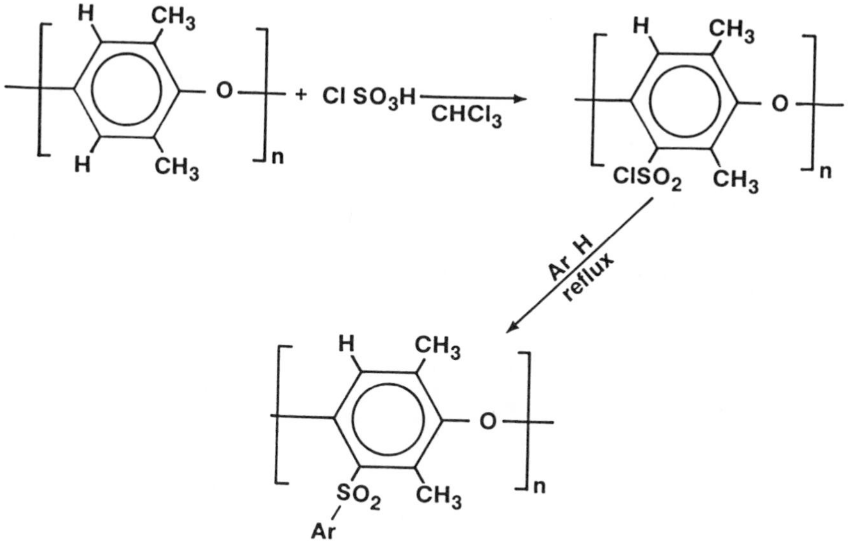

Scheme 1. Two step procedure of sulfonylation reaction on PPO.

Table II. The Permeability to Gases of Sulfonylated PPO

No.	R	Substitution Degree[a] (mol%)	P_{CO_2} (barrer)	α_{CO_2/CH_4}
1.	H	0	64	16.4
2.	$-SO_2-C_6H_4CH_3$	60	89.60	21.81
3.	$-SO_2-C_6H_3(CH_3)_2$	23	72.96	20.00
4.	$-SO_2-C_6H_5$	38	99.20	27.38
5.	$-SO_2-C_6H_4C_2H_5$	31	73.60	16.40
6.	$-SO_2-OH$	—	19.84	32.80

a) Determined by elemental analysis.

Friedel-Crafts Reactions on PPO and Properties of the Resulting Polymers. There are two hydrogens on the aromatic ring of PPO which can react through Friedel-Crafts reactions. The substitution of the first available position from the aromatic ring occurs easily by the treatment of the PPO with sulfonyl chloride or acid chlorides in the presence of a Friedel-Crafts catalyst. The remaining aromatic hydrogen could not be removed by a second abstraction reaction, and consequently only monosubstitution was achieved.

Table III presents the experimental reaction conditions used in the sulfonylation and acylation of the PPO backbone and the degrees of substitution obtained.

The substitution degree determined by ^1H NMR spectroscopy was between 17.5% and 79.4%, depending on the reaction conditions and sulfonylating or acylating agent. A complete substitution of the PPO aromatic rings was not obtained under these experimental conditions. The presence of one carbonyl or sulfonyl group attached to a PPO phenyl ring, drastically decreases the nucleophilicity of the remaining unsubstituted position and, therefore, no disubstitution of the phenylenic units was realized. This is due to the strong electron withdrawing character of the carbonyl and sulfonyl groups. At the same time, when bulky substituents are attached to the PPO backbone, steric hindrance also decreases the accessability of nearby unsubstituted positions (16). Consequently, electronic factors retard the second electrophilic substitution on the same PPO phenyl ring while steric factors also contribute to limit the degree of monosubstitution to about 80%.

Typical ^1H NMR spectra of the sulfonylated PPO and acylated PPO are presented in Figures 1-2 together with the assignment of their protonic resonances (17). Multiple resonances from the aliphatic region in Figure 1 are due to the sequence distribution of the structural units. It is difficult to provide an exact assignment for this region at this time. The percent of substitution of the PPO aromatic units was determined from the ratio of the integrals of proton resonances due to unsubstituted 2,6-dimethyl-1,4-phenylene units (signal H_a at δ=6.5 ppm) and substituted units (signal H_b at δ=6.0 ppm).

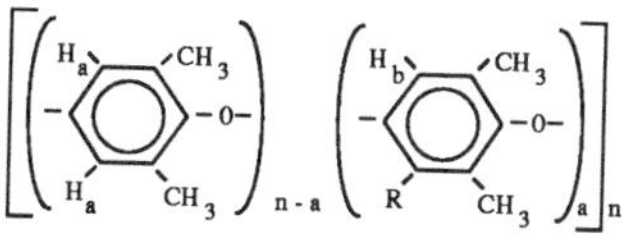

In all cases monosubstitution was demonstrated by measuring the ratio between the integrals of the unsubstituted aromatic proton of the substituted phenylenic units and some representative protons from the newly attached pendant groups. For example, the ratio between the signals c and e from spectra II and III (Figure 1), the ratio between the signals c and f from the spectra IV and VI (Figure 2), and c and e from the spectra VI (Figure 2) were used to demonstrate monosubstitution.

Both thermogravimetric analysis and differential scanning calorimetric studies were carried out on modified and unmodified PPO samples. Table IV presents the weight losses and the glass transition temperatures of the most representative polymers.

As was expected, the substitution of PPO with rigid and bulky side groups decreases the flexibility of the polymer chain and the glass transition temperatures of modified polymers increases. This

Table III. Modification of PPO under Friedel-Crafts Conditions (PPO 5% Solution in Nitrobenzene)

No.	Modifying Agent	Reaction Conditions			Substitution degree[a] (mol %)
		PPO:M.A.:$AlCl_3$ (mol:mol:mol)	Time (hours)	Temperature (°C)	
	Sulfonyl Chlorides:				
1.	$N(CH_3)_2SO_2Cl$	1.0:1.5:0.55	5.0	70	32.3
2.	$C_6H_5SO_2Cl$	1.0:1.0:1.10	7.0	80	60.7
3.	$BrC_6H_4SO_2Cl$	1.0:1.0:1.10	9.5	80	59.0
4.	$CH_3C_6H_4SO_2Cl$	1.0:1.0:1.10	7.5	80	66.6
5.	$(CH_3)_3C_6H_2SO_2Cl$	1.0:1.0:1.10	7.0	60	46.3
6.	(naphthyl)-SO_2Cl	1.0:1.0:1.10	8.5	80	63.5
	Acid Chlorides:				
7.	C_2H_5COCl	1.0:1.0:1.10	4.0	50	42.7
8.	$CH_3(CH_2)_2COCl$	1.0:1.0:1.10	3.0	50	58.6
9.	$CH_3(CH_2)_{10}COCl$	1.0:1.0:1.10	6.0	50	59.0
10.	$CH_3(CH_2)_{12}COCl$	1.0:0.5:0.55	3.0	50	17.5
11.	$CH_3(CH_2)_{12}COCl$	1.0:1.0:1.10	6.0	50	55.0
12.	$CH_3(CH_2)_{14}COCl$	1.0:1.0:1.10	6.5	50	50.0
13.	$C_6H_5CH_2COCl$	1.0:0.5:0.55	6.0	50	29.2
14.	$CH_3C_6H_4COCl$	1.0:1.0:1.10	8.0	60	79.4

a) determined by ^1H NMR
M.A. - modifying agent

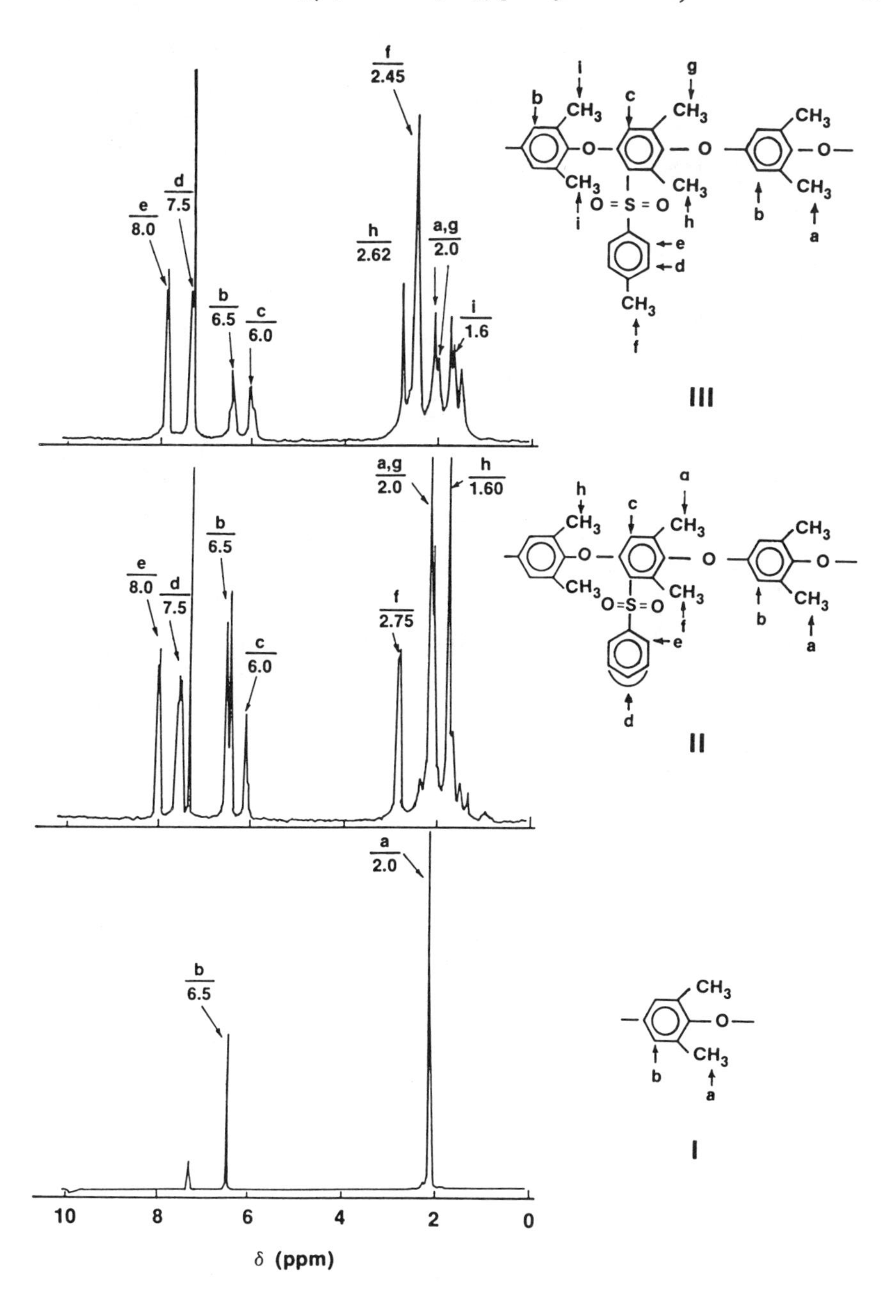

Figure 1. ^{1}H NMR spectra of PPO (I), PPO modified with benzenesulfonyl chloride (Sample No. 2, Table III) (II), and PPO modified with p-toluenesulfonyl chloride (Sample No. 4, Table III) (III) Reproduced with permission from Ref. 17. Copyright 1987, Wiley.)

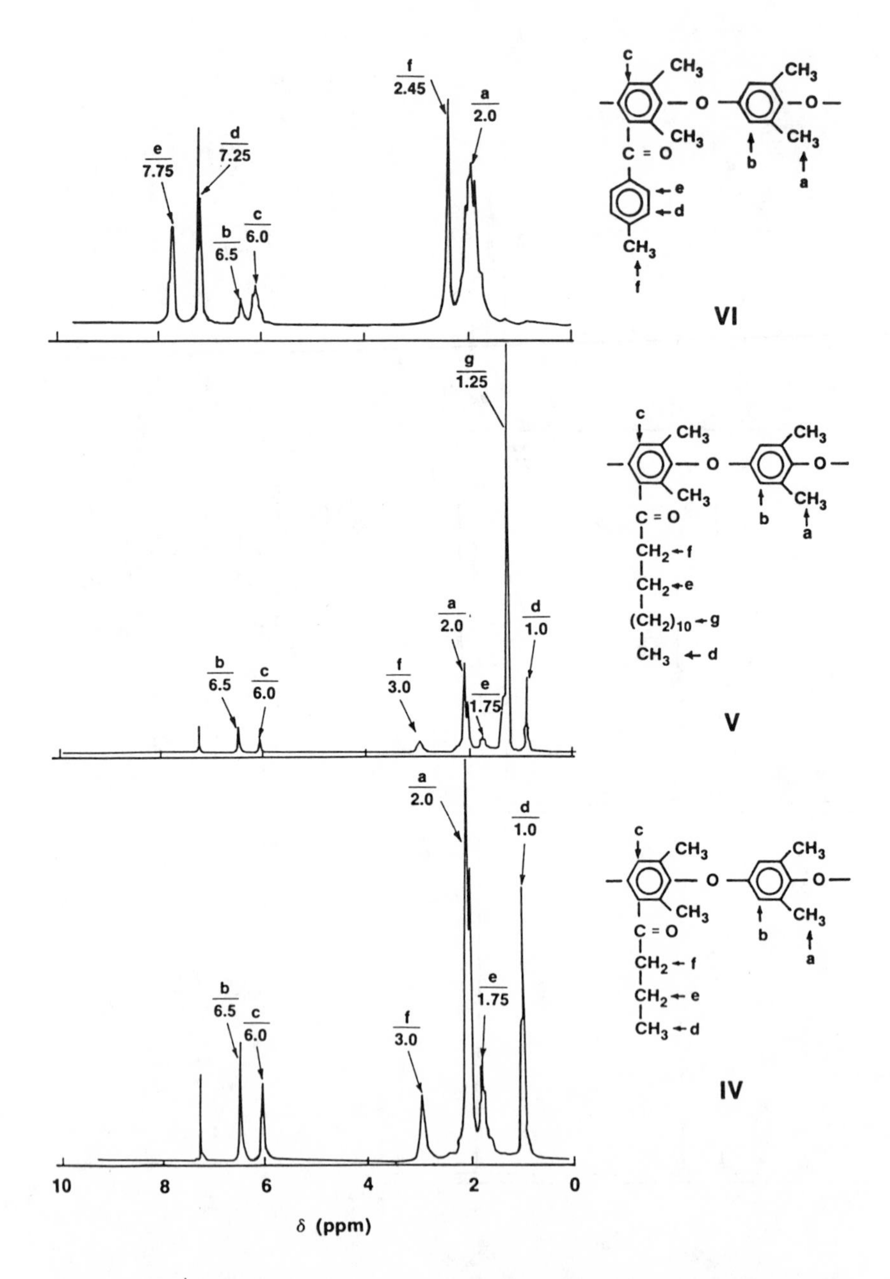

Figure 2. ^{1}H NMR spectra of PPO acylated with propionyl chloride (Sample No. 8, Table III) (IV), PPO acylated with myristoyl chloride (Sample No. 11, Table III) (V), and PPO acylated with p-toluoyl chloride (Sample No. 14, Table III) (VI). (Reproduced with permission from ref. 17. Copyright 1987 Wiley.)

Table IV. Thermal Characterization of PPO Modified by Friedel-Crafts Reactions

Polymer	Number Corresponding to Table III	Zone I		Zone II		Zone III		T_G (°C)
		Temp. Range (°C)	Weight Loss (%)	Temp. Range (°C)	Weight Loss (%)	Temp. Range (°C)	Weight Loss (%)	
PPO		75-175	0.3	175-300	0.6	300-400	0.2	212
PPO sulfonylated with p-toluene-sulfonyl chloride	4	100-200	0.4	200-300	0.2	300-400	3.5	225
PPO sulfonylated with dimethyl sulfamoyl chloride	1	0-250	0.0	250-325	1.9	325-400	11.7	245
PPO acylated with myristoyl chloride	10	0-150	0.0	150-300	0.7	300-400	2.5	116
PPO acylated with myristoyl chloride	11	0-200	0.0	200-300	0.1	300-400	3.0	165

Source: Reproduced with permission from Ref. 17. Copyright 1987, Wiley.

effect was exemplified by the PPO modified with p-toluenesulfonyl chloride which has a T_G of 265°C. Similarly an increase in the value of the T_G was also observed when PPO was modified with polar substituents such as dimethylsulfamoyl group. In this case the T_G value was 245°C.

The increase in the length of the side chain results normally in an internal plasticization effect caused by a lower polarity of the main chain and an increase in the configurational entropy. Both effects result in a lower activation energy of segmental motion and consequently a lower glass transition temperature. The modification of PPO with myristoyl chloride offers the best example. No side chain crystallization was detected by DSC for these polymers.

The sulfonylated and acylated PPO presents solubility characteristics which are completely different from those of the parent PPO. Table V presents the solubility of some modified structures compared to those of unmodified PPO. It is very important to note that, after sulfonylation, most of the polymers become soluble in dipolar aprotic solvents like dimethyl sulfoxide (DMSO), N,N- dimethylformamide (DMF) and N,N-dimethylacetamide (DMAC). At the same time it is interesting to mention that, while PPO crystallizes from methylene chloride solution, all the sulfonylated polymers do not crystallize and form indefinitely stable solutions in methylene chloride. Only some of the acetylated polymers become soluble in DMF and DMAC, and none are soluble in DMSO. The polymers acetylated with aliphatic acid chlorides such as propionyl chloride are also soluble in acetone.

The permeation properties of substituted PPO to a carbon dioxide, methane, nitrogen mixture were studied for several systems. The results are presented in Table VI.

An increase in permeability together with an increase in selectivity for CO_2/CH_4 was clearly demonstrated for PPO substituted with p-toluenesulfonyl chloride and for PPO substituted with dimethylsulfamoyl chloride (Table VI). An interesting example is PPO modified with myristoyl chloride. This long side chain substituted polymer showed an increase in permeability for methane of 5.19 times while its permeability for carbon dioxide remained almost unchanged. Therefore its selectivity to CO_2/CH_4 decreases dramatically compared to that of the parent polymer. However it can be speculated that by varying the length of the side chain of the modified structures, certain differences in permeability among different hydrocarbon gases are likely to occur. This is an important aspect which may prove useful for hydrocarbon/ hydrocarbon separations.

Conclusions

Chemical modifications of PPO by electrophilic substitution of the aromatic backbone provided a variety of new structures with improved gas permeation characteristics. It was found that the substitution degree, main chain rigidity, the bulkiness and flexibility of the side chains and the polarity of the side chains are major parameters controlling the gas permeation properties of the polymer membrane. The broad range of solvents available for the modified structures enhances the possibility of facile preparation of PPO based membrane systems for use in gas separations.

Table V. Solubility of PPO Modified by Friedel-Crafts Reactions

No.	Modifying Agent	Solvent						
		Chloroform	Methylene Chloride	DMF	DMSO	DMAC	THF	Acetone
1.	None	+	±	–	–	–	–	–
2.	Benzenesulfonyl chloride	+	+	+	±	+	+	–
3.	p-Toluenesulfonyl chloride	+	+	+	±	+	+	–
4.	Naphthalenesulfonyl chloride	+	+	+	+	+	+	–
5.	Dimethylsulfamoyl chloride	+	+	+	+	+	+	–
6.	p-Toluoyl chloride	+	+	+	–	+	+	–
7.	Propionyl chloride	+	+	+	–	+	+	+
8.	Lauroyl chloride	+	+	–	–	–	+	–
9.	Palmitoyl chloride	+	+	–	–	–	+	–
10.	Phenyl acetyl chloride	+	+	+	–	+	+	–

\+ Soluble $\geq$ 1 g/100 ml
– Insoluble
± Tendency to crystallize from solution

Source: Reproduced with permission from Ref. 17.

Table VI. The Permeability to Gases of PPO Modified by Friedel-Crafts Reactions

No.	Modifying Agent	P_{CH_4} (barrer)	P_{CO_2} (barrer)	P_{N_2} (barrer)	α_{CO_2/CH_4}	α_{CH_4/N_2}
1.	None	3.90	64.00	3.30	16.40	1.18
2.	Toluenesulfonyl chloride	3.20	75.00	1.55	23.40	2.06
3.	Dimethylsulfamoyl chloride	3.31	80.99	2.45	24.47	1.35
4.	Myristoyl chloride (Polymer 11, Table III)	20.26	58.09	11.84	2.87	1.71

Testing temperature = 25°C

References

1. A.J. Erb and D.R. Paul, J. Membr. Sci., 8, 11 (1981).
2. D.R. Paul, J. Membr. Sci., 18, 75 (1984).
3. J.W. Barlow and D.R. Paul, J. Appl. Polym. Sci., 29, 845 (1984).
4. K. Toi, G. Morel and D.R. Paul, J. Appl. Polym. Sci., 27, 2997 (1982).
5. L. Verdet and J.K. Stille, Organometallics, 1, 380 (1982).
6. R. P. Kambour, J. T. Bendler and R. C. Bopp, Macromolecules, 16, 753 (1983).
7. A.J. Chalk and A.S. Hay, J. Polym. Sci., Part A-1, 7, 691 (1968).
8. S. Xie, W.J. MacKnight and F.E. Karasz, J. Appl. Polym. Sci., 29, 1678 (1984).
9. I. Cabasso, J.J. Grodzinski and D. Vofsi, J. Appl. Polym. Sci., 18, 1969 (1974).
10. D.M. White and C.M. Orlando, ACS Symp. Series 6, 178 (1975).
11. W.J. Ward III and Robert M. Salemne (to General Electric) U.S. Pat. 3,780,496 (1973).
12. G. Li, U.S. Pat. 4,521,224 (1985).
13. E.S. Percec and G. Li, U.S. Pat. 4,596,860 (1986).
14. R.A. Pasternak, J.F. Schimscheimer and J. Heller, J. Polym. Sci., Part A-2, 8, 467 (1970).
15. C.E. Rogers, in Recent Development in Separation Science, Vol. II, N.N. Li ed., CRC Press, p. 107, 1975.
16. G.A. Olaf, S. Kobayashi and J. Nishimura, J. Am. Chem. Soc., 95, 564 (1973).
17. S. Percec, J. Appl. Polym. Sci., 33, 191 (1987).

RECEIVED August 27, 1987

Chapter 5

Synthesis and Reactions of Halogenated Polyethers and Polysulfides

Melvin P. Zussman [1], Jenn S. Shih [2], Douglas A. Wicks, and David A. Tirrell

Polymer Science and Engineering Department, University of Massachusetts, Amherst, MA 01003

Halogenated polyethers and polysulfides present a number of interesting possibilities in investigations of the reactivity of polymer chains and in the design of new reactive polymers. In this chapter, we explore the consequences of side chain extension, leaving group variation, and neighboring group participation in determining the stability and reactivity of these two classes of polymeric materials.

Polyepichlorohydrin (PECH) is well known as a reactive elastomer. Displacement at the carbon-chlorine bond of PECH has been accomplished with a wide variety of nucleophilic reagents, for the purposes of polymer modification, grafting and crosslinking (1,2). On the other hand, the PECH structure (1) is hardly optimal from the point of view of its reactivity as a substrate for nucleophilic

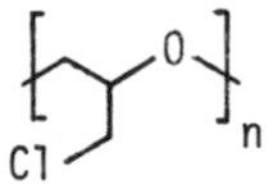

$$\left[\text{--CH}_2\text{CH(CH}_2\text{Cl)O--}\right]_n \qquad \textbf{1}$$

substitution: chloride is modest in its leaving group ability, and the β-branch point (i.e. the chain backbone) would be expected to depress reaction rates by a factor of 10 or so (3).

Within the past several years, we have examined the synthesis and reactions of several classes of polymers related to PECH. We have adopted three simple approaches to the preparation of polymeric substrates more reactive than PECH toward nucleophilic substitution. We have: i). removed the β-branch point by extension of the side chain, ii). replaced the chloride leaving group by a more reactive bromide and iii). replaced the backbone oxygen atom by a sulfur atom that offers substantial anchimeric assistance to nucleophilic

[1]Current address: E. I du Pont de Nemours and Company, Experimental Station, Wilmington, DE 19898

[2]Current address: GAF Corporation, Wayne, NJ 07470

0097-6156/88/0364-0060$06.00/0

displacement. Each of these simple structural changes affords elastomeric polymers more reactive than PECH toward attack by nucleophilic species. We present in this chapter an overview of the synthesis and reactions of these new reactive polymers.

Side Chain Extension

The repeating unit structure of PECH would suggest that the reactivity of the polymer would be rather similar to that of isobutyl chloride. The β-oxygen atom would not be expected to offer substantial anchimeric assistance in nucleophilic displacement, and should perhaps be mildly deactivating as a result of a small inductive effect (4). In general, the reactivity of isobutyl halides toward nucleophilic reagents is depressed in comparison with the reactivity of primary, straight-chain analogues. Streitwieser lists comparative data for nine reactions of isobutyl and 1-propyl halides (3); within this set, $k_{propyl}/k_{isobutyl}$ ranges from 4.3 (for RBr + Cl^- in acetone) to 33.8 (for RI + Br^- in acetone).

The β-branch point present in PECH is absent in polymers of higher 1,2-epoxy-ω-chloroalkanes. Such polymers are readily prepared by treatment of the neat monomers with the modified triethylaluminum catalyst introduced by Vandenberg (5,6); results for (2-chloroethyl)oxirane, (3-chloropropyl)oxirane and (4-chlorobutyl)-oxirane (**2a**-**c**) are summarized in Table I (7,8).

2a: n = 2
b: n = 3
c: n = 4

$(CH_2)_{n-1}CH_2Cl$ → $[\ldots O \ldots]_n$ with $(CH_2)_{n-1}CH_2Cl$

Table I. Homopolymerization of (Chloroalkyl)oxiranes

Monomer	Polymerization[a] Time (Days)	Yield (%)	η_{inh}[b] (dL/g)	Tg(°C)
2a	14	89	4.12[c]	-27
2b	14	78	3.44[c]	-42
2c	10	77	1.21	-50

[a] Polymerization at room temperature in presence of 5 mol-% $AlEt_3/H_2O$/acetylacetone (1/0.5/0.5)

[b] In $CHCl_3$, 0.5 wt %, 35°C

[c] GPC peak molecular weight > 10^6

In each case, the polymer is obtained as a white elastomer of high molecular weight (>10^6 in some experiments). Each of these polymers is soluble in benzene, in chlorinated hydrocarbons and in dipolar aprotic solvents. Figure 1 shows the kinetics of chloride substitution by tetra-n-butylammonium benzoate in N,N-dimethylacetamide at 50°C, for PECH and for the polymers of **2a**-**c**. Under these conditions, each of the higher homologues is about equally reactive, and all are converted to the benzoate more rapidly than PECH. Each

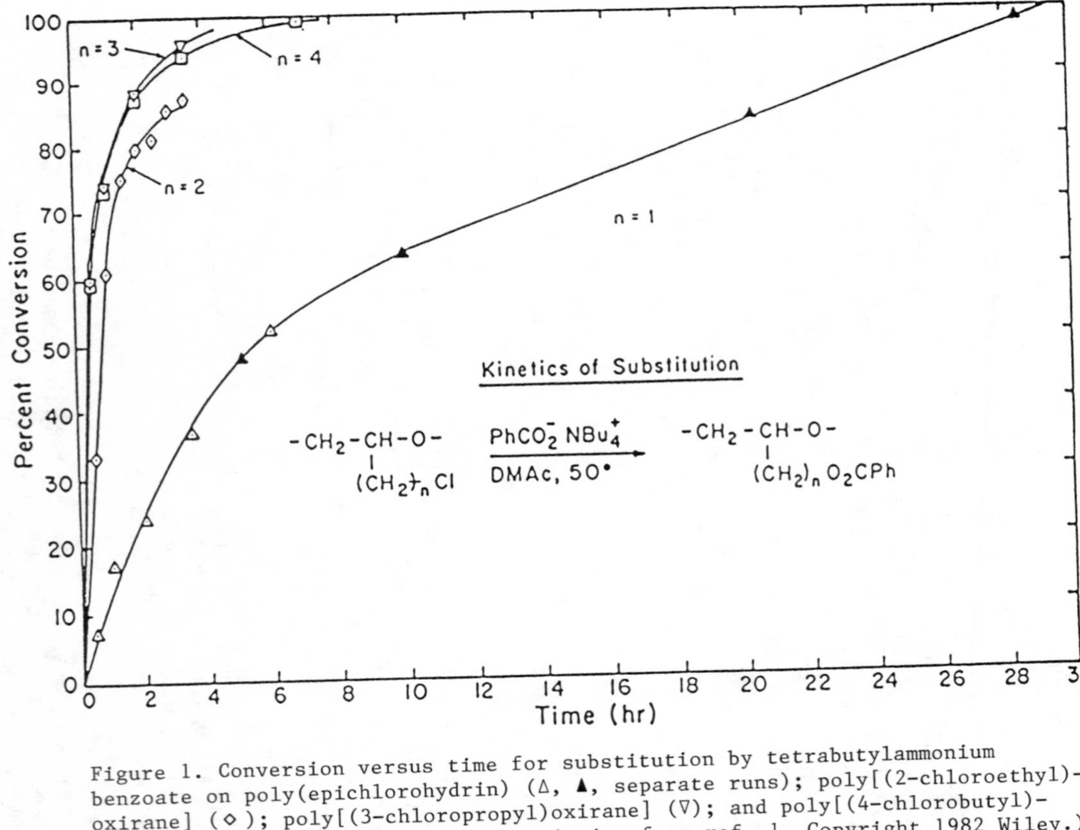

Figure 1. Conversion versus time for substitution by tetrabutylammonium benzoate on poly(epichlorohydrin) (Δ, ▲, separate runs); poly[(2-chloroethyl)-oxirane] (◇); poly[(3-chloropropyl)oxirane] (∇); and poly[(4-chlorobutyl)-oxirane] (□). (Reproduced with permission from ref. 1. Copyright 1982 Wiley.)

of these reactions shows negative deviations from second-order kinetics, but second-order rate constants can be estimated from the initial reaction rates. This procedure gives a second-order rate constant of 3.9 x 10^{-4} M^{-1} s^{-1} for poly[(2-chloroethyl)oxirane] and a value of 0.63 x 10^{-4} M^{-1} s^{-1} for PECH. Thus the higher homologues are more reactive than PECH by a factor of 6, an observation completely consistent with known structure-reactivity relations for simple alkyl halides.

The reactive carbon-chlorine bond in these polyethers allows their conversion to other interesting materials. For example, quantitative chloride displacement by benzoate anion followed by basic methanolysis affords hydrophilic polyether elastomers that are water-soluble or water-swellable, depending on side chain length (9). Poly[(3-hydroxypropyl)oxirane] is a colorless elastomer that can be cast into tough, clear films from water or methanol. Poly-[(4-hydroxybutyl)oxirane] is insoluble in water, but quite hydrophilic; a film immersed in water gains 46% in weight in 90 min at room temperature, while retaining good mechanical integrity. Uses for such materials as adhesives, as biomaterials and as contact lenses may be anticipated.

Bromide as Leaving Group

Perhaps the most straightforward approach to increasing the reactivity of halogenated substrates--polymeric or otherwise--is to exploit the well known order of leaving group ability, i.e., I > Br > Cl > F. For example, data provided by Streitwieser suggest that replacement of chloride by bromide will lead to rate enhancements of 50- to 200-fold, depending on reaction conditions (3). One cannot take this approach too far in the design of reactive polymeric substrates, however, since the leaving group must be inert to the conditions of polymerization if protection-deprotection schemes are to be avoided. The ideal leaving group would be completely unreactive toward the nucleophilic species involved in chain growth, but readily displaced in subsequent polymer modification reactions.

Bromide appears to be useful in this regard. Table II summarizes our homopolymerization experiments on (2-bromoethyl)oxirane, (3-bromopropyl)oxirane and (4-bromobutyl)oxirane (3a-c) (10-12).

3a: n = 2
b: n = 3
c: n = 4

$(CH_2)_{n-1}CH_2Br$ oxirane $\longrightarrow$ $[\,-CH_2-CH(-(CH_2)_{n-1}CH_2Br)-O-\,]_n$

Table II. Homopolymerization of (Bromoalkyl)oxiranes

Monomer	Catalyst	Polymerization Time (hr)	Yield (%)	η_{inh}[c] (dL/g)	Tg (°C)
3a	a	24	67	3.5	-35
3a	b	24	63	0.6	
3b	a	3	35	0.6	-35
3b	b	24	74	0.5	
3c	a	24	65	1.7	-34
3c	b	24	68	0.6	

[a] Room temperature, 6-7 mol-% $AlEt_3/H_2O$/acetylacetone (1/0.5/1)

[b] Room temperature, 6-7 mol-% $AlEt_3/H_2O$/acetylacetone (1/0.5/0.5)

[c] In $CHCl_3$, 0.5 wt %, 30°C

Polymerization of each of these oxiranes is effectively accomplished by treatment of the neat monomer with the chelated aluminum catalyst. It is interesting to note that higher molecular weights are generally realized by using the catalyst mixture that contains a 1:1 ratio of triethylaluminum and acetylacetone; this may be a result of chelation of the most acidic catalyst sites and a consequent reduction in chain transfer associated with Al-assisted ionization of the C-Br bond. The poly[(ω-bromoalkyl)oxirane]s are slightly tacky elastomers at room temperature; all undergo a glass transition at approximately -35°C.

Figure 2 shows the kinetics of bromide substitution by tetra-n-butylammonium benzoate in $CDCl_3$ at 45°C, for poly(epibromohydrin) (PEBH) and for the polymers of 3a-c. Although these results are not directly comparable to those in Figure 1 because of differences in solvents and temperatures, qualitative consistency is apparent. First of all, each of the higher homologues is about equally reactive, and all react more rapidly than PEBH. The kinetics show negative deviation from second-order behavior, as before, but the initial slopes of the second-order plots are again useful for comparative purposes. Table III lists the rate constants obtained in this way. As in the chloride series, extension of the side chain by a single carbon atom removes the β-branch point and accelerates the reaction by a factor of about 10. (The significance of the apparent two-fold higher reactivity of the 3-bromopropyl side chain compared to its 2- and 4-carbon homologues has not been determined.) Finally, the bromides are indeed more reactive than the chlorides; despite the use of $CDCl_3$ as the reaction solvent, the bromides react more than an order of magnitude faster than the chlorides, even though the latter were converted in DMAc, a superior solvent for nucleophilic substitutions of this kind.

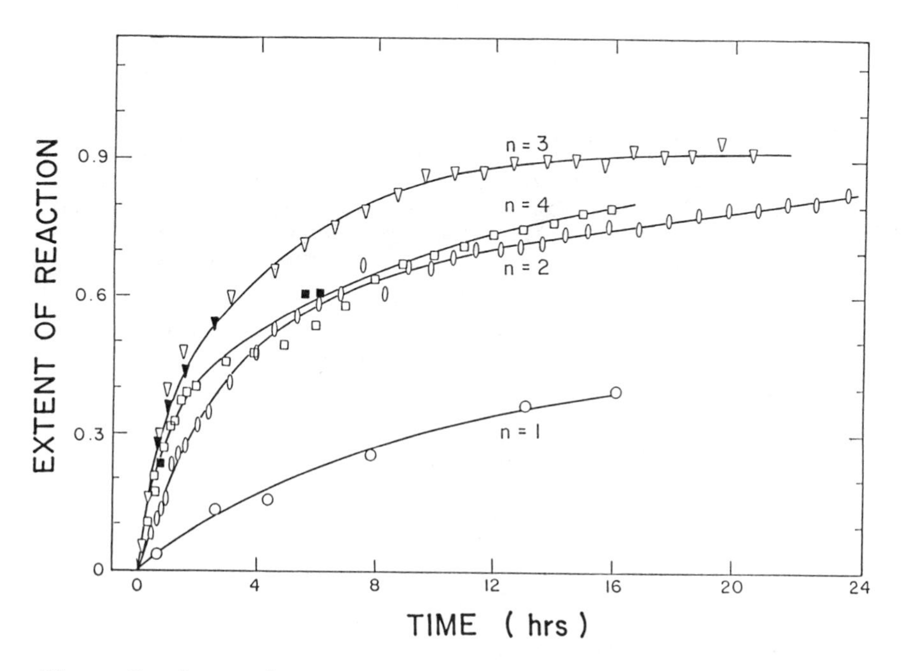

Figure 2. Conversion versus time for substitution by tetrabutylammonium benzoate on poly(epibromohydrin) (O); poly[(2-bromoethyl)oxirane] (0); poly[(3-bromopropyl)oxirane] (▽, ▼, separate runs); and poly[(4-bromobutyl)oxirane] (□, ■, separate runs).

Table III. Kinetics of Substitution on Poly(ω-bromoalkyl)oxiranes

Side Chain Length	$[-CH_2Br]_0$[b] (M)	$[Benzoate]_0$[b] (M)	k[c] (M^{-1} s^{-1})
1	8.57×10^{-3}	1.74×10^{-2}	0.8×10^{-3}
2	6.97×10^{-3}	1.40×10^{-2}	8.2×10^{-3}
3	4.53×10^{-3}	9.11×10^{-3}	19×10^{-3}
4	3.88×10^{-3}	8.10×10^{-3}	7.9×10^{-3}

a 45°C, $CDCl_3$
b Initial Concentrations
c Initial slope of second-order kinetic plot

Anchimeric Assistance by Backbone Sulfur

The acceleration of halide displacement by neighboring sulfides is among the most familiar examples of anchimeric assistance in organic chemistry. Table IV illustrates the magnitude of the effect for primary alkyl chlorides (4).

Table IV. Relative Rate Constants for Hydrolysis of Alkyl Chlorides[a]

RCl	k_{rel}[b]
$CH_3CH_2CH_2CH_2Cl$	1.00
$CH_3CH_2OCH_2CH_2Cl$	0.18
$CH_3CH_2SCH_2CH_2Cl$	2750
$CH_3CH_2OCH_2CH_2CH_2Cl$	0.77
$CH_3CH_2SCH_2CH_2CH_2Cl$	1.00

a Aqueous dioxane, $[H_2O]$ = 20 M, 100°C
b Rate relative to 1-chlorobutane

A β-sulfide accelerates the solvolysis nearly 3000-fold compared to the simple alkyl halide and approximately 15,000-fold compared to the β-chloroether. On the other hand, the γ-sulfide offers no assistance; apparently closure to the four-membered cyclic sulfonium ion cannot compete with direct displacement by the external nucleophile.

These results suggest that poly[(chloromethyl)thiirane] (PCMT, **4**) should be substantially more reactive than polyepichlorohydrin

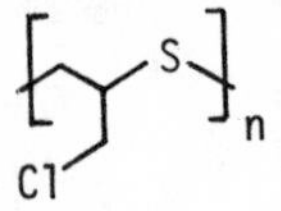

4

toward nucleophilic substitutions, and indeed it is (13). In addition, the formation of the intermediate episulfonium ion by neighboring group displacement leads to a rearrangement of the CMT repeating unit to one (5) derived from the ring- opening polymerization of 3-chlorothietane (3CT, Scheme I) (13-17).

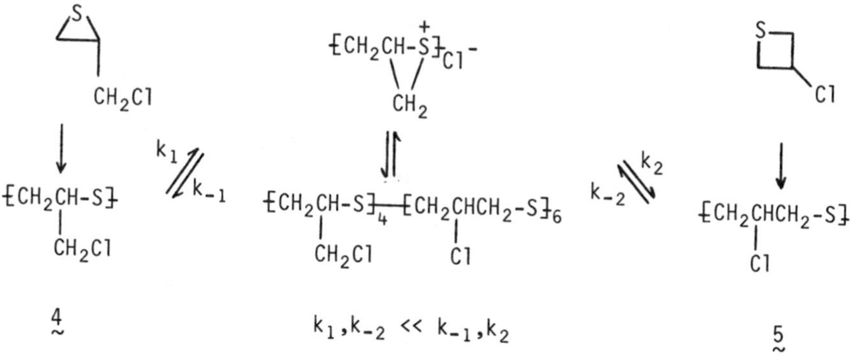

$$-[CH_2CH(CH_2Cl)-S]- \; (4) \underset{k_{-1}}{\overset{k_1}{\rightleftharpoons}} -[CH_2CH(CH_2Cl)-S]_4-[CH_2CH(Cl)CH_2-S]_6- \underset{k_{-2}}{\overset{k_2}{\rightleftharpoons}} -[CH_2CH(Cl)CH_2-S]- \; (5)$$

$$k_1, k_{-2} \ll k_{-1}, k_2$$

Scheme I

We have examined the effects of concentration, temperature, solvent and added electrolyte on the kinetics of this structural interconversion. In all instances, the kinetics are well described by the rate law for a reversible first-order reaction [Equation 1]:

$$(K + K^-)t = \ln \left[\frac{f_0(3CT) - f_\infty(3CT)}{f(3CT) - f_\infty(3CT)} \right] \quad (1)$$

where f(3CT) is the fraction of 3-chlorothietane repeating units in the copolymer and the subscripts 0 and ∞ refer to initial and equilibrium copolymer structures, respectively; K and K^- are combinations of the elementary rate constants k_1, k_2, k_{-1} and k_{-2} (cf. Scheme I) such that

$$K = \frac{k_1 k_2}{k_{-1} + k_2} \quad (2)$$

and

$$K^- = \frac{k_{-1} k_{-2}}{k_{-1} + k_2} \quad (3)$$

The quantity $(K + K^-)$ is thus obtained as the slope of a plot of the right side of Equation 1 versus time, and because

$$f(3CT)_\infty = \frac{K}{K + K^-} \quad (4)$$

the composite rate constants K and K^- can be determined individually. Each of these quantities (K and K^-) may be viewed as a rate constant for cyclization, multiplied by a factor that describes the partitioning of the sulfonium ion intermediate between the two isomeric products of chloride ion attack. Table V summarizes the values of ($K + K^-$) obtained in this way:

Table V. Kinetics of Repeating Unit Isomerization in Poly[(chloromethyl)thiirane] and Poly(3-chlorothietane)

Temperature (°C)	Solvent	Equilibrium Fraction 3CT Units	$K + K^-$ ($10^{-6}s^{-1}$)
48	none[a]	0.685 ± .04	9.4 ± 0.9
40	none[a]	0.690[d]	6.3 ± 1.6
35	none[a]	0.692[d]	3.6 ± 0.9
20.5	none[a]	0.700[d]	0.34 ± 0.03
35.5	none[b]	0.675 ± .03	1.2 ± 0.14
35.5	$CHCl_3$[b]	0.59 ± .03	0.33 ± 0.04
35.5	NO_2Ph[b]	0.62 ± .03	0.47 ± 0.09
35.5	CD_2Cl_2[a]	0.55 ± .02[e]	0.42 ± 0.06
35.5	CD_2Cl_2[c]	0.55 ± .02[e]	0.36 ± 0.07

[a] PCMT-C (ref 15)
[b] PCMT-E (ref 15)
[c] P3CT (ref 15)
[d] Calculated from result at 48°C
[e] Based on rearrangement of P3CT

These data reveal several interesting features of the repeating unit isomerization. First of all, the rate constants are of roughly the same order of magnitude as those observed in solvolyses and rearrangements of β-chlorosulfides of low molecular weight when differences in solvents and temperatures are taken into account. Table VI provides a summary of comparative data from the literature. We have also determined the apparent rate constant for the analogous isomerization of 1-chloro-2-ethylthiopropane (7) to 2-chloro-1-ethylthiopropane (8) to be 4 x $10^{-6}s^{-1}$ at 45°C in the absence of solvent. The latter is of course a composite rate constant of the kind defined by Equations 2 and 3.

S Cl → S Cl

7 8

Table VI. Rates and Activation Energies of Reactions of β-Chlorosulfides

Substrate	Temp (°C)	Solvent	$k(s^{-1})$	E_a(kcal/mol)	Ref
$CH_3PhSCH_2CH_2Cl$	40	$EtOH/H_2O$	1.2×10^{-5}	18.6	18
$PhSCH_2CH_2Cl$	40	MeOH	2.5×10^{-4}	18	19
$CH_3SCH_2CH_2Cl$	55	AcOH	4.8×10^{-4}	--	20
2-endo-chloro-7-thiabicyclo-[2.2.1]heptane	25	AcOH	1.8×10^{-4}	16.4	21
erythro-Ph(ClPhS)CHCHClCH$_3$	70	$Cl_2CHCHCl_2$	4.5×10^{-6}[a]	23	22
threo-Ph(ClPhS)CHCHClCH$_3$ (6)	50	$Cl_2CHCHCl_2$	9.2×10^{-6}[a]	9	22

[a] These are composite rate constants of the kind defined by Equations 2 and 3.

A second point of interest is the apparent activation energy of the isomerization. From an examination of the temperature dependences of K and K^- for reaction in the bulk polymer, we find apparent activation energies of 22-23 kcal/mol for each of the forward and reverse processes. Given the composite nature of K and K^{-1}, these are true activation energies only if the partitioning of the sulfonium ion intermediate is insensitive to temperature. Nonetheless, the similarity of these parameters to those listed in Table VI argues for a common mechanism that involves in all instances rate-determining ring-closure to the episulfonium ion, as suggested by Scheme I. The sole exception in Table VI is the low activation energy for rearrangement of threo-6; this behavior appears to be anomalous, and no explanation is provided by the original authors (22).

The solvent dependence of the reaction rate is also consistent with this mechanistic scheme. Comparison of the rate constants for isomerizations of PCMT in chloroform and in nitrobenzene shows a small (ca. 40%) rate enhancement in the latter solvent. Simple electrostatic theory predicts that nucleophilic substitutions in which neutral reactants are converted to ionic products should be accelerated in polar solvents (23), so that a rate increase in nitrobenzene is to be expected. In fact, this effect is often very small (24). For example, Parker and co-workers (25) report that the S_N2 reaction of methyl bromide and dimethyl sulfide is accelerated by only 50% on changing the solvent from 88% (w/w) methanol-water to N,N-dimethylacetamide (DMAc) at low ionic strength; this is a far greater change in solvent properties than that investigated in the present work. Thus a small, positive dependence of reaction rate on solvent polarity is implicit in the sulfonium ion mechanism.

A final observation consistent with rate-determining cyclization is that the reaction rate is relatively insensitive to added electrolyte. Addition of 0.5 equivalents of tetra-n-butylammonium chloride or tetra-n-butylammonium azide to chloroform solutions of

PCMT produces very small, and approximately equal, increases in rate. Alternative reaction mechanisms that invoke as rate-determining steps either i). attack by chloride ion, or ii). unassisted S_N1 dissociation of the carbon-chlorine bond are inconsistent with this result.

Repeating Unit Isomerization vs. Isomerization Polymerization

Repeating unit isomerization is similar in several respects to isomerization polymerization (26,27). Isomerization polymerization may be defined as a process whereby a monomer of structure A is converted to a polymer of repeating unit structure B, wherein the conversion of A to B represents a structural change which is not a simple ring opening or double bond addition:

$$nA \longrightarrow \left(B\right)_n$$

The product of an isomerization polymerization is thus determined by the relative rates of the propagation and isomerization steps; i.e., it is kinetically determined. If isomerization is much faster than propagation, the homopolymer of B is obtained; competitive rates will lead to A-B copolymers.

We define repeating unit isomerization as a process subsequent to polymerization, in which an intramolecular rearrangement of the repeating unit leads to a thermodynamically preferred structure:

$$nA \longrightarrow \left(A\right)_n \longrightarrow \left(A\right)_x\left(B\right)_y$$

Thus propagation must be much faster than isomerization, and the product will be determined by thermodynamics, rather than by reaction kinetics. The net results of the two processes may be quite similar, however, in that polymers of unexpected structures may be obtained, and copolymers may be prepared by polymerization of a single monomer.

Acknowledgments

This work was supported by grants from the Polymers Program of the National Science Foundation. Support of our research by an NSF Presidential Young Investigator Award is also gratefully acknowledged.

Literature Cited

1. E. Schacht, D. Bailey and O. Vogl, J. Polym. Sci. Polym. Chem. Ed. 16, 2343 (1978) and references therein.
2. E.J. Vandenberg, in C.C. Price and E.J. Vandenberg, Eds. Coordination Polymerization, Plenum, 1983, p. 11.
3. A. Streitwieser, Jr., Solvolytic Displacement Reactions, McGraw-Hill, New York, 1962.
4. H. Bohme and K. Sell, Chem. Ber., 81, 123 (1948).
5. E.J. Vandenberg, J. Polym. Sci. 47, 486 (1960).
6. E.J. Vandenberg, J. Polym. Sci. A17, 525 (1969).

7. J.S. Shih, J.F. Brandt, M.P. Zussman and D.A. Tirrell, J. Polym. Sci. Polym. Chem. Ed. 20, 2839 (1982).
8. J.S. Shih, Ph.D. Dissertation, Carnegie-Mellon University, 1984.
9. J.S. Shih and D.A. Tirrell, J. Polym. Sci. Polym. Chem. Ed. 22, 781 (1984).
10. J.S. Shih and D.A. Tirrell, J. Macromol. Sci. Chem. A21, 1013 (1984).
11. S. Vaze and D.A. Tirrell, J. Bioactive and Compatible Polym. 1, 79 (1986).
12. D.A. Wicks and D.A. Tirrell, Polym. Prepr. 27(2), 21 (1986).
13. M.P. Zussman and D.A. Tirrell, Macromolecules 14, 1148 (1981).
14. M.P. Zussman and D.A. Tirrell, Polymer Bull. 7, 439 (1982).
15. M.P. Zussman and D.A. Tirrell, J. Polym. Sci. Polym. Chem. Ed. 21, 1417 (1983).
16. M.P. Zussman, Ph.D. dissertation, Carnegie-Mellon University, 1982.
17. M.P. Zussman and D.A. Tirrell, J. Polym. Sci. Polym. Chem. Ed., in press.
18. R. Bird and C.J.M. Stirling, J. Chem. Soc. Perkin II, 1221, 1973.
19. F.G. Bordwell and W.T. Brannen, Jr., J. Am. Chem. Soc. 86, 4645 (1964).
20. I. Tabushi, Y. Tamaru and Z. Yoshida, Bull. Chem. Soc. Japan 47, 1455 (1974).
21. I. Tabushi, Y. Tamaru, Z. Yoshida and T. Sugimoto, J. Am. Chem. Soc. 97, 2886 (1975).
22. G.H. Schmid and V. Csizmadia, Can. J. Chem. 50, 2465 (1972).
23. R.W. Alder, R. Baker and J.M. Brown, Mechanism in Organic Chemistry, Wiley, New York, 1971, p. 43.
24. A.J. Parker, Chem. Rev. 69, 1 (1969).
25. Y.C. Mac, W.A. Millen, A.J. Parker and D.W. Watts, J. Chem. Soc. (B), 525 (1967).
26. J.P. Kennedy and R.M. Thomas, Makromol. Chem. 53, 28 (1962).
27. J.P. Kennedy in H.F. Mark, N.G. Gaylord and N.M. Bikales, Eds., Encyclopedia of Polymer Science and Technology, Wiley, New York, 1967, Vol. 7, p. 754.

RECEIVED September 11, 1987

Chapter 6

Polymeric Dialkylaminopyridines as Supernucleophilic Catalysts

Subhash C. Narang[1], Susanna Ventura[1], and Roopram Ramharack[2]

[1]**SRI International, 333 Ravenswood Avenue, Menlo Park, CA 94025**
[2]**3M Company, St. Paul, MN 55144**

Supernucleophilic polymers containing the 4-(pyrrolidino)pyridine group were synthesized from the corresponding maleic anhydride copolymers and also by cyclopolymerization of N-4-pyridyl bis(methacrylimide). The resulting polymers were examined for their kinetics of quaternization with benzyl chloride and hydrolysis of p-nitrophenylacetate. In both instances, the polymer bound 4-(dialkylamino)pyridine was found to be a superior catalyst than the corresponding low molecular weight analog.

4-(Dimethylamino)pyridine (1, DMAP) and its analogs, particularly 4-(pyrrolidino)pyridine (2, PPY) have acquired enormous importance and utility as supernucleophilic catalysts in synthetic organic and polymer chemistry (1,2).

In 1967, (3) it was discovered that DMAP catalyzes the benzoylation of m-chloroaniline 10^4 times faster than pyridine. This enormous increase in reaction rate is unmatched by any other nucleophilic acylation catalyst (3). It was shown that the catalytic action of DMAP and PPY is not primarily due to their larger pK_a's with respect to pyridine, but is a result of enhanced nucleophilic catalysis.

0097-6156/88/0364-0072$06.00/0

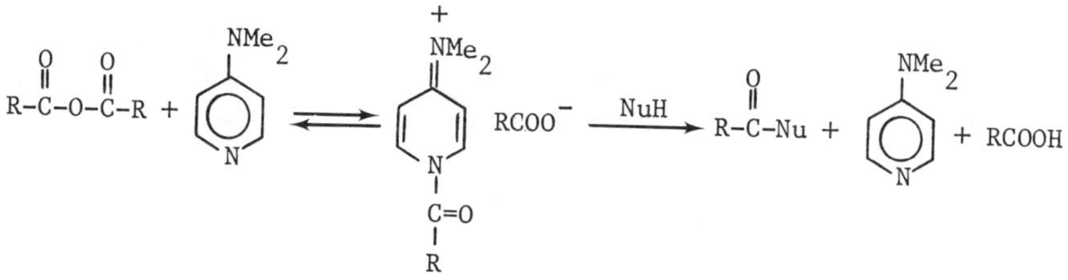

$$R\text{-}\overset{O}{\overset{\|}{C}}\text{-}O\text{-}\overset{O}{\overset{\|}{C}}\text{-}R + \text{DMAP} \rightleftharpoons \text{N-acyl-4-dimethylaminopyridinium } (\text{C=O, R}) \; RCOO^{-} \xrightarrow{NuH} R\text{-}\overset{O}{\overset{\|}{C}}\text{-}Nu + \text{DMAP} + RCOOH$$

The intermediate N-acylpyridinium salt is highly stabilized by the electron donating ability of the dimethylamino group. The increased stability of the N-acylpyridinium ion has been postulated to lead to increased separation of the ion pair resulting in an easier attack by the nucleophile with general base catalysis provided by the loosely bound carboxylate anion. Dialkylaminopyridines have been shown to be excellent catalysts for acylation (of amines, alcohols, phenols, enolates), tritylation, silylation, lactonization, phosphonylation, and carbomylation and as transfer agents of cyano, arylsulfonyl, and arylsulfinyl groups (1-3). DMAP has been found to improve the coupling reaction in solid phase peptide synthesis (4). Polypeptides synthesized via the DMAP-DCC method were found to be of higher purity compared to other methods of synthesis. In polymer chemistry, DMAP has been used for hardening of resins and in the synthesis of polyurethanes, (5) polycarbonates, (6) and polyphenyleneoxides (7).

The tremendous scope of utilization of DMAP and PPY as catalysts has led to an active interest in the development of their polymeric analogs. The pioneering work was carried out by Hierl et al (8) and Delaney et al. (9). They attached 4-dialkylaminopyridine derivatives to poly(ethyleneimine) and found the modified polymers to be highly active catalysts for hydrolysis of p-nitrophenylcarboxylates. Since then, many research groups have reported the synthesis of polymers functionalized with 4-dialkylaminopyridine (10-18).

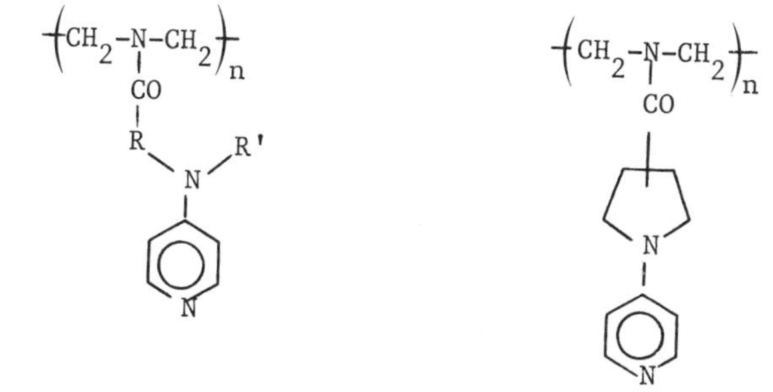

$$\left(CH_2\text{-}N\text{-}CH_2 \right)_n \text{ with N-CO-R-N(R')-(4-pyridyl)} \qquad \left(CH_2\text{-}N\text{-}CH_2 \right)_n \text{ with N-CO-(3-pyrrolidinyl)-N-(4-pyridyl)}$$

EXPERIMENTAL

Materials

All reagents and chemicals were commercially available materials purified by standard laboratory procedures.

N-4-(Pyridyl) Methacrylamide

A solution of methacryloyl chloride (44.8 g, 0.43 mol) in tetrahydrofuran (100 ml) was added dropwise to a stirred solution of 4-aminopyridine (20 g, 0.21 mol) dissolved in a two phase system of 10% aq. NaOH (300 ml) and tetrahydrofuran (50 ml) at 15-20°C. The reaction mixture was stirred for 2 hrs. Conc. HCl was added to the reaction mixture to lower the pH to 8 and the reaction mixture extracted with dichloromethane (4 x 100 ml). The organic extract was dried and the solvent evaporated to give the desired product in 80% yield. ^{1}H NMR ($CDCl_3$, TMS) δ 2.05 (s, 3H), 5.6 (s, 1H), 5.9 (s, 1H), 7.8 (m, 2H), 8.6 (m, 2H), 9.6 ppm (s, 1H); IR (KBr) 3200, 3140, 3040, 2980, 1675, 1590, 1510, 1415, 1330, 1290, 1200, 1155, 1000, 940, 860 cm^{-1}; mass spectrum (70 eV) m/e 162 (molecular ion), 147, 134, 119, 93, 78, 69.

N-4(Pyridyl) bis(methacrylimide)

A solution of N-4-(pyridyl)methacrylamide (27 g, 0.17 mol) in tetrahydrofuran (250 ml) was added to a stirred slurry of sodium hydride (8 g of a 60% dispersion in mineral oil, 0.2 mol) in tetrahydrofuran (150 ml) at 15-20°C under a dry nitrogen atmosphere. After stirring for 2 hrs, a solution of methacryloyl chloride (21 g, 0.20 mol) in anhydrous tetrahydrofuran (25 ml) was added dropwise over one hour and the reaction mixture stirred for another hour. The reaction mixture was quenched by addition of a few ml of methanol. The solvent was evaporated under vacuum and the crude product crystallized from ether in 60% yield, m.p. 113-114°C; IR (KBr) 3000, 2980, 1720, 1680, 1635, 1595, 1460, 1355, 1310, 1280, 1160 cm^{-1}; ^{1}H NMR ($CDCl_3$, TMS) δ 2.0 (s, 6H), 5.6 (d, 4H), 7.1 (m, 2H), 8.7 ppm (m, 2H); ^{13}C NMR($CDCl_3$) δ 18.5, 120.9, 122.5, 143.1, 146.6, 150.8, 173.3 ppm, mass spectrum (70 eV) m/e 230 (molecular ion), 229, 202, 189, 187, 174, 161, 145, 78, 70, 69. Anal. Calcd. for $C_{13}H_{14}N_2O_2$: C 67.8; H 6.1; N 12.2%. Found: C 67.8; H 6.13; N 12.16%.

Cyclopolymerization of N-4-(Pyridyl) bis(methacrylimide)

N-4-(Pyridyl) bis(methacrylimide) (8.6 g, 37 mmol) was dissolved in dimethylformamide (35 ml). The solution was degassed under argon and azobis(isobutyronitrile) (0.4 g, 2 mmol) was added. The reaction mixture was heated to 75°C under argon. After 24 hrs, another portion of AIBN (0.4 g, 2 mmol) was added and the reaction continued for another 24 hrs. The resulting polymer was isolated by precipitation from ether in 65% yield. DSC of the polymer indicated a Tg at 110°C. TGA showed onset of decomposition at 400°C under argon. The intrinsic viscosity determined in chloroform solution at 30°C, was 0.05 dlg^{-1} indicating a low molecular weight polymer; UV spectrum in CH_2Cl_2 showed a λ_{max} at 230 nm. IR showed bands at 1790 and 1730 cm^{-1} corresponding to the five membered imide ring; ^1H NMR ($CDCl_3$, TMS) δ 1.24 (s, 6H), 1.70 (s, 4H), 8.24 (broad, 2H) 8.67 ppm (broad, 2H); ^{13}C NMR ($CDCl_3$) δ 15.9, 31.1, 50.7, 119.4, 139.2, 150.7, 178.8 ppm.

Reduction of Poly(N-4-(pyridyl) bis(methacrylimide))

To a stirred solution of the polymer (2 g, 17 meq of C=O) in diglyme (100 ml), boron trifluoride etherate (1.2 ml, 10 mmol) was added under a nitrogen atmosphere. The reaction mixture was cooled to 0°C and a solution of sodium borohydride (1.32 g, 35 mmol) in diglyme (80 ml) was added dropwise over 30 minutes. The reaction mixture was heated to 70°C for 16 hrs and then quenched with methanol (2-3 ml). The solvent was evaporated under vacuum and the crude product was stirred with aq. sodium hydroxide (5%, 100 ml). The polymer was filtered and the residue washed with water. The polymer was dried in a vacuum oven at 90°C. The degree of functionalization was 67% as calculated by UV and ^1H NMR spectroscopy. TGA of the pyrrolidinopyrdine polymer showed the onset of weight loss at 350°C under argon. UV λ_{max} 261 nm (basic form, $CH_3OH/H_2O/NaOH$); 281 nm (acidic form, $CH_3OH/H_2O/HCl$); ^1H NMR (d_6-DMSO, TMS) δ 0.95 (broad, 10H), 3.2 (broad, 4H), 6.4 (broad, 2H), 8.1 ppm (broad, 2H); ^{13}C NMR (d_6-DMSO) δ 20, 29, 45, 59, 107, 149, 151, 152 ppm.

Kinetics Experimental

The quaternization kinetics were followed by ^1H NMR spectroscopy using a JEOL FX-90Q NMR spectrometer. Solvolysis of p-nitrophenylacetate was followed by UV spectroscopy using a Hewlett Packard 8450 A diode array spectrophotometer.

The kinetics were followed by measuring the increase in absorbance at 400 nm due to the formation of the p-nitrophenoxide anion in a tris buffer solution at pH 8.5. The substrate was used in excess over the free base catalyst, whose concentration was calculated from spectrophotometric data.

Pseudo-first-order rate constants (k_{obs}) determined from the linear relationship of ln ($A_\infty - A_t$) with time, were calculated for different concentrations of catalyst. A pseudo-first-order rate constant was also calculated for the uncatalyzed hydrolysis at pH 8.5, and was substracted from the values found for the catalyzed hydrolysis.

From the slope of the linear relationship between k_{obs} and the base concentration, the second-order-rate constants were calculated.

RESULTS AND DISCUSSION

We have previously (13) reported a rapid two step synthesis of 4-(pyrrolidino)pyridine copolymers via the reaction of commercially available maleic anhydride copolymers with 4-aminopyridine followed by reduction with $LiAlH_4$, yielding polymers with a high degree of functionalization.

-CH—CH-CH$_2$-CH- / O=C, C=O, O / R

$\xrightarrow{H_2N\text{-}C_5H_4N}$

-CH—CH-CH$_2$-CH- / O=C, C=O, N-(4-pyridyl) / R

$\xrightarrow{LiAlH_4}$

-CH—CH-CH$_2$-CH- / H_2C, CH_2, N-(4-pyridyl) / R

In order to determine the efficiency of the polymers as reagents in nucleophilic catalysis, it was decided to study the rate of quaternization with benzyl chloride. Table I shows the second-order-rate constants for the benzylation reaction in ethanol. Comparison with DMAP indicates that poly(butadiene-co-pyrrolidinopyridine) is the most reactive of all the polymers examined and is even more reactive than the monomeric model. This enhanced reactivity is probably due to the enhanced hydrophobicity of the polymer chain in the vicinity of the reactive sites.

Table I

KINETICS OF QUATERNIZATION IN ETHANOL[a]

Nucleophile	k (mol^{-1} ℓ min^{-1})
4-Dimethylaminopyridine	0.0010
Poly(butadiene-co-pyrrolidinopyridine)	0.0014
Poly(methyl vinyl ether-co-pyrrolidinopyridine)	0.0009
Poly(octadec-1-ene-co-pyrrolidinopyridine)	b

[a]The kinetics were studied at 25.4°C.
[b]The rate of reaction was too small to measure, presumably because of steric hindrance due to the long alkyl chain.
Source: Reproduced with permission from Ref. 13. Copyright 1985, Wiley.

The dielectric constant of the solvent in the microenvironment of the polymer chain has been shown to be different from that in the bulk solvent (19). This change in dielectric constant might enhance the nucleophilicity of the pyridine ring and therefore increase the rate of quaternization. The kinetic results are consistent with the observations of Overberger et al., (20), who showed that increased hydrophobic nature of the substrate led to faster reaction rates in nucleophilic catalysis. In the present case one would expect the butadiene copolymer to be more hydrophobic than the methylvinylether copolymer. An alternative synthesis of supernucleophilic polymers has been achieved using the following reaction sequence.

NH_2 + Cl, O — $NaOH/H_2O$, THF →

1. NaH/THF
2. Cl, O

DMF/AIBN
75°C

$NaBH_4/BF_3 \cdot Et_2O$

The monomer synthesis and cyclopolymerization were carried out following the procedure of Butler et al. (21). The resulting polyimide was shown to possess primarily pyrrolidine rings as indicated by infrared spectroscopy (21). Initially, the reduction was carried out with $LiAlH_4$ in tetrahydrofuran. However, the reaction always led to some overreduction of the pyridine ring. Reduction with BH_3/THF was unsuccessful. However, $NaBH_4/BF_3 \cdot Et_2O$ proved to be an excellent reagent for the reduction step, resulting in the formation of the desired polymer. The polymer was shown by UV and NMR spectroscopy to be 67% functionalized primarily with 4-pyrrolidinopyridine moieties by UV and NMR spectroscopy; according to these data no starting polymer is present anymore, and the nature of the remaining 23% of functionalization is unknown.

Kinetic Studies. The pioneering work of Hierl et al. (8) and Delaney et al. (9) had established that hydrolysis of *p*-nitrophenylcarboxylates was an excellent means of observing the nucleophilic catalysis by 4-(dialkylamino) pyridine functionalized polymers. Hydrolysis of p-nitrophenylacetate in a buffer at pH 8.5 showed that the polymer was a slightly better catalyst than the monomeric analog PPY (Table II). However, preliminary results indicate that the polymer bound 4-(dialkylamino) pyridine is more effective as a catalyst than the monomeric analog in the hydrolysis of longer carbon chain p-nitrophenylcarboxylates, such as p-nitrophenylcaproate.

Figures I and II show a comparison of the reaction profile for PPY and polymer catalyzed hydrolysis for p-nitrophenylacetate and p-nitrophenylcaproate monitored by the appearance of p-nitrophenoxide absorption by UV-VIS spectroscopy. These results confirm the effectiveness of the interactions between the hydrophobic polymer chain and the hydrocarbon portion of the substrate, as it was previously mentioned, in accordance with the observations of Overberger et al (20).

Table II

KINETICS OF HYDROLYSIS OF P-NITROPHENYLACETATE AT 25°C

Nucleophile	k (mol^{-1} ℓ sec^{-1})
4-Pyrrolidinopyridine	428
Poly(4-pyrrolidinopyridine)	493

Conclusions

4-Pyrrolidinopyridine based polymers react faster with benzyl chloride than low molecular weight analogs. The polymer synthesized by cyclopolymerization of N-4-pyridyl bis(methacrylimide)

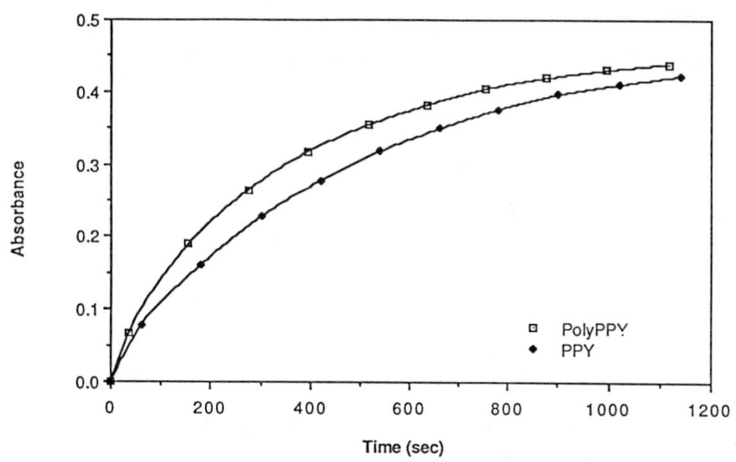

Figure 1. Absorbance at 400 nm vs. time for the poly(4-pyrrolidinopyridine) and 4-pyrrolidinopyridine catalyzed hydrolysis of p-nitrophenylacetate at pH 8.5.

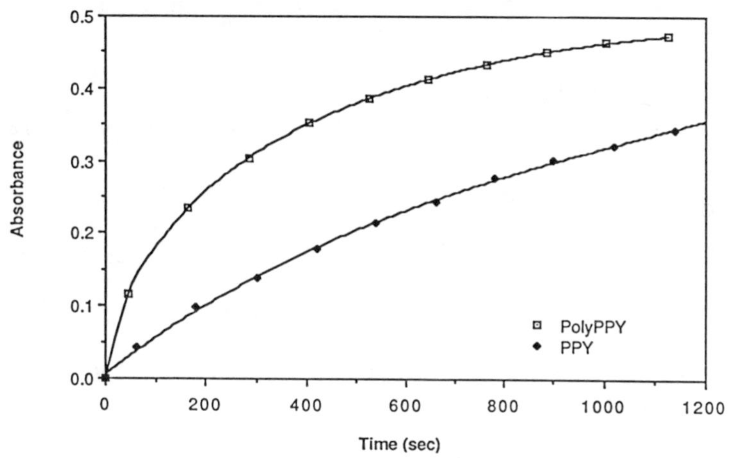

Figure 2. Absorbance at 400 nm vs. time for the poly(4-pyrrolidinopyridine) and 4-pyrrolidinopyridine catalyzed hydrolysis of p-nitrophenylcaproate at pH 8.5.

followed by reduction is a supernucleophilic polymer which shows higher catalytic activity compared to the low molecular weight analogous catalyst, PPY.

Acknowledgments

We are grateful to the Army Research Office for their generous support of this work.

References

† Abstracted in part from the Ph.D. thesis of R. Ramharack, submitted to Polytechnic Institute of New York.

1. Hofle, G., Steglich, W. and Vorbruggen, V. Angew. Chem. Int. Ed. Engl. 1978, 17, 569 and references therein.
2. Scriven, E.F.V. Chem. Soc. Rev. 1983, 12, 129 and references therein.
3. (a) Litvinenko, L. M. and Kirichenko, A. I. Dokl, Akad. Nauk, SSSR Ser. Khim., 1967, 176, 97.
   (b) Steglich, W. and Hofle, G. Angew, Chem. Int. Ed. Engl., 1969, 8, 981.
4. Wang, S. S., Tam, J. P., and Merrifield, R. B. Int. J. Peptide Protein Res., 1981, 18, 459.
5. Wild, J. H. and Tate, F.E.G. U.S. Patent 3144 452, 1964.
6. Jaquiss, D.B.G., Mark, V. and Mitchell, L. C. U.S. Patent 4286 085.
7. Verlaan, J.P.L., Alferink, P.J.T. and Challa, G. J. Mol. Catal., 1984, 24, 235.
8. Hierl, M. A., Gamson, E. P. and Klotz, I. M. J. Am. Chem. Soc., 1979, 101, 6020.
9. Delaney, E. J., Wood, L. E. and Klotz, I. M. J. Am. Chem. Soc., 1982, 104, 799.
10. Shinkai, S., Tsuji, H., Hara, Y. and Manabe, O. Bull. Chem. Soc. Jpn., 1981, 54, 631.
11. Tomoi, M., Akada, Y. and Hakiuchi, H. Makromol. Chem. Rapid Commun., 1982, 3, 537.
12. Guendouz, F., Jacquier, R. and Verducci, J. Tetrahedron. Lett., 1984, 25, 4521.
13. Narang, S. C. and Ramharack, R. J. Polym. Sci. Polym. Lett. Ed., 1985, 23, 147.
14. Mathias, L. J., Vaidya, R. A. and Bloodworth, R. H. J. Polym. Sci. Polym. Lett. Ed., 1985, 23, 289.
15. Storck, W. and Manecke, G. J. Mol. Catal., 1985, 30, 145.
16. Menger, F. M. and McCann, D. J. J. Org. Chem., 1985, 50, 3928.
17. Tomoi, M., Goto, M., Kaikuchi, H. Makromol. Chem. Rapid Commun., 1985, 6, 397.
18. Klotz, I. M., Massil, S. E. and Wood, L. E., J. Polym. Sci., Polym. Chem. Ed., 1985, 23, 575.
19. Mikes, F., Strop P. and Kalal, J. Makromol. Chem, 1984, 175, 2375.
20. Overberger, C. G. and Guterl, A. C. J. Polym. Sci., Polym. Symp., 1978, 62, 13.
21. Butler, G. B. and Myers, Myers J. Macromol. Sci. Chem., 1971, A5, 135, references therein.

RECEIVED November 6, 1987

NEW SYNTHESIS ROUTES

Introduction to New Synthesis Routes

Chemical reactions on polymers provide a broad array of new synthesis routes to novel polymers with unique properties. The variety of polymer subtrates is nearly limitless, and includes addition polymers, e.g. polyolefins and halogenated polyolefins, polydienes, polystyrenes, and acrylic polymers, as well as condensation polymers, e.g. polyesters, polyamides, epoxies, polyurethanes, polyamines and polyimines, and silicones. The reactions which can be used on the polymers are also highly varied, including oxidation, reduction, addition, elimination, cyclization or cycloaddition, substitution, and condensation. The reactions can be used to modify the polymer backbone, to introduce or modify branched side chains, or to change functional substituents. The level of activity has been high and continually increasing, and has been reflected by the publication of numerous patents and research articles in recent years.

Graft polymers represent one of the most important areas for new synthsis routes, since very unique properties can be obtained via introduction of a branched side chain on a polymer backbone. Polymerization can be initiated by a functional group on the polymer backbone, and the graft chains formed from a variety of addition or condensation monomers. The effect of the graft on the polymer properties depends on the chemical nature, molecular weight and polydispersity of both the graft and the polymer backbone. However, a complex mixture of products is often formed from the grafting reaction, making it difficult to obtain the optimal properties from the new material.

One of the newer routes to graft polymers employs the macromonomer method to introduce well defined side chains onto polymers. Polymers with controlled molecular weight, narrow polydispersity, and functional (polymerizable) termination can be prepared using anionic polymerization techniques. These macromonomers can be derived from a variety of polymer families, including polydienes, polystyrenes, and silicones. The functional termination has most frequently been allyl, vinyl, acrylate, or methacrylate. The macromonomer can then be copolymerized with a variety of addition monomers, e.g. styrene, acrylate, to produce the graft polymer. The graft polymers prepared using macromonomers frequently possess phase separated structures, with the nature of the phase separation determined by the molecular weight of the macromonomer. Bulk properties, e.g. permeability, modulus, and thermal stability, as well as surface properties, e.g. adhesion, wettability, and chemical resistance, can be influenced by the nature of the graft.

Side-chain liquid crystalline polymers, e.g. where the mesogen is present in the side chain rather than the polymer backbone, represent another type of graft polymer. Many publications exist on these new and interesting materials. Chemical reactions on polymers to introduce the mesogenic unit have been an attractive alternative to direct polymerization of the mesogenic monomer. Side-chain liquid crystalline polymers have been prepared by reaction of mesogenic alcohols or amines with polyacrylates, polymethacrylates, and polyacrylamides. Liquid crystalline polysiloxanes have also been prepared via hydrosilation using a vinyl substituted mesogen. More recently, liquid crystalline polymers were prepared using chemical reactions to introduce the mesogenic side chains on poly(2,6-dimethyl-1,4-phenylene oxide) and poly(epichlorohydrin).

Novel heterocyclic polymers have been prepared using chemical reactions on polymers. For example, polyvinyl amine can be modified to introduce a pyridine-containing side chain, and the resulting polymer is a selective macromolecular complexing agent. Cyclization reactions on linear nitrogen-containing polymers to produce polymers with heterocyclic backbones have also been reported.

Many other types of novel polymers have been produced using chemical reactions on polymers. Examples include ring opening reactions on polymers containing cyclic moieties in the polymer backbone or polymer side chain, and coupling of reactive monomers, e.g. isocyanotoethyl methacrylate, with functional groups on the polymer to produce radiation curable polymers.

The use of chemical reactions on polymers to provide new synthesis routes to unique new polymers is a growing area of research. The papers in this section of the book illustrate the diversity of new materials and new synthesis techniques which are becoming available.

Chapter 7

Styrenic- and Acrylic-Siloxane Block and Graft Copolymers

S. D. Smith, G. York, D. W. Dwight[1], and J. E. McGrath

Department of Chemistry and Polymer Materials and Interfaces Laboratory, Virginia Polytechnic Institute and State University, Blacksburg, VA 24061-8699

The synthesis and characterization of graft copolymers, where the backbones were either acrylic or styrenic with siloxane grafts, were prepared and characterized in order to elucidate the molecular and solid state structures, as well as general physical behavior. Although two microphases could be achieved, the results showed that miscibility (as measured by several techniques) increased with lower graft molecular weights and as the solubility parameters of the backbone were designed to be closer to that of the poly(dimethylsiloxane) (PSX) grafts.

An important polymer modification reaction is the grafting to or from a polymer backbone by some chemical method to produce a branched structure (1). The characterization of the products of these reactions is often somewhat less well defined than block copolymers (2) due to the complexity of the mixture of products formed. It is therefore useful to prepare and characterize more well defined branched systems as models for the less well defined copolymers. The macromonomer method (3) allows for the preparation of more well defined copolymers than previously available.

Anionic polymerization in suitable systems allows the preparation of polymers with controlled molecular weight, narrow molecular weight distributions and functional termination. The functional termination of a living anionic polymerization with a polymerizable group has been used frequently in the preparation of macromonomers (4). Our research has encompassed the anionic homo and block copolymerizations of D_3 or hexamethyl cyclotrisiloxane with organolithiums to prepare well defined polymers. As early as 1962 PSX macromonomers were reported in the literature by Greber (5) but the copolymerization of these macromonomers did not become accepted technique until their value was demonstrated by Milkovich and

[1]Current address: P.O. Box 160, Sumneytown, PA 18084

0097-6156/88/0364-0085$06.00/0

coworkers (6). Since then, considerable work has been dedicated to the macromonomer technique in many laboratories around the world.

In our own research, the functional termination of the living siloxanolate with a chlorosilane functional methacrylate leading to siloxane macromonomers with number average molecular weights from 1000 to 20,000 g/mole has been emphasized. Methacrylic and styrenic monomers were then copolymerized with these macromonomers to produce graft copolymers where the styrenic or acrylic monomers comprise the backbone, and the siloxane chains are pendant as grafts as depicted in Scheme 1. Copolymers were prepared with siloxane contents from 5 to 50 weight percent.

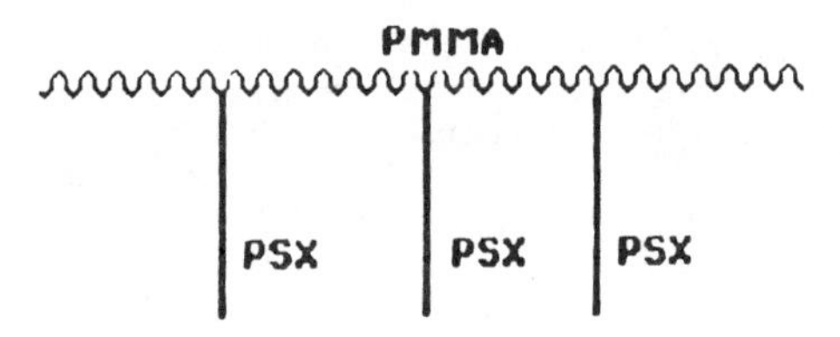

The initiation of the cyclic siloxane monomers with a living polymeric lithium species such as polystyryl lithium leads to block copolymers, as outlined in Scheme 2, were also of interest. These styrenic-siloxane block copolymers were prepared with siloxane contents from 10 to 50 weight percent.

Scheme 2.

$$—(CH_2-\underset{\underset{R}{|}}{\overset{}{CH}}(C_6H_4))_A—|\sim\sim(\overset{CH_3}{\underset{CH_3}{Si}}-O)_B\sim\sim\overset{CH_3}{\underset{CH_3}{Si}}-CH_3$$

$R = H, CH_3, t\text{-}C_4H_9$

Recently, the importance of incorporating silicon containing segments into copolymers has become increasingly well appreciated. The desirable properties of silicon include ion etch resistance for microlithography (7,8), and higher oxygen permeability in membrane materials (9), as well as biocompatability. Unfortunately, homo- or random copolymers of polysiloxanes often have Tg's too low to be useful as film forming materials. In contrast, the phase separation of these graft and block materials discussed in this paper allow the incorporation of silicon moieties at relatively high levels while maintaining the solubility and integrity of the glassy polymer phase. Alloys of these multiphase copolymers are also being investigated.

Experimental

Cyclohexane (Phillips 99+ mole %) was purified by stirring over H_2SO_4 then distilled from a sodium dispersion. Tetrahydrofuran (Fisher) was distilled from a sodium/benzophenone complex. Toluene (Fisher) was degassed thoroughly and used directly in radical polymerizations.

D_3 (Petrarch) was purified by melting and stirring in the presence of finely divided calcium hydride then sublimed under vacuum. The D_3 was then dissolved in purified cyclohexane and stored under N_2.

Styrene (Fisher), p-methylstyrene (Mobil), and t-butylstyrene (DOW) were purified by passing through a column of activated alumina and then carefully degassed to remove all traces of O_2. Further purification by vacuum distillation from dibutyl magnesium resulted in anionically pure monomers. Methacrylate monomers were distilled from calcium hydride and stored in a freezer until use.

Macromonomers were prepared by initiation of the D_3 in cyclohexane with s-BuLi after which purified tetrahydrofuran was added to promote propagation of the siloxanolate species. Termination with 3-methacryloxypropyl dimethylchlorosilane (Petrarch, Inc.) afforded the macromonomer in high yields which was then precipitated in methanol and dried under vacuum. Macromonomers were analyzed by several techniques to confirm not only their molecular weight and molecular weight distribution but also their methacrylate end group functionality. Vapor phase osmometry (VPO), run in toluene using sucrose octacetate as a standard, was used as a method for determination of number average molecular weight. NMR analysis (IBM 270 MHz SL instrument) allowed a ratioing of alpha methyl protons on the methacrylate group to the silicon methyls as a determination of functionality. NMR was not sufficiently precise for molecular weights above approximately 2000 and another technique based on the UV analysis of the methacrylate functionality was developed. UV spectra run on a Perkin Elmer 552 instrument (scanned from 350 nm to 190 nm at 20 nm/minute) established a wavelength maximum at 214 nm for the carbonyl group of our methacrylate functional endgroup (Figure 1). Then, absorbance measurements were determined at 214 nm and a calibration plot using Beers law with methyl methacrylate as a standard in THF was prepared. The end group analysis of known macromonomer concentrations thus was used to establish the functionality.

Free radical copolymerizations of the alkyl methacrylates were carried out in toluene at 60°C with 0.1 weight percent (based on monomer) AIBN initiator, while the styrenic systems were polymerized in cyclohexane. The solvent choices were primarily based on systems which would be homogeneous but also show low chain transfer constants. Methacrylate polymerizations were carried out at 20 weight percent solids and the styrenic systems were started at 50 weight percent solids then diluted with time to prevent solidification of the mixture. Copolymers after workup were extracted extensively with hexanes or isopropanol, depending on polymer solubility, to remove any polysiloxane homopolymer. Under appropriate conditions, these amounts were quite low.

The anionic polymerization of the styrene monomers was conducted at -78°C in THF with s-BuLi as the initiator. The second block was prepared by adding to this living polymer a calculated amount of D_3/cyclohexane mixture. Siloxanolate initiation was evidenced by loss of the characteristic orange color of the living styrene anion. The reaction was allowed to warm to room temperature and given adequate time (5-20 hours) for propagation to near quantitative conversion. Termination was achieved by reaction with trimethyl chlorosilane in slight molar excess at room temperature. The copolymers were then precipitated, dried and extracted to remove any siloxane homopolymer. GPC indicates (Figure 2) a typically narrow distribution for these copolymers which were also noted to be free of styrenic homopolymer. Typical number average molecular weights prepared were 100-150,000 for the first block. The polysiloxane constituted 10 to 60 weight percent of the overall copolymers. Thus, the total molecular weight was as high as 300,000 gm/mole. Viscosities become very high and must be controlled to keep the molecular weight distributions from broadening.

Polymers were analyzed by Gel Permeation Chromatography (GPC) with a Waters 150-C GPC fitted with micro styragel columns of 500Å, 10^3, 10^4, 10^5, and 10^6 Å porosities in THF. Poly(methyl methacrylate) (PMMA) and polystyrene standards (Polysciences) were generally used to estimate number average molecular weights. Number average molecular weights of greater than 150,000 were desired for optimal polymer properties.

NMR spectra on an IBM 270 MHz SL instrument were used to determine the siloxane incorporation into the copolymer. FTIR spectra were run on a Nicolet MX-1 spectrophotometer.

Differential Scanning Calorimeter (DSC) thermograms were obtained on a Perkin Elmer DSC-2 run at 10°C per minutes. Dynamic Mechanical Thermal Analysis (DMTA) spectra were obtained on a Polymer Labs DMTA at a frequency of 1Hz with a temperature range from -150°C to +150°C at a scan rate of 5°C per minute.

Contact angle measurements were obtained using a goniometer, measuring the advancing angle from 2 to 20 microliter drop sizes, of purified water upon polymer films at room temperature. Films were cast on metal plates and allowed to dry slowly from chloroform solutions. Several spots were measured on each film and the results averaged.

ESCA spectra were obtained with a Kratos XSAM -800 instrument with Mg anode 200 watts under a vacuum of 10-9 torr. TEM analysis was done on a Phillips IL 420 T STEM at 100 KV in the TEM mode.

Morphological studies were conducted on model copolymers containing a methyl methacrylate backbone and approximately

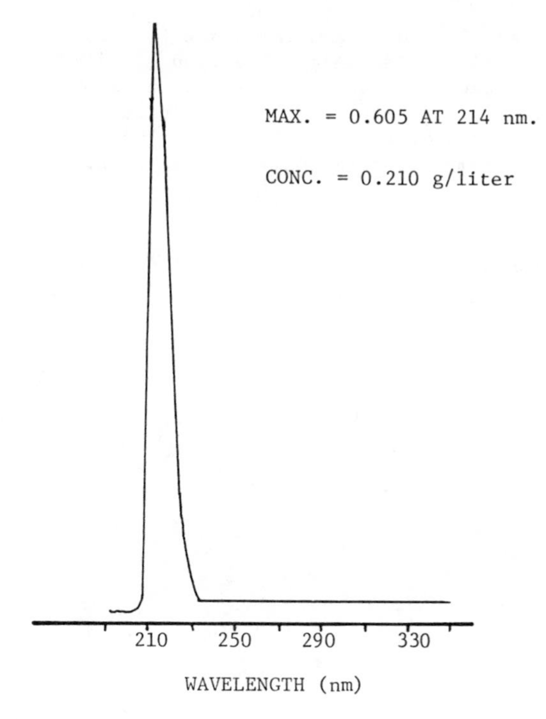

Figure 1. UV spectrum of methacrylate functional PSX macromonomer with M_n 1600.

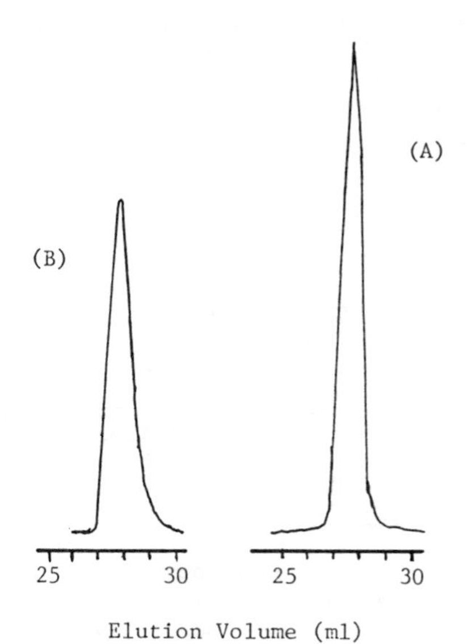

Figure 2. Size exclusion chromatograms of polystyrene standard (A) and p(t-butyl styrene)-b-PSX (B).

5,000, 10,000 and 20,000 molecular weight ($\bar{M}n$) polydimethyl siloxane side chains. The weight fraction of siloxane was held approximately constant at 20% and 50%. Thin (approximately 100 nm) films were cast on water and thick (approximately 1 mm) films were cast on stainless steel. Specimens from the latter films were microtomed at cryoscopic temperatures.

Results and Discussion

The preparation of methacrylate functional polydimethyl siloxane (10,11,12) of high functionality via Equation 1 was an

$$\underset{1}{RLi} + \underset{2}{D_3} \xrightarrow{THF} \underset{3}{R\text{-}(\text{-}Si(CH_3)_2\text{-}O\text{-})_n^- \; Li^+} + \underset{4}{Cl\text{-}Si(CH_3)_2\text{-}(CH_2)_3\text{-}O\text{-}C(=O)\text{-}C(CH_3)=CH_2}$$

$$\longrightarrow \underset{5}{\boxed{PSX}\text{—}O\text{-}C(=O)\text{-}C(CH_3)=CH_2} \qquad (1)$$

important first step in the preparation of the well defined graft copolymers. The macromonomers prepared were characterized for number average molecular weights via VPO. The functionality of the end groups were also determined by NMR or UV analysis which should provide identical molecular weights for perfectly monofunctional materials. As can be seen in Table I, a good correspondence was obtained. Incorporation of the macromonomers into copolymers via free radical copolymerization can also be used as a check on functionality since nonfunctional materials obviously will not be incorporated. Proton NMR was used to confirm the amount of PSX

Table I. Mn Values of Methacrylate Functional Poly(Dimethyl Siloxane) via Spectroscopic and Osmotic Techniques

SAMPLE	1H NMR[a]	VPO[b]	UV[c]
1	---	6500	6000
2	1600	1500	1700
3	0900	1030	0900

[a] Comparison of silicon methyl/acrylate methyl
[b] Solvent: Toluene
[c] From ester carbonyl at absorbance 214 nm (UV)

incorporation (Tables II and III) into the copolymers. In general, the incorporated amounts are relatively close to the charged amounts, indicating both high macromonomer functionality and appropriate copolymerization conditions.

Table II. NMR Analysis of Poly(Dimethyl Siloxane) Content in Styrenic Graft and Block Copolymers

Sample Description	Charged	Found
Grafts:		
P(Styrene)-g-PSX	30	27
P(p-Methyl Styrene)-g-PSX	10	9
P(t-Butyl Styrene)-g-PSX	20	20
Blocks:		
P(Styrene)-b-PSX	20	20
P(p-Methyl Styrene)-b-PSX	20	19
P(t-Butyl Styrene)-b-PSX	10	10
P(t-Butyl Styrene)-b-PSX	40	42

Table III. Influence of Siloxane Content and Molecular Weight on the Water Contact Angles of Poly(Methyl Methacrylate)-Poly(Dimethyl Siloxane) Graft Copolymers[a]

Macro Monomer M_n	Weight % Charged	Weight % via NMR	Contact Angle
PMMA Homopolymer Control			74°
1,000	5	4.0	97°
	10	7.6	
	15	9.0	
	20	15.0	105°
5,000	5	4.3	99°
	10	6.8	
	15	12.4	
	20	16.8	107°
10,000	5	5.8	108°
	10	9.6	
	15	14.0	
	20	15.9	109°
20,000	5	4.0	109°
	10	7.0	
	20	12.0	109°

[a] Copolymer work-up included extraction with hexane to remove any polysiloxane homopolymer.

Very well defined block copolymers of styrene, p-methyl styrene, p-tert-butyl styrene and PSX were prepared, as judged by GPC, with molecular weights of over 100,000 g/mole. Impurity levels at this molecular weight become extremely critical and careful vacuum techniques must be used where appropriate to exclude all contaminations.

As mentioned earlier, the graft molecular weight and copolymer compositions dictate the properties of the system. DSC measurements indicate partial phase mixing for low molecular weight grafts, as evidenced by a shifting of the Tg of PMMA to lower temperatures (Table IV). Although the Tg of the polymethylmethacrylate systems with graft molecular weights of 5K, 10K and 20K are roughly equivalent, changes in the breadth of the transition indicate differences in the phase mixing, which is to be expected. Similar results are observed by Dynamic Mechanical Thermal Analysis (Figure 3, Table IV).

Table IV. Influence of Graft Molecular Weight on Glass Transition Temperature for PMMA-g-PSX Copolymers (~16 wt.% PSX)

Macronomer Mn	Tg by DSC (°C)	Tg by DMTA[a] (°C)
1000	111	110
5000	123	123
10000	125	126
20000	127	---
Control PMMA	127	

[a]Determined by Mechanical Loss Plot (1 Hz)

Since poly(dimethyl siloxane) is well known to display a low surface energy, it was expected to dominate the surface of the microphase separated copolymers. Contact angle analysis indeed indicated a change in surface composition with molecular weight and composition of the grafts. Copolymers containing higher molecular weight grafts, which are believed to phase separate to a higher extent, have higher contact angles due to the presence of the higher concentration of the polysiloxane at the surface (Table III). This agrees with several results from our laboratory and elsewhere (13). Again the 5000 mw graft system does not dominate the surface at low percent siloxane to the same extent, indicating a partial phase mixing. ESCA was used as a tool to quantify the surface composition. Through the use of angular dependent depth profiling the domination of the surface was observed to be greater for higher molecular weight grafts. It is noteworthy to point out that higher angles penetrate deeper into the surface and as an approximation, an angle of 10 degrees measures about the top 5-10Å of the surface, while the 90 degree measurement indicates the composition over the top 60 Å. Another factor is that the number of electrons escaping follows an inverse exponential relationship with depth, so that surface composition has a large effect on even the large angles. In Table V we can see the trends in composition with graft molecular weight and

Table V. Influence of Siloxane Graft Molecular Weight and Composition on Surface Composition of PMMA-g-PSX Copolymers by Variable Angle XPS (ESCA) (Copolymers ~5 wt.% PSX)

	% Poly(Dimethyl Siloxane) Detected		
Macromonomer Mn / Exit Angle	10°	30°	90°
1000	79	52	41
5000	86	86	69
10000	100	97	72
20000	100	100	90

depth. In our styrenic block copolymers, typical PSX block molecular weights were greater than 10K and high surface concentrations of PSX were again observed. Results from the styrenic block and graft copolymers are somewhat similar and are presented in Table VI.

Table VI. ESCA Study of Styrenic-Siloxane Block and Graft Copolymers (Copolymer Composition ~10 wt.% PSX)

	% PSX		
Copolymer / Angle	10°	30°	90°
P(Styrene)-b-PSX	91	80	52
P(Styrene)-g-PSX 1K	79	52	32
P(p-Methyl Styrene)-g-PSX 1K	48	20	12
P(t-Butyl Styrene)-b-PSX	82	56	40
P(t-Butyl Styrene)-g-PSX 1K	39	24	29

From Transmission Electron Microscopy (TEM), the morphology of the poly(methyl methacrylate) graft system changed significantly as a function of architecture and composition (Figure 4). The average domain sizes vary from 12 nm for the 5K grafts to 21 nm for the 20K grafts, while the 1K graft systems showed no apparent phase separation. Further details of this interesting morphology will be reported later. However, it is already clear that remarkably well defined solid state graft structures can be developed. As a comparison to the TEM measurements of domain size, Small Angle X-ray scattering (SAXS) (courtesy of Prof. G. L. Wilkes) was also run on the PMMA-g-PSX copolymers (Table VII). Two trends are

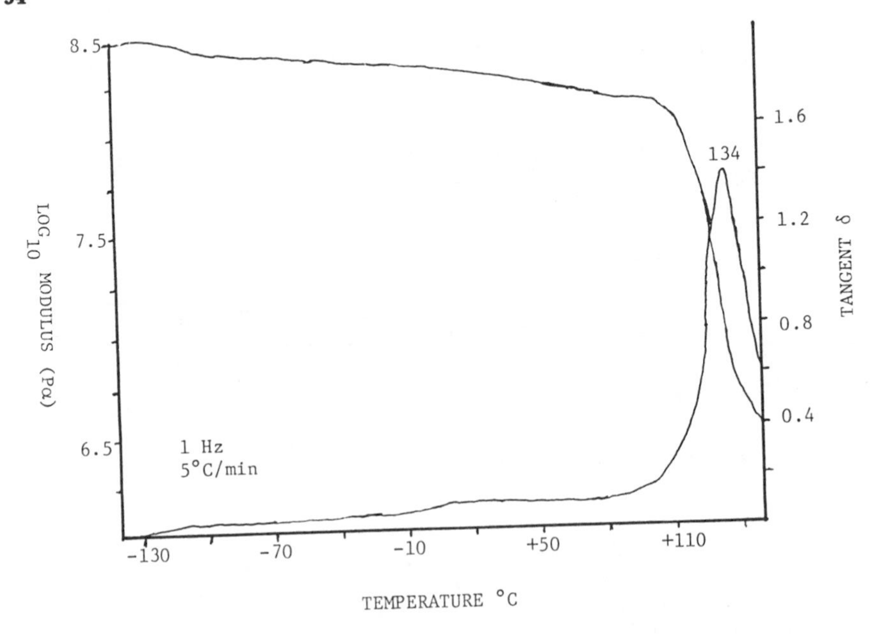

Figure 3. Dynamic mechanical behavior of PMMA-g-PSX copolymer: 20 wt.% siloxane of M_n 10,000.

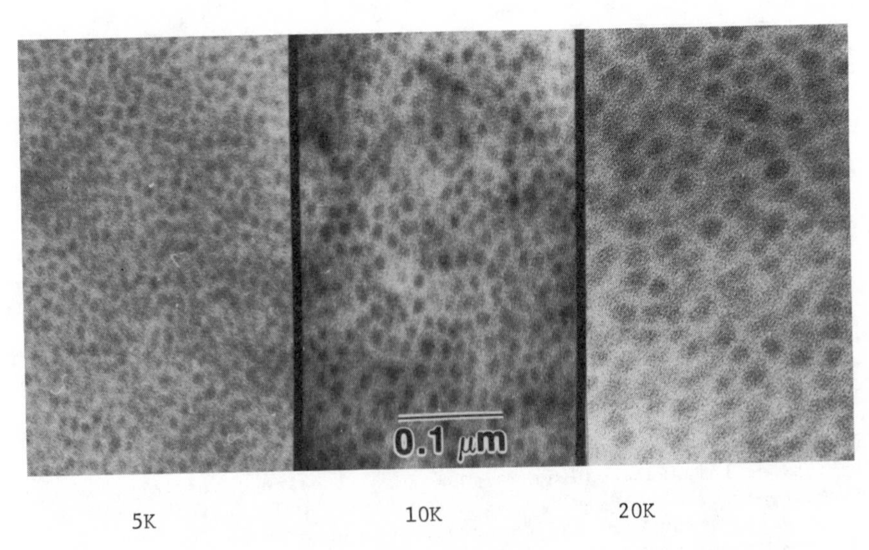

Figure 4. Comparison of the morphologies of PMMA-g-PSX (16 wt.% PSX) of 5000, 10000 and 20000 g/mole siloxane grafts by TEM.

Table VII. Comparison of Small Angle X-Ray Analysis of PMMA-g-PSX Copolymers with TEM Domain Size Data (Copolymers ~16 wt.% PSX)

Macromonomer Mn	Domain Size (TEM)	Interdomain Spacing	Max. Scattering Intensity
1000	___	12.4 nm.	30
5000	8 nm.	18.8 nm.	110
10000	15 nm.	28.6 nm.	260
20000	20 nm.	37.6 nm.	540

apparent, one being that there is a corresponding increase in interdomain spacing with graft molecular weight. The other trend is that the normalized scattering intensity also varies with graft molecular weight, indicating that better phase separation is occurring at least up until 10^4 g/mole. Also SAXS confirms that even at graft lengths of as low as 1000 there is a phase separation although domain size is small and phase definition is probably too weak to be defined in the TEM technique. Further analysis of these phenomena will be reported in forthcoming papers.

Similar studies were undertaken as a comparison in the styrenic systems, although in these systems we also were able to compare the graft systems to diblock systems of similar composition. DSC also shows a depression of the high temperature Tg for all of the styrenes for the 1K graft systems. The 5K systems showed a smaller depression but still a slight lowering of the Tg (Table VIII). Clearly, the t-butylstyrene system is most influenced, as anticipated.

Table VIII. Dependence of Architecture and Molecular Weight of Styrenic-Siloxane Block[a] and Graft Copolymers on Tg (Copolymers ~20 wt.% PSX)

Copolymer Type	Tg (DSC), °C
P(Styrene)-b-PSX	106
P(Styrene)-g-PSX 1K	74
P(Styrene)-g-PSX 5K	100
P(Styrene)-g-PSX 10K	103
P(p-Methyl Styrene)-b-PSX	115
P(p-Methyl Styrene)-g-PSX 1K	80
P(p-Methyl Styrene)-g-PSX 5K	107
P(p-Methyl Styrene)-g-PSX 10K	110
P(t-Butyl Styrene)-b-PSX	149
P(t-Butyl Styrene)-g-PSX 1K	100
P(t-Butyl Styrene)-g-PSX 5K	139
P(t-Butyl Styrene)-g-PSX 10K	143

[a] Block Copolymers are of ~100K-20K block molecular weights and are diblocks.

ESCA analysis showed a similar trend of incomplete surface coverage for the 1K systems. Also, no domains were visible in any of the 1K styrenic graft systems by TEM. There is an expected trend with respect to solubility parameter, p(t-butyl styrene) ($\delta \cong 8.1$) has a solubility parameter much closer to that of polydimethylsiloxane ($\delta \cong 7.3$) than does p(p-methyl styrene) which is closer than p(styrene) ($\delta \cong 9.1$). Preliminary data does indicate that for a copolymer of similar composition the p(t-butylstyrene) polymer has smaller domains and, in general, a partially phase mixed surface and solid state structure.

Conclusions

The macromonomer technique provides the possibility of preparing graft copolymers of copolymers not synthetically possible previously. The usefulness of the macromonomer technique was demonstrated and some interesting properties of the acrylic-siloxane and styrenic siloxane graft polymers prepared by this technique were reported. Complimentary very well defined block copolymers of styrene, p-methyl styrene and p-tert-butyl styrene with polydimethyl siloxane were also prepared. Preliminary characterization shows that block lengths and compositions were close to the desired values and that narrow molecular weight distributions were achieved. We have illustrated the utility of the technique and shown the need for further investigations which are continuing in our laboratories.

Literature Cited

1. Battaerd, H.; Tregear, G. W. Graft Copolymers; Wiley: New York, 1967.
2. Noshay, A.; McGrath, J. E. Block Copolymers: Overview and Critical Survey; Academic Press: New York, 1977.
3. Milkovich, R. In Anionic Polymerization: Kinetics, Mechanisms and Synthesis; ACS Symposium Series No. 166; American Chemical Society: Washington, DC, 1981.
4. Rempp, P.; Lutz, P.; Masson, P.; Franta, E. Makromol. Chem. 1984, Suppl. 8, 3.
5. Greber, G.; Reese, E. Makromol. Chem. 1962, 55, 96.
6. Milkovich, R.; Chiang, M. U.S. Patent 3 786 116, 1974.
7. Reichmanis, E.; Smolinsky, G. J. Electrochem. Soc.: Solid State Science and Tech., 1985, 132, 1178.
8. Bowden, M.; et al. PMSE Preprints, ACS Anaheim, September 1986.
9. Kawakami, Y.; Aoki, T.; Hisada, H.; Yamamura, Y.; Yamashita, Y. Polymer Comm., 1985, 26(5), 133.
10. Rempp, P.; Franta, E. Advances in Polymer Science, 1984 58.
11. Kawakami, Y.; Yamashita, Y. In Ring Opening Polymerization, Kinetics, Mechanisms and Synthesis; McGrath, J. E., Ed.; ACS 286, 1985, Chapter 19.
12. Cameron, G. G.; Chisholm, M. S. Polymer, 1985, 26, 437.
13. Gaines Jr., G. L. Macromolecules, 1981, 14, 208.

RECEIVED August 27, 1987

Chapter 8

Effect of the Polymer Backbone on the Thermotropic Behavior of Side-Chain Liquid Crystalline Polymers

Coleen Pugh and Virgil Percec

Department of Macromolecular Science, Case Western Reserve University, Cleveland, OH 44106

Poly(epichlorohydrin) (PECH) and poly(2,6-dimethyl-1,4-phenylene oxide) (PPO) containing pendant mesogenic units separated form the main chain through spacers of zero to ten methylene units were synthesized and characterized in order to test the "spacer concept." Both polymers were modified by phase transfer catalyzed esterifications of the chloromethyl groups (PECH) or the bromobenzyl groups (brominated PPO) with potassium ω-(4-oxybiphenyl) alkanoates and potassium ω-(4-methoxy-4'-oxybiphenyl)-alkanoates. While PPO required ten methylene units as a spacer and 4,4'-methoxybiphenyl as mesogen to present thermotropic liquid crystalline mesomorphism, PECH required no spacer.

Since Finkelmann and Ringsdorf introduced the spacer concept, it has been well accepted that a requirement for obtaining thermotropic side-chain liquid crystalline polymers is that a flexible spacer must be introduced to partially decouple the mobility of the main chain from that of the mesogenic groups (1-4). Without a spacer, mesophase formation would require that the polymer backbone be significantly distorted from its normal random coil conformation. At the same time, the backbone imposes a steric hindrance on the packing of the mesogens. For this reason, most polymers with the mesogenic groups directly attached to the backbone are amorphous (5). There are however several exceptions to this rule, most of which are listed in Table I, together with their phase transitions. Why is this? It must certainly be true that the motions of the mesogen must be decoupled from those of a rigid polymer backbone. But what about very flexible polymer backbones? We propose that such a backbone may itself act as a flexible spacer.

Table I demonstrates that most liquid crystalline polymers lacking a spacer are formed from a flexible polyacrylate backbone. In contrast, the methyl substituent in polymethacrylate backbones both reduce main chain mobility and imposes additional steric barriers to mesophase formation. Therefore, successful liquid crystalline formation of polymethacrylates has been achieved only

0097-6156/88/0364-0097$06.50/0

Table I. Polymers with Direct Attachment of Mesogens

Polymer	X	R	Phase Transitions	Reference
—CH_2-CH— O=C-O— (biphenyl)			S 285 I	7-11
—CH_2-CH— O=C-O— (phenyl)(H)			S 205 I	7,8
—CH_2-CH— O=C-O— (phenyl)—N=N—(phenyl)			K 81 S 222 I	13
—CH_2-CX— O=C-O— (phenyl)(phenyl)—R	X = H	R = CN	S 270 I	12
	X = H	R = C_5H_{11}	S 303 I	12
	X = CH_3	R = CN	S 240 I	12
	X = CH_3	R = C_5H_{11}	S 232 I	12
—CH_2-CH— (phenyl) N=CH—(phenyl)—R		R = OC_4H_9	K 88.3 N 120.6 I	13
		R = OC_6H_{13}	K 94.5 S 97.7 N 116 I	13
		R = CN	K 113.8 N 140.5 I	14
—CH-CX— O=C-O—(phenyl)—O-C(=O)—(phenyl)—OR	X = H	R = $C_{16}H_{33}$	S 200 I	5
	X = CH_3	R = C_4H_9	S 270 I	5
	X = CH_3	R = C_9H_{19}	S 220 I	5
	X = CH_3	R = $C_{12}H_{25}$	S 215 I	5
	X = CH_3	R = $C_{16}H_{33}$	S 220 I	5
—CH_2-CX— O=C-O—(phenyl)—CH=N—(phenyl)—R	X = H	R = OC_2H_5	K 78 N 136.5 I	15
	X = H	R = COOH	K 201 S 226 I	14
	X = CH_3	R = COOH	K_1 182 K_2 201 S 205 N	16
—CH_2-CX— O=C-O—(phenyl)—N=N—(phenyl)—R	X = CH_3	R = CH_3	S 270 I	5
	X = H	R = OC_5H_{11}	S >270 I	5
	X = CH_3	R = OC_5H_{11}	S >270 I	5
—CH_2-CX— O=C-O-[Chol]	X = H		K 125.8 I (124.8 C 91 K)	17
	X = CH_3		K 114.8 (111.8 C)	17

when the anisotropy of the mesogen is increased by increasing the number of stiff and highly polarizable units that it contains. This in turn enlarges the mesogen, and is most frequently achieved by attaching a polar substituent para to the polymerizable group of the monomer.

A review of the literature demonstrates some trends concerning the effect of the polymer backbone on the thermotropic behavior of side-chain liquid crystalline polymers. In comparison to low molar mass liquid crystals, the thermal stability of the mesophase increases upon polymerization (3,5,18). However, due to increasing viscosity as the degree of polymerization increases, structural rearrangements are slowed down. Perhaps this is why the isotropization temperature increases up to a critical value as the degree of polymerization increases (18).

There are also some trends when looking at the main chain flexibility. Table II demonstrates that when the main chain flexibility decreases from cyanobiphenyl containing polymethacrylates to polysiloxanes, not only does the Tg drop, but the isotropization temperature increases. However, the trend is the opposite when the mesogen is methoxyphenyl benzoate (18). Therefore, this effect of the main chain flexibility is still ambiguous.

In order to determine the necessity and/or the length of the spacer that is required to achieve liquid crystalline behavior from flexible vs. rigid polymers, we have introduced mesogenic units to the backbones of a rigid [poly(2,6-dimethyl-1,4-phenylene oxide) (PPO)] and a flexible [poly(epichlorohydrin) (PECH)] polymer through spacers of from 0 to 10 methylene groups via polymer analogous reactions.

The synthetic procedure used for the chemical modification of PPO involved in the first step the radical bromination of PPO methyl groups to provide a polymer containing bromobenzyl groups. The bromobenzyl groups were then esterified under phase-transfer-catalyzed (PTC) reaction conditions with potassium 4-(4-oxybiphenyl)butyrate (Ph3COOK, Ph3COO-PPO), potassium 4-(4-methoxy-4'-oxybiphenyl)butyrate (Me3COOK, Me3COO-PPO), potassium 5-(4-oxybiphenyl)valerate (Ph4COOK, Ph4COO-PPO), potassium 5-(4-methoxy-4'-oxybiphenyl)valerate (Me4COOK, Me4COO-PPO), potassium 11-(4-oxybiphenyl)undecanoate (Ph10COOK, Ph10COO-PPO) and potassium 11-(4-methoxy-4'-oxybiphenyl) undecanoate (Me10COOK, Me10COO-PPO). The notations between parentheses correspond to the starting potassium salt and the resulting functionalized PPO.

PECH was modified under similar reaction conditions, except that dimethylformamide (DMF) was used as the reaction solvent. In addition, the phase-transfer-catalyzed etherification of the chloromethyl groups of PECH with sodium 4-methoxy -4'-biphenoxide was used to synthesize PECH with direct attachment of the mesogen to the polymer backbone. Similar notations to those used to describe the functionalized PPO are used for functionalized PECH. In this last case, PPO was replaced with PECH. Esterification routes of both PPO and PECH are presented in Scheme I.

The attachment of mesogenic units to a polymer backbone via polymer analogous reactions is not a new concept, although they are much less frequently used than the polymerization of mesogen containing monomers. Liquid crystalline polyacrylates,

Table II. Influence of the Main Chain Flexibility on Liquid Crystalline Phase Transitions for Polymers with Cyano-biphenyl as Mesogen

Main chain	n[a]	Tg (C)	Ti (C)	Phase	Reference
Polysiloxane	6	14	166	S	18
	9	-1	157	N	19
Polyacrylate	2	50	112	N	20
	5	40	120	N	20
	6	35	125	S	18
	11	25	145	S	20
Polymethacrylate	2	95	—	—	20
	5	60	121	S	20
	6	55	100	S	18
	11	40	121	S	20

[a] length of methylenic units between mesogen and main chain

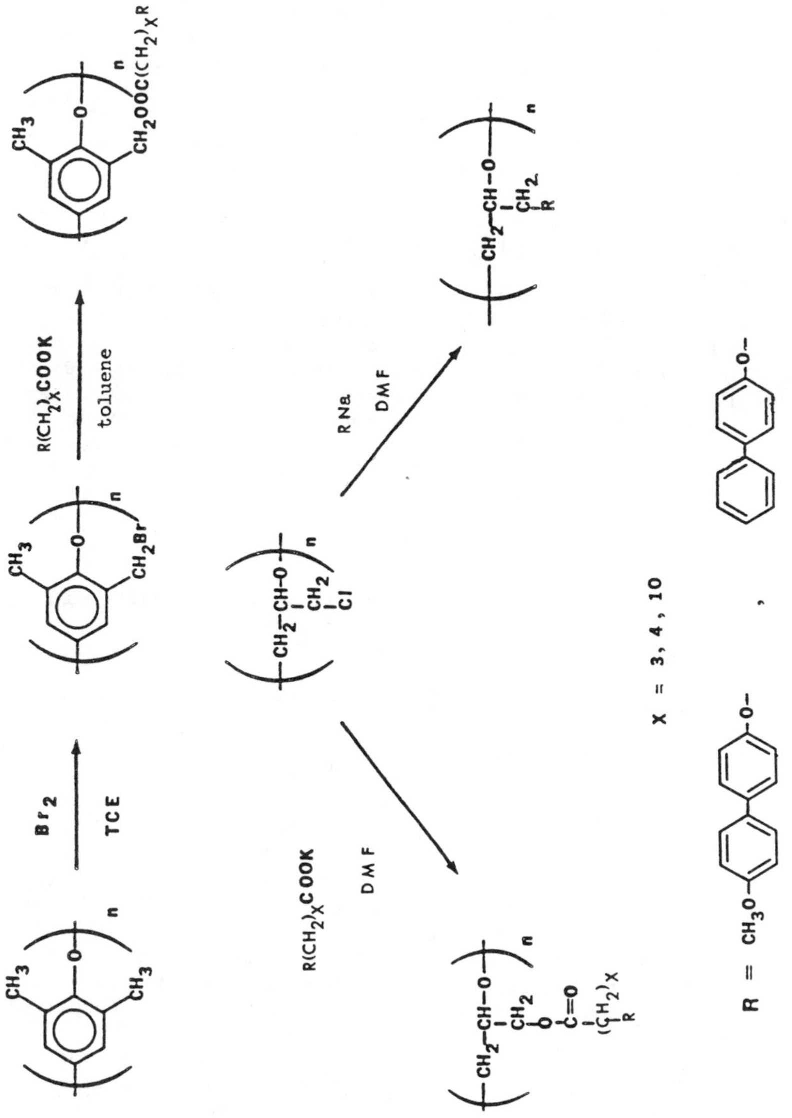

Br_2 / TCE

$R(CH_2)_XCOOK$ / toluene

RNa / DMF

$R(CH_2)_XCOOK$ / DMF

X = 3, 4, 10

R = CH_3O–(biphenyl)–O– , (biphenyl)–O–

Scheme 1. Synthetic routes used for the modification of PPO and PECH

polymethacrylates, and polyacrylamides have been prepared by conventional esterification or amidation of poly(acryloyl chloride) and poly(methacryloyl chloride) with a mesogenic alcohol or amine in the presence of triethylamine (8,21,22). Similarly, liquid crystalline polyacrylates, polymethacrylates, and polyitaconates have been prepared by phase-transfer-catalyzed reactions on the sodium salts of the corresponding polycarboxylates (23-25). In addition, alternating poly(methylvinylether-co-maleate) copolymers were prepared by the PTC reactions of poly(methylvinylether-co-disodium maleate) with mesogen containing bromoalkylesters (26). The most important use of this class of reactions, however, is in the preparation of liquid crystalline polysiloxanes, which cannot be obtained by any other method. We have recently summarized the work on the preparation of liquid crystalline polysiloxanes (19), which involves the platinum catalyzed hydrosilation reaction of vinyl substituted mesogenic molecules with poly(hydrogen methylsiloxane) or its copolymers.

Lastly, it was demonstrated with PPO substituted with a series of alkyl side-chains as we have here, that the glass transition temperature decreases with an increase in the side-chain length (28). At the same time, the Tg's of the more flexible side-chain liquid crystalline polymers investigated to date are always much higher than those of the corresponding polymers without the mesogenic side-chains (3). Therefore, it is quite likely that we may obtain side-chain liquid crystalline polymers of approximately the same Tg from PPO and PECH.

Experimental

The starting polymers were commercial products purified by precipitation with methanol from chloroform solutions: PECH (B.F. Goodrich, $\bar{M}n$ = 873,000, $\bar{M}w$ = 10,250,000); PPO (Aldrich, $\bar{M}n$ = 19,000m $\bar{M}w$ = 49,000).

The mesogenic units with methylenic spacers were prepared by reacting the sodium salt of either 4-methoxy-4'-hydroxybiphenyl or 4-phenylphenol with a bromoester in DMF at 82° C for at least 4 hours in the presence of tetrabutylammonium hydrogen sulfate (TBAH) as phase transfer catalyst. In this way, ethyl 4-(4-oxybiphenyl)butyrate, ethyl 4-(4-methoxy-4'-oxybiphenyl)butyrate, ethyl 4-(4-oxybiphenyl)valerate, ethyl 4-(4-methoxy-4'-oxybiphenyl)-valerate, n-propyl 4-(4-oxybiphenyl)undecanoate and n-propyl 4-(4-methoxy-4'-oxybiphenyl)undecanoate were obtained. These esters were hydrolyzed with base and acidified to obtain the carboxylic acids. The corresponding potassium carboxylates were obtained by reaction with approximately stoichiometric amounts of potassium hydroxide. Experimental details of these syntheses were described elsewhere (27).

Bromobenzyl groups were introduced into PPO by radical bromination of the methyl groups. The PPO bromobenzyl groups and PECH chloromethyl groups were then esterified under phase-transfer-catalyzed reaction conditions with the potassium carboxylates just described. This procedure has been described previously (29). The sodium salt of 4-methoxy-4'-hydroxybiphenyl was also reacted with PECH (no spacer).

Thermal analysis was performed with a Perkin-Elmer DSC-4

differential scanning calorimeter equipped with a Perkin-Elmer TADS thermal analysis data station. Heating and cooling rates were 20° C/min, and Indium was used as the calibration standard. All samples were heated to just above Tg and quenched before the first heating scan was recorded. A Carl-Zeiss optical polarizing microscope equipped with a Mettler FP82 hot stage and FP80 central processor was used to analyze the anisotropic textures.

Table III summarizes the reaction conditions and the results of the substitution of PPO for all reactions performed, while Table IV presents the results for the modification of PECH.

Results and Discussion

The synthetic route used for the chemical modification of PPO and PECH is outlined in Scheme 1. The results of the esterification reactions of PPO and of PECH are summarized in Tables III and IV respectively. It was previously shown that the only available procedure for the nucleophilic substitution of bromomethylated PPO was by solid-liquid phase-transfer reactions in aprotic nonpolar solvents (29), since PPO is not soluble in aprotic dipolar solvents which are required for conventional nucleophilic substitutions. In both cases, it was necessary to use elevated temperature (60°C) to obtain good halide displacement. However, it is again demonstrated that PECH is less reactive toward nucleophilic displacement than PPO. Although Me4COOK seem to result in the most efficient substitution, and although the carboxylate nucleophilicities would be expected to change with different spacers, these results are not completely comparable due to different amounts of excess hydroxide present in each displacement reaction. In addition, we found that sodium 4-phenylphenoxide degraded PECH to a number average molecular weight of approximately 3000, rather than strictly displacing halide as was the case with the other nucleophiles.

Table V summarizes the thermal characterization of substituted PPO, and demonstrates that although the Tg is easily dropped with any of the substituents, liquid crystalline behavior is not observed until 4-methoxy-4'-hydroxybiphenyl is decoupled from backbone PPO by ten methylenic units. Therefore, liquid crystalline behavior can be obtained from very rigid polymers, provided a long enough spacer is employed. A smectic liquid crystalline mesophase was confirmed by polarized optical microscopy for Me10COO-PPO, but the exact smectic phase could not be determined because some thermal crosslinking takes place during extensive annealing. This could be the result of the presence of unreacted bromobenzyl groups. In contrast to Me10COO-PPO, Ph10COO-PPO is not liquid crystalline. although there is comparatively little substitution in this case, p-biphenyl itself would not be expected to act as a mesogen in such a rigid polymer due to its low degree of anisotropy and polarizability.

Comparison ot Tables V and VI demonstrates that the thermal behavior of Me10COO-PPO and Me10COO-PECH are very similar, with the glass transition temperatures converging with substitution. Therefore, it appears that when very long spacers are used with the same mesogen, the polymer backbone has little effect on the thermotropic phases formed. This conclusion is supported by additional unpublished experiments performed in our laboratory.

Table III. Reaction Conditions and Results of Synthesis of PPO Containing Biphenyl Groups*

#	Nucleophile	Mole Fraction Structural Units Containing $-CH_2Br$	Moles per Mole $-CH_2Br$		Reaction		$\%CH_2Br$ Substituted
			Nucleophile	TBAH	Temp. (^{o}C)	Time (hr)	
1	biPhO-$(CH_2)_3$COOK	0.52	2.2	0.22	25	45	25
2		0.74	1.8	0.24	25	45	26
3		0.74	1.8	0.12	60	46	100
4	MeO-biPhO-$(CH_2)_3$COOK	0.52	2.0	0.19	25	40	15
5		0.74	2.0	0.19	25	40	12
6		0.74	1.7	0.39	60	61.5	100
7	biPhO-$(CH_2)_4$COOK	0.52	2.0	0.21	25	62	87
8		0.74	1.9	0.07	25	62	93
9		0.74	1.9	0.14	60	46	100
10	MeO-biPhO-$(CH_2)_4$COOK	0.52	2.1	0.14	25	62	50
11		0.74	2.1	0.10	25	62	61
12		0.74	2.0	0.39	60	61.5	100
13	biPhO-$(CH_2)_{10}$COOK	0.52	2.0	0.12	25	96	25
14		0.74	1.9	0.21	25	96	24
15	MeO-biPhO-$(CH_2)_{10}$COOK	0.74	0.9	0.24	60	54.5	71
16		0.74	0.9	0.24	60	139.5	75

* Solvent = toluene

Table IV. Reaction Conditions and Results of Synthesis of PECH Containing Methoxybiphenyl Groups*

#	Nucleophile	Moles per Mole $-CH_2Cl$		Reaction Time (hr)	$\%CH_2Cl$ Substituted
		Nucleophile	TBAH		
1	MeO-biPhONa	0.76	0.10	92	29
2		0.76	0.10	140	36
3	MeO-biPhO-$(CH_2)_3$COOK	0.69	0.13	91	22
4		0.75	0.12	108.5	26
5		0.75	0.12	139.5	30
6		0.75	0.12	158.5	34
7	MeO-biPhO-$(CH_2)_4$COOK	0.75	0.12	19	23
8		0.75	0.12	49	42
9		0.75	0.12	83	52
10		0.73	0.11	91	51
11		0.75	0.12	115.5	62
12		0.75	0.12	140	65
13		0.75	0.12	159.5	65
14		0.75	0.12	181.5	65
15	MeO-biPHO-$(CH_2)_{10}$COOK	0.64	0.17	54	17

* Solvent = DMF; 60 ^{o}C

Table V. Thermal Characterization of PPO Containing Biphenyl Groups

#[a]	Polymer Sample	Temperature (°C)			
		Heating		Cooling	
		Tg	Endotherms	Tg	Exotherms
1	0.13 Ph3COO-PPO[b]	141.7	—		
2	0.19 Ph3COO-PPO	130.0	—		
3	0.74 Ph3COO-PPO	69.5	—		
4	0.08 Me3COO-PPO	172.5	—		
5	0.09 Me3COO-PPO	155.2	—		
6	0.74 Me3COO-PPO	69.4	—	58.4	—
7	0.43 Ph4COO-PPO	80.8	—		
8	0.69 Ph4COO-PPO	67.0	—		
9	0.74 Ph4COO-PPO	59.6	—		
10	0.26 Me4COO-PPO	106.2	—		
11	0.45 Me4COO-PPO	89.2	—	85.6	—
12	0.74 Me4COO-PPO	62.5	—	53.1	—
13	0.13 Ph10COO-PPO	105.5	—		
14	0.18 Ph10COO-PPO	83.7	—		
15	0.53 Me10COO-PPO	54.2	114.5	39.2	77.5
16	0.56 Me10COO-PPO	38.6	72.4, 116.8, 129.0	c	40.4, 101.4

a) from Table III; b) mole fraction of mesogen substituted structural units; c) buried in peak

Table VI. Thermal Characterization of PECH Containing Methoxybiphenyl Groups

#[a]			Temperature (°C) 1st Heat		1st Cool		2nd Heat	
			Tg	Endotherms	Tg	Exotherms	Tg	Endotherms
1	0.29	MeO-PECH	9.2	66.6, 82.9	-1.6	65.7	7.0	81.2
2	0.36	MeO-PECH	17.3	60.1, 80.5	-3.9	62.9	12.4	81.6
3	0.22	Me3COO-PECH	-4.1	39.8, 49.6	-9.8	----	-2.5	----
4	0.26	Me3COO-PECH	-0.9	7.9, 52.3	-11.7	13.9	-0.4	43.1
5	0.30	Me3COO-PECH	0.1	51.6, shoulder	-7.3	26.7	1.8	51.7
6	0.34	Me3COO-PECH	3.7	50.0, shoulder	-6.9	31.6	3.7	56.8
7	0.23	Me4COO-PECH	-9.1	shoulder, 44.7	-13.6	----	-7.5	34.1
8	0.42	Me4COO-PECH	7.7	55.0, 70.3	2.0	42.0, 55.2	6.2	70.7
9	0.52	Me4COO-PECH	18.6	58.8, 85.4, 107.2	4.7	70.6	17.3	88.1
10	0.51	Me4COO-PECH	17.0	57.3, 83.5, 103.2	-5.2	65.7	16.0	83.4
11	0.62	Me4COO-PECH	34.3	73.5,100.5, 114.4	----	82.9	37.6	101.7
12	0.65	Me4COO-PECH	35.2	72.0,105.0, 117.3	----	89.0	40.2	107.1
13	0.65	Me4COO-PECH	36.0	67.0,107.9, 119.2	----	91.4	47.7	110.4
14	0.65	Me4COO-PECH	34.1	62.4,107.8, 118.1	----	91.1	44.1	109.8
15	0.17	Me10COO-PECH			-20.3	34.7, 88.6	-12.9	45.7,74.1,105.4

[a] from Table IV; [b] mole fraction of mesogen substituted structural units

In contrast to the substituted PPO's, it is theoretically possible to obtain the same substituted PECH's by homopolymerization of the corresponding mesogenic oxirane, or by its copolymerization with epichlorohydrin. We have attempted these polymerizations in order to better interpret the thermal behavior of the more complicated copolymers that we have obtained by polymer analogous reactions. Homopolymerization would be instructive because the incorporation of nonmesogenic units into liquid crystalline homopolymers doesn't as a rule change the type of mesophase obtained (5).

However, functional oxiranes are generally difficult to polymerize. One possibility is cationic ring opening polymerization. This however results in high conversion to cyclic oligomers due to backbiting of the oxonium ion chain end to produce stable six-membered dioxane molecules (30). In order to obtain linear polymers, the activated monomer mechanism can be used. Because of the addition of an alcohol to the cationic system, propagation occurs by attack of a neutral growing hydroxyl terminated chain end on a protonated (i.e. activated) monomer, thereby eliminating the possibility of backbiting (30-32). However, it is usually not possible to obtain high molecular weight polymers because as the ratio of monomer to initiator increases, so does the possibility for competition by conventional cationic propagation increase. This results in both linear and cyclic oligomers. To prepare linear high polymers from oxiranes, ionic coordinative catalysts are usually employed. The Vandenberg catalyst, $R_3Al.0.6\ H_2O$, is the most frequently used coordinate catalyst ((33-35). The obtained polymers are, however, highly crystalline (isotactic) and therefore difficult to characterize by techniques requiring solubility.

A fourth alternative is $AlEt_3$/metal chelate polymerizations, such as the $AlEt_3$/bis(dimethylglyoxime)nickel system. The problem with these catalysts is that they result in either high yields of low molecular weight polymers, or low yields of high molecular weight polymers, but not both high yield and high molecular weight (36,37). A more promising method allowing the control of molecular is a living polymerization initiated by metalloporphyrins. The initiator is generated by the reaction of diethylaluminum chloride and 5,10,15,20-tetraphenylporphyrin, which is used either by itself (38) or in combination with a quaternized ammonium or phosphonium salt (39) to generate a single or two growing chains, respectively, by insertion of the monomer in the aluminum-halide bonds. So far, none of these polymerization methods could lead to a convenient method for the polymerization and copolymerization of mesogenic oxiranes. Work on this line is in progress in our laboratory.

In contrast to PPO, Table V demonstrates that no spacer is required to obtain liquid crystalline behavior from PECH. Figures 1-8 show some representative DSC traces for Me10COO-PECH, Me3COO-Pech and Me4COO-PECH. In addidtion to the first cooling and second heating scans, the first heating scan is also shown. This is because the multiple endotherms observed with all spacers in this scan can be reproduced simply by annealing above Tg for several hours. Therefore, these endotherms cannot be dismised as traces of solvent or the result of thermal history, and the phases are evidently kinetically controlled. It must be noted that this is the first time liquid crystalline polymers of such high molecular weight have

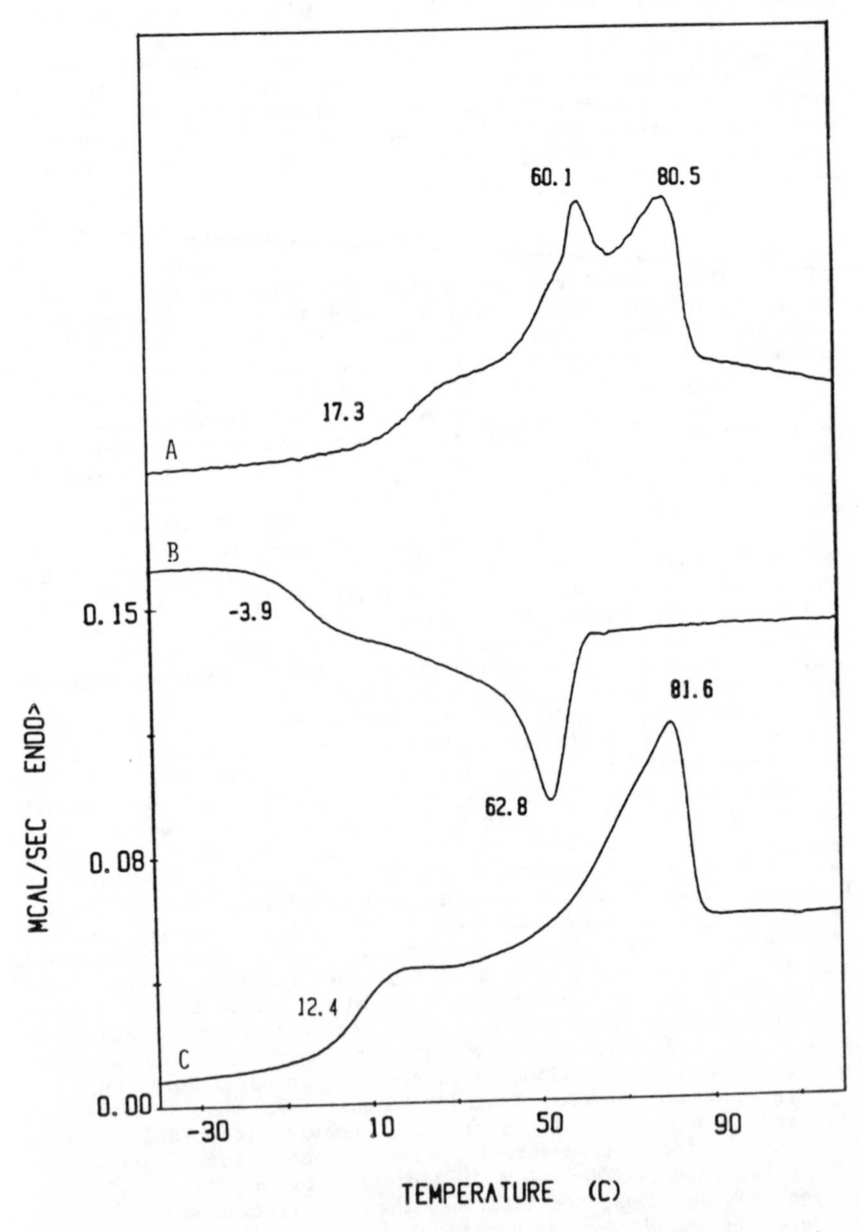

Figure 1. DSC traces of 0.36 MeO-PECH (#2/Table IV): (A) first heat; (B) first cool; (C) second heat.

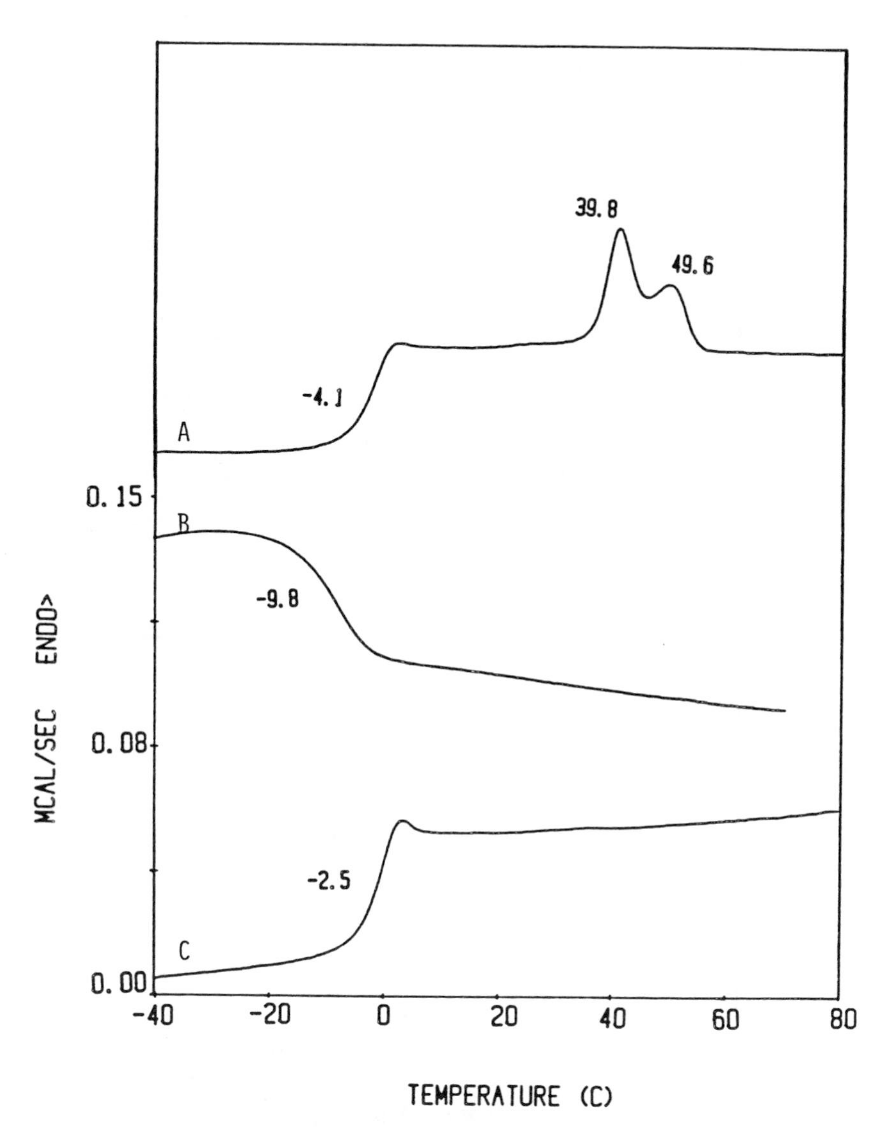

Figure 2. DSC traces of 0.22 Me3COO-PECH (#3/Table IV): (A) first heat: (B) first cool; (C) second heat.

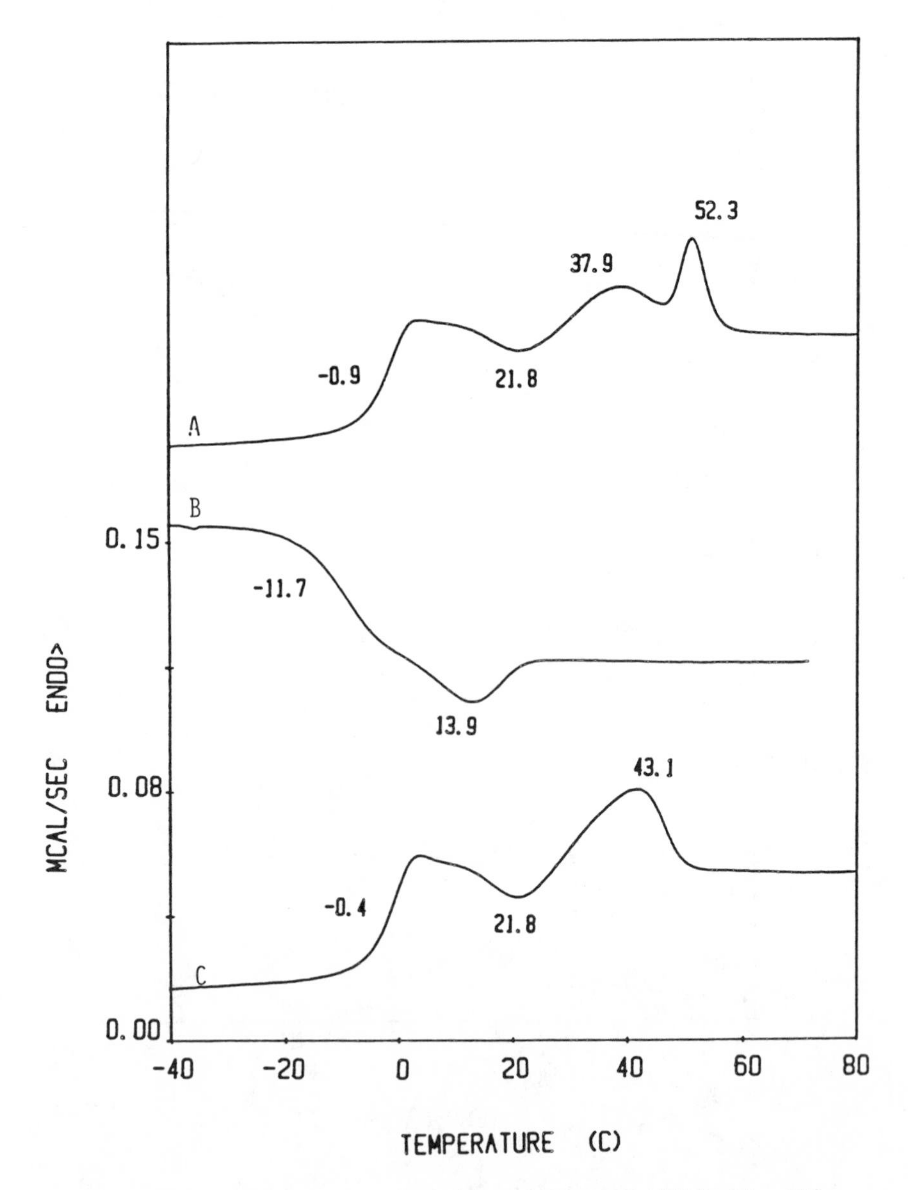

Figure 3. DSC traces of 0.26 Me3COO-PECH (#4/Table IV): (A) first heat; (B) first cool; (C) second heat.

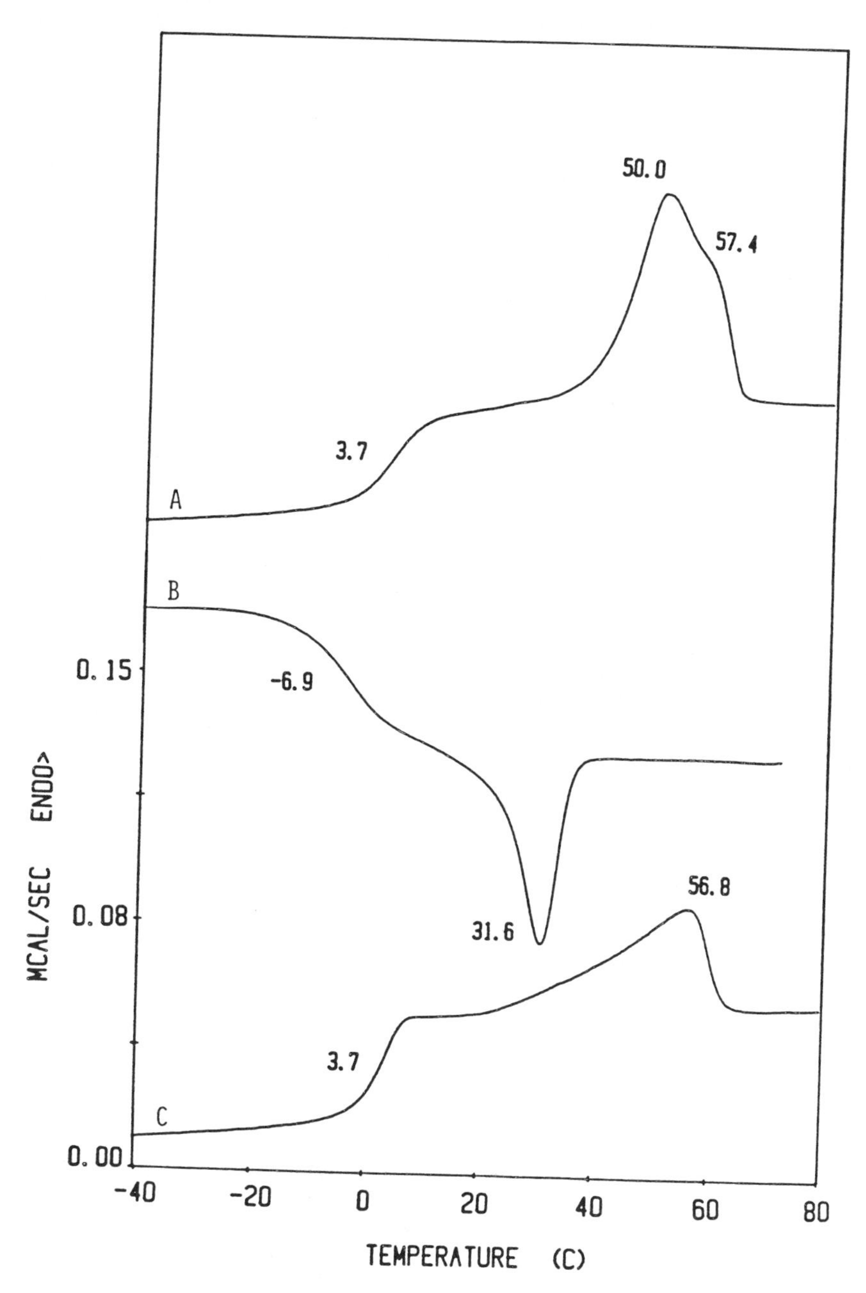

Figure 4. DSC traces of 0.34 Me3COO-PECH (#6/Table IV): (A) first heat; (B) first cool; (C) second heat.

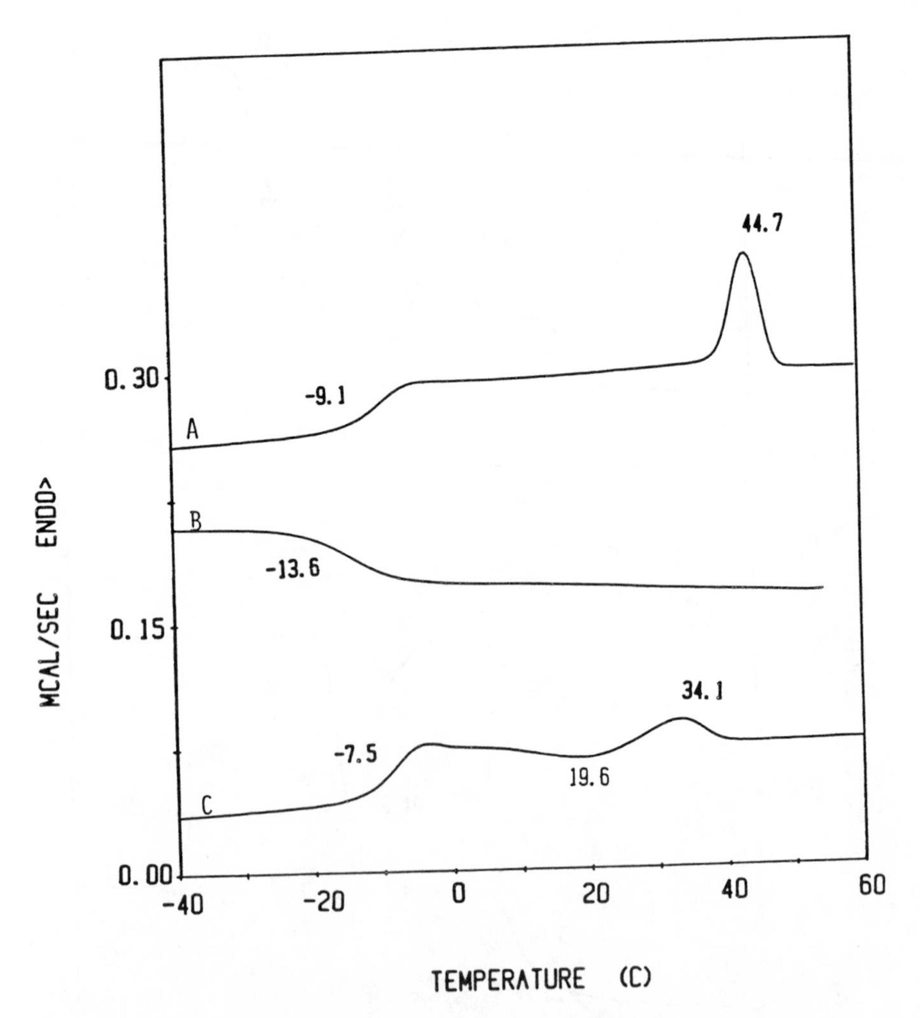

Figure 5. DSC traces of 0.23 Me4COO-PECH (#7/Table IV): (A) first heat; (B) first cool; (C) second heat.

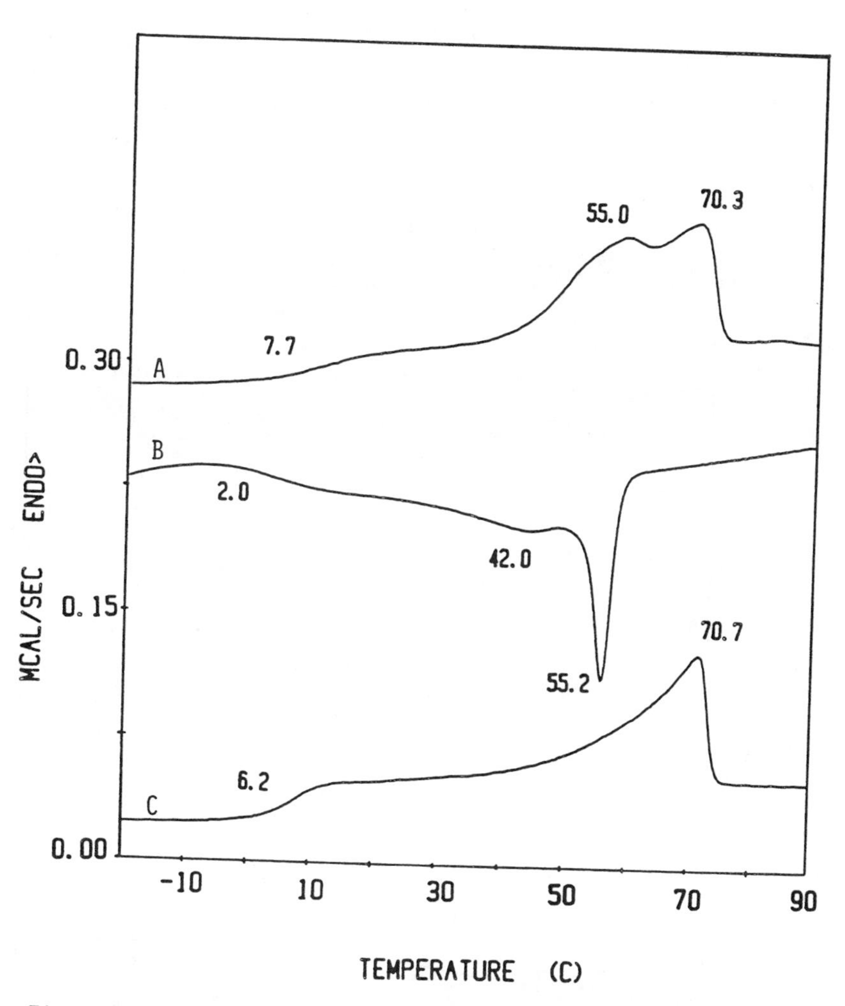

Figure 6. DSC traces of 0.42 Me4COO-PECH (#8/Table IV): (A) first heat; (B) first cool; (C) second heat.

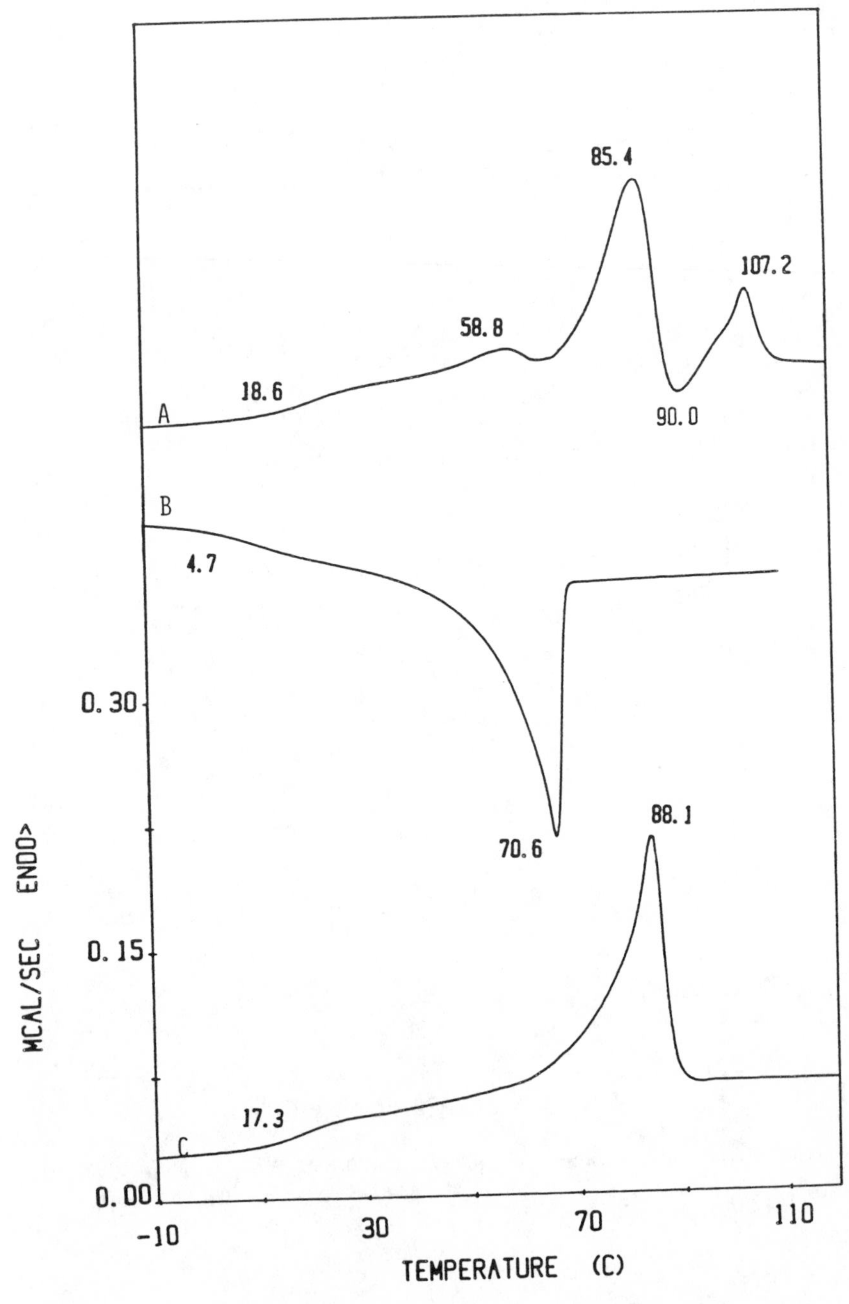

Figure 7. DSC traces of 0.51 Me4COO-PECH (#10/Table IV): (A) first heat; (B) first cool; (C) second heat.

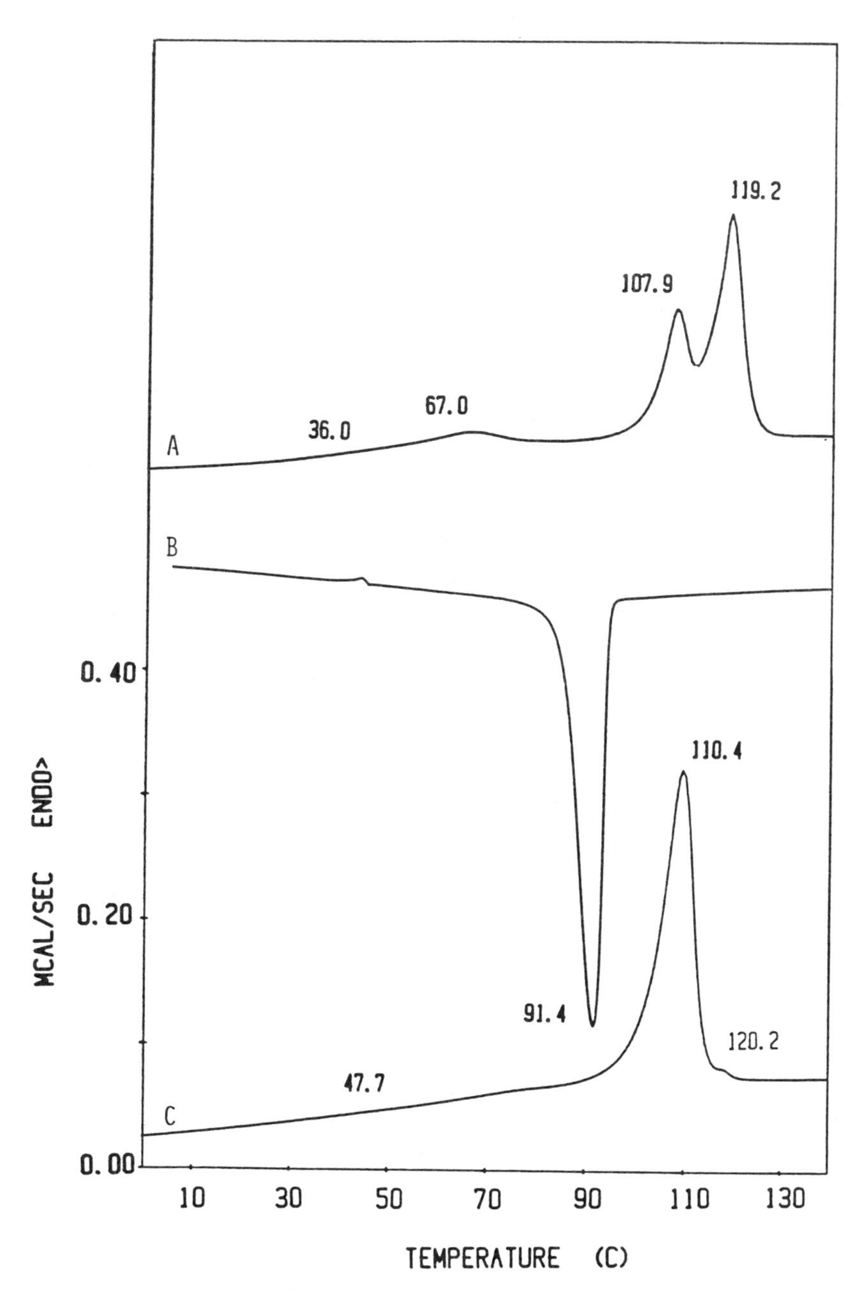

Figure 8. DSC traces of 0.65 Me4COO-PECH (#13/Table IV): (A) first heat; (B) first cool; (C) second heat.

been obtained. We have found with main-chain liquid crystalline polymers (40) that beyond a critical molecular weight, not only are the crystalline transitions kinetically controlled, but the liquid crystalline transitions also become kinetically rather than thermodynamically controlled, presumably as a result of chain entanglements. This results in a higher degree of supercooling than is normally observed for liquid crystalline phases, especially for smectic mesophases, as is seen here.

Figures 2 and 5 demonstrate that there is a minimum concentration limit necessary for mesophase formation in copolymers of mesogenic monomers with nonmesogenic monomers (3), since the liquid crystalline endotherm is not observed in the second heating scan. In addition, Figure 5 demonstrates that the first endotherm in each sample may be a crystalline melting. The first heating scan of 0.23 M34COO-PECH shows a strong liquid crystalline endotherm at 44.7°C, and a weak endotherm in the form of a tail at the temperature corresponding to the weak melting endotherm in the second scan. This weaker first endotherm is obviously crystalline melting since cold crystallization of the same enthalpic content is required just prior to the melting. This, and the lack of a crystallization exotherm on cooling demonstrates that crystallization is slow and occurs over a large range of temperatures with these samples. This is probably due to the proximity of the Tg and the temperature of melting (Tm), combined with a great deal of crystalline supercooling. Apparently, structural ordering in these PECH systems is much easier in solution (i.e., prior to precipitation in the nonsolvent methanol) than by thermal rearrangements.

Beyond the minimum mesogenic concentration limit of approximately 26% with these shorter spacers (or no spacer), isotropization is also seen in the second heating scans with melting seen only as a tail to isotropization. Once 51% substitution is reached with Me4COO-PECH, we see, in addition to melting and the "normal" liquid crystalline endotherm (Ti), crystallization followed by a second endotherm at higher temperature. However, it is not observed again on reheating scans unless the sample is annealed, and the enthalpy of the endotherm in the second heating scan is equal to the total enthalpy content of the first two peaks in the first scan. When substitution is increased further, this third endotherm is seen as a minor peak in additional heating scans and therefore represents the true isotropization temperature.

Preliminary room temperature x-ray data of 0.65 Me4COO-PECH indicates that the sample presents a highly ordered smectic mesophase which was not yet completely assigned. The textures seen by polarized optical microscopy are also typical of smectic phases. Due to the very high molecular weights involved, textures specific to mesophase in thermodynamic equilibrium could not be developed within a reasonable amount of time by annealing.

Conclusions

These results demonstrate that side-chain liquid crystalline polymers can be synthesized by polymer analogous reactions from theoretically any polymer backbone. When the polymer backbone is rigid, as in the case of PPO, a long spacer is required both to decrease the Tg of the parent polymer and to partially decouple the mobility of the mesogens from that of the main chain. Flexible backbones such as PECH do not require a long spacer between the mesogen and the polymer main chain. Liquid crystallinity can be induced by directly attaching the mesogen to the polymer backbone providing that the Tg of the resulting polymer is below the isotropization temperature. This can easily be accomplished with PECH copolymers where the flexible nonsubstituted structural units may behave as a spacer.

Acknowledgments

We thank the Center for Adhesives, Sealants, and Coatings (CASC) of Case Western Reserve University and the Office of Naval Research for generous financial support.

Literature Cited

1. H. Finkelmann, H. Ringsdorf and J. H. Wendorff, Makromol. Chem., 1978 179, 273 .
2. H. Finkelmann, M. Happ, M. Portugal and H. Ringsdorf, Makromol. Chem., 1978, 179, 2541 .
3. H. Finkelmann and G. Rehage, Adv. Polym. Sci., 1984, 60/61, 99.
4. M. Engel, B. Hisgen, R. Keller, W. Kreuder, B. Reck, H. Ringsdorf, H. W. Schmidt and P. Tschirner, Pure Appl. Chem., 1985,57, 1009.
5. V. P. Shibaev and N. A. Plate, Adv. Polym. Sci., 1984, 60/61, 173.
6. M. Bacceredda, P. L. Magagnini, G. Pizzirani and P. Guisti, J. Polym. Chem. Polym. Lett. Ed., 1971, 9, 303.
7. P. L. Magagnini, A. Marchetti, F. Matera, G. Pizzirani and G. Turchi, Eur. Polym. J., 1974,10, 585.
8. V. Frosini, P. L. Magagnini and B. A. Newman, J. Polym. Sci. Polym. Phys. Ed.,1974, 12, 23.
9. B. Bresci, V. Frosini, D. Luppinaci and P. L. Magagnini, Makromol. Chem. Rapid Comun., 1980, 1, 183.
10. V. Frosini, G. Levita, D. Lupinacci and P. L. Magagnini, Mol. Cryst. Liq. Cryst., 1981, 66, 21.
11. C. M. Paleos, G. Margomenou-Leonidopoulou, S. E. Filipakis, and A. Malliaris, J. Polym. Sci. Polym. Chem. Ed., 1982, 20, 2267.
12. A. K. Alimoglu, A. Ledwith, P. A. Gemmell, G. W. Gray, F. R. S. Lacy and D. Lacy, Polymer, 1984, 25, 1342.
13. A. Blumstein, R. B. Blumstein, S. B. Clough and E. C. Hsu, Macromolecules, 1975, 8, 73.
14. S. B. Clough, A. Blumstein and A. DeVries, A.C.S. Polym. Prepr., 1977, 18(2), 1.
15. E. Perplies, H. Ringsdorf and J. Wendorff, Makromol. Chem., 1974, 175, 553.
16. A. Blumstein, S. B. Clough, L. Patel, L. K. Kim, E. C. Hsu and R. B. Blumstein, A.C.S. Polym. Prepr., 1975, 16(2), 241.
17. E. C. Hsu, S. B. Clough and A. Blumstein, A.C.S. Polym. Prepr., 1977, 18(1), 709.

18. V. Shibaev and N. Plate, Pure Appl. Chem., 1985, 57(11) 1589; H. Stevens, G. Rehage and H. Finkelmann, Macromolecules, 1986, 17, 851.
19. C. S. Hsu, J. M. Rodriguez-Parada and V. Percec, J. Polym. Sci., Polym. Chem. Ed., in press.
20. R. V. Tal'Roze, V. P. Shibaev and N. A. Plate, Polym. Sci. U.S.S.R., 1983, 25, 2863.
21. C. M. Paleos, S. E. Filippakis and G. Margomenou-Leonidopoulou, J. Polym. Sci. Polym. Chem. Ed., 1981, 19, 1427.
22. H. Kamogawa, J. Polym. Sci. Polym. Lett. Ed., 1972, 10, 7.
23. P. Keller, Macromolecules, 1984, 17, 2937.
24. P. Keller, Macromolecules, 1985, 18, 2337.
25. P. Keller, Mol. Cryst. Liq. Cryst. Lett., 1985, 2(3-4), 101.
26. P. Keller, Makromol. Chem. Rapid Commun., 1985, 6, 707
27. C. Pugh and V. Percec, Polym. Bull., 1986, 16, 513; C. Pugh and V. Percec, Polym. Bull., 1986, 16, 521.
28. B. Cayrol, A. Eisenberg,J. F. Harrod and P. Rocaniere, Macromolecules, 1972, 5, 676.
29. C. Pugh and V. Percec, Macromolecules, 1986, 19, 65.
30. S. Penczek, P. Kubisa and K. Matyjaszewski, Adv. Polym. Sci., 1980, 37, and 1985, 68/69.
31. K. Brezezinska, R. Szymanski, P. Kubisa and S. Penczek, Makromol. Chem., Rapid Commun., 1986, 7, 1.
32. Y. Okamoto, in Ring-Opening Polymerization. Kinetics, Mechanisms, and Synthesis, J. E. McGrath Ed., A.C.S. Symposium Series, 1985, 286, p. 361.
33. E. J. Vandenberg, Macromol. Synthesis, 1972, 4, 49.
34. E. J. Vandenberg, Pure Appl. Chem., 1976, 48, 295.
35. E. J. Vandenberg, J. Polym. Sci. Polym. Chem. Ed., 1985, 23, 951.
36. S. Kambara, M. Hatano and K. Sakaguchi, J. Polym. Sci., 1961, 51, S7.
37. S. Kambara and A. Takahashi, Makromol. Chem., 1963, 63, 89.
38. T. Aida and S. Inoue, Macromolecules, 1982, 14, 1162, and references therein.
39. T. Yasuda, T. Aidas and S. Inoue, Macromolecules, 1984, 17, 2217, and references therein.
40. V. Percec and H. Nava, J. Polym. Sci. Part A: Polym. Chem., 1987, 25, 405.

RECEIVED March 5, 1987

Chapter 9

Polyimidazolinones via Thermal Cyclodehydration of Polyamides Containing α-Amino Acid Units

Jerald K. Rasmussen [1], Larry R. Krepski [1], Steven M. Heilmann [1], Kumars Sakizadeh [1], Dean M. Moren[1], Howell K. Smith, II[1], and Alan R. Katritzky [2]

[1]**Corporate Research Laboratories, 3M Company, 201–2N–20, 3M Center, St. Paul, MN 55144**
[2]**Department of Chemistry, University of Florida, Gainesville, FL 32611**

Polyamides containing α-aminoacid units are readily obtained by reaction of bisazlactones (2-oxazolin-5-ones) with diamines. When polyamines such as diethylenetriamine (DETA) or triethylenetetramine (TETA) are used as the diamine component, the resultant polyamides readily cyclodehydrate above 200°C to produce polymers containing 2-imidazolin-5-one units in the backbone. Polyamides derived from simple diamines (e.g. 1,6-hexanediamine) cyclodehydrate only in the presence of a suitable catalyst. Carboxylate salts and certain Lewis acids have been found to be efficient catalysts for this transformation.

Polyamides and polyesteramides having a regular arrangement of diacid and α-aminoacid units derived from bisazlactones **1** were initially reported in 1943 (1) (Equation 1). Although relatively little information on polymer properties was given, bis(secondary amines) (**2**, X = NR^5) were recommended as monomers because of the occurrance of secondary reactions with primary amines. Cleaver and Pratt subsequently described (2) solution polymerizations using several diamines, including diprimary amines, in greater detail. They characterized the polyamides **3** as being amorphous, high melting, difficultly soluble materials which decomposed at their melting points. Korshak and coworkers later reported (3) the preparation of several polyamides **3** displaying crystallinities of from 10-55%. Ueda and coworkers found (4) that solubilities of the resultant polyamides were improved when R^2 and R^3 were phenyl groups. A series of polyamides **3** were all found to decompose at around 250°C as measured by TGA. More recently, a variety of film- and fiber-forming polyamides **3** based upon naturally occurring α-aminoacids have been described (5) which are biodegradable.

Polyamides containing β-aminoamide linkages have been

0097–6156/88/0364–0119$06.00/0

reported to cyclodehydrate to the corresponding 2-imidazoline units upon heating at temperatures in excess of 275°C (6) (Equation 2).

The current work was initiated as part of a program designed to develop new polymers incorporating heterocyclic nuclei. Our continuing interest in the utilization of azlactones in polymer chemistry (7-9) prompted us to combine the concepts outlined in Equations 1 and 2 in an attempt to prepare polymers containing a regular arrangement of amide and imidazoline linkages.

Experimental

All solvents and reagents were reagent grade or were purified before use. IR spectra were recorded on a Perkin-Elmer 983; NMR spectra were recorded at 200 MHz on a Varian XL-200 using tetramethylsilane as an internal reference. Inherent viscosities were determined at a concentration of 0.5g/dL using a Canon-Fenske viscometer at 30°C. Monomer and polymer synthesis has recently been described (10).

Catalyst Study. Equivalent amounts of p-phenylenebis(4,4-dimethyl-2-oxazolin-5-one) (2) and Jeffamine D-2000 (polyoxypropylenediamine from Texaco Chemical Co., amine equiv. weight 1023) were mixed with 5 mole % of the desired catalyst. The stirred mixture was heated at 240°C under argon for 30 minutes, then an additional 1.5 hours under vacuum (<1 torr) and collected. The amount of cyclization was estimated by ^{1}H-NMR in $CDCl_3$ by comparison of the integrated intensities of the absorptions due to the gem-dimethyl substituents. These absorptions appeared at 1.39 ppm in the cyclic form and at 1.73 ppm in the open-chain form of the polymer (see Scheme 4). Results are listed in Table I.

Results and Discussion

Initial polymerization studies were conducted with 2,2'-tetramethylenebis(4,4-dimethyl-2-oxazolin-5-one) **4** and triethylenetetraamine **5** (Scheme 1) (10). Polyamide **6** was isolated by precipitation into acetone from DMF solution. Heating in the melt at 180-190°C under vacuum resulted in the elimination of water and production of a new hygroscopic, glassy polymer. Spectral differences between the two polymers were quite dramatic. In the IR, bands associated with the amide carbonyl at 1645 and 1520 cm^{-1} were replaced by two new absorptions at 1725 and 1630 cm^{-1}. In the ^{13}C-NMR, the new polymer displayed absorptions at 161.6 and 187.2 ppm as compared to the amide carbonyls at 171.6 and 173.9 ppm for polyamide **6**. These results seemed to be inconsistent with the anticipated cyclization to imidazoline units, and model reactions verified that, in fact, cyclization to imidazolinone units was occurring (Scheme 1). Subsequently, additional polyimidazolinones were prepared utilizing diethylenetriamine or N-methyliminobispropylamine (**9**) as the polyamine component (10).

Reported (2) difficulties encountered in the preparation and

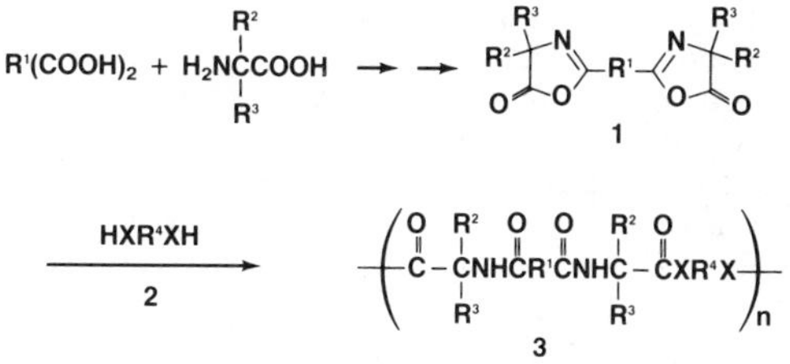

$$R^1(COOH)_2 + H_2NC(R^2)(R^3)COOH \rightarrow \rightarrow \mathbf{1} \quad (1)$$

$$\xrightarrow[\mathbf{2}]{HXR^4XH} \left(\overset{O}{\overset{\|}{C}} - C(R^2)(R^3)NH\overset{O}{\overset{\|}{C}}R^1\overset{O}{\overset{\|}{C}}NHC(R^2)(R^3) - \overset{O}{\overset{\|}{C}}XR^4X \right)_n \quad \mathbf{3}$$

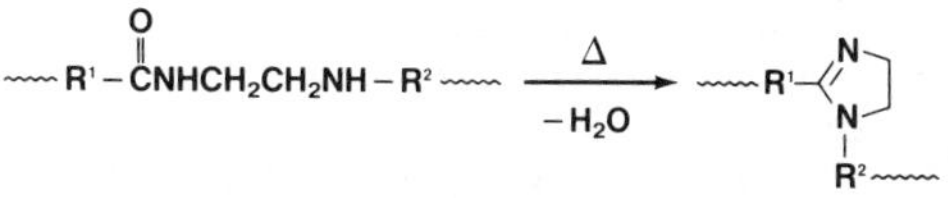

$$\sim R^1 - \overset{O}{\overset{\|}{C}}NHCH_2CH_2NH - R^2 \sim \xrightarrow[-H_2O]{\Delta} \sim R^1\text{-imidazoline-}N\text{-}R^2 \sim \quad (2)$$

$$\mathbf{4} + H_2N(CH_2)_2NH(CH_2)_2NH(CH_2)_2NH_2 \; (\mathbf{5})$$

$$\downarrow$$

$$\left[CH_2NH(CH_2)_2\,NH\overset{O}{\overset{\|}{C}} - C(CH_3)_2 - NH\overset{O}{\overset{\|}{C}} - (CH_2)_4\overset{O}{\overset{\|}{C}}NHC(CH_3)_2 - \overset{O}{\overset{\|}{C}}NH(CH_2)_2NHCH_2 \right]_n \quad \mathbf{6}$$

$$\downarrow$$

$$\left[CH_2NH(CH_2)_2\,N - C - (CH_2)_4 - C - N(CH_2)_2NHCH_2 \right]_n \quad \mathbf{7}$$

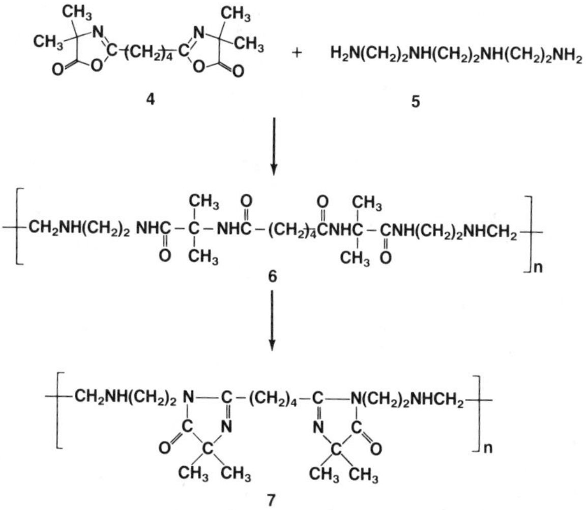

Scheme 1

reactions of certain bisazlactones, especially those having hydrogen atoms in the 4-position, led us to consider preparation of polyimidazolinones from the acyclic precursors of the bisazlactones (Scheme 2). Initially, N,N'-azeleoylbis- (α-aminoisobutyric acid) (**8a**) was condensed with amine **9** at elevated temperature and reduced pressure. Although spectral analysis indicated that the desired polyimidazolinone did form, the transformation was found to be substantially slower than when starting with bisazlactone **4**. Extension to other diacids **8**, however, proved to be difficult. For example, attempted condensation of adipoylbisglycine (**8b**) (2) with amine **9** led to a crosslinked, insoluble mass. The IR spectrum of this material indicated the presence of amide functionality but no imidazolinone absorptions.

Ueda and coworkers reported, in 1975, the preparation of polyamides **11** from unsaturated bisazlactones **10** and a variety of diamines (Scheme 3) (11). In a DTA study of one of these polyamides (**11**, R = CH_3, R^1 = $(CH_2)_6$), an endotherm was noted near 200^oC. It was suggested, without supporting evidence, that this endotherm might be due to a cyclodehydration reaction leading to polyimidazolinone **12**.

Encouraged by Ueda's report (11) and our own work, we next attempted the cyclodehydration of polyamides derived from simple diamines. Bisazlactone **4** was reacted with 1,6-hexanediamine in the melt to give polyamide **13** (Scheme 4). Continued heating (200-220^oC, 3 hrs) did not result in any detectable reaction. Similarly, other polyamides remained unchanged even after prolonged heating under vacuum at 250^oC. In contrast to these results, addition of sodium acetate to the polymer melt did catalyze the slow formation of polyimidazolinone **14**. Since it was previously known that hot aqueous sodium hydroxide would promote imidazolinone formation in model systems (12), we initiated a study to identify other possible catalysts for polyamide cyclodehydration. For this study we chose the polyamide derived from p-phenylenebis(4,4-dimethylazlactone) and Jeffamine D-2000 (Scheme 5). Steric hindrance due to the α-methyl substituent of the Jeffamine component in this polyamide leads to a reduced cyclization rate, thus providing a practical model for catalyst screening. Under a standard set of conditions (see Experimental), the catalyst effectiveness was determined by measuring the amount of cyclization by 200 MHz ^{1}H-NMR. Results listed in the Table show that carboxylate salts and certain Lewis acids are quite efficient catalysts.

In summary, the preparation of polyimidazolinones from polyamides containing α-aminoacid units (**3**, X = NH) can now be considered to be a general reaction provided that R^2 and/or R^3 are not hydrogen. When the polyamide has additional secondary or tertiary amine functionality in the backbone, cyclodehydration appears to be exceptionally facile. In the absence of amine functionality however, a catalyst is necessary to promote cyclization. Further studies of this new heterocyclic polymer system are ongoing in our laboratories.

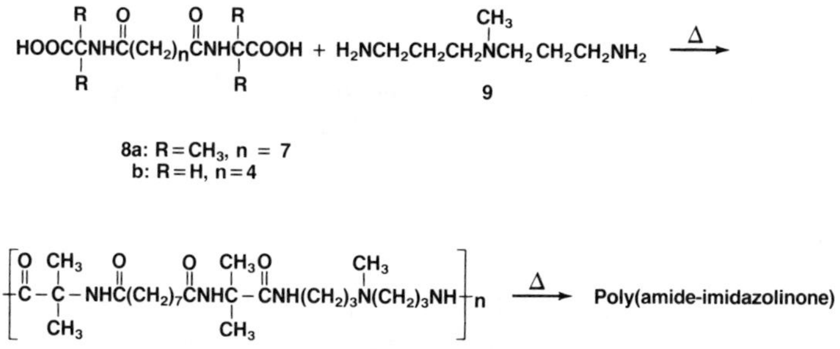

$$HOOCC(R)_2NHC(=O)(CH_2)_nC(=O)NHC(R)_2COOH + H_2NCH_2CH_2CH_2N(CH_3)CH_2CH_2CH_2NH_2 \xrightarrow{\Delta}$$

9

8a: $R = CH_3$, $n = 7$
b: $R = H$, $n = 4$

$$\left[C(=O)-C(CH_3)_2-NHC(=O)(CH_2)_7C(=O)NHC(CH_3)_2-C(=O)NH(CH_2)_3N(CH_3)(CH_2)_3NH \right]_n \xrightarrow{\Delta} \text{Poly(amide-imidazolinone)}$$

Scheme 2

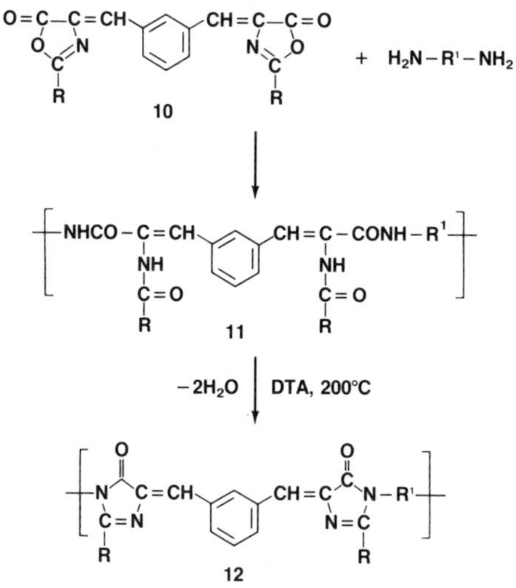

10 + $H_2N-R^1-NH_2$

↓

$$\left[NHCO-C(NHC(=O)R)=CH-C_6H_4-CH=C(NHC(=O)R)-CONH-R^1 \right]$$

11

$-2H_2O$ | DTA, 200°C

↓

12

Scheme 3

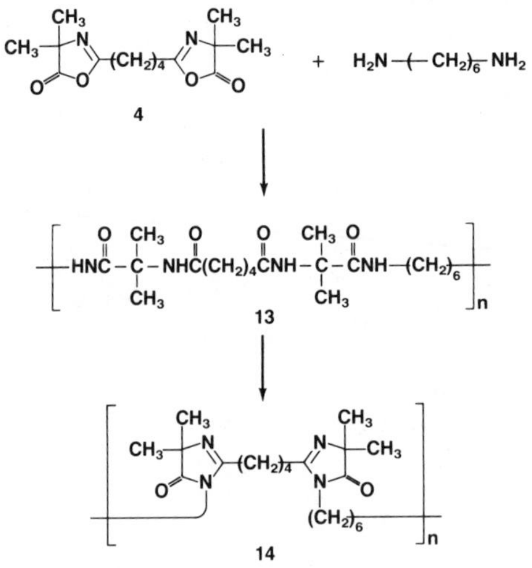

4 + $H_2N-(CH_2)_6-NH_2$

13

14

Scheme 4

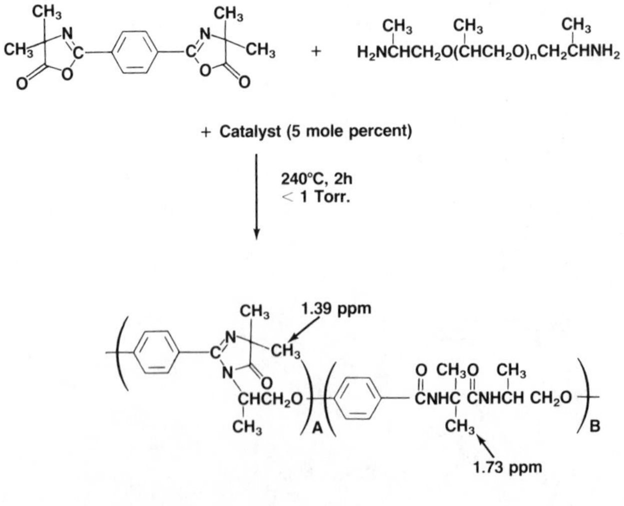

$H_2NCHCH_2O(CHCH_2O)_nCH_2CHNH_2$

+ Catalyst (5 mole percent)

240°C, 2h
< 1 Torr.

1.39 ppm

1.73 ppm

$$\% \text{ Cyclization} = \frac{A}{A + B}$$

Scheme 5

Table I. Catalyst Effectiveness

Catalyst (5 mole percent)	% Cyclization
None	0
H_2SO_4	0
Sb_2O_3	0
NaOPh	5
Na_2CO_3	11
NaOAc	12.5
$AlCl_3$	16
KOt-Bu	19
$Zn(CN)_2$	21
Na Benzoate	24
K Cyclohexanebutyrate	27.5
Sn $(BU)_2(laurate)_2$	28
$ZnCl_2$	32
$Zn(OAc)_2$	34
CsOAc	37
Na Pivalate	44
$FeCl_3$	56

Acknowledgments

The authors would like to acknowledge the assistance of the following members of the Corporate Research Analytical and Properties Research Laboratory, 3M, for help in obtaining and interpreting some of the NMR spectra: Mr. James R. Hill and Dr. Sadanand V. Pathre.

Literature Cited

1. L. Chassevent, P. Brot, Fr. Patent 887,530, Nov. 16, 1943.
2. C.S. Cleaver, B.C. Pratt, J. Am. Chem. Soc., 1955, 77, 1541-1543, 1544-1546.
3. T.M. Frunze, V.V. Korshak, L.V. Kozlov, Izv. Akad. Nauk SSSR,, 1959, 535-539.
4. M. Ueda, K. Kino, K. Yamaki, Y. Imai, J. Polym. Sci., Polym. Chem., 1978, 16, 155-162.
5. R.D. Katsarava, D.P. Kharadze, L.I. Kirmelashvili, M.M. Zaalishvili, Acta Polym., 1985, 36, 29-38.
6. D.E. Peerman, D.G. Swan, H.G. Kanten, U.S. Patent 4,049,598, Sept. 20, 1977.
7. S.M. Heilmann, J.K. Rasmussen, F.J. Palensky, J. Polym. Sci., Polym. Chem., 1984, 22, 1179-1186.
8. S.M. Heilmann, J.K. Rasmussen, L.R. Krepski, H.K. Smith II, J. Polym. Sci., Polym. Chem., 1984, 22, 3149-3160.
9. S.M. Heilmann, J.K. Rasmussen, L.R. Krepski, H.K. Smith II, J. Polym. Sci., Polym. Chem., 1986, 24, 1-14.
10. J.K. Rasmussen, S.M. Heilmann, L.R. Krepski, H.K. Smith II, A.R. Katritzky, K. Sakizadeh, J. Polym. Sci., Polym. Chem., 1986, 24, 2739-2747.

11. M. Ueda, K. Kino, Y. Imai, J. Polym. Sci., Polym. Chem., 1975, 13, 659-667.
12. A. Kjaer, Acta Chem. Scand., 1953, 7, 889-899.

RECEIVED September 23, 1987

Chapter 10

Synthesis and Characterization of *N*-Phenyl-3,4-dimethylenepyrrolidine Polymers

Raphael M. Ottenbrite and Herbert Chen

Department of Chemistry, Virginia Commonwealth University, Richmond, VA 23284

This study involved the preparation and characterization of poly(N-phenyl 3,4-dimethylenepyrrolidine) and the subsequent oxidation and reduction of this polymer. The parent polymer was not very soluble, so it was difficult to characterize. However, after oxidizing in the presence of palladium on carbon in nitrobenzene, the resultant poly(N-phenyl 3,4-dimethylenepyrrole) was soluble in several organic solvents. Attempts to reduce the original polymer to the pyrrolidone were unsuccessful.

Polyamines and their ammonium salts have been of interest because they are known to have potential applications as chelating agents (1-3), ion exchange resins (4-6), flocculants (7,8), and other industrial uses (9). Recent biomedical applications have constituted another important use of polymeric amines; they have been investigated for use as biocompatable materials, polymeric drugs, immobilization of enzymes, cell-culture substratum and cancer chemotherapeutic agents (10-12).

Polymeric amines may be broadly classified into three main categories according to the location of amino group(s) in the structure. They are namely polymers having;

I. pendent amine functions attached to the main polymer chain

II. amine nitrogen incorporated in the backbone of the polymer chain

III. cyclic amine structures incorporated into the polymer chain

The most important polyamine with cyclic amine ring on the main chain, is poly(dimethyldiallyammonium chloride) (DMDAC) (13). Due to its cationic charges, poly(dimethyldiallylammonium chloride) can

0097-6156/88/0364-0127$06.00/0

interact with ions and colloids of opposite charges. Therefore, DMDAC is being effectively used as a flocculating agent, retention agent, antistatic agent, and as an electroconductive coating (14). Because of its quaternary ammonium groups, poly(dimethyldiallylammonium chloride) is also a biocide against bacteria, algae and fungi.

To explain the formation of non-crosslinked polymers from the diallyl quaternary ammonium system, Butler and Angelo proposed a chain growth mechanism which involved a series of intra- and intermolecular propagation steps (15). This type of polymerization was subsequently shown to occur in a wide variety of symmetrical diene systems which cyclize to form five or six-membered ring structures. This mode of propagation of a non-conjugated diene with subsequent ring formation was later called cyclopolymerization.

We also studied the structure of poly(N,N-dimethyl-diallylammonium bromide) using poly(N,N-dimethyl-3,4-dimethylenepyrrolidinium bromide) as a model system (16). These studies unequivocally confirmed that polydiallyl quaternary ammonium system consisted predominantly, if not exclusively, of five-membered rings linked mainly in a 3,4-cis configuration. By investigating synthetic polymers with defined structures and composition, it is hoped that some relationship between the polymeric structure and properties could be clarified. We now wish to report the 1,4-polymerization of N-phenyl-3,4-dimethylene pyrrolidine and the effects of oxidation and reduction of this polymer.

Synthesis of N-Phenyl-1,3,4,6-Tetrahydrothieno(3,4-c)-Pyrrole-2,2-Dioxide

The synthesis of N-phenyl-1,3,4,6-tetrahydrothieno(3,4-c)-pyrrole-2,2-dioxide (II) was carried out by reacting aniline with 3,4-bis(bromomethyl)-2,5-dihydrothiophene-1,1-dioxide (I). The latter compound was synthesized by the bromination of the cycloaddition product, prepared from 2,3-dimethyl-1,3-butadiene and sulfur dioxide (17).

The yield of N-phenyl-1,3,4,6-tetrahydrothieno(3,4-c)-pyrrole-2,2-dioxide (II) was increased by gently flowing N_2 over the solution after the reaction was over. The N_2 sweep appeared to serve two purposes; first, it retarded the oxidation of the product (II) and, secondly, it cooled down the reaction mixture by evaporation of methanol thus causing more product to precipitate and improving the yield. The precipitated products including NaBr and unreacted carbonate salt were filtered immediately after the solution was cooled to room temperature, because on standing, the product would become strongly colored and difficult to purify. The reaction yield ranged from 50-75%.

When p-ethylaminobenzoate and N,N-dimethyl-p-phenylene diamine sulfate were similarly treated with the dibromosulfone (I) for 2 h, the yield was only 32% and 15% respectively (Figure 1). A possible explanation is that the nucleophilicity of these monosubstituted anilines is weaker than that of aniline while a Na_2CO_3 1,4-HBr elimination reaction could be competing with the substitution reaction, leading to the lower yield (18).

Synthesis of N-Phenyl-3,4-Dimethylenepyrrolidine

The decomposition of N-phenyl-1,3,4,6-tetrahydrothieno(3,4-c)-pyrrole-2,2-dioxide (II) was carried out in a sublimator at elevated temperatures (~160°C) to expel SO_2 and to give the exocyclic diene (III).

In addition to monomer III two major by products (IV) and (V) were detected (18). Compound (IV) was a dimer which resulted from Diels-Alder reaction between two exocyclic dienes (III) and was found at the bottom of the sublimator after thermal decomposition of the sulfone II. The GC/mass spectrum of this sample showed a molecular ion peak of 342 m/e that was in agreement with dimer molecular weight. The compound (V) had a pink tint and was found in a trace quantity in the crude N-phenyl-3,4-dimethylenepyrrolidine (III) by NMR (δ 1.9 - 1.8, $-\underline{CH}_3$, s). It could also be readily formed when N-phenyl-3,4-dimethylenepyrrolidine (III) solution was exposed to air for a period of time.

Compound (V) was readily soluble in acetone and methanol. The GC/mass spectrum of this sample showed a molecular ion peak of 171 which was in agreement with its proposed molecular weight. Other major relavent peaks were at m/e 170, 156, 141, and 77. The peak at m/e 170 was due to the loss of one allylic hydrogen, peaks at 156 and 141 were due to the loss of methyl groups (M-15) and (M-30). The strong peak at m/e 77 was characteristic of $C_6H_5^+$ ion. The amount of the undesirable oxidation product (V) formed during the recrystallization of the exocyclic diene (III) from ether could be significantly reduced if a trace quantity of an antioxidant such as N-phenyl-1-naphthylamine was added.

When the decomposition of N-phenyl-1,3,4,6-tetrahydrothieno-(3,4-c)-pyrrole-2,2-dioxide (II) was carried out in a sublimator a relatively high yield (80-95%) was obtained. However, under identical conditions, the decomposition of 5-(carboethoxyphenyl)-1,3,4,6-tetrahydrothieno(3,4-c)-pyrrole-2,2-dioxide yielded only 15% of diene product. This observation was found in agreement with the results reported by Alston (18). It was suggested that the yield from these sulfones depended on the relative volatility of the exocyclic diene formed since these dienes could undergo dimerization readily at the decomposition temperature of 160°C.

An alternative preparation of the exocyclic diene (III) was attempted by first thermally decomposing the bromosulfone (I) to yield 2,3-bis(bromomethyl)-1,3-butadiene VI, which, in turn, was allowed to react with aniline following the procedure described by Gaoni (19).

Diene (VI) was reacted with aniline in methanol in the presence of sodium carbonate to yield N-phenyl-3,4-dimethylenepyrrolidine. However, the purified yield was poor (~20%). The alternate route was a more direct method to prepare the monomer (III) without the side reactions on elimination that could take place in the first synthetic approach. Nevertheless, due to the poor yield, all the monomer (III) used for this work was prepared via the first route. The alternative method does have applicability in the preparation of substituted N-phenyl derivatives as these could be less volatile and more difficult to recover by sublimation.

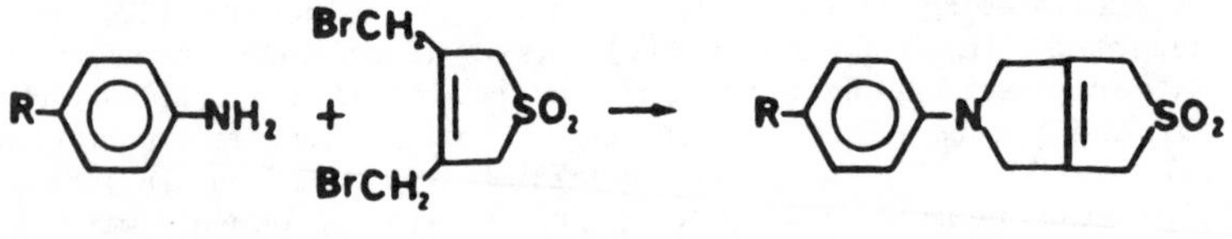

I II

where R = -H; $-COOC_2H_5$; $-N(CH_3)_2$

Figure 1. Reaction of 3,4-Bis(bromomethyl)-2,5-Dihydrothiophene-1, 1-Dioxide with Monosubstituted Aniline.

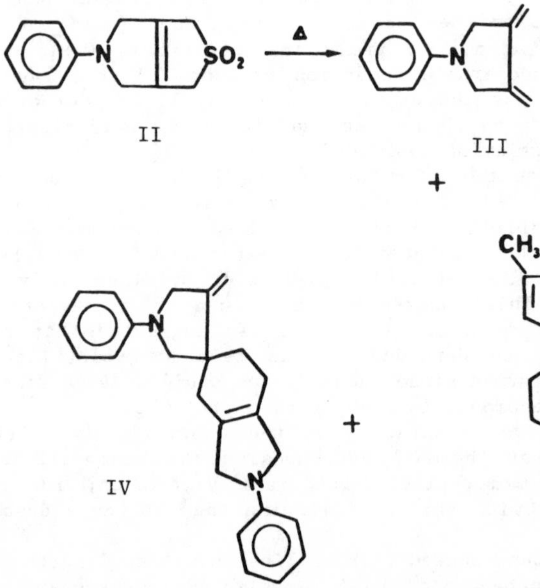

II III IV V

Figure 2. Thermal Decomposition of 1,3,4,6-Tetrahydrothieno-(3,4-C)-Pyrrole-2,2-Dioxide.

Polymer Synthesis and Characterization of Poly(N-Phenyl-3,4-Dimethylenepyrroline) (VI)

Due to the low solubility of the monomer (III) in benzene, the polymerization had to be carried out at less than 10% (w/v) monomer concentration. A yield of 92% was obtained by AIBN initiation at 60°C. Ammonium persulfate and benzoyl peroxide initiators were found to be ineffective. The solubility characteristics of poly(N-phenyl-3,4-dimethylenepyrroline) are listed in Table I. The polymer was insoluble in most common solvents except for formic acid and trifluoroacetic acid. The polymer was characterized by C,H elemental analysis, IR and ^{1}H NMR.

The viscosities were measured with an Ubbelohde Cannon 75-L, 655 viscometer. Formic acid was chosen as the solvent for the viscosity measurement because the polymer (VII) showed very low or no solubility in other common solvents. In a salt free solution, a plot of the reduced viscosity against the concentration of the polymer showed polyelectrolytic behavior, that is, the reduced viscosity η sp/c increased with dilution (Figure 4). This plot passed through a maximum at 0.25 g/dL indicating that the expansion of the polyions reached an upper limit, and the effects observed on further dilution merely reflected the decreasing interference between the expanded polyions.

The addition of potassium bromide (0.01M) suppressed the sharp rise in reduced viscosity at low concentration and a linear relationship was obtained. By extrapolation of the line, the intrinsic viscosity of the polymer was found to be 0.15 dL/g.

Poly(N-Phenyl-3,4-Dimethylenepyrrole) (VIII)

This polymer (VIII) was prepared by the oxidation of poly(N-phenyl-3,4-dimethylenepyrroline) (VII). Catalytic oxidation with Pd/C at 180°C in nitrobenzene gave the best yield with 100% conversion within 2 h. At temperatures higher than 180°C, the cleavage of the carbon-nitrogen bond was detected. At lower temperatures, the reaction time was much longer, and 100% conversion was more difficult to obtain. Other solvents such as DMF, monochlorobenzene and cumene were also tested, however, nitrobenzene appeared to be the best. Oxidation using DDQ in benzene did not appear to be as effective. The low yield (~20%) was probably caused by the poor solubility of the starting polymer (VII).

The pyrrole polymer VIII was characterized by C,H elemental analysis, IR, and ^{1}H NMR.

The viscosity of the oxidized polymer (VIII) was determined using DMF as a solvent. Chloroform was not a good solvent because it was too volatile and resulted in poor reproducibility. The reduced viscosities are plotted against polymer concentration (Figure 6). Polymer VIII behaved like a polyelectrolyte, the reduced viscosities increased sharply on dilution in a salt free solution. The addition of 0.01 M KBr did not completely suppress the loss of mobile ions; however, at 0.03 M KBr addition a linear relationship between the reduced viscosities and concentration was established.

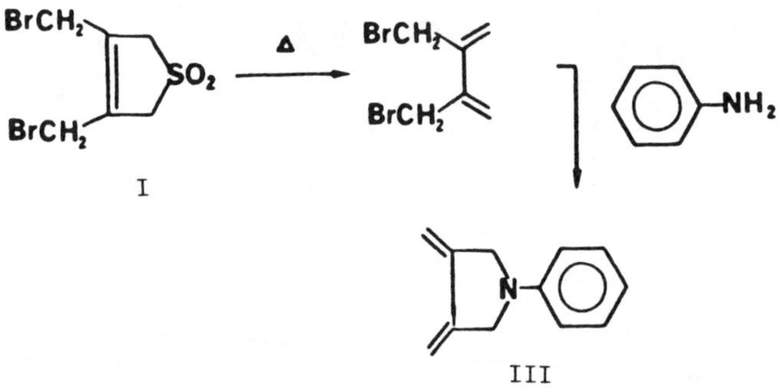

Figure 3. Alternative Preparation of Exocyclic Diene (III) Monomer.

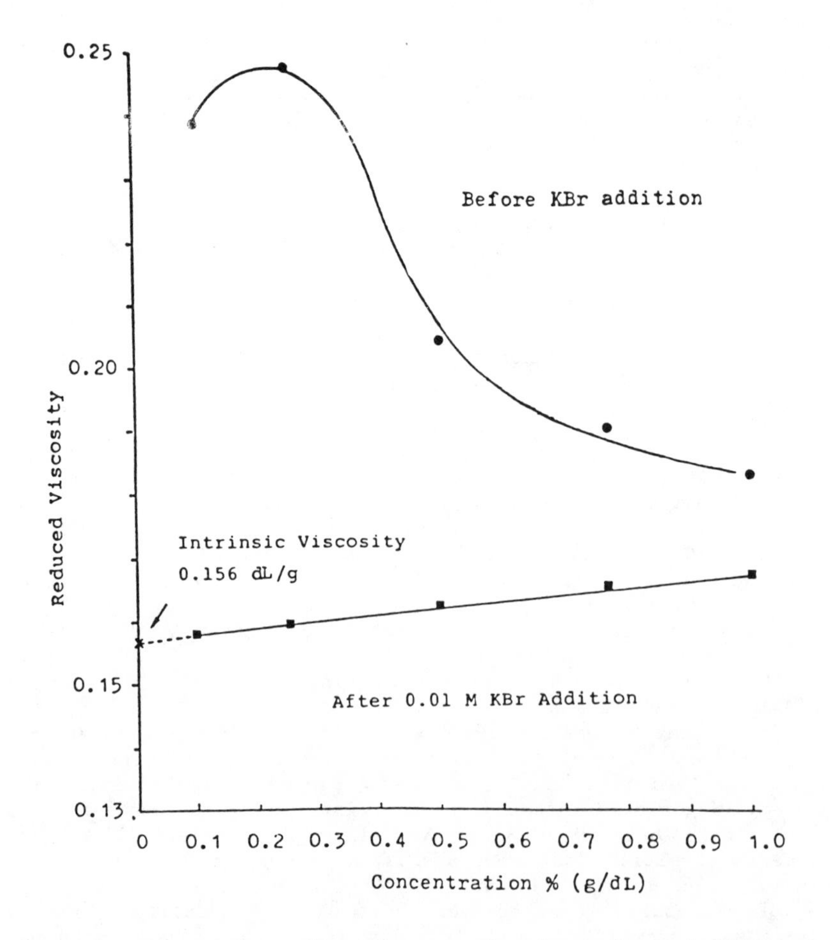

Figure 4. Reduced Viscosity-Concentration Curve for Poly(N-Phenyl-3,4-Dimethylenepyrroline).

VI $\xrightarrow[\text{or DDQ}]{\text{Pd/C}}$ VII

Figure 5. Oxidation of Poly(N-phenyl-3,4-dimethylenepyrroline).

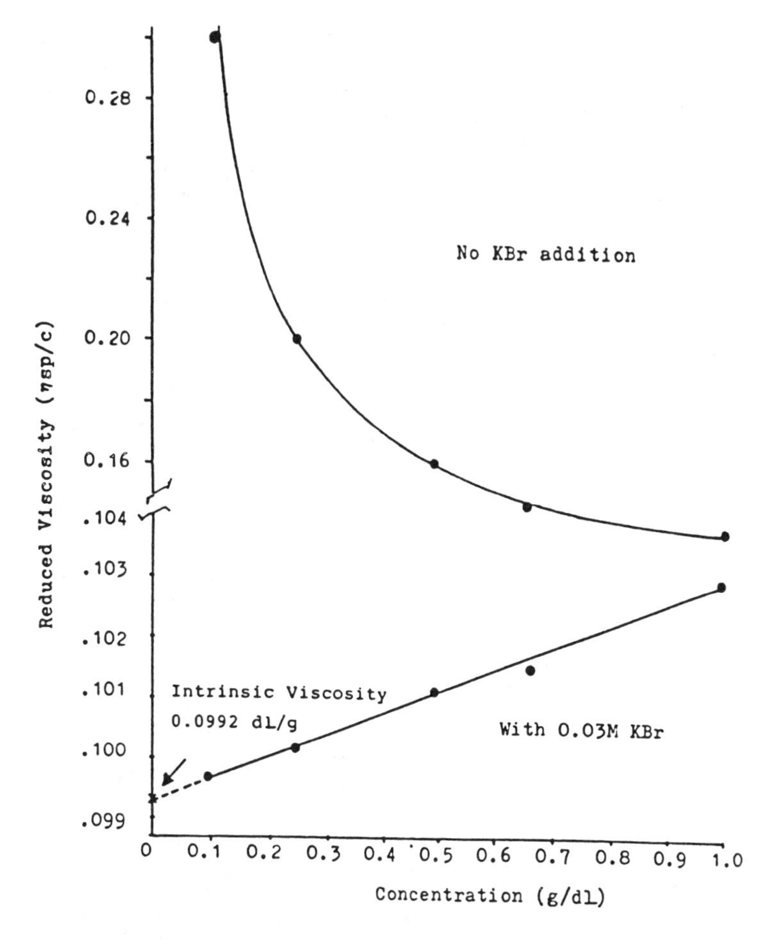

Figure 6. Reduced Viscosity-Concentration Curve for Poly(N-Phenyl-3,4-Dimethylenepyrrole).

Attempts to Prepare Poly(N-Phenyl-3,4-Dimethylenepyrrolidine) IX

The most widely used method of reducing a double bond is by catalytic hydrogenation. Attempts were made to prepare the saturated pyrrolidine polymer (IX) by the hydrogenation of poly(N-phenyl-3,4-dimethylene-pyrroline) with PtO_2/H_2. The catalytical reduction was carried out in a specially built stainless steel hydrogenator in which gaseous hydrogen was introduced from a commercial high-pressure cylinder. The catalyst concentration used varied from 10 to 50% of the substrate by weight. At 100°C under a pressure of 800 psi for 7 days, no appreciable reduction was observed. Under more severe conditions of 200°C with 1500 psi for 3 days, a substantial amount of C-N bond cleavage in the polymer was detected in the ^1H NMR spectrum, but no apparent reduction had taken place. Other transition metal catalysts such Pd/C, Raney nickel were tried and found equally ineffective. Several solvents were used including, formic acid, acetic acid and methanol. The type of solvent used did not appear to make much difference. One possible explanation for this poor hydrogenation was that the polymer molecules, being extremely long, existed mostly in a coil configuration. Therefore, it was very difficult for the catalyst to interact with the double bond and facilitate reduction.

An alternative method of preparing the saturated cyclic amines via cyclopolymerization of diallylamine or diallylammonium chloride was unsuccessful. Common free radical initiators such as 2,2'-azobisisobutyronitrile, ammonium persulfate, benzoyl peroxide were found to be ineffective. Several procedures reported in the literature were followed, and unfortunately all of them have resulted only a small amount of low molecular weight oligomers. Further research for polymerization conditions and types of initiation is still required.

Comparison of the Properties of N-Phenylpyrrolidine (VII) and N-Phenyl-Pyrrole Polymers (VIII)

The elemental analysis, IR and ^1H NMR spectra of polymers VII and VIII are in agreement with the proposed chemical structure for the 1,4-polymerization of N-phenyl-3,4-dimethylenepyrrolidine.

Solubility - The oxidized polymer (VIII) has a greater solubility than the original polymer (VII). It was found to be soluble in acetone, chloroform, benzene, DMF and DMSO. Unlike the polymer (VII), (VIII) was not soluble in formic acid or trifluoroacetic acid; that was expected since the pyrrole moiety is less basic than pyrrolidine. In the oxidized polymer, the pair of unshared electrons on the nitrogen atom are contributing to the pyrrole ring aromaticity, therefore, unavailable for protonation as in the case of polymer (VII). A comparison of the solubilities is given in Table I.

Thermal Analysis - Differential Scanning Calorimetry (DSC) and thermal gravimetric analysis (TGA) were used to characterize the thermal properties of the polymers synthesized. DSC analysis was performed on a Perkin-Elmer Differential Scanning Calorimeter, Model 2C with a thermal analysis data station. Thermal gravimetric analysis (TGA) was carried out on a DuPont thermal gravimeter, Model 951. From the DSC and TGA plots of poly(N-phenyl-3,4-dimethylene-

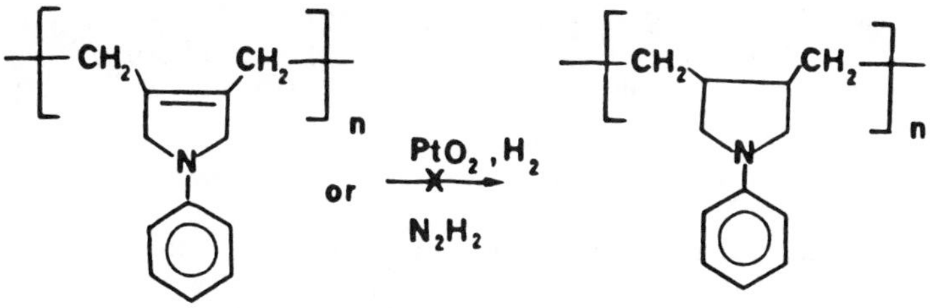

Figure 7. Attempted Hydrogenation of Poly(N-Phenyl-3,4-Dimethylenepyrroline).

TABLE I. Solubility[a] Of Some Poly(N-Phenyl-3,4-Dimethylenepyrroles)

Sample	HCL	Acetone	Methanol	Chloroform	Benzene	Formic Acid
N-Phenyl-3,4-dimethylene-pyrrolidine monomer(III)	s	s	p	s	s	s
Poly(N-phenyl-3,4-dimethylene-pyrroline) (VII)	s	i	i	i	i	s
Poly(N-phenyl-3,4-dimethylene-pyrrole) (VIII)	i	s	i	s	s	i

[a]Solubility: s = Soluble; greater than 1% (w/v)
p = Partly soluble; between 1% - 0.5% (w/v)
i = Insoluble; less than 0.1% (w/v)

pyrroline) (VII) the Tm (melting temperature) was determined to be 297.2°C with a heat of fusion 8.88 cal/g. The polymer was thermally stable with only 20% weight loss at 400°C and the complete decomposition occurred at 482°C.

The DSC and TGA plots of the oxidized polymer (VIII) showed that the Tm is 130°C and the weight loss of 20% and 80% was observed at 455°C and 600°C, respectively, compared to 400° and 482°C for the original polymer VII indicating the oxidized polymer was more stable to heat. This observation was consistent with the chemical structure of the oxidized polymer, which consisted of a repeating aromatic pyrrole structure and, therefore, should be more thermodynamically stable. The thermal data of the polymers are tabulated in Table II.

IR Spectra - Infrared spectral characteristics of poly(N-phenyl-3,4-dimethylenepyrroline) is given in Figure 8. The broad band at 3410 cm^{-1} was likely caused by absorbed water. The bands at 3040 and 3020 cm^{-1} were assigned to aromatic C-H stretch and olefinic C-H stretch. The two strong bands at 2960 and 2800 cm^{-1} were due to the aliphatic C-H stretch. The sharp intense band at 1680 cm^{-1} was attributed to the C=C stretching of pyrroline ring. The substitution of methylene groups for an α-hydrogen atom in the ring system resulted an increase in the frequency of C=C absorption. The medium to weak absorption bands at 1370 cm^{-1} was assigned to the C-N stretch. The two strong bands at 740 and 660 cm^{-1} were indicative

TABLE II. Thermal Properties of Polymers

Polymer	Tm* (°C)	Heat of Fusion (cal/g)	T(20% wt. loss)(°C)	T(80%wt. loss)(°C)
poly(N-phenyl-3,4-dimethylene-pyrroline) (VII)	297.2	8.88	350	465
poly(N-phenyl-3,4-dimethylene-pyrrole) (VIII)	130.3	7.73	455	600**

* Heating Rate: 10°C/min.
** Only about 80% loss weight. No more weight loss after that temperature.

of a monosubstituted benzene ring. The IR differences between the oxidized polymer VIII (Figure 9) and its original polymer (VII) are not significant. The broad band at 3410 cm^{-1} was likely caused by occluded water. The C=C absorption at 1680 cm^{-1} disappeared. Strong absorptions at 1600-1505 cm^{-1} were due to the benzene ring. The medium absorption band at 1390 cm^{-1} was designated to C-N stretch. The two strong bands at 754 and 690 cm^{-1} were indicative of monosubstituted aromatic.

NMR Spectra - The proton NMR spectrum of poly(N-phenyl-3,4-dimethylenepyrroline) (VII) had three singlet absorptions at δ 2.56, 4.81 and 7.60 respectively (Figure 10). The integration of these peaks showed a ratio of 4:4:5. The presence of exocyclic olefinic protons was not observed, indicating that 1,4- addition was predominant in the polymerization with little or no 1,2 addition taking place.

The ^{1}H NMR spectrum of the oxidized polymer (VIII) had three singlet absorption peaks at δ 2.81, 6.86 and 7.25, respectively, as shown in Figure 11. The peak at δ 2.81 was due to the methylene protons at the backbone of the polymer and the chemical shift of this peak remained very similar to that of the original polymer (VII). The second peak was due to the protons adjacent to nitrogen atom. That peak was shifted downfield to δ 6.86 because of the aromaticity within the heterocyclic ring. The phenyl protons appeared as a singlet in δ 7.25 and remained unchanged. The integration of the peak areas showed a correct ratio of 4:2:5 for the proposed structure.

CONCLUSIONS

This study has demonstrated that a cyclic pyrroline polymer could be prepared by a free radical initiation of the corresponding exocyclic diene monomer. The polymerization was shown to proceed predominantly by 1,4-addition as expected from a free radical initiator with diene monomers.

The oxidation of poly(N-phenyl-3,4-dimethylenepyrroline) with DDQ or Pd/C in nitrobenzene gave in a cyclic aromatic amine polymer with repeating pyrrole rings in the polymer backbone. Using Pd/C in

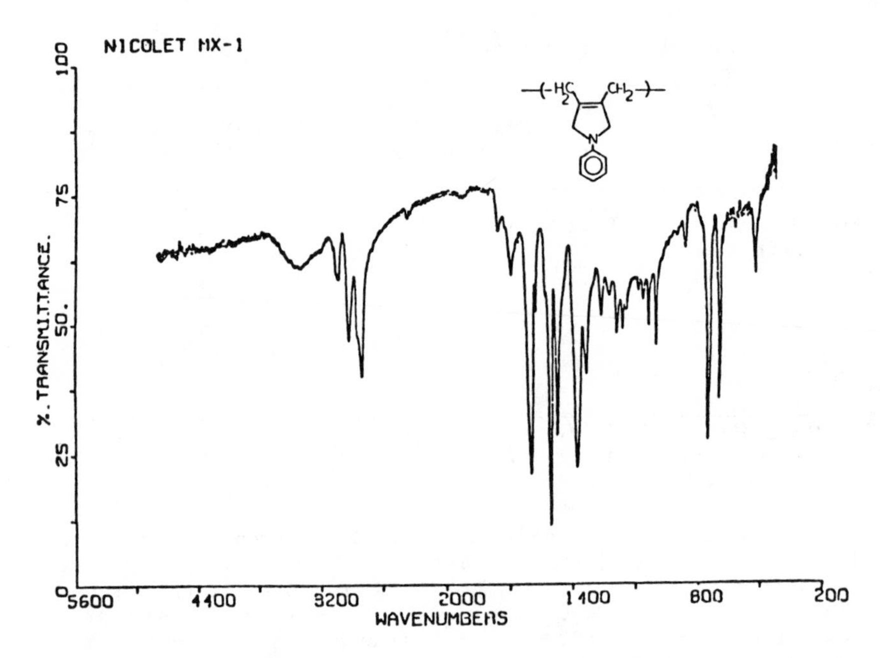

Figure 8. IR Spectrum of Poly(N-Phenyl-3,4-Dimethylenepyrroline).

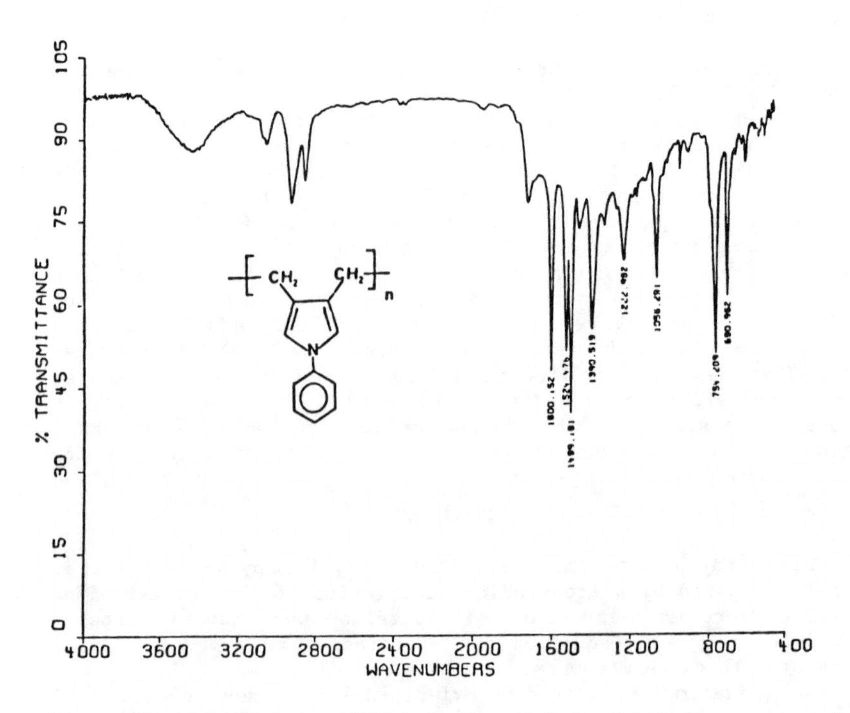

Figure 9. IR Spectrum of Poly(N-Phenyl-3,4-Dimethylene pyrrole).

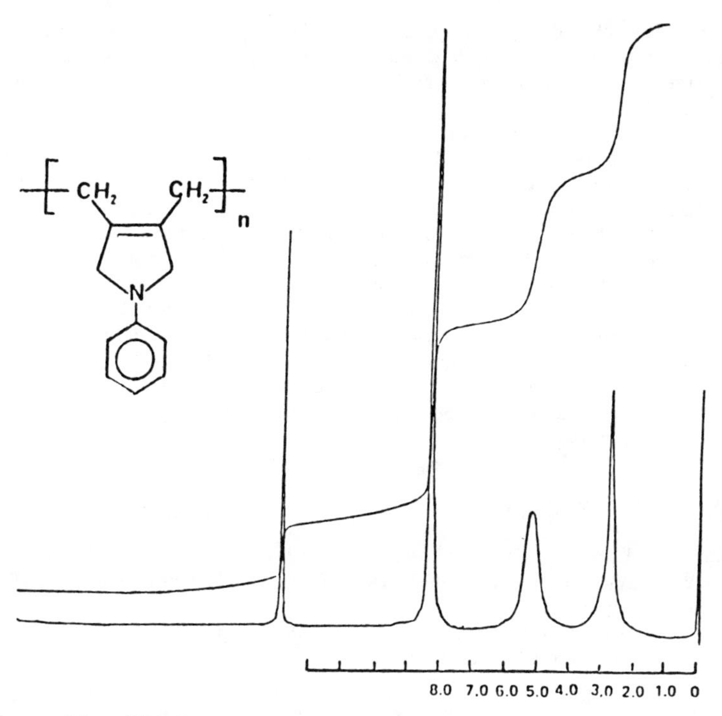

Figure 10. NMR Spectrum of Poly(N-Phenyl-3,4-Dimethylenepyrroline).

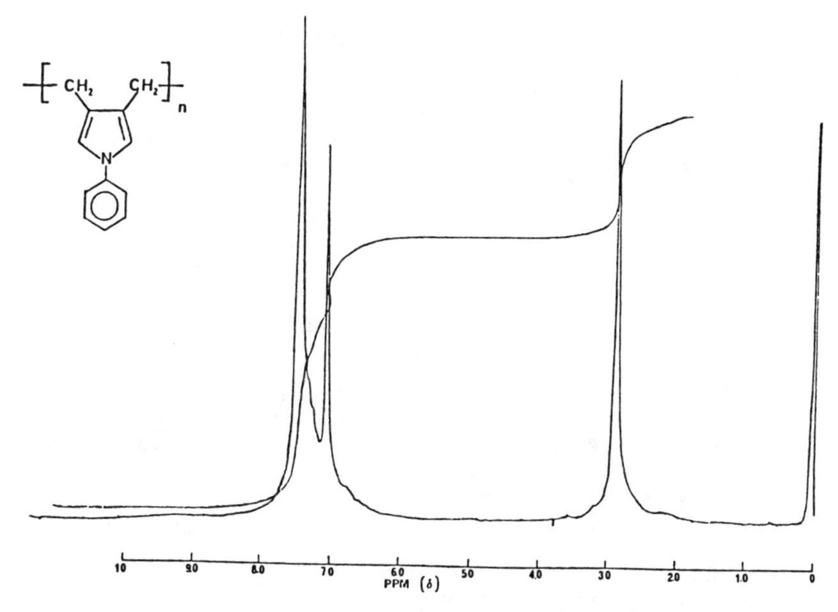

Figure 11. NMR Spectrum of Poly(N-Phenyl-3,4-Dimethylenepyrrole).

nitrobenzene for the dehydrogenation at 180°C for 2 hours gave the best yield. However, the double bond in poly(N-phenyl-3,4-dimethylenepyrroline) was found inert to both catalytic hydrogenation and diimide reduction.

Poly(N-phenyl-3,4-dimethylenepyrroline) had a higher melting point than poly(N-phenyl-3,4-dimethylenepyrrole) (171° vs 130°C). However, the oxidized polymer showed a better heat stability in the thermogravimetric analysis. This may be attributed to the aromatic pyrrole ring structures present in the oxidized polymer, because the oxidized polymer was thermodynamically more stable than the original polymer. Poly(N-phenyl-3,4-dimethylenepyrroline) behaved as a polyelectrolyte in formic acid and had an intrinsic viscosity of 0.157 (dL/g) whereas, poly(N-phenyl-3,4-dimethylenepyrrole) behaved as a polyelectrolyte in DMF and had an intrinsic viscosity of 0.099 (dL/g). No common solvent for these two polymers could be found, therefore, a comparison of the viscosities before and after the oxidation was not possible.

REFERENCES

1. Inaki, Y.; Kumura, K. and Takemoto, E. Macromol. Chem. 1973, 19, 171.
2. Geckeler, K.; Weingartner, K. and Bayer, E. "Polymeric Amines and Ammonium Salts", Goethals, E.J., Ed. Pergamon Press: Oxford, 1980, p. 67.
3. Small, H. Ing. Eng. Chem. Prod. Res. Develop. 1976, 6, 147.
4. Biswas, M. and Packirisawy, S. J. Applied Polym. Sci. 1980, 25, 511.
5. Boloto, B.A. "Polymeric Amines and Ammonium Salts", Goethals, E.J., Ed.; Pergamon Press: Oxford, 1980, p. 365.
6. Bolto, B.A.; Eppinger, K.H. and Jackson, M.B. J. Macromol. Sci. - Chem. 1982, A17 (1), 153.
7. Black, A.P.; Birkner, F.B. and Morgan, J.J. J. Am. Water Works Assoc. 1966, 57 (12), 1547.
8. Butler, G.B. "Polymeric Amines and Ammonium Salts", Goethals, E.J., Ed.; Pergamon Press: Oxford, 1980, p. 125.
9. Horn, D. "Polymeric Amines and Ammonium Salts", Goethal, E.J., Ed.; Pergamon Press: Oxford, 1980, p. 305.
10. Tsuruta, T. Am. Chem. Soc. Polym. Reprints 1979, 20, 350.
11. Ferruti, P. "Polymeric Amines and Ammonium Salts", Goethal, E.J., Ed.; Pergamon Press: Oxford, 1980, p. 305.
12. Tsuchida, E. "Polymeric Amines and Ammonium Salts", Goethal, E.J., Ed.; Pergamon Press: Oxford, 1980, p. 193.
13. Ottenbrite, R.M., W. Ryan, Ind. and Eng. Res and Dew. 19, 528 (1980).
14. Moroson, H. and Rotman, M. "Polyelectrolytes and their application", Rembaum, A. and Selegny, E. Holland, 1975.
15. Butler, G.B. and Angelo, R.J. J. Am. Chem. Soc. 1957, 79, 3128.
16. Ottenbrite, R.M. and Shillady, D.D., "Ring Size on Cyclopolymerization" Polymeric Amines and Ammonium Salts, E. Goethals, Ed. Pergamon Press Oxford, 1980.
17. Butler, G.B. and Ottenbrite, R.M. Tetrahedron Letters 1967, 48, 4873.
18. Ottenbrite, R.M. and Alston, P.V. J. Org. Chem. 1972, 37, 3360.
19. Gaoni, Y. Tetrahetron Letters 1973, 26, 2361.

RECEIVED September 29, 1987

Chapter 11

Polycyclopropanone Synthesis and Hydrogenolysis

Roderic P. Quirk and James H. Dunaway[1]

Institute of Polymer Science, University of Akron, Akron, OH 44325

Cyclopropanone was polymerized at -78°C to polymers with $\bar{M}_n$=4,000-20,000 and $\bar{M}_w/\bar{M}_n$=1.9-2.2 using triethylamine as initiator. Infrared, ^{1}H NMR, and ^{13}C NMR spectral analyses of the polymers are consistent with a polyacetal structure for the polymer. Hydrogenolysis reactions of the cyclopropane rings in the polymer were carried out using Pt/carbon and Pd/carbon catalysts. Infrared, ^{1}H NMR, and ^{13}C NMR spectral analyses of the hydrogenolysis product are consistent with selective ring-opening of the least substituted cyclopropane carbon-carbon bond to form gem-dimethyl groups. Hydrogenolysis was accompanied by some degradation as shown by size exclusion chromatographic analysis.

Ketones are generally not polymerizable, despite claims that acetone can be polymerized at low temperatures (1). A simple explanation for the lack of polymerizability of ketones compared to vinyl monomers can be deduced from consideration of Pauling (2) average bond energies as shown in Equations 1 and 2, where ΔH^{pol}(est) is the estimated enthalpy of polymerization based upon the difference in bond energies of the two single bonds formed in the polymer compared to the double bond in the monomer:

Pauling Bond Energies		ΔH^{pol}(est)	
C=C ⟶	2C-C		
147 kcal/mol	2x83.1 kcal/mol	-19.2 kcal/mol	(1)
C=O ⟶	2C-O		
174 kcal/mol	2x84 kcal/mol	+6 kcal/mol	(2)

[1]Current address: PPG Industries, Glass Research and Development, Harmar Township, PA 15238

0097-6156/88/0364-0141$06.00/0

Thus, estimated bond energy changes for polymerization indicate that in contrast to vinyl monomers carbonyl groups have unfavorable enthalpies of polymerization, because of the high bond energy of the carbonyl group. The contribution from the enthalpy of polymerization would be expected to dominate the free energy of polymerization since entropies of polymerization are generally negative and unfavorable for all monomers (3). This suggests that ketones which have ring strain energy which is decreased upon carbonyl addition might have more favorable thermodynamic parameters for polymerization. Cyclopropanone is an interesting monomer in this respect since there is much more bond angle strain (4) in the monomer compared to the polymer as shown in Equation 3. Therefore, it would be predicted that cyclopropanone, unlike unstrained ketones, might have a favorable free energy of polymerization. Indeed, it has been reported that

Polymerization (3)

C-C(=O)-C Bond Angle Strain: 120-60°=60°

$C-C(O_2)-C$ Bond Angle Strain: 109-60°=49°

cyclopropanone undergoes polymerization, presumably catalyzed by adventitious moisture, when warmed to room temperature (5,6). Given this encouraging information, it was important to investigate this interesting monomer and to determine not only its range of polymerizability (e.g., anionic, cationic, etc.), but also the properties of the resulting polymer.

Another intriguing aspect of the polymerization of cyclopropanone was the possibility that the resulting polymer could provide an indirect route to the hithertofore unreported polymer of acetone. It was envisioned that selective hydrogenolysis of the cyclopropane ring in polycyclopropanone could produce polyacetone as shown in Equation 4. There is precedent for the selective cyclopropane ring-opening of the least substituted cyclopropane carbon-carbon bonds

H_2 catalyst (4)

CH_3 CH_3

upon hydrogenation (7,8). Herein are reported results of investigations into the nucleophile-initiated polymerizations of cyclopropanone.

Experimental

Materials. Ketene was synthesized as described by Andreades and Carlson (9) from diketene. Diazomethane was synthesized from N-methyl-N-nitroso-p-toluenesulfonamide as outlined by Hudlicky (10). A typical synthesis of cyclopropanone involved the slow addition of 400 mL of a 1.0 M diethyl ether solution diazomethane to a 2-3-fold molar excess of ketene at -78°C. This synthesis was based on the procedure described by van Tilborg et al. (11). The reaction between diazomethane and ketene was essentially instantaneous as evidenced by the immediate disappearance of the characteristic yellow color of diazomethane and the evolution of nitrogen. After all of the diazomethane had been added, the excess ketene was removed by vacuum distillation at -55 to -65°C. Infrared spectra of diethyl ether solutions of cyclopropanone showed a strong absorption at 1820 cm^{-1} characteristic of the cyclopropanone carbonyl (5,6). Yields for this synthesis as determined by the analytical method described by van Tilborg et al. (11) ranged between 50-75%. Triethylamine was purified by distillation from freshly crushed calcium hydride and stored under argon in a refrigerator.

Characterization. Infrared spectra were obtained with a Beckman Model FT-2100 FTIR spectrometer. ^{1}H-NMR spectra were obtained on a Varian Model T-60 spectrometer (60 MHz) with deuterated chloroform as the solvent and $(CH_3)_4Si$ as the internal standard. ^{13}C NMR spectra were obtained using a Varian XL400 (100 MHz) spectrometer in deuterated chloroform which also served as the internal standard. Glass transition temperatures and melting points were determined with a du Pont 1090 Thermal Analyzer using the DSC mode. Samples were sealed in aluminum dishes and the analyses were carried out under an inert atmosphere (nitrogen) at heating rates of 5°C and 10°C per minute. Size exclusion chromatographic analyses were obtained with a Waters 150C GPC using a six μ-Styragel columns (10^2, $5x10^2$, 10^3, 10^4, 10^5, 10^6 Å) after calibration with standard polystyrene samples. The carrier solvent was THF at a flow rate of 1 mL per minute at 30°C. Vapor Pressure Osmometry (VPO) measurements were made on a Knauer Type 11.00 Vapor Pressure Osmometer using toluene as the solvent at 44°C. Sucrose octaacetate, recrystallized twice from ethanol, was used as the standard for VPO.

Polymerizations. Polymerization was typically achieved by the addition of 40 μL ($28.8x10^{-5}$ mol) of triethylamine to 350 mL of a cold (-78°C) 0.8 M diethyl ether solution of cyclopropanone. Initiation was usually accompanied by a 10°C increase in reaction temperature and the formation of a white precipitate within 2-3 minutes. The polymerization solution was kept at -78°C for an additional two hours and then allowed to warm up slowly to room temperature overnight. The polymer was purified by precipitation into methanol and dried under vacuum. The ^{1}H NMR spectrum was characterized by a singlet at δ1.0 ppm. The ^{13}C NMR spectrum was characterized by singlets at δ12.24 ppm and δ89.26 ppm in a roughly 2:1 area ratio. The molecular weights of the polymers were determined by a combination of vapor phase osmometry and size exclusion chromatography. Molecular weights ranged between 4,000 g/mol to >15,000 g/mol. A DSC trace of a 13,000 g/mol sample showed a sharp crystalline melting point at 167°C and a discontinuity between 0-5°C which might have been due to a glass transition. The low molecular weight polymers (<7,000 g/mol) were soluble in methylene chloride, chloroform, toluene and THF while the higher molecular weight polymers were much less soluble.

Hydrogenations. Hydrogenation was accomplished using a Series 4000 Parr High Pressure Reactor. The polymers were end-capped before hydrogenation by reacting them with a refluxing 70:30 (by volume) mixture of pyridine/acetic anhydride for one hour. Hydrogenation catalysts used were 10% Pt/C and 10% Pd/C. Methylene chloride was used as the solvent. Reaction temperatures were varied between 20°C to 150°C and hydrogen pressures were varied between 100 to 1500 psi. Reaction times ranged from a few hours to a few days.

Results and Discussion

Cyclopropanone Synthesis. The literature procedures (5,6) for the synthesis of cyclopropanone utilize the reaction of a 2-3 fold excess of ketene with diazomethane at -78°C as shown in Equation 5.

$$CH_2N_2 + CH_2{=}C{=}O \xrightarrow[Et_2O]{-78^\circ C} \text{cyclopropanone} + N_2 \quad (5)$$

Since the ketene will copolymerize with cyclopropanone, excess ketene was removed by addition of 3Å molecular sieves followed by evacuation at 1 mm Hg for 2-3 hours at -70°C. The amount of ketene was monitored by FTIR analysis. Ketene has a distinct strong absorption between 2130 and 2150 cm^{-1} (5,6). The solution FTIR spectrum of cyclopropanone in Figure 1 was taken before ketene removal.

Cyclopropanone Polymerization. Triethylamine is an efficient initiator for the polymerization of cyclopropanone. This initiator caused polymerization to start almost immediately as evidenced by the rapid increase in temperature and the formation of a precipitate within 2-3 minutes. From the data in Table 1 there does not appear to be any correlation between the amount of initiator added and the molecular weight of the resultant polymer. One possible explanation for this is that the polymer was synthesized under heterogeneous conditions thus limiting the access of monomer to growing polymer chains.

Table 1. Anionic Polymerization of Cyclopropanone

Sample	Moles Initiator	Grams Polymer	% Yield[a]	Theor. $\bar{M}_n$[b] (g/mol)	$\bar{M}_n$[c] (g/mol)	$\bar{M}_w/\bar{M}_n$[d]
1	0.0001787	8.10	95	45,500	4,600	1.82
2	0.000072	4.8	64	66,800	13,000	2.80
3	0.000180	5.7	42	32,000	24,000	1.82

a. based on titrated amount of cyclopropanone.
b. Grams of polymer/moles initiator.
c. Sample 1 determined by VPO; samples 2 and 3 determined by GPC using a polycyclopropanone calibration curve.
d. Based on GPC analysis using polystyrene standards.

FTIR, ^1H NMR and ^{13}C NMR analysis support the polyacetal structure containing intact cyclopropyl rings cited by previous workers (5,6) for the polymer of cyclopropanone (Equation 6). Evidence supporting this conclusion includes polymer infrared absorptions characteristic

$$Et_3N + n \text{ cyclopropanone} \xrightarrow{-78°C} \left[\text{C(cyclopropylidene)} - O \right]_n \qquad (6)$$

of the cyclopropyl ring at 3020, 1010, 975 and 950 cm^{-1}, and absorptions characteristic of a C-O-C linkage at 1130 and 1310 cm^{-1} as shown in Figure 2. One anomolous absorption is present at 1765 cm^{-1}. This absorption was present to a greater or lesser degree in all of the polymers synthesized for this study and is characteristic of a carbonyl group. One explanation for this absorption is that ketene was still present in the cyclopropanone solution when polymerization was initiated and was subsequently incorporated into the polymer. The attack of a nucleophile on ketene is shown in Equation 7, which shows that the anion of ketene has two resonance structures. In an

$$B^- + CH_2{=}C{=}O \rightleftharpoons [B{-}C(O^-){=}CH_2 \longleftrightarrow B{-}C(=O){-}CH_2^-] \qquad (7)$$

anionic polymerization of ketene, therefore, it would not be unreasonable to obtain a polymer containing two types of repeat units in the chain as shown in structure A below. Ketene polymers do contain

$$-\!\!\left[C(=O){-}CH_2 \right]\!\!\left[C(=CH_2){-}O \right]\!-$$

A

a mixture of these two repeat units and their relative proportions can be controlled by varying reaction conditions (14). Some of the polymers of cyclopropanone which had even larger carbonyl absorption than in the spectrum shown in Figure 2 also had absorptions in the 1625-1675 cm^{-1} range which are characteristic of carbon-carbon double bonds. Furukawa et al. (15) synthesized a polymer of ketene containing both types of repeat units as shown in structure A. The infrared spectrum they reported contained absorptions at 1770 cm^{-1} and 1650 cm^{-1} which they assigned to the carbonyl and the carbon-carbon double bonds, respectively. This fact, coupled with the observation that the carbonyl absorption was less intense when the ketene distillation was carried out for longer periods of time at -70°C, indicates that the carbonyl group in the polymer is most likely due to traces of ketene being copolymerized with cyclopropanone.

The ^1H NMR spectrum (Figure 3) of this polymer consists of a singlet at $\delta 1.0$ ppm which is consistent with the intact cyclopropyl rings in the polymer. The ^{13}C NMR spectrum (Figure 4) consists of two singlets, one at $\delta 12.24$ ppm and the other at $\delta 89.26$ ppm in an approximately 2:1 area ratio. From chemical shift calculations and model compound comparisons, the singlet at $\delta 12.24$ ppm can be assigned to the cyclopropyl methylene carbons while the singlet at $\delta 89.26$ ppm can be assigned to the backbone carbons (16).

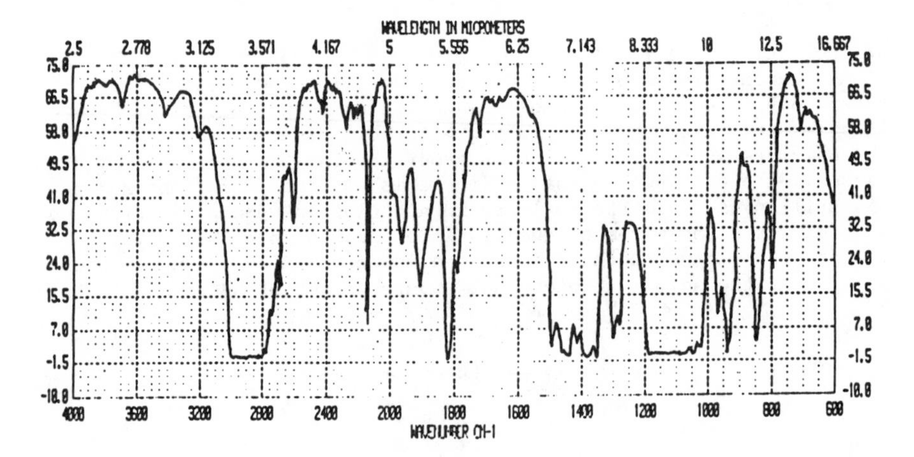

Figure 1 FTIR spectrum of a diethyl ether solution of cyclopropanone.

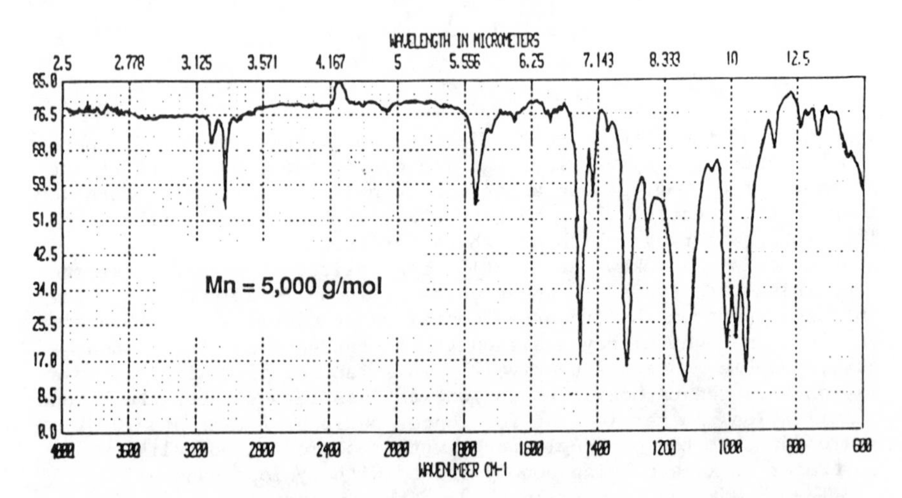

Figure 2 FTIR spectrum of a thin film of polycyclopropanone.

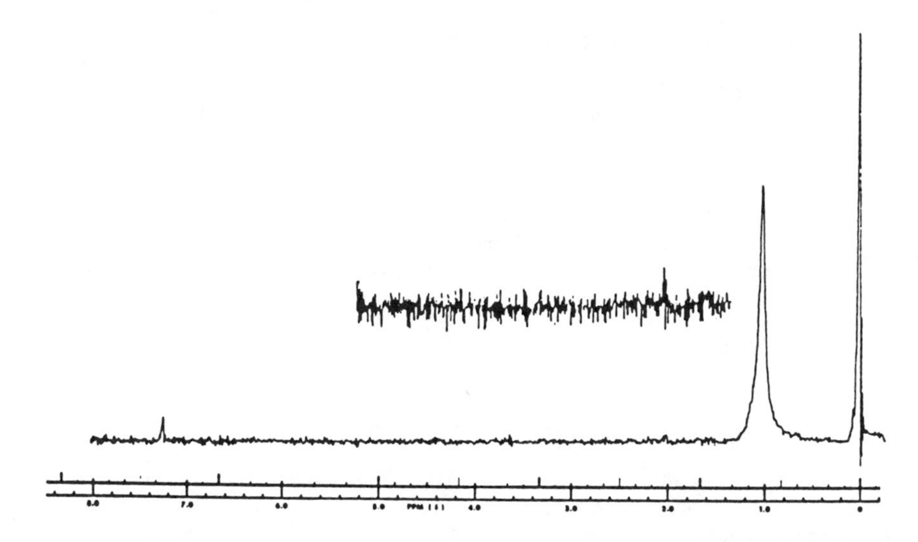

Figure 3 60 MHz ^{1}H NMR spectrum of polycyclopropanone in $CDCl_3$.

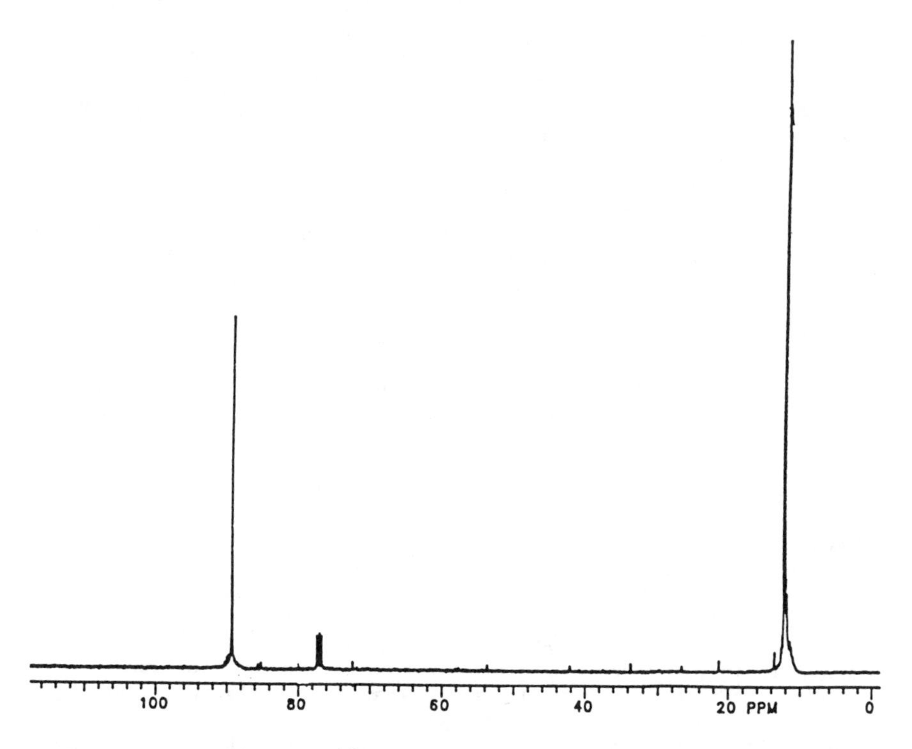

Figure 4 100 MHz ^{13}C NMR spectrum of polycyclopropanone in $CDCl_3$.

The sharp DSC melting point of high molecular weight polycyclopropanone at 167° to 170°C (see Figure 5) indicates that these polymers are highly crystalline. This is further supported by the relative insolubility of the higher molecular weight polymers. The lower molecular weight polymers melted at lower temperatures (130-150°C) and their melting points were not sharp.

Hydrogenolysis. Hydrogenation of polycyclopropanone was carried out using supported catalysts. Before hydrogenation was attempted the polymers were end-capped in a refluxing mixture of acetic anhydride and pyridine. This was done to convert thermally unstable hydroxyl end groups, which may have been present in the polymer, to more thermally stable acetyl end groups as shown in Equation 8. This was

$$HO\sim\sim\sim\sim OH \xrightarrow[\text{pyridine}]{\text{acetic anhydride}} CH_3\text{-}\overset{O}{\overset{\|}{C}}\text{-}O\sim\sim\sim\sim O\text{-}\overset{O}{\overset{\|}{C}}\text{-}CH_3 \qquad (8)$$

done to minimize the chance of depolymerization during hydrogenolysis at elevated temperatures. This precaution may or may not have been necessary since intact cyclopropanone repeat units in the chain should act as zipper stoppers if depolymerization starts.

Using 10% Pt/C and 10% Pd/C catalysts it was possible to achieve extents of hydrogenolysis of 2% to 50%. There appeared to be a direct correlation between temperature and rate of hydrogenolysis with the highest rate achieved with the highest temperatures. The extent of hydrogenolysis could be followed qualitatively by FTIR spectroscopy and quantitatively using proton NMR spectroscopy. An FTIR spectrum of partially hydrogenated polycyclopropanone is shown in Figure 6 which can be compared with the spectrum of polycyclopropanone in Figure 2. The major difference between these two spectra is the large aliphatic C-H stretch absorption between 2900 and 3000 cm^{-1} in Figure 5. Also in Figure 5 the cyclopropyl C-H stretch at 3020 cm^{-1} is much less intense, an absorption at 1375 cm^{-1} characteristic of methyl groups becomes much more pronounced and the absorptions at 1010, 975 and 950 cm^{-1} (cyclopropyl C-C) are less intense. All of these differences between the spectra in Figures 2 and 5 indicate that hydrogenolysis has occurred as indicated in Equation 4.

Spectra obtained by ^1H NMR allowed the progress of the hydrogenolysis reaction to be followed quantitatively. The cyclopropyl protons in the polymer have an ^1H NMR peak at $\delta 1.0$ ppm, while the protons on the ring-opened, gem-dimethyl repeat unit (Equation 4) have an ^1H NMR peak at $\delta 1.35$ ppm. The gem-dimethyl peak assignment was based on chemical shift calculations (17). Also the dimethyl acetal of acetone has protons in a similar chemical environment to those on the gem-dimethyl ketal repeat unit and they have a chemical shift between $\delta 1.3$ ppm and $\delta 1.4$ ppm (18). A proton NMR spectrum of partially hydrogenated polycyclopropanone is shown in Figure 7. As shown in Figure 6, the peaks at $\delta 1.0$ ppm and $\delta 1.35$ ppm do not overlap appreciably. Because of this it is possible to calculate the extent of hydrogenolysis from the integrated area of these peaks. ^1H NMR also provides evidence that the predominant mode of bond cleavage is at the bond opposite to the backbone carbon in the polymer. If the bond adjacent to the backbone carbon had been broken this

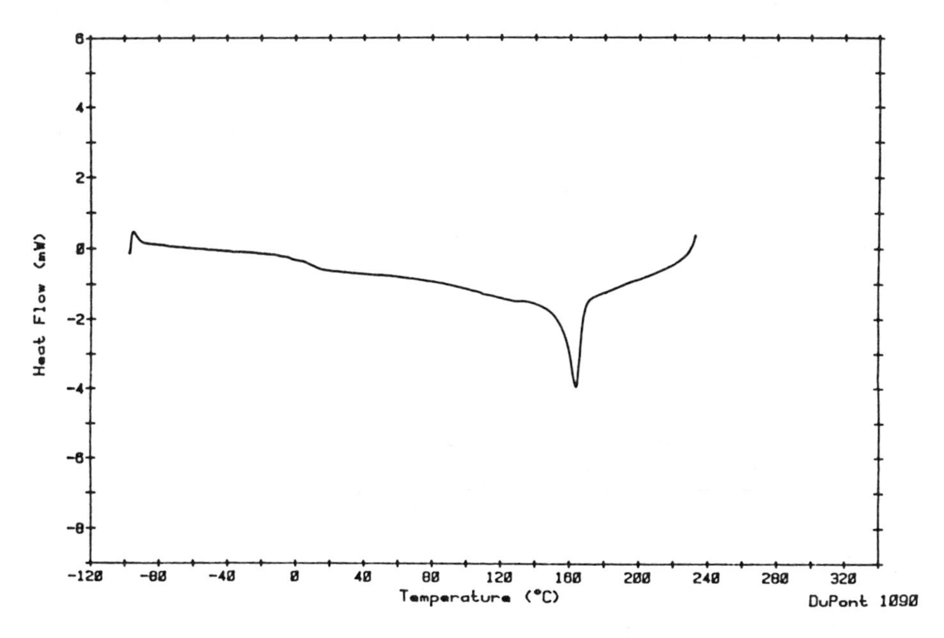

Figure 5 DSC trace for polycyclopropanone ($\overline{M}_n$=13,000 g/mol).

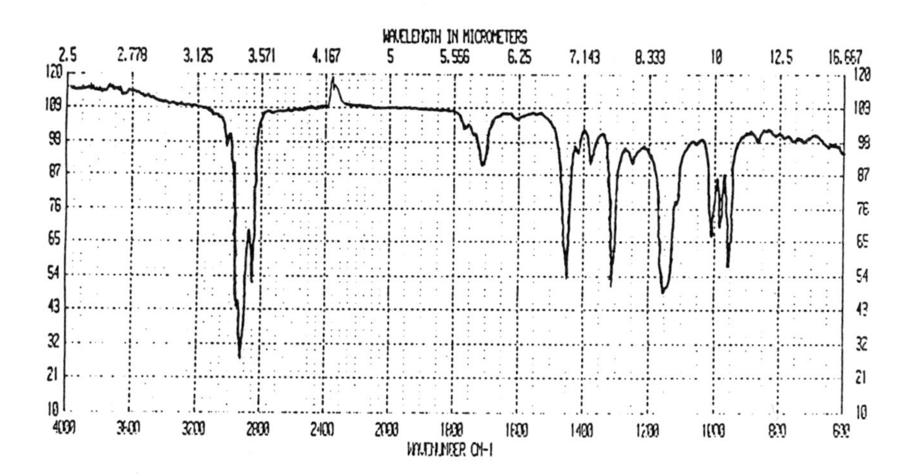

Figure 6 FTIR spectrum of a thin film of partially hydrogenated polycyclopropanone.

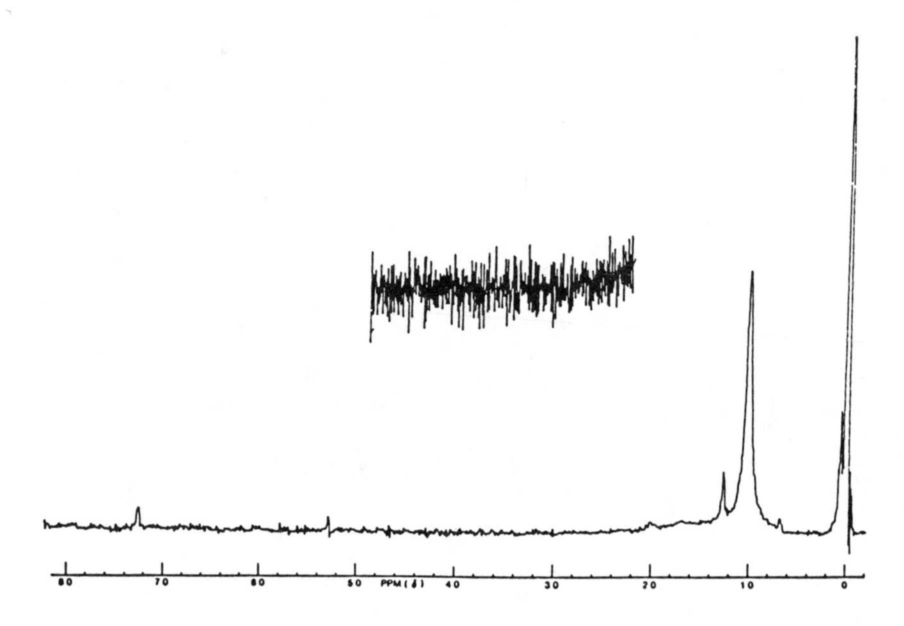

Figure 7 60 MHz ^{1}H NMR spectrum of partially hydrogenated polycyclopropane in $CDCl_3$.

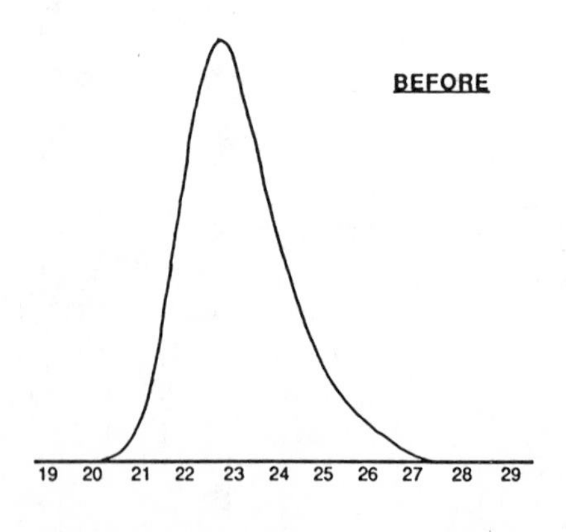

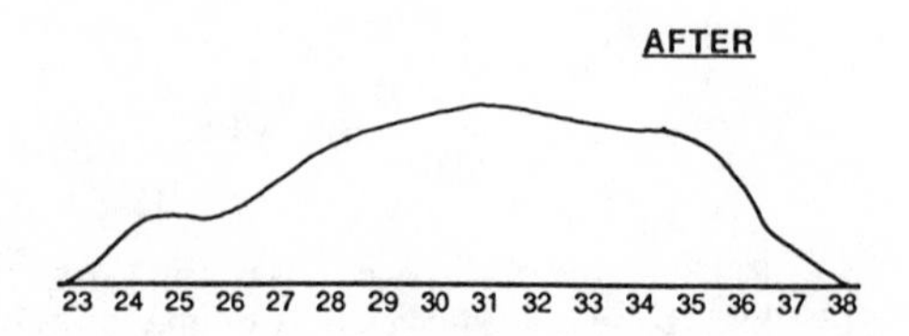

Figure 8 Size exclusion chromatograms of polycyclopropanone before and after hydrogenolysis.

would have left a unique proton attached to the backbone carbon which would have a chemical shift of approximately δ3.5 ppm based on chemical shift calculations (17). As can be seen from Figure 6, no noticeable peak appears at δ3.5 ppm after partial hydrogenolysis (approximately 10%) but a substantial peak does appear at δ1.3 ppm to δ1.4 ppm. This ^{1}H NMR evidence indicates that the predominant mode of bond cleavage during hydrogenolysis is opposite to the backbone carbon to give a gem-dimethyl structure which is identical to the structure expected for the repeat unit in a polymer of acetone (Equation 4).

While it was possible to obtain up to 50% hydrogenolysis using elevated temperature and long reaction times, this was obtained at the expense of molecular weight degradation. Evidence for this was a significant broadening of the molecular weight distribution ($\overline{M}_w/\overline{M}_n$) of the polymer after approximately 15% hydrogenolysis. Molecular weight distributions of the polymers before hydrogenolysis typically ranged between 1.8 to 2.2 (as determined by GPC using polystyrene standards). After about 15% hydrogenolysis the molecular weight distributions typically increased to over 14. Figure 8 is a GPC chromatogram of a polycyclopropanone ($\overline{M}_n$=24,000 g/mol, $\overline{M}_w/\overline{M}_n$=1.82) before and after hydrogenolysis using a 10% Pt/C catalyst for 48 hours at 60°C and 750 psi hydrogen pressure. Considering the low estimated ceiling temperature of polyacetone it is not surprising that polymer degradation occurs during hydrogenolysis, especially at elevated temperatures. It might be possible to avoid or at least minimize polymer degradation during hydrogenolysis by using a catalyst which is effective under milder conditions, thus limiting the exposure to high temperatures of the polymer containing acetone-type repeat units.

Conclusion

Using triethylamine as the initiator it is possible to synthesize polymers of cyclopropanone having molecular weights of between 4,000 g/mol to over 15,000 g/mol and having molecular weight distributions between 1.8 to 2.2. Spectral analyses of these polymers show that they have a polyacetal-type structure containing intact cyclopropyl rings. Polymers having molecular weights over 7,000 g/mol exhibit sharp, crystalline melting points between 167°C to 170°C.

Hydrogenolysis of end-capped samples of polycyclopropanone yield the gem-dimethyl repeat unit which is identical to the repeat unit for a polymer of acetone. It was possible to achieve extents of hydrogenolysis of up to 50% but after approximately 15% hydrogenolysis appreciable molecular weight degradation appears to occur. It might be possible to avoid molecular weight degradation during hydrogenolysis if a catalyst which is effective under milder conditions can be found.

Acknowledgments

We gratefully acknowledge the generous support of this research by the Exxon Education Foundation and Owens-Corning Fiberglas Corporation.

Literature Cited

1. Ivin, K. J. in Encyclopedia of Polymer Sci. & Eng., Supplement Vol. 2, Interscience, New York, N.Y., 1977, p 700.
2. Pauling, L. The Nature of The Chemical Bond, 3rd ed., Cornell University Press, Ithaca, N.Y., 1960, p 85, 189.
3. Sawada, H. Thermodynamics of Polymerization, Marcel Dekker, Inc., New York, N.Y., 1976.
4. Dauben, W. G.; Pitzer, K. S. in Steric Effects in Organic Chemistry, Newman, M. S., Ed., 156, p 1.
5. Schaafsma, S. E.; Steinberg, H.; DeBoer, T. J. Rec. Trav. Chim., Pays-Bas, 1966, 85, 1170.
6. Turro, N. J.; Hammond, W. B. J. Am. Chem. Soc., 1966, 88, 3672.
7. Rylander, P. N. Hydrogenation Methods, Academic Press, Orlando, Florida, 1985, p 174.
8. Newman, J. Chem. Rev., 1963, 63, 123.
9. Andreades, S.; Carlson, H. D. Org. Syn., 1986, 45, 50.
10. Hudlicky, M. J. Org. Chem., 1980, 45, 5377.
11. van Tilborg, W. J.; Steinberg, H.; DeBoer, T. J. Synth. Commun., 1973, 3, 189.
12. Whiffen, D. H.; Thompson, H. W. J. Chem. Soc., 1946, 1005.
13. France, G; Binaghi, M. Ketene Polymers, in Encyclopedia of Polymer Science and Technology, Vol. 8, Mark, H. F., Ed.; John Wiley & Sons, Inc., New York, N.Y., 1968.
14. Franco, G.; Binaghi, M. Ketene Polymers, in Encyclopedia of Polymer Science and Technology, Vol. 8, Mark, H. F., Ed.; John Wiley and Sons, Inc., New York, N.Y., 1968.
15. Furukawa, J.; Saegusa, T.; Mise, N.; Kawasaki, A. Makromol. Chem., 1960, 39, 243.
16. Brown, D. W. J. Chem. Ed., 1985, 62, 209.
17. Jackman, L. M. Applications of Nuclear Magnetic Resoance Spectroscopy in Organic Chemistry, Pergamon Press, New York, N.Y., 1963, p 59.
18. Pouchert, C. J. The Aldrich Library of NMR Spectra, Vol. II, 2nd ed., Aldrich Chemical Company, Inc., Milwaukee, WI., 1983, p 187.

RECEIVED August 27,1987

Chapter 12

Chain-Propagation and Step-Propagation Polymerization

Synthesis and Characterization of Poly(oxyethylene)-*b*-poly(pivalolactone) Telechelomer

K. B. Wagener and S. Wanigatunga

Department of Chemistry and Center for Macromolecular Science and Engineering, University of Florida, Gainesville, FL 32611

Anionic polymerization and polymer modification coupled with step growth polymerization is being used to prepare segmented copolymers containing polyether and polyester segments having narrow molecular weight distributions for each segment. In this method an alkoxide initiator having a masked alcohol group initiates ethylene oxide polymerization. The resulting polyether alkoxide is modified with succinic anhydride to give a carboxylate anion which polymerizes pivalolactone. When the resulting macroion is hydrolyzed selectively, a telechelomer, poly(oxyethylene)-b-poly(pivalolactone) containing hydroxy and carboxylic functionalities at its two ends, is formed. Synthesis of the telechelomer and the reaction intermediates will be discussed. NMR spectra and gel permeation chromatograms will also be presented to show that the products are pure and have narrow molecular weight distributions.

The field of multiphase segmented copolymers, although some 30 years old, remains quite active (1-4) in part due to the possibility of obtaining a variety of useful materials ranging from impact resistant plastics to elastomers (5,6). Poly(urethane ethers) and particularly poly(ester ethers) lend themselves to melt extrusion such that these polymers can be shaped easily, an extremely valuable feature. As true elastomers, however, these copolymers lack the physical properties that are demanded in many applications, particularly when they are melt processed (5). Specifically, the elastic properties that are deficient in melt extruded segmented copolymers are stress and recovery related, like immediate elastic recovery (often less than 95%), stress decay (often greater than 15%) and compression set (more than 10%). These properties are direct manifestations of the inefficiency of the physical crosslinks in the copolymer, which in turn are a function of how well the phase separation in the copolymer is achieved. Phase mixing - even to a small degree - adversely affects the properties of the copolymers.

Phase mixing can be attributed partly to the irregularity of the copolymer chain. In the case of poly(ester ether) copolymers synthesized by normal step polymerization reactions, the hard segment has a

NOTE: This chapter is part II in a series.

0097-6156/88/0364-0153$06.00/0

polydispersity ratio approaching 2, and this distribution of chain lengths within the hard segment adversely affects phase separation (7). Phase mixing is also induced by broad distributions in the soft segment. The low molecular weight fraction of either the hard or soft segment especially promote phase mixing (7).

Droscher and coworkers' support the hypothesis that high regularity within segments, i.e., monodispersity within each segment, leads to enhanced phase separation and crystallization phenomena (8). Their investigations involved monodisperse aromatic poly(ester ether) copolymers synthesized in a multistep scheme of nucleophilic substitution reactions (8-11). Inoue and coworkers (12) also have prepared low molecular weight versions of poly(ester ether) block copolymers having monodisperse segments using porphyrin catalysts. The molecular weights of these polymers are up to about 5000.

Our goal is to synthesize highly regular poly(ester ether) segmented copolymers, to investigate their phase separation behavior, and to see how the phase separation behavior affects mechanical and surface properties. The polymers will be synthesized using a novel strategy where both chain and step polymerizations are combined into one scheme. The synthetic scheme was outlined in a previous communication (13). When the two segments have narrow molecular weight distributions and when they are incompatible, the polymers will exhibit good phase separation and the surface of the polymer will be richer in the segment which has a lower surface energy. Proper selection of monomers, therefore, will lead to polymers having controlled surface properties such as biocompatibility, adhesion and weathering. If one segment is hard and the other is soft, the segment lengths could be varied to obtain thermoplastic elastomers having good mechanical properties. The surface and mechanical properties of these copolymers could then be compared with those of ABA and AB type block copolymers.

In this research we have set out to prepare a series of highly regular poly(oxyethylene-co-pivalolactone) segmented copolymers. Our strategy involves synthesizing a highly regular telechelomer (A telechelomer is a high molecular weight monomer, that self polymerizes via step polymerization, whereas a macromer is a high molecular weight monomer, that self polymerizes by chain polymerization.), then converting it to high polymer. In order to obtain narrow molecular weight distributions for both segments, the telechelomer has been prepared by anionic polymerization; the telechelomer then can be converted to high polymer by step polymerization. This paper discusses the synthetic steps involved up to the preparation of poly(oxyethylene)-b-poly(pivalolactone) telechelomer using anionic polymerization. ^{1}H and ^{13}C NMR spectra are presented to support the structure and purity of the telechelomer and the reaction intermediates. Gel permeation chromatograms (GPC) are also shown to confirm their purity and to indicate that narrow molecular weight distributions exist.

Experimental

Materials. Ethylene Oxide (EO) and pivalolactone (PVL) were dried over calcium hydride for a day before use. EO was further dried on a sodium mirror prior to passing into the reaction vessel. Dried PVL was sealed into ampules under high vacuum (10^{-6} mm Hg). Succinic anhydride (SA) was dried in a vacuum oven for a day and sublimed under high vacuum into ampules. Dry THF was obtained by refluxing THF on Na/K (1:1) alloy for six hours, distilling over the alloy and degassing under high vacuum in the presence of the Na/K alloy and benzophenone which is an indicator for dryness.

Synthesis of Masked Poly(oxyethylene)-b-poly(pivalolactone) Copolymeric Salt, 4. The synthesis of the initiator, 1, was reported in a previous communication (13). Figure 3 shows a typical apparatus used for the synthesis of 4. The reaction vessel "A" contains an ampule with initiator (approximately 0.005 mole) in THF and an ampule with excess SA. A coarse filter F connects flasks "A" and "B". "B" contains an ampule of PVL (2.5 mL, .04 mole). The apparatus was brought to 10^{-6} mm Hg. The initiator was added to the reaction vessel followed by THF (90 mL), then stirring was started and EO (5.5 mL, 0.11 mol) was distilled over the Na mirror through the main line, and the mixture was stirred at 25 °C under dry argon for three days. A mercury valve was used to release any sudden pressure. SA (2.0 gms, .02 mol) was sublimed into the reaction flask, "A", under vacuum and the mixture was stirred for six hours at 25°C. The flask "A" was then sealed off from the vacuum line, the contents filtered through F into flask "B", and "B" sealed off at R. PVL was then added and the mixture was stirred vigorously for three hours at 25°C. The product was precipitated in cold ether and purified thrice by dissolving in methylene chloride and reprecipitating in cold ether. (Elemental analysis, Calculated for $C_{112}H_{199}O_{50}K$: C 56.39, H 8.42; Found: C 56.50, H 8.43).

Synthesis of the Telechelomer, 5. The copolymeric salt, 4, (2.28 gms) was dissolved in methylene chloride (12.6 mL) and was shaken with 3N HCl (25.6 mL) in a separatory funnel for 30 minutes. The organic layer was separated and the product was precipitated in cold ether. The product was purified thrice by dissolving in methylene chloride and reprecipitating in cold ether. (Elemental analysis, Calculated for $C_{108}H_{192}O_{49}$: C 57.01, H 8.52; Found: C 56.94, H 8.62).

Instrumentation. 300 MHz ^{1}H NMR and 75 MHz ^{13}C NMR spectra were obtained using Nicolet NT-300 Spectrometer operating at a field of 7 telsa. 50 MHz ^{13}C NMR were obtained using Varian XL-Series NMR Superconducting Spectrometer System. All spectra were taken at room temperature in $CDCl_3$. Sufficient relaxation times were given in taking ^{1}H spectra so that the integrations are accurate. A Waters GPC System containing a Model 6000A solvent delivery system coupled with a R-401 Differential Refractomer was used to obtain GPC chromatograms (solvent: CH_2Cl_2, flow rate: 1 mL/min, 10 A μ styragel column at 25°C). Elemental analysis of products were done by Atlantic Microlab, Inc.

Results and Discussion

Figure 1 indicates the approach used to synthesize poly(oxyethylene)-b-poly(pivalolactone) telechelomers. An acetal capped anionic initiator, 1, (13) polymerizes ethylene oxide (EO) to give 2, a potassium alkoxide of a masked polyether, and this "new" initiator is to be used to polymerize pivalolactone (PVL). Since potassium alkoxides are strong nucleophiles, they can randomly attack at both the carbonyl carbon and the β-methylene carbon in lactones, (Figure 2); such a random attack would result in a pivalolactone segment containing irregularities. Lenz (15), and Hall (16), and Beaman (17) have investigated PVL polymerization and have shown that the less nucleophilic carboxylate anion is preferable in polymerizing PVL smoothly. The weaker carboxylate anion will attack only at the methylene

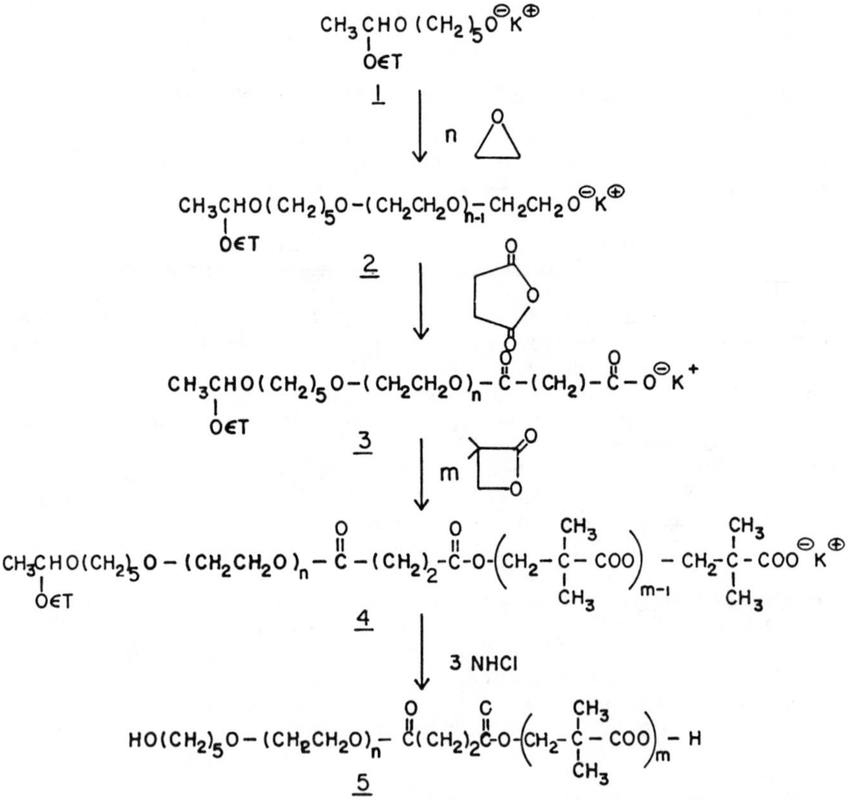

$$CH_3CH(O\epsilon T)O(CH_2)_5O^{\ominus}K^{\oplus}$$

1

$\downarrow$ n (ethylene oxide)

$$CH_3CH(O\epsilon T)O(CH_2)_5O-(CH_2CH_2O)_{n-1}-CH_2CH_2O^{\ominus}K^{\oplus}$$

2

$\downarrow$ (succinic anhydride)

$$CH_3CH(O\epsilon T)O(CH_2)_5O-(CH_2CH_2O)_n-\overset{O}{\overset{\|}{C}}-(CH_2)-\overset{O}{\overset{\|}{C}}-O^{\ominus}K^{+}$$

3

$\downarrow$ m (pivalolactone)

$$CH_3CH(O\epsilon T)O(CH_2)_5O-(CH_2CH_2O)_n-\overset{O}{\overset{\|}{C}}-(CH_2)_2-\overset{O}{\overset{\|}{C}}-O-\left(CH_2-\underset{CH_3}{\overset{CH_3}{C}}-COO\right)_{m-1}-CH_2-\underset{CH_3}{\overset{CH_3}{C}}-COO^{\ominus}K^{\oplus}$$

4

$\downarrow$ 3 NHCl

$$HO(CH_2)_5O-(CH_2CH_2O)_n-\overset{O}{\overset{\|}{C}}(CH_2)_2\overset{C}{\overset{\|}{C}}-O-\left(CH_2-\underset{CH_3}{\overset{CH_3}{C}}-COO\right)_m-H$$

5

Figure 1. Synthetic scheme for poly(oxyethylene)-b-poly(pivalolactone) telechelomer.

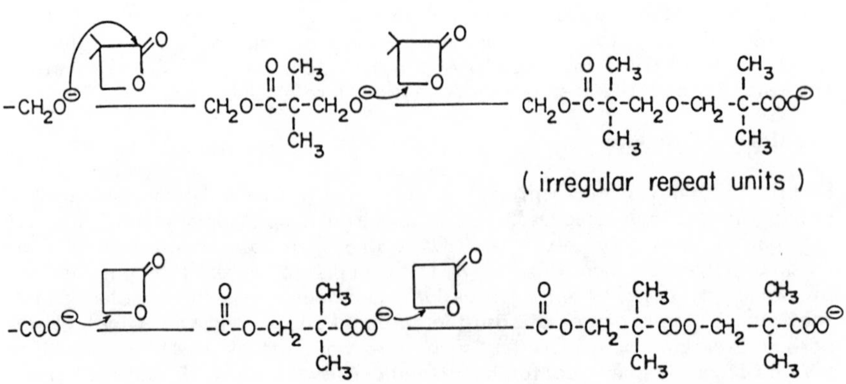

$$-CH_2O^{\ominus} + \text{(pivalolactone)} \longrightarrow CH_2O-\overset{O}{\overset{\|}{C}}-\underset{CH_3}{\overset{CH_3}{C}}-CH_2O^{\ominus} + \text{(pivalolactone)} \longrightarrow CH_2O-\overset{O}{\overset{\|}{C}}-\underset{CH_3}{\overset{CH_3}{C}}-CH_2O-CH_2-\underset{CH_3}{\overset{CH_3}{C}}-COO^{\ominus}$$

(irregular repeat units)

$$-COO^{\ominus} + \text{(pivalolactone)} \longrightarrow -\overset{O}{\overset{\|}{C}}-O-CH_2-\underset{CH_3}{\overset{CH_3}{C}}-COO^{\ominus} + \text{(pivalolactone)} \longrightarrow -\overset{O}{\overset{\|}{C}}-O-CH_2-\underset{CH_3}{\overset{CH_3}{C}}-COO-CH_2-\underset{CH_3}{\overset{CH_3}{C}}-COO^{\ominus}$$

Figure 2. Unsuitability of alkoxide to polymerize pivalolactone.

carbon adjacent to oxygen and as such the PVL segment will be devoid of any defects (Figure 2). Consequently, it became necessary for us to convert the alkoxide anion of the masked polyether into a carboxylate anion. Another problem caused by alkoxide anions has been reported by Yamashita and Hane (14). They polymerized PVL using polystyrene containing an alkoxide ion end group and observed that the homopolymer of pivalolactone was also formed in addition to poly(styrene)-b-poly(pivalolactone) block copolymer. The mechanism for the formation of homopolymer was unclear.

A variety of reagents could be used to carry out such a conversion (18,19). We chose to react the alkoxide ion with succinic anhydride (SA), because the alkoxide ion could be converted quantitatively to the carboxylate ion when excess of SA is used, and also because no side reactions are reported (19). The carboxylate anion, 3, thus formed was used to polymerize PVL giving the masked poly(oxyethylene)-b-poly(pivalolactone) copolymeric salt, 4. The salt, 4, was converted to the telechelomer, 5, by acid hydrolysis.

^{13}C and ^{1}H NMR of the initiator, 1, and the masked polyether, 2, have been presented in a previous communication (13). ^{13}C NMR of the masked polyether with a succinic anhydride end group, 3, is shown in Figure 4, which is an attached proton test (APT) spectrum with primary and tertiary carbons appearing as upward signals and secondary and quarternary carbons appearing as downward signals. Three upward signals at 15.3, 19.3 and 99.5 ppm are due to two methyl carbons "a" and "b" and the methine carbon "g" coming from the initiator, 1. Downward signals at 22.5, 29.5, and 60.5-65.2 ppm are due to "c", "d" and "e" carbons which also come from the initiator, 1. The major downward signal at 69.9 ppm is assigned to the ethylene oxide repeat units; the small signals surrounding the major signal is assigned to the end ethylene oxide units (20). These signals were present in the ^{13}C NMR spectrum of the masked polyether, 2 (13). The incorporation of the succinic acid end group is clearly seen by the new signals "r" and "s" appearing at 173.2 and 176.1 ppm. A new set of downward signals also appear between 30.2-30.6 ppm; these are assigned to the methylene carbons "i" of the succinic acid moiety.

The ^{1}H NMR of this carboxylate capped prepolymer 3, (Figure 5) confirm the carbon assignments. The "a", "b", "c", "d" and "g" protons of the initiator are centered around 1.2, 1.3, 1.4, 1.6 and 4.6 ppm respectively as a triplet, doublet, multiplet, and a quartet. The ethylene oxide repeat units give a major signal at 3.71 ppm. The two methylene hydrogens "i" of the succinic acid moeity occur at 2.6 ppm. The ratio of "i" protons to "g" proton in the spectra is 2, showing that the functionalization of the masked polyether by succinic anhydride is quantitative. The signal at 4.2 ppm is due to the "j" methylene protons that became different from "(h')" methylene protons due to the incorporation of the succinic acid moeity. Protons "(h')" and "(h')" of the end ethylene oxide units give signals close to the major signal at 3.7 ppm (20). The broad signal at 5.4 ppm is due to the presence of some moisture in the sample which is very hygroscopic.

This carboxylated capped prepolymer, 3, which acts as a new initiator smoothly polymerizes pivalolactone giving poly(oxyethylene)-b-poly(pivalolactone) copolymeric salt, 4. ^{13}C APT NMR spectrum of 4 (Figure 6) shows the presence of all the signals of the ^{13}C APT NMR of 3 plus a number of signals coming from the newly incorporated PVL units. The new signals "l" and "k" appearing at 42.9 and 23.0 ppm are due to methylene and methyl carbons. Also the two new signals "t" and "u" appearing at 174.1 and 178.2 ppm are due to carbonyl carbons of pivalolactone units and the

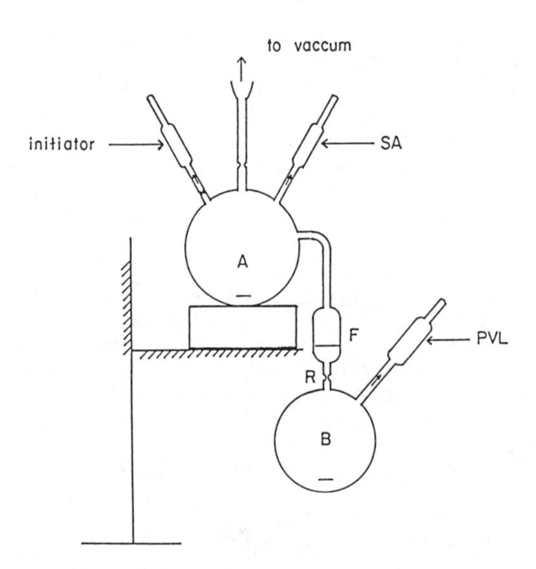

Figure 3. Apparatus for the synthesis of poly(oxyethylene)-b-poly(pivalolactone) salt, 4.

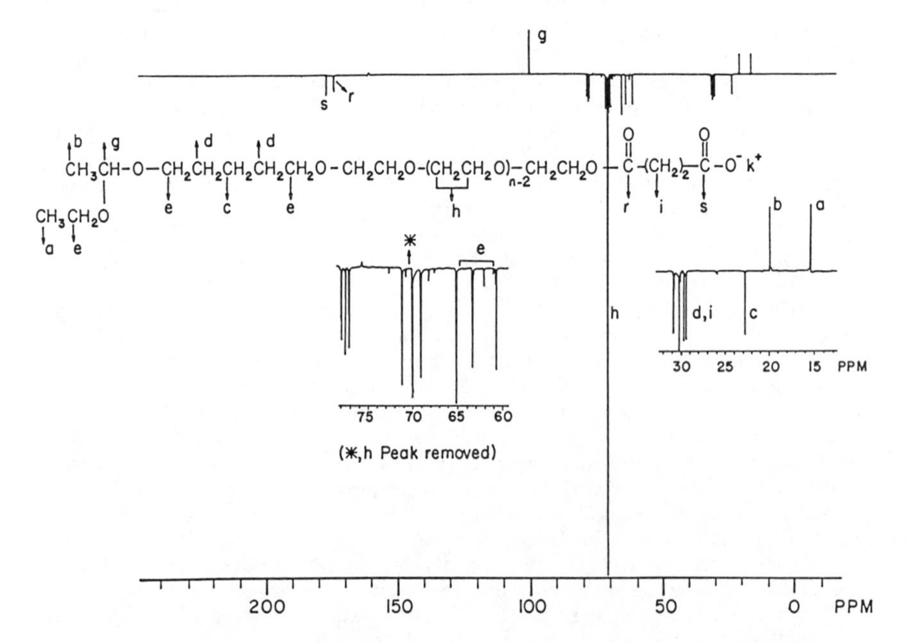

Figure 4. 75 MHz ^{13}C NMR spectrum of the masked polyether with succinic acid end group, 3, using attached proton test sequence (CH, CH_3, Pos.; CH_2, C, neg.) in $CDCl_3$ at 25°C.

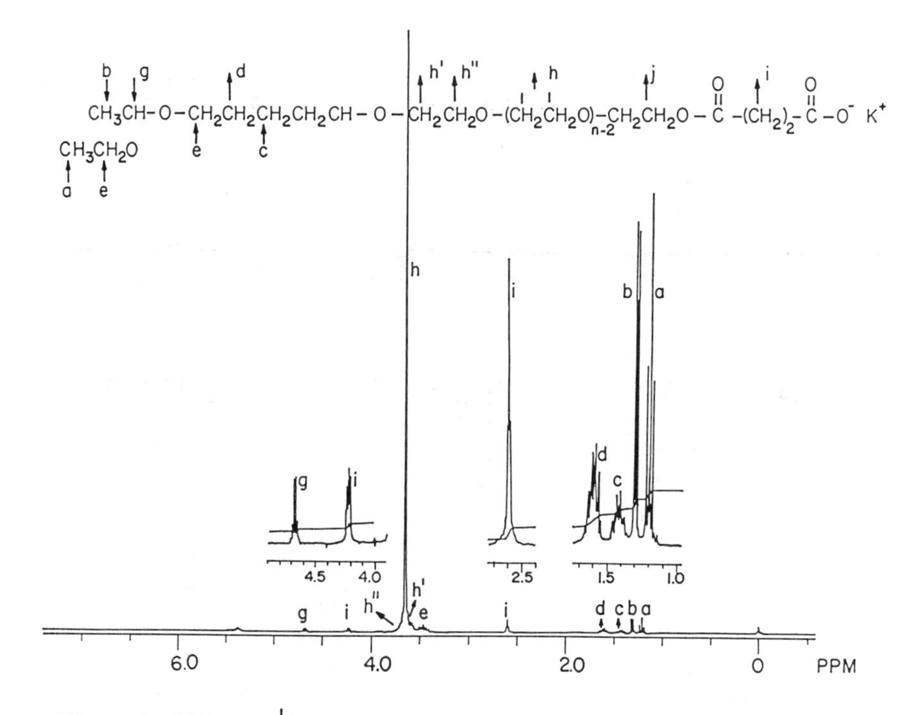

Figure 5. 300 MHz ^1H NMR spectrum of the masked polyether with succinic acid end group, 3.

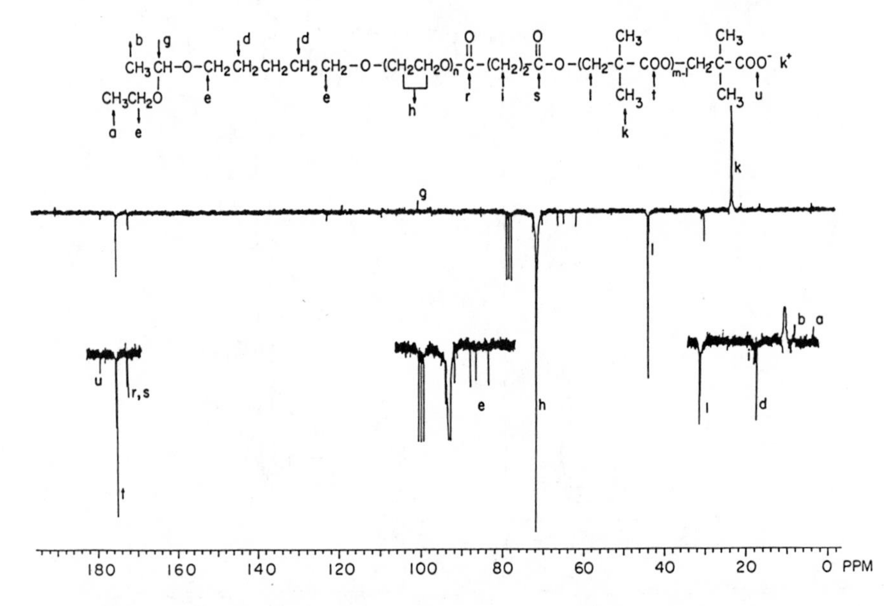

Figure 6. 50 MHz ^{13}C NMR spectrum of the masked poly(oxyethylene)-b-poly(pivalolactone) copolymeric salt, 4, using attached proton test sequence (CH, CH_3, Pos.; CH_2, C, neg.) in $CDCl_3$ at 25°C.

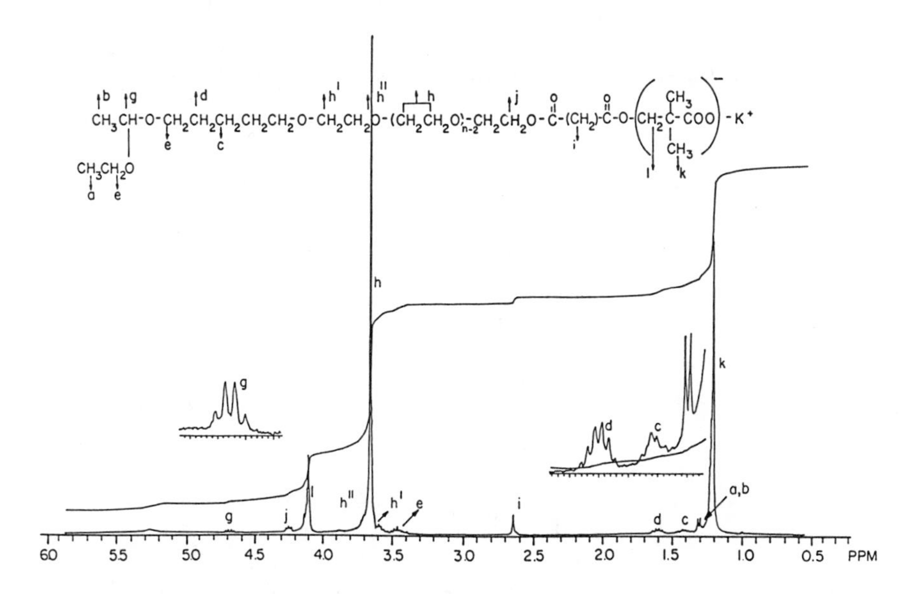

Figure 7. 300 MHz ^{1}H NMR spectrum of the masked poly(oxyethylene)-b-poly(pivalolactone) copolymeric salt, 4.

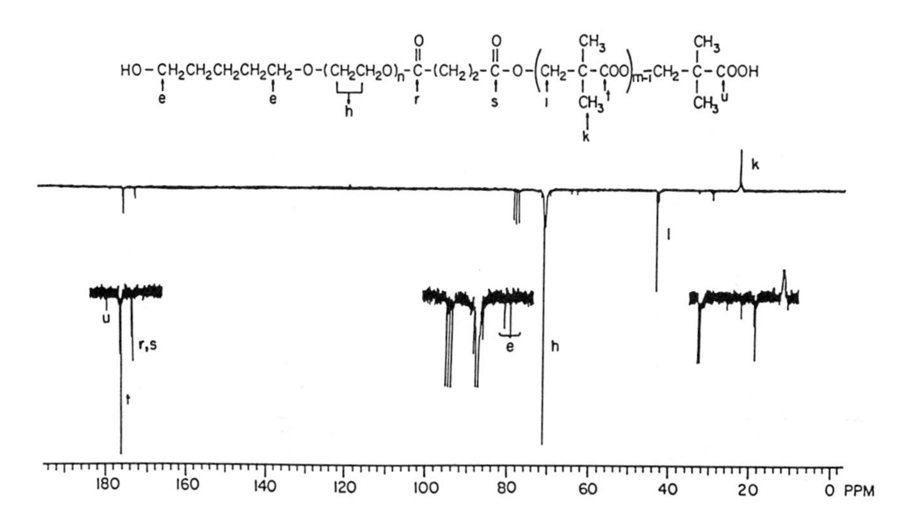

Figure 8. 50 MHz ^{13}C NMR spectrum of the telechelomer, 5, using attached proton test sequence (CH, CH_3, Pos.; CH_2, C, neg.) in $CDCl_3$ at 25°C.

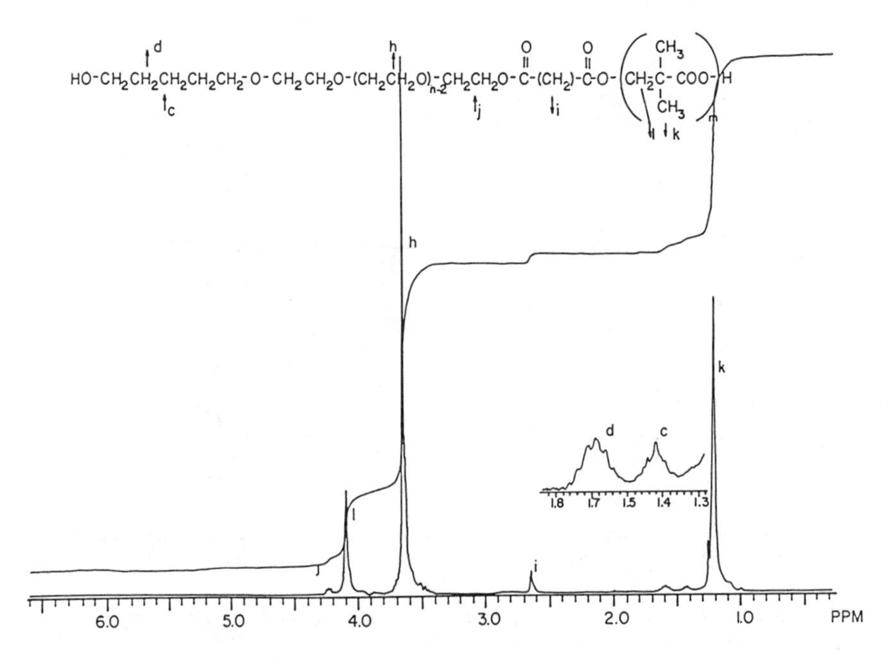

Figure 9. 300 MHz ^{1}H NMR spectrum of the telechelomer.

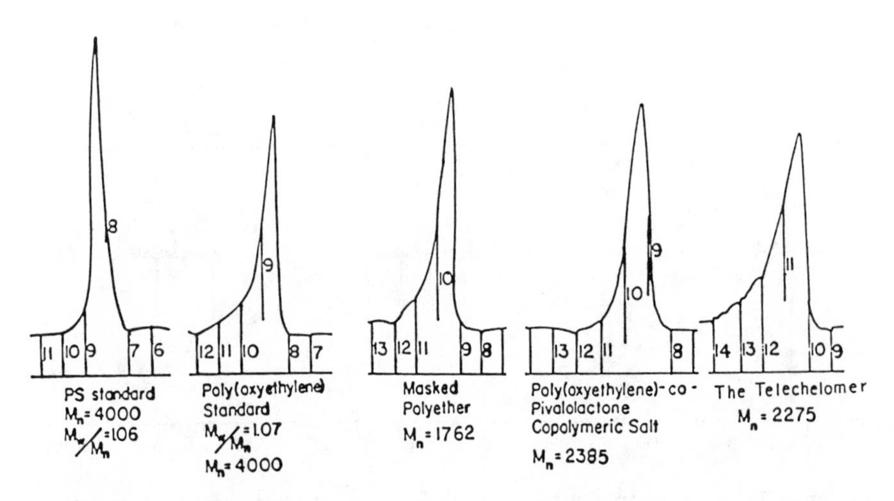

Figure 10. GPC curves (in CH_2Cl_2, at 25°C, flow rate: 1mL/min., using 10^3A μ styragel column).

end carboxylate anion, respectively. ^{1}H NMR spectrum (Figure 7) confirms the incorporation of PVL into 3 while forming 4. The only differences between the proton spectra of 3 and 4 are the new signals "k" and "l" appearing at 1.2 and 4.6 ppm in the spectra of 4; these are assigned to the methylene and methyl protons of the pivalolactone units in 4.

The conversion of the salt, 4, to the telechelomer, 5, was done using aqueous HCl (3M) under homogeneous conditions. Here the polymer dissolved in methylene chloride was shaken with aqueous HCl (3M) for 30 minutes. This process should remove the acetal mask, and this is what is observed in the ^{13}C APT NMR spectra of 5 (Figure 8). The signals "g", "a" and "b" coming from the acetal group are absent here. It is also interesting to note the changes in "e" signals. In 4, three "e" signals were present at 65.1, 63.8 and 60.6 ppm. In structure 5, however, only two signals are seen in this region, one at 63.6 ppm and the other at 62.3 ppm. The signal at 65.1 ppm has disappeared and the signal at 60.6 ppm has moved to 62.3 ppm as a result of hydrolysis. On this basis, the signal at 62.3 ppm is assigned to the end alcohol carbon in the telechelomer. Proton NMR of 5 (Figure 9) confirms the loss of acetal mask; here the quartet at 4.7 ppm and the doublet at 1.3 ppm corresponding to the "g" and "b" protons have disappeared. The loss of "a" protons could not be seen due to the large "k" signal at 1.2 ppm.

The possibility exists that the acid used in hydrolysis could catalyze the hydrolysis of the succinic ester group in the middle of the telechelomer itself. Even though NMR (^{1}H and ^{13}C) cannot easily eliminate this possibility, we have evidence that such a hydrolysis did not take place. For instance, hydrolysis should result in the formation of poly(pivalolactone) which is insoluble both in methylene chloride and water, but no insolubles were evident. Also, the Gel Permeation Chromatograms do not show impurities in the product.

Gel Permeation Chromatograms of the telechelomer and the intermediates are shown in Figure 10. The number average molecular weights (M_n) were determined by ^{1}H NMR. Sufficiently large relaxation times were used in obtaining spectra so that integrations are reasonably accurate. The molecular weight distributions of the polystyrene and poly(oxyethylene) standards are 1.06 and 1.07 respectively. Some absorptive effective effects are seen in the polymers including the polyoxyethylene standard; but they show narrow molecular weight distributions. The telechelomer has a carboxylic acid end group in addition to the alcohol end group and therefore the absorptive effects are more evident. Based on the atomic ratios obtained from NMR, percent atomic ratios were calculated. These values agree with those obtained from elemental analysis.

Our future work includes the synthesis of highly regular segmented copolymers via step polymerization of the telechelomer and investigating their structure-property relationships.

Acknowledgments

We would like to thank Army Research Office for its financial support. We also thank Dr. T. E. Hogen-Esch for helpful suggestions and Dr. H. E. Hall for sending us a pivalolactone sample.

Literature Cited

(1) Cooper, S. L.; Miller, J. A.; Lin, S. E.; Hwang, K. K. S.; Wu, K. S.; Gibson, P. E. Macromolecules, 1985, 18, 32.

(2) McGrath, J. E.; Sheridan, M. M.; Hoover, J. M.; Ward, T. C. Polymer Preprints, 1985, 26(1), 186.

(3) Franta, E.; Reibel, L. Polymer Preprints, 1985, 26(1), 55.

(4) Ogata, S.; Kakimoto, M.; Imai, Y. Macromolecules, 1985, 18, 851.

(5) Wolfe, J., Jr. Polymer Preprints, 1978, 19(1), 5.

(6) Wolfe, J., Jr. Rubber Chem. Technol. 1977, 50(1), 230.

(7) Wegner, G.; Fujii, T.; Meyer, W. H.; Lieser, G. Angew. Makromol. Chem., 1978, 74, 295.

(8) Droescher, M.; Bandara, U.; Schmidt, F. Macromol. Chem. Phys. Suppl., 1984, 7, 107.

(9) Droescher, M.; Schmidt, F. Makromol. Chem., 1983, 184, 2669.

(10) Droescher, M.; Hasslin, H. Makromol. Chem., 1980, 181, 301.

(11) Droescher, M.; Bill, R.; Wegner, G. Makromol. Chem., 1981, 182, 1033.

(12) Inoue, S.; Yasuda, T.; Aida, T. Macromolecules, 1984, 17, 2217.

(13) Wagener, K. B.; Wanigatunga, S. Polymer Preprints, 1986, 27(1), 105.

(14) Yamashita, Y.; Hane, T. J. Polym. Sci. Chem. Ed., 1973, 11, 425-434.

(15) Lenz, R.; Bigdelli, E. Macromolecules, 1978, 11, 493.

(16) Hall, H. Macromolecules, 1969, 2, 488.

(17) Wilson, D. R.; Beaman, R. G. J. Polym. Sci., A-1, 1970, 8, 2161.

(18) Young, R. N.; Quirk, R. P.; Fetters, L. J. Adv. Polym. Sci., 1984, 56, 70.

(19) Harris, J. M. Macromol. Chem. Phys. Rev., 1985, C25(3), 341-345.

(20) Hashimoto, K.; Sumitomo, H.; Yamanori, H. Polymer J., 1985, 17(5), 682.

RECEIVED August 27, 1987

SURFACE MODIFICATION OF POLYMERS

Introduction to Surface Modification of Polymers

Surface modification of polymers is a major and growing area of research, since the surface properties of a polymer are often critical to the use or application of the polymer. In many instances, the surface properties may even be more important than the bulk properties of the polymer. Examples of this include biocompatibility, adhesion, wettability, and weatherability. As a consequence, surface modification is one of the active research areas in the field of chemical reactions on polymers. Chemical approaches to surface modification include plasma treatment, irradiation, often to initiate graft copolymerization, and a broad array of chemical transformations, including oxidation or reduction, on hydrocarbon or functionalized materials.

Radiation induced surface modification is one of the most thoroughly investigated areas, and can result in both physically and chemically altered surfaces. Both high energy radiation, such as gamma or electron beam radiation, and low energy radiation, such as ultraviolet light, have been used to initiate surface modification. Initially, research was concentrated on the use of high energy sources; more recently there has been an increase in activity using low energy radiation sources. For the purposes of this symposium and book, radiation effects resulting in physical change (e.g. microstructuring) were excluded. Instead, a focus on new chemical entities via chemical reaction at the polymer surface was chosen.

Graft polymerization of monomers on the surface of polymers has been a very active area of radiation induced surface modification. In most instances, radiation is used to generate a free radical at the surface of the polymer in the presence of monomers capable of undergoing free radical polymerization, eg. acrylates, olefins, etc. The free radicals have been generated by approaches which include use of photoinitiators and sensitizers, as well as scission of either saturated or unsaturated carbon-carbon bonds. The monomers have been provided in solution or as pure gases/liquids. In this way, polymers such as polyolefins, polyesters, and polystyrene have been grafted with a variety of addition polymers.

Radiation has also been used to induce surface modification of polymers through the reaction of small molecules with a polymer substrate. The reaction may be addition, e.g. halogen, oxygen, ozone, or elimination, e.g. dehydrohalogenation. In some instances, combinations of chemical reactions occur, leading to complex surface modifications.

The array of non-radiation induced chemical reactions performed on the surface of polymers is extremely large and diverse. The variety of substrates includes polyesters, acrylics, epoxies, polyamides, polyurethanes, polyolefins and halogenated polyolefins, polyamines and polyimines, and polystyrenes. The chemical reactions have included oxidation, reduction, addition, elimination, cyclization, substitution, and condensation. A few examples will serve to demonstrate the numerous application areas for this field of research. Polyurethanes and silicones have been surface modified by grafting of hydrophilic groups, by attachment of alkyl groups, and by addition of pharmacologically active agents to improve blood compatibility of the polymer surface, i.e. to create an antithrombogenic surface. Polyethylene film has been chemically oxidized on the surface to improve adhesion properties. Poly(vinylidene fluoride) has been surface modified to produce perfluoroalkyl carboxylic acids resulting in a polymer with cation exchange properties. Chemical reactions can be used to modify many properties of a polymer surface, e.g. stability, wear, polarity/hydrophilicity, adhesion, and biocompatibility.

Surface modification of polymers by chemical reactions has been an extremely active area of research for many years. The level of activity has been demonstrated by the numerous publications, including review articles and monographs, in the past decade. Many opportunities exist for new, improved chemical transformations to produce desired changes in the surface properties of polymers. The papers which follow provide a good cross-section of recent activities in the field.

Chapter 13

Modification of Polymer Surfaces by Photoinduced Graft Copolymerization

B. Rånby, Z. M. Gao, A. Hult, and P. Y. Zhang

Department of Polymer Technology, The Royal Institute of Technology, S-100 44 Stockholm, Sweden

During the last few years, new developments in polymer photochemistry have made it possible to graft various functional monomers onto surfaces of inert polymers like polyethylene, polypropylene and polyethyleneterephthalate. In the first attempts, initiator and monomer were transferred in vapor phase into a "UV Cure" irradiator containing the polymer sheet to be surface grafted. After a few minutes irradiation with a high pressure mercury lamp at about 50°C, a rather complete cover of grafted acrylic acid, acrylamide and other vinyl monomers could be obtained. In later experiments a continuous grafting method has been developed where a tape or a fiber bundle after suitable pretreatment is grafted by UV irradiation for a few seconds. Homopolymer formed is removed by washing and grafted polymer analyzed by dye absorption, IR reflection and ESCA spectroscopy.

During the last 30 years radiation chemistry has been and still is an active area of polymer research. High energy radiation, e.g. electron beam and gamma radiation, initiates ionization and radical formation in polymers which may result in modification, crosslinking and/or degradation. An early studied and common modification involves ionizing irradiation followed by free radical initiated graft copolymerization of monomers present or subsequently added (1). Radiation induced free radical graft copolymerization has also been accomplished using UV light (2).

Since the pioneering work by Oster et al. in the 1950's (3,4) where monomers were photografted with UV light onto polymers blended with photoinitiators, only few researchers have more recently followed up this work. The interest has been focused on the use of high energy radiation.

In a study in the 1960's by Howard et al. (5) thin polymer films were saturated with monomers and then modified by irradiation with

0097-6156/88/0364-0168$06.00/0

UV light in a second step. In later work reported by Tazuke, Kimura et al. (6-8) successful photograftings on the polymer surface of thin films (polyethylene, polypropylene and polystyrene) were achieved with various monomers. In this work, grafting was induced by UV irradiation through the film which was in contact with a solution containing initiator and monomer. The method is slow and - in addition to surface grafting - large amounts of homopolymer are formed.

In our surface photografting research we have developed two new processes applied to sheets, films and fibers of polyethylene (PE), polypropylene (PP) and poly(ethylene-terephthalate) (PET).

1. UV-induced surface grafting of a polymer sheet using vapor phase transfer of sensitizer and monomer from a volatile solution (9). A very thin layer of grafted polymer is formed on both sides of the sheet.
2. UV-induced surface grafting of a polymer tape or a fiber bundle as substrate in a continuous operation with pretreatment of the substrate (presoaking) in a solution of sensitizer and monomer (10). The grafting takes place in a thin layer of solution on the surface of the moving substrate.

Experimental

1. In the vapor phase experiments, the photograftings are carried out in specially designed photoreactor constructed and built in our laboratory (Figure 1). The reactor is equipped with a 1 kW high pressure mercury UV lamp (HPM-15 from Philips) which can be moved to vary the distance to the substrate. The grafting takes place in an atmosphere of nitrogen in a thermostated chamber closed with a clear quartz window. Sensitizer and monomer evaporates from a solution of a volatile solvent in an open bucket which is shielded from the UV-irradiation with aluminium foil. HPM lamp and reaction chamber are enclosed and shielded in a house of aluminium sheet metal through which air is sucked for cooling.
2. For the continuous grafting process, the experiments are carried out in specially designed photografting device, constructed and built in our laboratory (Figure 2). The sample (film tape or fiber bundle) is pulled from a feed roll (1) through a presoaking solution (2) into the reaction chamber up and down on the running rolls (5) and out to a take-off roll (19) which is rotated by a driving motor (20) with gear to regulate the take-off speed. The reaction chamber (22) is closed, has nitrogen atmosphere through a gas inlet (6), is heated by a thermostated electric heater (8,9) and cooled by a water spiral (11). A thermocouple (3) shielded from UV-radiation measures the temperature (21) inside the reaction chamber which is separated from the lamp box (23) by a quartz window (7) with a shutter (16). The UV lamp (12) with reflection (13) is supported by a movable stand (17) and air-cooled by an air inlet (14) and a forced air outlet (15). During the experiments nitrogen is bubbled through a wash bottle (18) containing a solution of sensitizer and monomer, into the reaction chamber (4). The irradiation time is varied by using different speed of the driving motor (20).

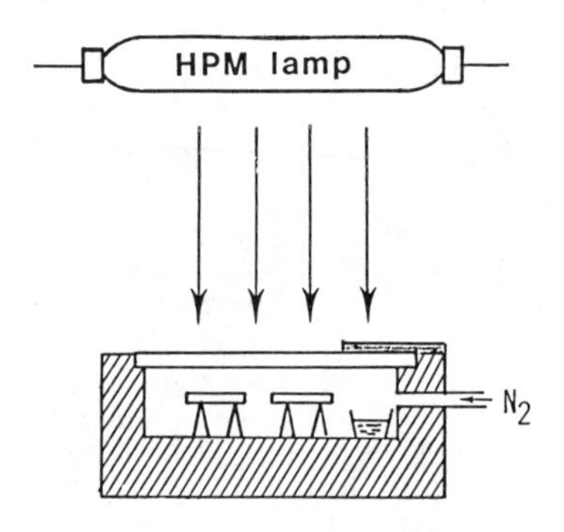

Figure 1. Surface grafting by vapor phase process.

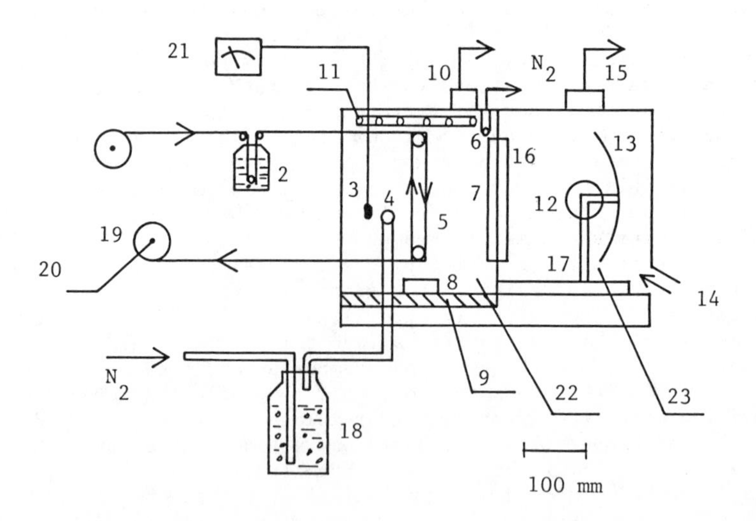

Figure 2. Surface grafting by continuous process.

Materials

Acrylic acid (AA) and methacrylic acid (MAA) (purchased from Merck) are freed from inhibitor on a neutral aluminium oxid column and distilled. Acrylamide (AM) from Kebo, Stockholm, is recrystallized once from chloroform solution before use. Other monomers of analytical grade were purchased from Merck and used as received: crotonic acid (CA), tiglic acid (TA), 3-methyl crotonic acid (3-MCA), and α-methyl cinnamic acid (α-MCia) (Table 1). Benzophenone (analytical grade, Kebo) and acetone (spectroscope grade, Merck) were used as supplied.

Polyethylene (PE) was a commercial LD type (without additives) with a density of 0.92 and polypropylene (PP) was also a commercial material with a density of 0.91. The polyolefin samples were melt pressed to 1 mm thick sheets (plates) which were wiped clean with acetone and used directly for the grafting experiments with the vapor-phase process.

For the continuous process, strips of commercial film and split-film fibers were used as described in the experiments reported.

Analysis

The results of the surface grafting reactions are analyzed with several different methods.

1. ESCA spectra to measure the relative intensity of Cls, Nls and Ols bands at about 285, 400 and 532 eV binding energy, respectively. The instrument used is a Leybold-Heraeus spectrometer.
2. Multiple internal reflection IR spectra (ATR-IR) to measure e.g. the relative absorption of COOH, NH_2 and CH_2 bands at 1730, 1667 and 1475 cm^{-1}, respectively. The instrument is a Perkin-Elmer spectrometer (type 580 B).
3. Light absorption of a dye, crystal violet, adsorbed from solution on the surface of film or fiber, before and after grafting, and measured by absorption of visible light at about 600 nm.

Results and Discussion

Photoinduced free radical graft copolymerization onto a polymer surface can be accomplished by several different techniques. The simplest method is to expose the polymer surface (P-RH) to UV light in the presence of a vinyl monomer (M). Alkyl radicals formed, e.g. due to main chain scission or other reactions at the polymer surface can then initiate graft polymerization by addition of monomer (Scheme 1). Homopolymer is also initiated (HRM·).

$$\text{P-RH} \xrightarrow[h\nu]{UV} \begin{cases} \text{P}\cdot + \cdot\text{RH} \\ \text{P-R}\cdot + \text{H}\cdot \end{cases} \xrightarrow{+\ M} \begin{cases} \text{PM}\cdot + \text{HRM}\cdot \\ \text{P-RM}\cdot \end{cases} \qquad \text{Scheme 1}$$

The limitation of this method is the relatively low quantum yield of radical formation by chain scission for most polymers. It will take high doses at short wavelengths (< 300 nm) to produce enough initiating radicals for a complete surface coverage of grafted polymer chains. Under such conditions there will also be a considerable amount of degradation of the bulk polymer. One way to get around this problem is to add a photosensitator to the polymer.

The advantage of using a sensitizer is twofold:

1. A lower exposure dose is required for polymer radical formation.
2. Wavelengths that are non-destructive to the bulk polymer can be used. The latter is an advantage only if the sensitizer is located at the polymer surface. If it is evenly distributed in the polymer matrix it will cause degradation of the bulk polymer. There are several ways to add the sensitizer to the polymer surface. In the present work we have used:
   a. a vapor phase technique for surface deposition and
   b. a very thin surface layer of sensitizer and monomer solution by presoaking and removal of excess solution.

The sensitizer in our experiments is benzophenone (BP) which reacts as shown in Scheme 2. UV light of 300 to 400 nm is absorbed and excites the aromatic ketone group to a singlet state which by intersystem crossing (ISC) reverts to a triplet state, abstracts a

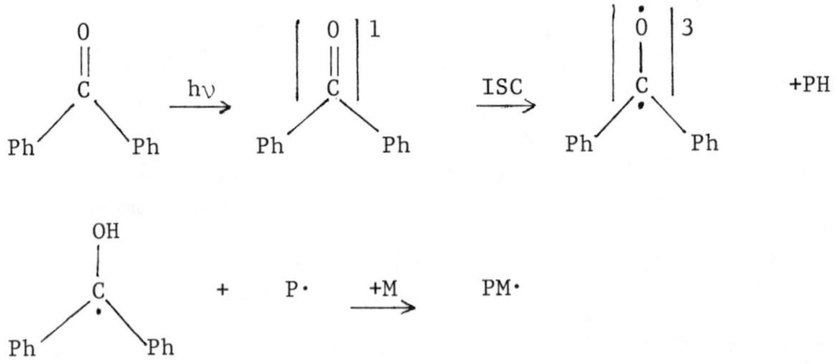

$$Ph_2C{=}O \xrightarrow{h\nu} [Ph_2C{=}O]^1 \xrightarrow{ISC} [Ph_2\dot{C}{-}\dot{O}]^3 \quad +PH$$

$$Ph_2\dot{C}{-}OH \quad + \quad P\cdot \xrightarrow{+M} PM\cdot$$

Scheme 2

hydrogen from the polymer and forms a polymer radical. The resulting ketyl radical is unreactive and has no UV absorption at 300 to 400 nm. The polymer radicals may combine (crosslink) or add monomer to form grafted chains.

Photografting of Acrylic Monomers onto Polyolefin Surfaces

The reason for surface modification of a polymer is in most cases a wish to improve wettability and adhesion towards more hydrophilic materials. Both adhesion and wettability are interfacial phenomena. It is therefore unnecessary to alter the chemical properties of the bulk material, e.g. by using vapor phase deposition of sensitizer and monomer. This can be seen by comparison of ATR-IR and ESCA spectra of grafted PP surfaces (Figure 3).

In the continuous process the solution of sensitizer and monomer forms a thin surface layer on the polymer in which the photoinitiated reaction takes place without affecting the bulk phase of the polymer.

In the vapor phase process, the grafting reaction starts at the

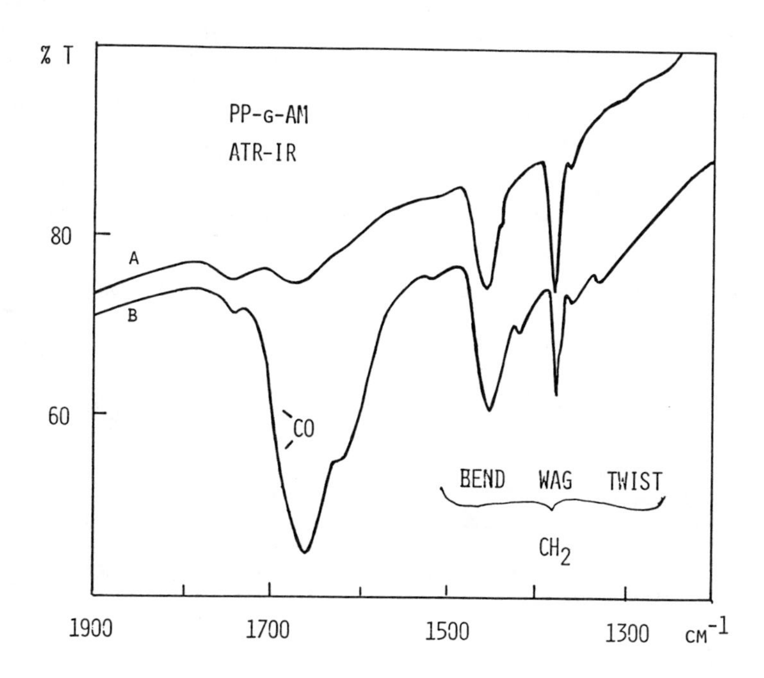

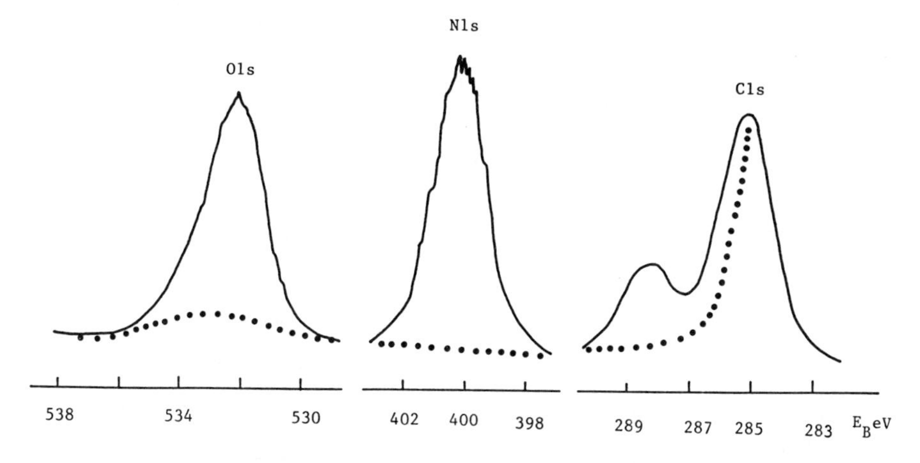

Figure 3. Reflection infrared spectra (ATR-IR) of a polypropylene surface before (A) and after (B) grafting with acrylamide (AM) by the vapor phase process (above). ESCA spectra of the same surface before (dotted lines) and after (full lines) the surface grafting (below).

polymer vapor interphase and grows out from this. After some time when the polymer surface is covered with grafted chains, there will be a competition between grafting and homopolymerization and some freshly deposited monomer will be grafted on already formed graft-co-polymer. Further, since deposition and irradiation takes place at the same time, there is a high probability that the grafted layer becomes crosslinked.

The growth of the grafted layer can be studied by following the dye adsorption on the grafted surface. Crystal violet is strongly adsorbed by the grafted AM layer. Figure 4 shows a comparison of the growth kinetics of grafting on the PE and PP surfaces. For both polymers a non-linear growth kinetics is observed. This is probably caused by competition between grafting and homopolymerization. After the initial graft-co-polymerization on the polymer surface, subsequently adsorbed monomer can either be grafted or homopolymerized. All grafted samples are extracted in water for 24h before they are analyzed and most homopolymer is then washed off. It is concluded from these experiments that graft-co-polymerization occurs predominantly during the early stages of the reaction. In the later stages (here after about 20 min.) we have mainly homopolymerization.

The grafting reaction seems to be more efficient on the PE surface than on the PP (Figure 4). This is consistent with observations made by Tazuke and coworkers (11). One explanation is that the primary formed radicals on PP are more likely to undergo rearrangement to chain end radicals than radicals on PE. For the same reason it is more difficult to crosslink PP than PE. The chain end radicals can easily diffuse into the polymer matrix and there be out of reach for deposited monomer.

It is well known from degradation studies (12) that photodegradation of semicrystalline polyolefins mainly occurs in the amorphous regions. To study if this was true also here a comparison was made between a PE plate and a highly oriented PE fiber (Figure 5). The ESCA spectra of the two surfaces show that the highly crystalline PE fiber exhibits a significantly lower degree of grafting than the PE plate. More experiments have to be done here to understand better the grafting mechanism. These results indicate that the solubility of the sensitizer in the polymer substrate is important in the grafting reactions. BP is more soluble in the amorphous regions of the polyethylene than in the crystalline lamellae.

Grafting of Acrylic Acid Derivatives onto Polyethylene and Polypropylene Surfaces

Various mono-, di- and tri-substituted acrylic monomers (Table 1) were used as grafting monomers. Only mono- and 1,1-disubstituted monomers, e.g. acrylic and methacrylic acid, can undergo homopolymerization. The reluctance of 1,2-disubstituted and higher substituted monomers to form honompolymers by free radical mechanism has been attributed to steric interaction between the monomer and the endgroup radical. This type of steric hinderance would increase the activation energy required for the propagation reaction to such an extent as to favour chain transfer or termination reactions. As a consequence, the higher substituted monomers only form thin grafted layers on the substrate surface. This can be demonstrated with ESCA and dye adsorp-

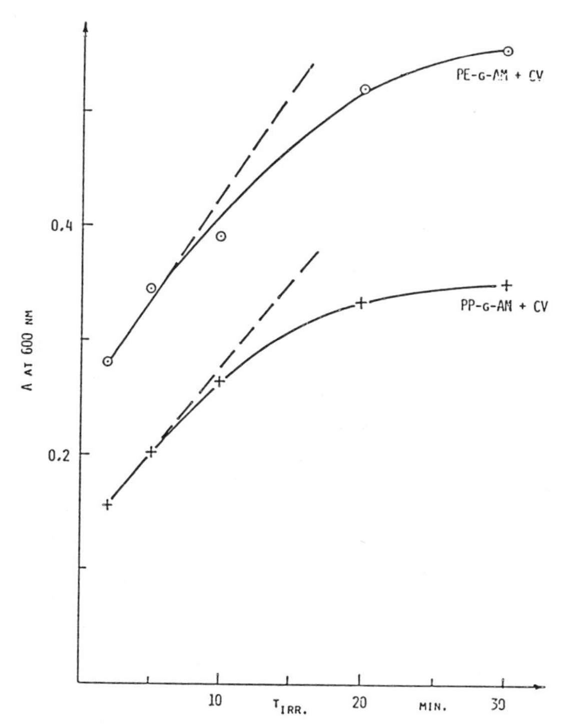

Figure 4. Kinetics of surface grafting of polyethylene (PE) and polypropylene (PP) films with acrylamide (AM) by the vapor phase method measured as light absorption at 600 nm after dipping in an aqueous solution of crystal violet (CV).

Table 1. Monomers Used for Surface Graft Copolymerization onto Polyolefin Surfaces

Monomer	Structure	Bp °C	Mp °C
Acrylamide (AM)	$CH_2 = CH - CONH_2$	125/25	84-86
Acrylic Acid (AA)	$CH_2 = CH - COOH$	139	13
Methacrylic Acid (MAA)	$CH_2 = C(CH_3) - COOH$	163	16
Crotonic Acid (CA)	$(CH_3)CH = CH - COOH$	180	71-73
Tiglic Acid (TA)	$(CH_3)CH = C(CH_3) - COOH$	198	61-64
3-Methyl Crotonic Acid (3-MCA)	$(CH_3)_2 C = CH - COOH$	~200	~40
α-Methyl Cinnamic Acid (α-MCiA)	$PhCH = C(CH_3) - COOH$	228	74

The methyl groups of TA are trans, the phenyl and carboxyl groups of CA are trans and the phenyl and methyl groups of α-MCiA are trans.

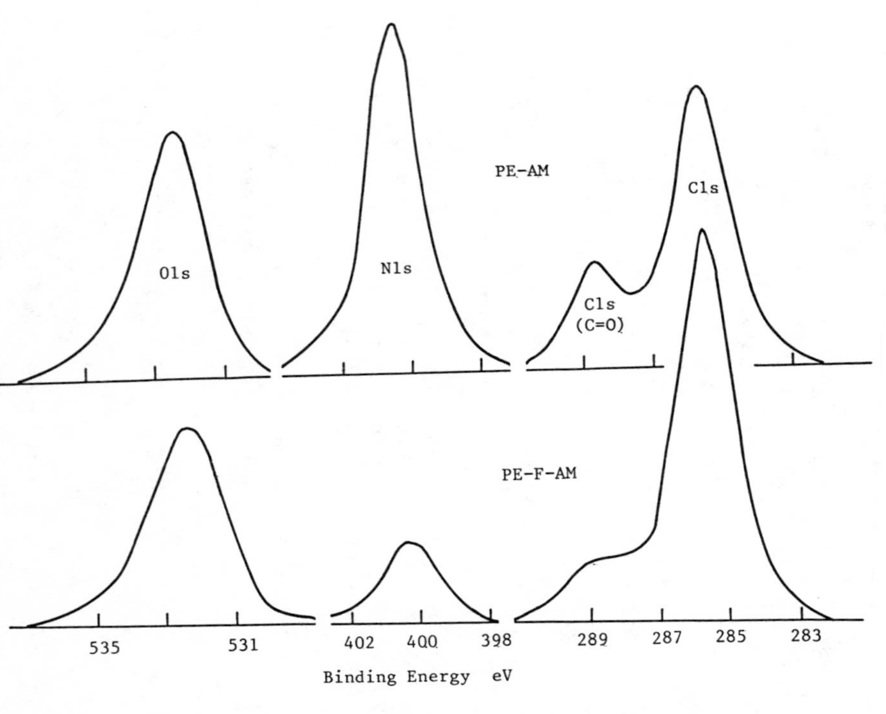

Figure 5. ESCA spectra of polyethylene film (PE) and polyethylene fiber (split film PE-F) after grafting for 2 min. with the vapor phase method.

tion for PE (Figure 6) and PP (Figure 7). The polyolefin plates of 1 mm thickness are grafted on both sides when irradiated in the vapor phase process, e.g. PP grafted with methacrylic acid (Figure 8). The reflection IR analysis is penetrating deeper (about 1 μm) than the ESCA analysis (5-10 nm).

Surface Grafting of Acrylamide onto a Polyester Sheet

The surface grafting of a commercial sheet of poly(ethylene terephtalate) was analyzed by ESCA spectra and dye adsorption. The sheets were washed with acetone to remove impurities and then grafted by irradiation for 2 and 20 min. (Figure 9). The blank sample showed no N1s peak in the ESCA spectrum. After grafting the N1s peak increased with increasing grafting time.

The more detailed ESCA spectra show resolved peaks (Figure 10). The blank sample (A) has a C1s peak containing three components: a main peak at 285 eV referred to carbon bonded to other carbons and hydrogens only, and two minor peaks due to C-O and O=C-O at 287 and 289 eV, respectively. The O1s peak has two components of equal intensity referred to O(C=O) and O(C-O-C)groups at 533 and 535 eV, respectively. The peak intensities correspond very well to the chemical structure of the polyester $+ OC-C_6H_4-CO-O-(CH_2)_2-O +_n$.

After grafting with AM the N1s peak appears, the C1s peak has only one side peak (C=O) at about 286 eV and the O1s peak is dominated by O(C=O) which indicates an almost complete cover of grafted poly(acrylamide) chains $+ CH_2-CH(CO-NH_2) +_n$. The grafting kinetics (Figure 11) is similar to that ot PE and PP substrates (comp. Figure 4).

Some Results of the Continuous Method

The continuous process has been applied to thin strips of PP film (5 mm wide), PP fiber, PET fiber (yarn), and split film of high density PE. On all substrates efficient grafting of AM and AA could be obtained after short irradiation times (5-10 sec.). Only a few results can be presented in this review. The PE split film blank, presoaked (no irradiation) and irradiated for 6 and 8 sec. show extensive grafting of AM and AA (Figure 12 and 13). The kinetics of the grafting process shows a very rapid initial reaction to a sharp maximum for AM (Figure 14) and a broad maximum for AA (Figure 15). These results are interpreted as due to evaporation of the solvent (acetone) in the presoaking solution of initiator and monomer. Grafting in a dry layer of initiator and monomer is apparently slow. At the long irradiation times for AM, the acetone may have evaporated before the samples enters the reaction chamber. AA is a liquid and can react also without solvent present, i.e. during long irradiation times. Separate measurements of ESCA spectra before and after washing the grafted samples of PE and PP have shown that the amount of homopolymer in this process usually is not more than 30-40% of the total amount of polymer formed, i.e. 60-70% of the polymer is grafted. The adsorption of dye increases rapidly with increasing amounts of grafted AM on PP film: the AM concentration in the presoaking solution is 0.2, 0.5, 0.8 and 1.3 M in these experiments and the UV irradiation time 8 sec. in all experiments (Figure 16). The continuous grafting method is a very efficient and rapid process.

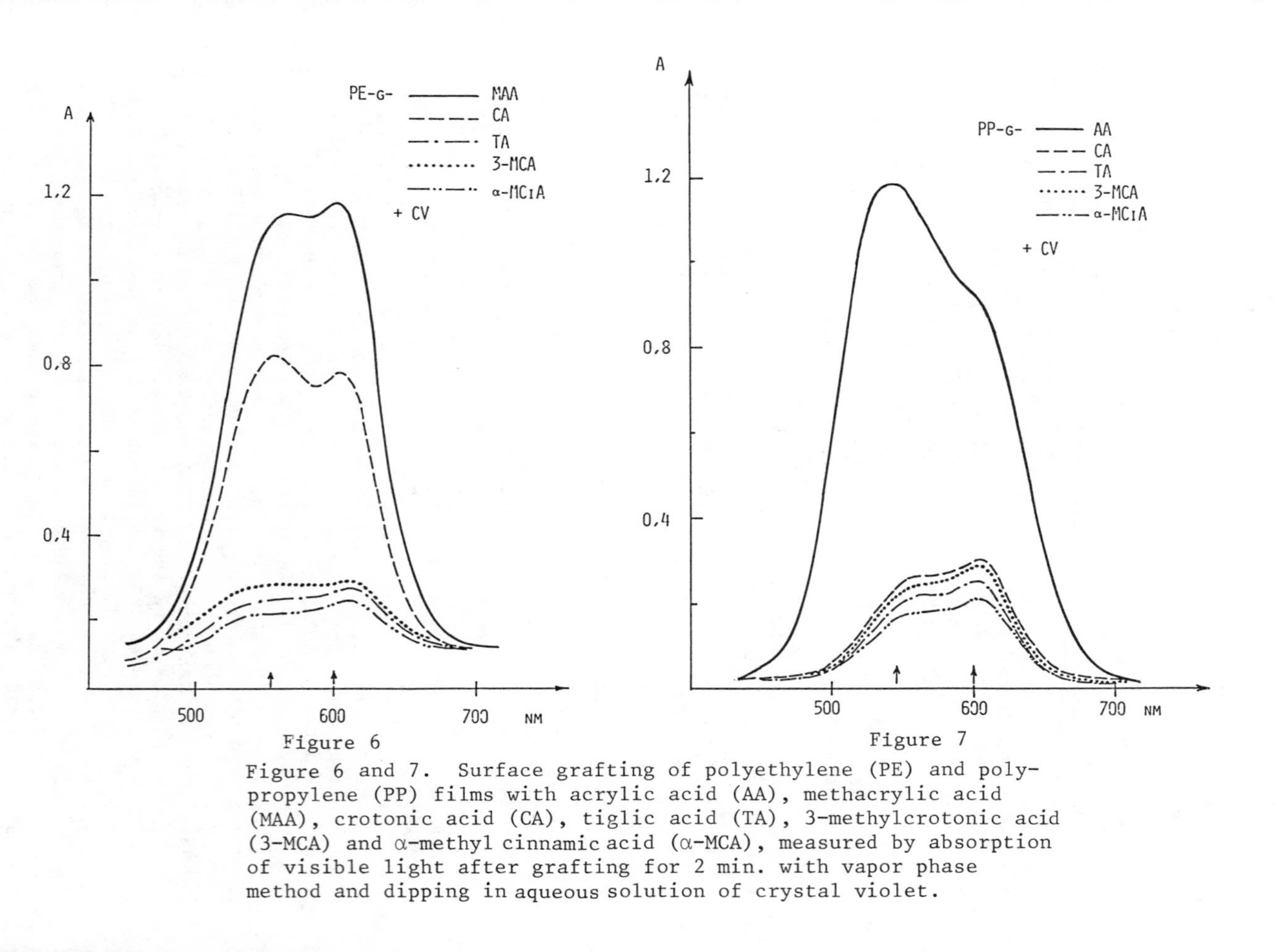

Figure 6

Figure 7

Figure 6 and 7. Surface grafting of polyethylene (PE) and polypropylene (PP) films with acrylic acid (AA), methacrylic acid (MAA), crotonic acid (CA), tiglic acid (TA), 3-methylcrotonic acid (3-MCA) and α-methyl cinnamic acid (α-MCA), measured by absorption of visible light after grafting for 2 min. with vapor phase method and dipping in aqueous solution of crystal violet.

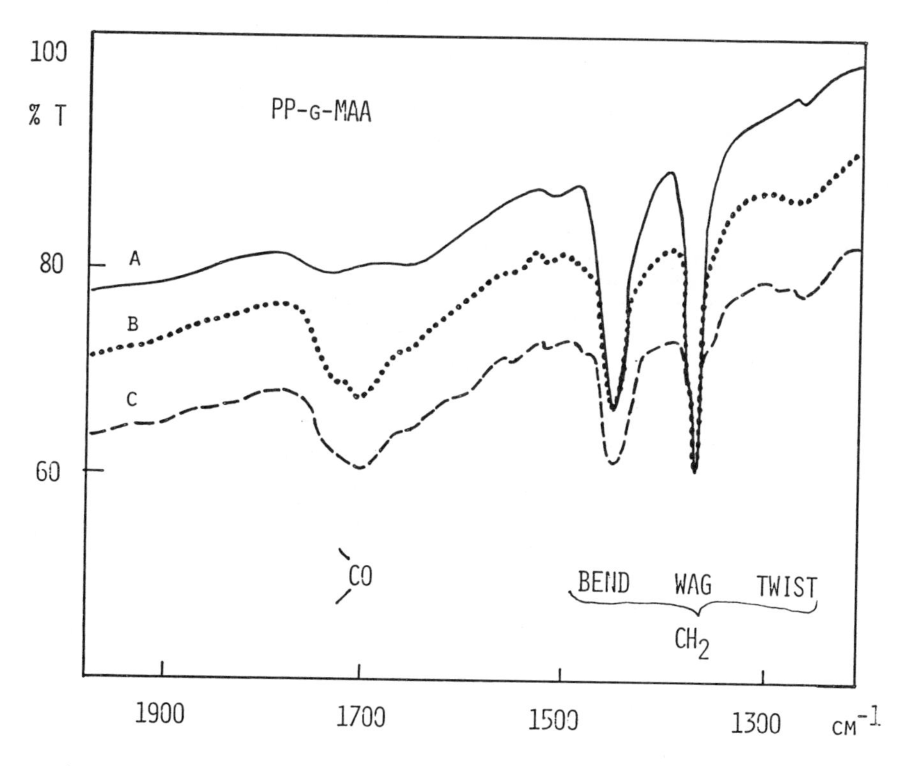

Figure 8. Reflection infrared spectra (ATR-IR) of polypropylene film surface before (A) and after grafting with methacrylic acid (MAA) measured at the top (B) and the bottom (C) surface.

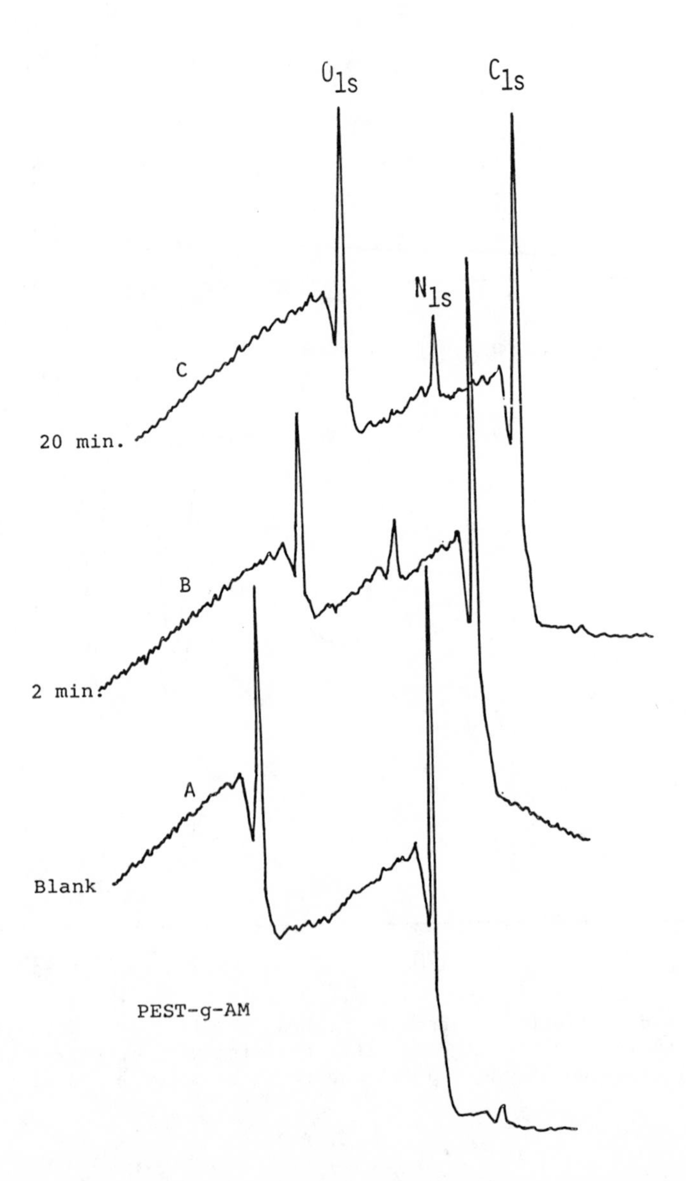

Figure 9. Wide scan ESCA spectra of polyester film surface (PET) before grafting (A) and after grafting for 2 and 20 min. (B and C) with acrylamide (AM) using the vapor phase method.

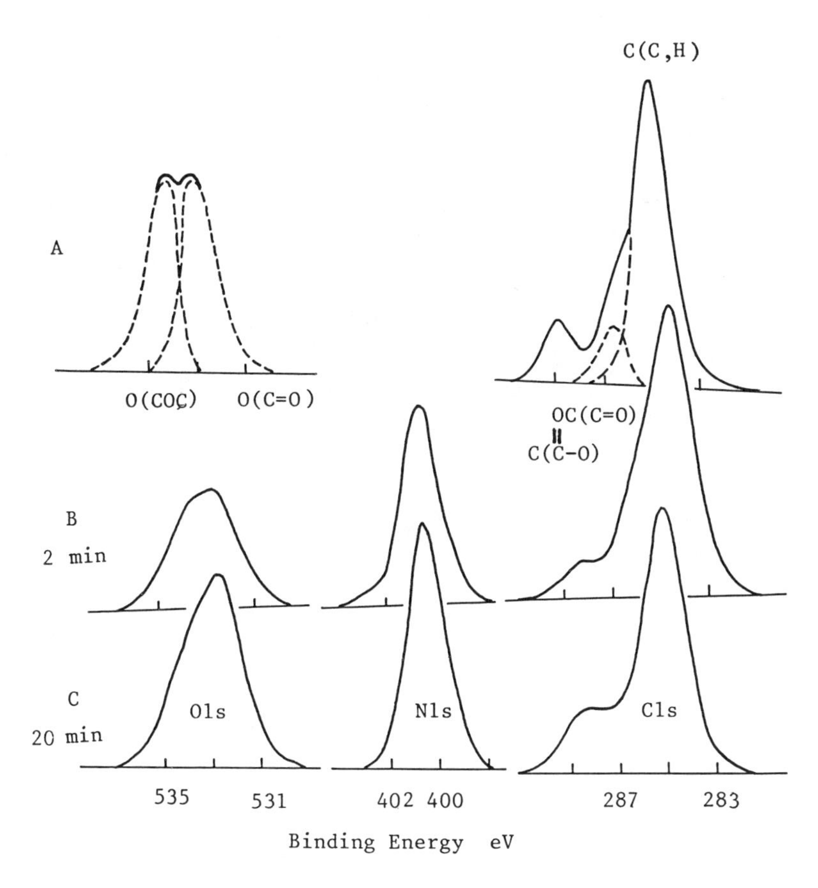

Figure 10. Detailed ESCA spectra of the polyester film surface as described in Figure 9.

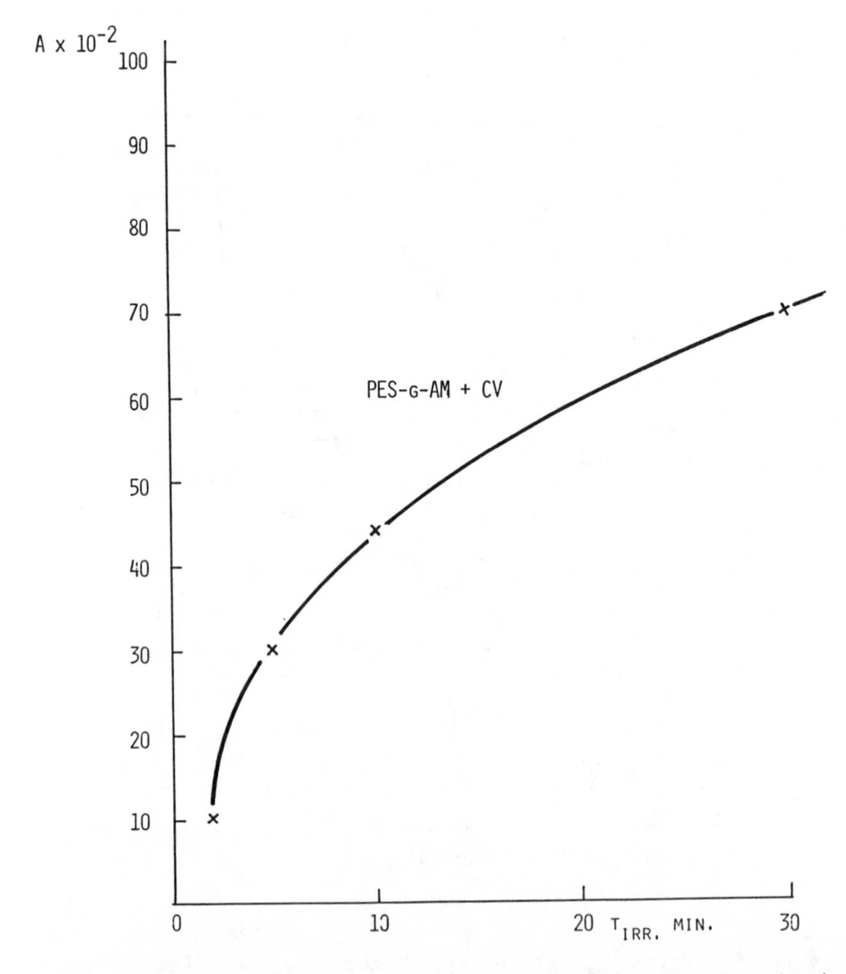

Figure 11. Kinetics of surface grafting of polyester film (PET) with acrylamide (AM) by the vapor phase method, measured as light absorption at 600 nm after dipping in an aqueous solution of crystal violet (CV).

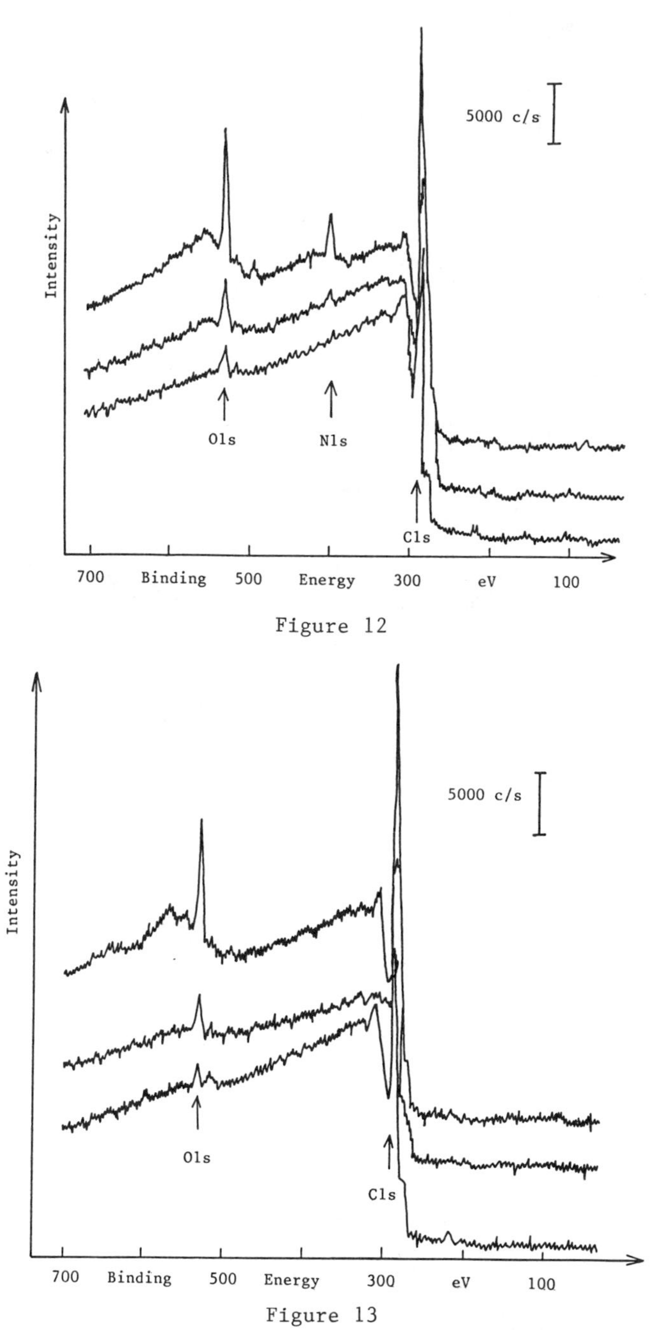

Figure 12

Figure 13

Figure 12 and 13. ESCA spectra of polyethylene split film before grafting (A), after presoaking in sensitizer-monomer solution (B) and after grafting with acrylic acid (8 sec., Figure 12) and acrylamide (6 sec., Figure 13) using the continuous method.

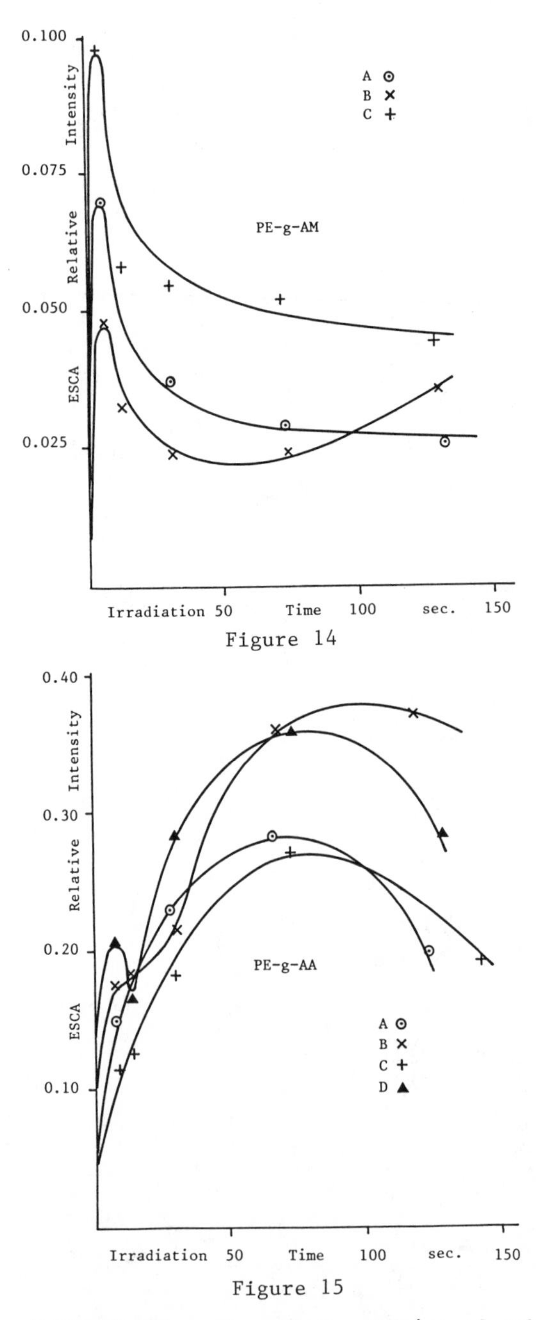

Figure 14 and 15. Kinetics of surface grafting of polyethylene split film, with acrylonitrile (AM, Figure 14) and acrylic acid (AA, Figure 15), using the continuous method and measured as relative intensity of N1s/C1s (Figure 14) and O1s/C1s peaks (Figure 15) in the ESCA spectra.

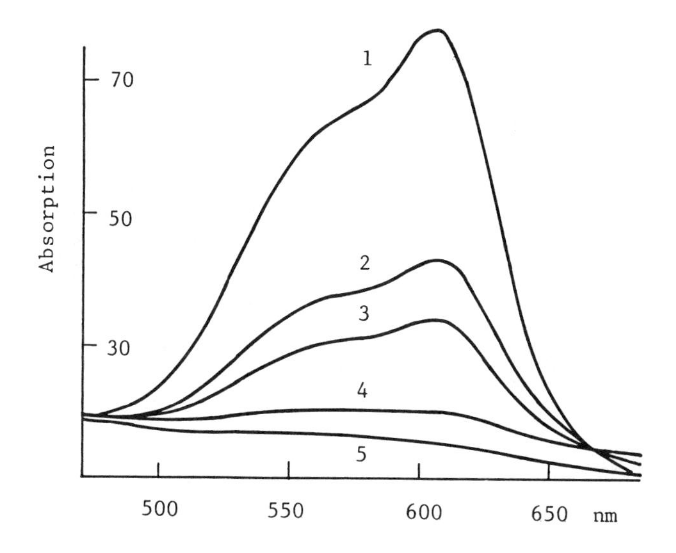

Figure 16. Surface grafting of polypropylene film strips after 10 sec. irradiation measured as light absorption after dipping in aqueous crystal violet solution. The presoaking solutions contain 0.2 M benzophenone (all) and 1.3 M acrylamide (1), 0.8 M (2), 0.5 M (3) and 0.2 M (4). Curve 5 refers to a blank sample (no irradiation).

Conclusions

Two methods for modification of polymer surfaces by photoinitiated graft copolymerization have been developed: a discontinuous method (1) with vapor phase transfer of initiator and monomer and a continuous method (2) with presoaking of a film strip or a fiber bundle in a solution of initiator and monomer. Both methods have been applied to polyelefins and linear polyester.

Method (1) is a slow process which requires several minutes of UV-irradiation for grafting a surface layer of an acrylic monomer onto the polymer substrate.

Method (2) is a fast and efficient process which requires only 5 to 10 sec. UV-irradiation for surface grafting of the polymer substrate. About 2/3 of the polymer formed is grafted (about 1/3 is homopolymer).

For the two methods, the resulting grafting of functional monomers, e.g. acrylic acid and acrylamide, has been measured by multiple reflection IR spectra, ESCA spectra, and dye adsorption from an aqueous solution of crystal violet. The measurements indicate that the inert surfaces of the polymer substrates are modified by a complete surface layer of the grafted monomers.

Acknowledgments

The research work is supported by grants from the National Swedish Board for Technical Development and the visiting scientists Gao Z M and Zhang P Y by fellowships from Academia Sinica and Ministry of Education, P.R. of China, respectively, which is gratefully acknowledged.

Literature Cited

1. Charlesby, A. Atomic Radiation and Polymers, Pergamon Press, New York, N.Y. 1960; Chapter 22.
2. Chapiro, A. Radiation Chemistry of Polymeric Systems, Interscience Publ., New York, N.Y. 1962; Chapters IV-VII.
3. Oster, G.; Shibata, O. J. Polymer Sci. 26, 233 (1957).
4. Oster, G.; Oster, G.K.; Moroson, H. J. Polymer Sci. 34, 671 (1959).
5. Howard, G.J.; Kim, S.R.; Peters, R.H. J. Soc. Dyers Colour, 85, 468 (1969).
6. Tazuke, S.; Kimura, H. J. Polymer Sci., Polymer Lett. Ed. 16 (10), 497 (1978).
7. Tazuke, S.; Kimura, H. Makromol. Chem. 179, 2603 (1978).
8. Tazuke, S.; Matoka, T.; Kimura, H.; Okada, T. ACS Symp. Ser. Nr 121, 217 (1980).
9. Gao Z.M.; Hult, A.; Rånby, B. Photoinitiated Grafting at Polymer Surfaces, paper presented at IBM Research Laboratory, San José, Calif., USA, April 16, 1984.
10. Zhang P.Y.; Rånby, B. Photoinitiated Grafting at Polymer Surfaces, paper presented at symposium on Polymer Surfaces and Interphases, Durham, England, April 17, 1985.
11. Cf. e.g. Tazuke, S. Polymer-Plast.Techol.& Eng.14 (2), 107 (1980).
12. Cf. Rånby, B.; Rabek, J.F. Photodegradation, Photo-oxidation and Photostabilization of Polymers, J. Wiley, London, 1975; p. 122.

RECEIVED September 19, 1987

Chapter 14

Photoozonization of Polypropylene

Oxidative Reactions Caused by Ozone and Atomic Oxygen on Polymer Surfaces

J. F. Rabek [1], J. Lucki [1], B. Rånby[1], Y. Watanabe[2], and B. J. Qu[3]

[1]Department of Polymer Technology, The Royal Institute of Technology, S-100 44 Stockholm, Sweden
[2]Research Institute for Polymers and Textiles, Tsukubo, Japan
[3]University of Science and Technology of China, Hefei, Anhui, China

Polypropylene films were oxidized by treatment with ozone and ozone- UV- light. The main products of photolysis of ozone are atomic oxygen, singlet oxygen and oxygen. All of these oxygen species such as ozone, atomic oxygen and singlet oxygen are very reactive with polymer surface. The changes that resulted in polypropylene surface oxidation were followed by IR(ATR), fluorescence, ESR and ESCA spectroscopy and the contact angle measurements. Polypropylene surface is oxidized more rapidly in the presence of atomic oxygen formed from photolysis of ozone than that of ozone,however, separation of both reactive oxygen species in this method is imposible. The surface oxidation products are mainly carbonyl and/or carboxyl groups, with lower level of hydro-peroxide groups. The mechanisms of oxidation by ozone and atomic oxygen have been proposed.

Ozone formed in atmosphere is a by-product of atmospheric photo-chemical reactions, mainly in photochemical smog incidents and its concentrations may attain as high as 50 ppm (1). Ordinarily, ozone in the lower atmosphere is present only in very low concentrations in the range of less than 1 to 2 or 3 ppm.High concentration of ozone is formed closely to high-voltage instalations and during high--voltage discharges in stormy weather.

The low concentrations of ozone normally present in the atmosphere are sufficient to cause severe oxidation and cracking in polyolefins (2-13) and many other polymers such as polystyrene (6,11,12,14,15), poly(vinyl chloride) (11,12,16) and rubbers (11,12,17-20). Where the ozone concentration is increased by air pollution, higher altitudes or the present of electrical machinery, the rate of degra-dation considerably increases.The prevention of such degradation represents a matter of considerable economic interest since it can greatly improve service life of polymers and plastics. An additional application of this research is to apprise the suitability of poly-mers for upper atmosphere application (Space Shutle flights) where a plentitude of ozone and atomic oxygen prevails.

0097-6156/88/0364-0187$06.00/0

Inspite of extensive and significant studies on the reaction of ozone with polymers, nothing has been published on the reaction of ozone photolysis products with polyolefins. Ozone has UV absorptions at 255 (strong, quantum yield of photolysis Ø =1.0), 305 (weak) and 340 nm (weak). Ozone absorbs also at longer wavelength (510,615,670 and 760 nm) (21,22). The main products of photolysis of ozone are atomic oxygen, singlet oxygen and oxygen (22). All of these three oxygen species, ozone, atomic oxygen and singlet oxygen, are very reactive with most of all polymers (23,24). UV-irradiation accelerates the ozonization of low molecular weight organic compounds (18). In our study we have found that UV-light greatly accelerates the rate of ozone attack at the polypropylene surface. Presented results were undertaken to determine mechanism of the photo-oxidation of polypropylene surface upon UV-irradiation of polymer films in ozone.

Experimental

Isotactic polypropylene (PP) was supplied by Polyscience Inc. Warrington (USA) and purified by a common method. Ozone was generated by Ozon Generator Model 502 (Fischer Labor und Verfahrentechnik, Germany) equipped with ozone concentration measuring device. A mixture of ozone and oxygen at the flow rate 500 ml/min and ozone concentration $4x10^{-3}$ g/l passed through a quartz cell, contained PP films (50-70 μm), which could be in addition irradiated with UV or UV/VIS light from lamps: low pressure mercury lamp, type HPK 125 W (L1) or high pressure mercury lamp, type HPK 15, 1000 W (L2), both Philips, Holland, from the distance 30 cm.

Atomic oxygen was generated by photolysis of ozone which passed through a quartz capillary tube (Ø 2 mm^2 and lehgth 60 mm), using the 253.7 nm line from a low-pressure mercury lamp (L1).

Fluorescence emission spectra were recorded with Perkin-Elmer LS Luminescence Spectrometer, whereas IR (transmission and ATR spectra) with Perkin-Elmer 1710 IR Fourier Transform Spectrometer.

ESR spectra were determined with a Bruker ER-420 ESR Spectrometer using accessories for solid samples, ultraviolet irradiation in a room temperature.

ESCA core-level spectra for C_{1s} and O_{1s} were recorded with a Leybold-Heraeus Spectrometer using $AlK\alpha_{1,2}$ excitation radiation. Typical operating conditions for the X-ray gun were 13 kV and 14 mA and a pressure of $3x10^{-8}$ mbar in the sample chamber.

The contact angle has been measured by a common method.

The hydroperoxide (POOH) concentration was determined iodometrically after decomposition of the ozonides with excess of alcoholic sodium hydroxide.

Results and Discussion

Being an endothermic allotrope of oxygen, ozone may serve as a precursor for reactive oxygen species such as atomic oxygen and singlet oxygen. The absoption of light by ozone consists of three bands: 200-320 nm (Hartley band), 300-360 nm (Huggins band) and 440-850 nm (Chappuis band). The primary photochemical processes differ considerably in each of these bands. The quantum yield of ozone photolysis at 254 nm (L1) is nearly unity (0.9± 0.2). The main photoproducts are atomic oxygen O and singlet oxygen $^1O_2$ ($^1\Delta$) according to the reactions (22):

$$O_3 \xrightarrow{+h\nu} O(^1D) + {}^1O_2(^1\Delta) \qquad (1)$$

$$O(^1D) + O_3 \longrightarrow 2O + O_2 \qquad (2)$$

$$^1O_2(^1\Delta) + O_3 \longrightarrow O + 2O_2 \qquad (3)$$

The quantum yield of ozone decomposition at 334 nm (L2) is 4, indicating that one of the products must be an excited species capable of decomposing O_3 further. The primary process of the O_3 photolysis at 334 nm occurs according to the reactions:

$$O_3 \xrightarrow{+h\nu} O(^3P) + {}^1O_2(\Sigma \text{ or } {}^1\Delta) \qquad (4)$$

$$O(^3P) + O_3 \longrightarrow O_2 + O_2 \qquad (5)$$

$$^1O_2(\Sigma \text{ or } {}^1\Delta) + O_3 \longrightarrow O(^3P) + 2O_2 \qquad (6)$$

From the mechanism the overall quantum yield of O_3 decomposition is 4. The results show that all reactions (1-6) must be more complicated by the fact that the ozone consumption occurs in two stages, a very fast process (less than 50 μsec) followed by a slow process lasting many milliseconds. Upon the exposure of polymer samples to ozone and light, four active oxygen species such as ozone itself, atomic oxygen, singlet oxygen and oxygen may simultaneously react with a polymer. In the case of PP films, singlet oxygen does not react at all with a polymer surface (25-27). However, ESCA measurements (9,28) support reaction of $^1O_2$ (Δ) with polyolefin surfaces.

The PP sample exposed to ozone shows for a rapid formation of polymer peroxy radicals. A singlet unsymmetric line ESR spectrum (Fig.1) is very characteristic for oxygenated polymer radicals (29).

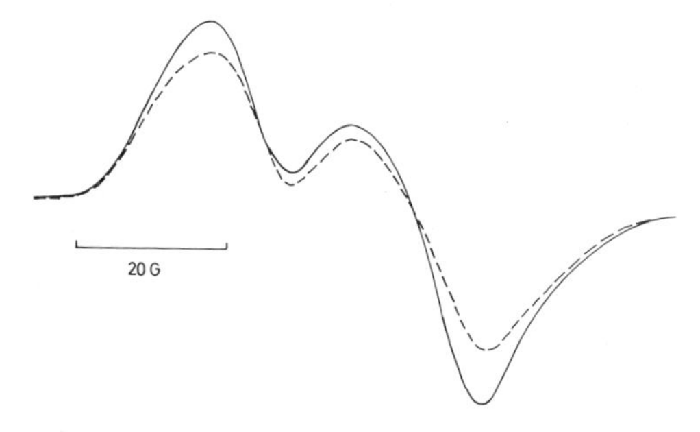

Figure 1. ESR spectra of POO$^\bullet$ radicals formed during exposure of polypropylene films to: (– –) ozone and (——) ozone and UV light (L1).

These ESR spectra are in good agreement with ESR spectra of ozonized PP published previously (30). The rapid formation of peroxy radicals indicates that ozone reacts with PP without induction period. In the initial stage of reaction the hydroperoxide groups (POOH) concentration increases and the rate of POOH formation is linearly dependent on the ozone concentration (Fig.2). After prolonged ozonization the concentration of POOH remains almost constant.

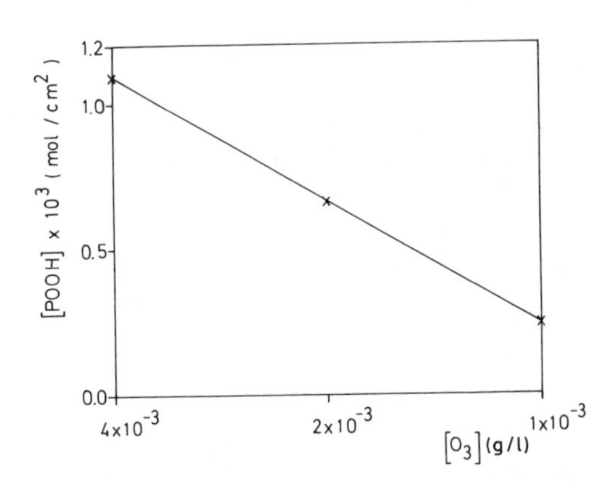

Figure 2. Kinetic curve of POOH groups formation in polypropylene film after 10 hours exposure to ozone.

The formation of POOH during simultaneous exposure of PP films to ozone and light (L1 or L2) can not be obtained kinetically. The experimental results show for rapid formation of hydroperoxide groups which are partially decomposed under UV-irradiation. There is no linear dependence on the ozone concentration.

The analysis of IR spectra of PP samples exposed to ozone or ozone-UV irradiation show for the formation of a band at 1714 cm^{-1} (Fig.3) attributed to carbonyl groups.

The carbonyl index obtained from transmission and ATR spectra of PP films UV-irradiated (L2) in ozone shows that the concentration of carbonyl groups at the surface is 8 times higher than in a bulk. ATR spectra show that oxygen groups are located at the surface to the depth of 0.6 μm. Kinetic curves of carbonyl groups formation in different experimental conditions are shown in Fig.4.

The highest carbonyl groups content is obtained during UV-irradiation of PP films in ozone. The concentration of carbonyl groups is 2 times higher than in the case where samples were exposed only in ozone. It is generally well known fact that the concentration of carbonyl and hydroperoxy groups in strongly oxidized polymer decreases rapidly with distance from the surface (31). The ESCA method allows for the penetration of a very thin layer of the order 1-3 nm. ESCA spectra of PP films treated with ozone and UV-irradiated in ozone are shown in Fig.5. Surface oxidation of PP films is indicated in the formation of the O_{1s} peak and the component in the C_{1s} band due to oxygen-containing groups. These results are similar to that reported previously for surface ozonization of polyethylene films (9).

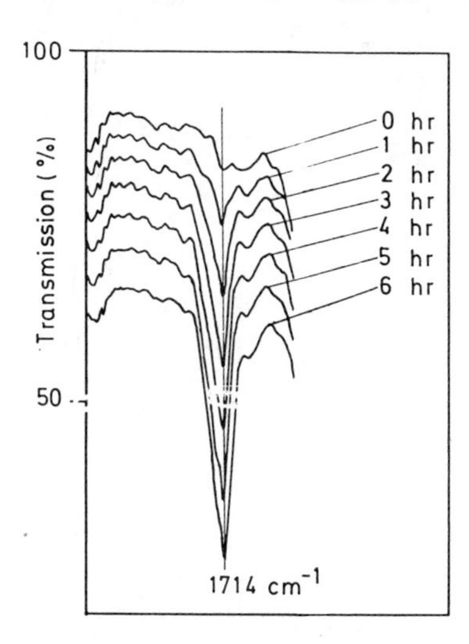

Figure 3. Formation of carbonyl group absorption at 1714 cm^{-1} in polypropylene samples exposed to ozone and UV light (L2).

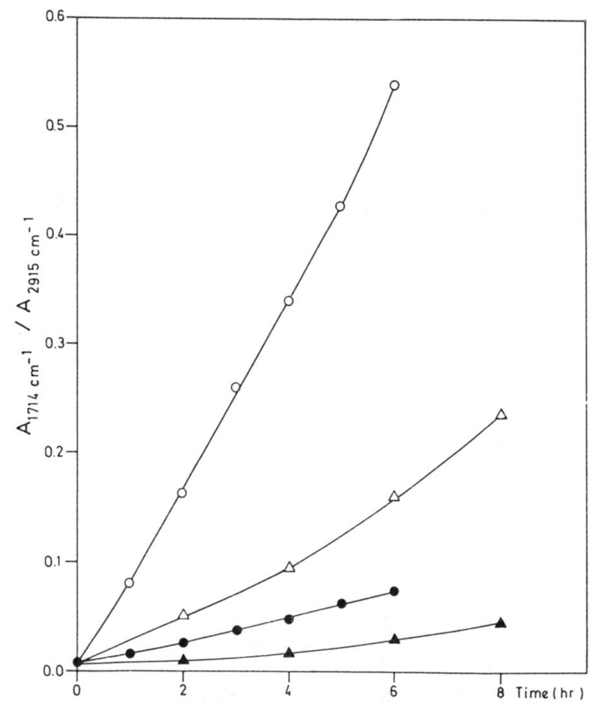

Figure 4. Kinetics of carbonyl group formation at 1714 cm^{-1} in polypropylene samples: (O) and (●) ozone and UV light (L2); (Δ) and (▲) ozone only; (O) and (●) ATR spectra; (Δ) and (▲) transmission spectra.

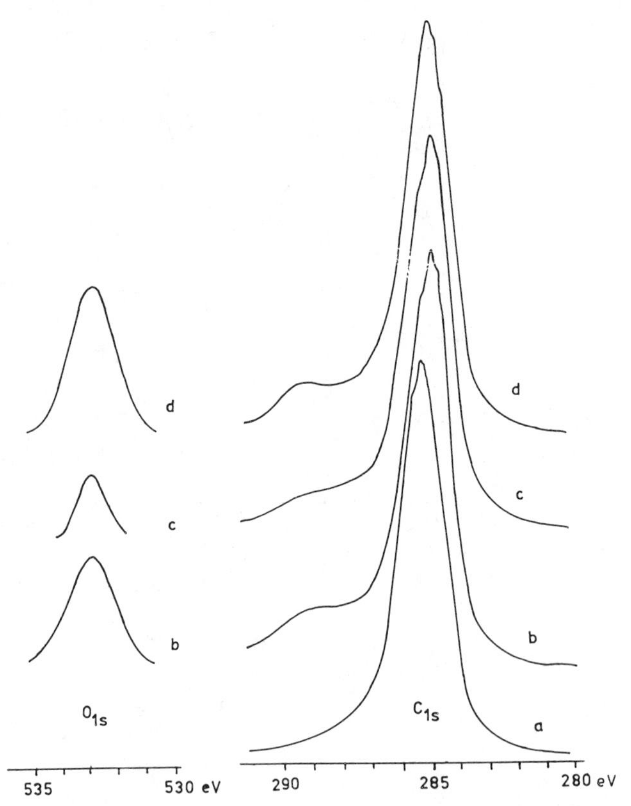

Figure 5. ESCA spectra of polypropylene films: a. pure sample; b. exposed to atomic oxygen from photolysis of ozone (L1); c. exposed to ozone only; d. exposed to ozone and UV light (L1). The O_{1s} band (10x) for samples b,c,d is presented after substraction of the O_{1s} band of the sample a.

The main difference is that the C_{1s}/O_{1s} intensity ratio for samples UV-irradiated in ozone is lower than that observed in samples exposed in ozone only, what indicate higher content of oxygen functional groups. The tails on the high bonding energy sides of the C_{1s} peaks show that oxygen incorporated into polymer surface is in the form of different functional groups. By analogy with surface oxidation of polyethylene films (9), these tails can be attributed to component peaks at binding energy of 286.5, 288.5 eV, corresponding to carbon in C-O, C=O and C(O)OH groups. The ozone oxidized products of polyethylene include ketone and/or ethers, esters and carboxylic acids (5,10,13,18). The shape of C_{1s} envelope in the case of PP irradiated in ozone (Fig.5d) is similar to that observed in other samples exposed to ozone only (Fig.5c) or to atomic oxygen (Fig.5b).

The most obvious change in an oxidation of the PP surface results in an increase in the wettability of the polymer. The PP shows a great decrease in contact angle with water or water/alcohol (3:1) after treatment with ozone or ozone-UV irradiation or atomic oxygen (Table 1).

Table 1. Change in the wettability of polypropylene film after exposure to ozone and ozone-UV irradiation

Experimental conditions	Contact angle (degrees)	
	H_2O	H_2O - C_2H_5OH (3:1)
Unexposed film	85.5	51.8
After ozone treatment, 8 hrs	54.4	26.7
After ozone - UV irradiation, 8 hrs	18.2	11.3

Ozone diffuses readily into amorphous region of the polyethylene (32) and oxidation probably occurs much deeper in the solid sample. Ozone also attacks the cryatalline part of polyethylene but it has a slow initiation stage followed by more rapid oxidation (13). Because ozone does not diffuse into the crystalline regions (13,32), oxidation is restricted to the surface. The resulting oxidized functional groups on the crystalline regions will remain at the surface, whereas those formed in the amorphous region can diffuse into the bulk.

Our results show that not only ozone but atomic oxygen formed of photolysis of ozone are responsible for oxidation of PP film surfaces. Preferential attack occurs in the case of ozone on double bonds in PP, but single bonds also yield oxygen containing groups. It is known that polyolefins always contain small amount of olefinic unsaturations of which vinylidene is more important than vinyl in polypropylene (33). These groups can be responsible for a fluorescence observed in PP (34) (Fig.6). It is generally accepted that these groups are involved in photo-oxidative reactions. Vinylidene decrease is associated with the rapid formation and decay of hydroperoxide groups. The formation of conjugated carbonyl groups in the latter stage of oxidation in place of vinylidene indicates that the vinylidene decay is associated with the oxidation of the allylic groups. Fluorescence observed in PP films disappears continuously under treatment with ozone (Fig.6a) and UV-irradiation in ozone (Fig.6b) or oxygen (Fig.6c). These results show that almost all of olefinic unsaturations are oxidized.

Kinetic curves of fluorescence disappearing (Fig. 7) allow for comparison of the oxidation effect, which is the highest in the case of UV-irradiation in ozone.

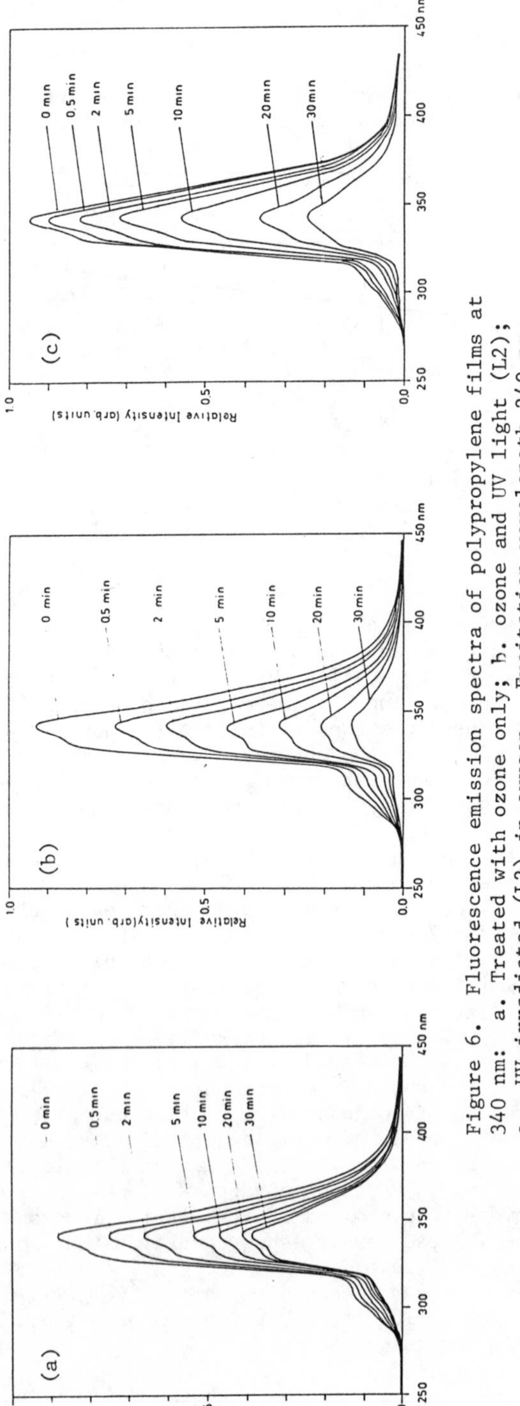

Figure 6. Fluorescence emission spectra of polypropylene films at 340 nm: a. Treated with ozone only; b. ozone and UV light (L2); c. UV-irradiated (L2) in oxygen. Excitation wavelength 240 nm.

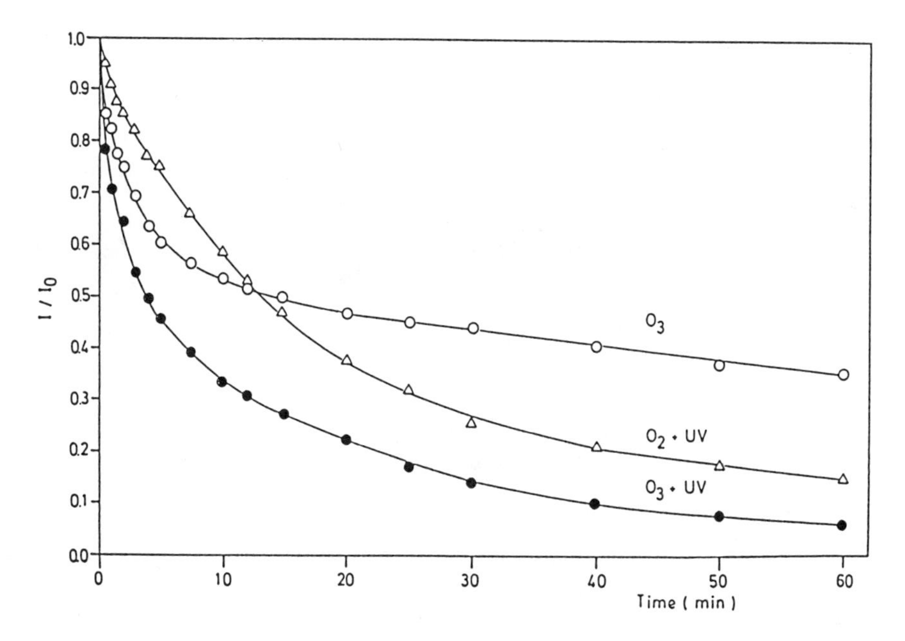

Figure 7. Kinetics of fluorescence disappearing at 340 nm of polypropylene films: (○) treated with ozone only; (●) ozone and UV light (L2); (△) UV irradiated (L2) in oxygen. Excitation wavelwngth 240 nm.

There have been numerous theoretical and experimental efforts to explain the mechanism by which ozone reacts with double bonds of unsaturated substances (11,12,35). Perhaps the more widely accepted reaction is the Criegee mechanism which produces the two groups A and B (as shown below) (36-42):

$$R_2C{=}CR_2 + O_3 \longrightarrow \underset{A}{R_2C(O_3)\text{-}CR_2} \longrightarrow \underset{B}{R_2C(\text{-}O^-\text{-}O\text{-}O^+)\text{-}CR_2} \longrightarrow \underset{C}{R_2\overset{+}{C}\text{-}OO^-} + \underset{D}{R_2C{=}O} \quad (7)$$

Product A, the structure of which Criegee leaves in doubt, is extremely unstable and quickly reverts into products C and D through the intermediate B.

The analysis of published data on reactions of ozone with low molecular hydrocarbons shows that double bonds react with ozone more quickly than saturated bonds (12). Ozone reacts with saturated hydrocarbons in reactions in which hydrogen abstraction is followed by re-hydridization of the carbon atom form sp^3 to sp^2 state (43,44):

$$\geqslant C\text{-}H + O_3 \longrightarrow \left[\geqslant C^{\bullet} + HO^{\bullet} + O_2\right] \longrightarrow \text{products} \quad (8)$$

It has been shown that transition of a backbone carbon from the sp^3 to sp^2 state is promoted by tensile stresses and inhibited by compressive strains (10,44). The acceleration of the process of ozone oxidation of the polymers under load is not associated with the changes in supramolecular structure or segmental mobility of the chain. The probably reason of this effect is a decreasing of the activation energy for hydrogen abstraction (44). The mechanism of initial stages of the reaction of ozone with PP can be represented as:

$$\text{-}CH_2\text{-}CH(CH_3)\text{-} + O_3 \longrightarrow \left[\text{-}\dot{C}H\text{-}CH(CH_3)\text{-} + O_2 + {}^{\bullet}OH\right] \longrightarrow \text{-}CH(OO^{\bullet})\text{-}CH(CH_3)\text{-} + {}^{\bullet}OH \text{ (rapid)} \quad (9)$$

$$\text{-}CH_2\text{-}CH(CH_3)\text{-} + O_3 \longrightarrow \left[\text{-}CH_2\text{-}\dot{C}(CH_3)\text{-} + O_2 + {}^{\bullet}OH\right] \longrightarrow \underset{(POO^{\bullet})}{\text{-}CH_2\text{-}C(OO^{\bullet})(CH_3)\text{-}} + {}^{\bullet}OH \text{ (rapid)} \quad (10)$$

The polymer peroxy radicals (POO•) are detected by ESR spectroscopy (Fig.1). This is particularly interesting since neither of the proposed products of ozone chain scission described previously in the Criegee mechanism are susceotible to detection by ESR. It can be, therefore, concluded that subsequent or additional reactions immediately occur after chain scission or the decomposition of the zwitterion through proton migration take place. The zwitterions (reaction 7) can be very reactive and react with themselves or other part of the chain (42). Another possible reaction which could be considered is that the zwitterion may strip a hydrogen atom from another chain. If the free radicals are formed by a direct hydrogen abstraction or by secondary reactions of these types, ESR does not provide a direct measure of bond rupture in PP.Formation of oxidized groups such as C=O, OOH or COOH is a result of secondary reactions in which formation of polymer hydroperoxy radical (POO•) seems to be essential:

$$\mathrm{POO^{\bullet}} \rightarrow \mathrm{-C(=O)-CH_2-} \quad \text{(slow)} \quad (11)$$

$$\mathrm{POO^{\bullet}} \rightarrow \mathrm{-C(=O)-OH} \quad \text{(slow)} \quad (12)$$

$$\mathrm{POO^{\bullet}} \rightarrow \mathrm{-CH(OOH)-CH_2-} \quad \text{(slow)} \quad (13)$$

All of the chemical changes that result from ozonization are oxidative reactions including the breaking of the polymer chain.
A PP sample after ozonization in the presence of UV-irradiation becomes brittle after 8 hrs of exposure, whereas the same effect in ozone is noticeable after 50-60 hours.Degradation of polymer chain occurs as a result of decomposition of peroxy radicals. The oxidation rapidly reaches saturation, suggesting the surface nature of ozone and atomic oxygen against of PP as a consequence of limited diffusion of both oxygen species into the polymer. Ozone reacts with PP mainly on the surface since the reaction rate and the concentration of intermediate peroxy radicals are proportional to the surface area and not the weight of the polymer. It has been found that polyethylene is attacked only to a depth of 5-7 microns (45).

Atomic oxygen oxidation of polymers has been reported by a few authors (46,50). Experiments were limited to the measurements of weight-loss data and changing of the wetteability (46-48), and only two papers were devoted to the study mechanism of atomic oxygen oxidation of polydienes (49,50).

The PP samples exposed to atomic oxygen show for the formation of polymer peroxy radicals (POO•), which give almost identical ESR spectrum as in the case of ozone reaction (Fig.1). The ESCA spectra (Fig.5) indicate that atomic oxygen oxidation is more effective than ozonization. These results support our assumption that ozonization of PP film in the presence of UV-irradiation is probably a result of simultaneous attack of atomic oxygen, molecular oxygen (which is present in excess in ozone-oxygen mixture) and ozone (which was not completely photolysed). Oxygen atoms were generated by photolysis of ozone, using the 253,7 nm line from L1 lamp. The initially formed excited oxygen atoms $O(^1D)$ are rapidly deactivated to the ground state atomic oxygen O, by collisions with the excess of ozone molecules present. In this method it is impossible to separate both species, atomic oxygen and ozone.

Mechanism of the reactions which are belived to occur when a polymer such as PP is exposed to atomic oxygen are following:

$$-CH_2-C(H)(CH_3)- \; + \; O \longrightarrow -CH_2-\dot{C}(CH_3)- \; + \; {}^{\bullet}OH \qquad \text{(rapid)} \qquad (14)$$

$$-CH_2-\dot{C}(CH_3)- \; + \; {}^{\bullet}OH \longrightarrow -CH_2-C(OH)(CH_3)- \qquad \text{(rapid)} \qquad (15)$$

$$-CH_2-\dot{C}(CH_3)- \; + \; {}^{\bullet}OH \longrightarrow -CH{=}C(CH_3)- \; + \; H_2O \qquad \text{(rapid)} \qquad (16)$$

$$-CH_2-\dot{C}(CH_3)- \; + \; O_2 \longrightarrow -CH_2-C(OO^{\bullet})(CH_3)- \qquad \text{(rapid)} \qquad (17)$$

$(POO^{\bullet})$- detected by ESR spectroscopy

$$-CH{=}C(CH_3)- \; + \; O_3 \longrightarrow \text{products according reaction 7} \qquad \text{(rapid)} \qquad (18)$$

$$POO^{\bullet} + PH \longrightarrow POOH + P^{\bullet} \qquad \text{(slow)} \qquad (19)$$

$$POOH \longrightarrow PO^{\bullet} + {}^{\bullet}OH \qquad \text{(slow)} \qquad (20)$$

Slow reaction which also occur during photo-oxidation and/or thermal oxidation can take place during oxidation with atomic oxygen, but these slow reactions are of little importance because of the rapid oxidation which usually occurs. More results which explain atomic oxygen oxidation mechanism of PP, will be published separately.

Acknowledgment. These investigations are part of a research program on the environmental degradation of polymers supported by the Swedish National Board for Technical Developments (STU), which we gratefully acknowledge.

Literature Cited

1. Finlayson-Pitts,B.J. and Pitts, J.N.Jr., "Atmospheric Chemistry: Fundamentals and Experimental Techniques",Wiley,New York,1986.
2. Boutevin, B., Pietrasanta, Y., Taha, M., and Sarraf, T., Europ.Polym.J.,20,875 (1984).
3. Giesler, G.,and Wergin, H., J.für Prakt.Chem.,25,135-140 (1964).
4. Giesler, G., and Wergin, H., J. für Prakt.Chem.,25,141 (1964).
5. Kefeli, A.A., Razumovskii, S.D., and Zaikov, G.E., Polym.Sci. USSR, 13,904 (1971).
6. Kefeli, A.A., Razumovskii, S.D., Markin, V.S., and Zaikov, G.E., Vysokomol.Soedin., A,14,2413 (1972).
7. Lazar, M., Rubb.Chem.Technol., 36,527 (1963).
8. Lebel, P.H., Rubb.Plast.Age, 45,297 (1964).
9. Peeling, J., and Clark, C.T., J.Polym.Sci., A1,21,2047 (1983).
10. Popov, A.A., Krisyuk, B.E., Blinov, N.N., and Zaikov, G.E., Europ.Polym.J., 17,169 (1981).
11. Razumovskii, S.D., and Zaikov, G.E., Develop.Polym.Stabil.,6,239 (1985).
12. Razumovskii, S.D., and Zaikov, G.E., "Ozone and Its Reactions with Organic Compounds",Elsevier, 1984.
13. Priest, D.J., J.Polym.Sci., A2,9,1771 (1971).
14. Kefeli, A.A., Rakovskii, S.K., Shopov, D.M., Razumovskii, S.D., Rakovskii, R.S., and Zaikov, G.E., J.Polym.Sci.,A1,19,2175 (1981).
15. Razumovskii, S.D., Karpukhin, O.N., Kefeli, A.A., Pokholok,T.V., and Zaikov, G.E., Vysokomol.Soedin.,A,13,782 (1971).
16. Abdullin, M.I., Gataullin, R.F., Minsker, K.S., Kefeli, A.A., Razumovskii, S.D., and Zaikov, G.E., Europ.Polym.J.,14,811 (1978).
17. Devries, K.L., and Simonson, E.R., Ozone Chem.Technol.,1975,257.
18. Razumovskii, S.D., Kefeli, A.A., and Zaikov, G.E., Europ.Polym. J.,7,275 (1971).
19. Tucker, H., Rubb.Chem.Technol.,32,269 (1959).
20. Yakubchik, A.I., Kasatkina, N.G., and Pavlovskaya, T.E., Rubb. Chem.Technol.,32,284 (1959).
21. Calvert, J.C., and Pitts, J.N.Jr., "Photochemistry", Wiley,New York,1967,p.209.
22. Okabe, H., "Photochemistry of Small Molecules", Wiley,New York, 1978,p.237.
23. Rånby, B., and Rabek, J.F., "Photodegradation, Photooxidation and Photostabilization of Polymers", Wiley, London, 1975,p.254 and 351.
24. Rånby, B., and Rabek, J.F., in "The Effects of Hostile Environments on Coatings and Plastics", (Garner, D.P., and Stahl, G.A., eds), ACS Symp.Ser.,No229,Washington, DC, 1983,p.291.
25. Carlsson, D.J., Suprunchuk, T., and Wiles, D.M., J.Polym.Sci., B,14,193 (1976).
26. Carlsson, D.J., and Wiles, D.M., J.Polym.Sci.,A1,14,493 (1976).
27. Rabek, J.F., in "Singlet Oxygen" (Frimer, A.A.,ed.),Vol.4,CRC Press,Boca Raton,FL,1985,p.1.
28. Dilks, A., J.Polym.Sci.,A1,19,1319 (1981).
29. Rånby, B., and Rabek, J.F., "ESR Spectroscopy in Polymer Research", Springer Verlag,Berlin,1977,p.257.

30. Yamauchi, J., Ikemoto, K., and Yamaoka, A., Makromol.Chem.,178, 2483 (1977).
31. Carlsson, D.J., and Wiles, D.M., Macromolecules,4,174 (1971).
32. Kefeli, A.A., Razumovskii, S.D., Markin, V.S., and Zaikov, G.E., Polym.Sci.,USSR,14,2812 (1972).
33. Amin, M.U., Tillekeratne, L.M.K., and Scott, G., Europ.Polym.J., 11,85 (1976).
34. McKellar, J.F., and Allen, N.S., "Photochemistry of Man-Made Polymers",Applied Science Publishers,London, 1979,p.10.
35. Bailey, P.S., "Ozonization in Organic Chemistry", Vol.1-2, Academic Press, New York, 1982.
36. Criegee, R., Record of Chemical Progress,18,111 (1957).
37. Criegee, R., Kerckov, A., and Zinke, H., Chem.Berichte,88,1878 (1955).
38. Criegee, R., and Wenner, G., Justus Liebigs Ann.Chem.,564,9 (1959).
39. Murray, R.W., Acc.Chem.Res.,1,313 (1968).
40. Gillies, C.W., and Kuczkowski, R.L., J.Am.Chem.Soc.,94,6337(1972).
41. Kuczkowski, R.L., Acc.Chem.Res.,16,42 (1983).
42. Bailey, P.S., Chem.Rev.,58,925 (1958).
43. Popov, A.A., Rakovskii, S.K., Shopov, D.M., and Ruban, Z.V., Izv. Akad.Nauk SSSR, 982,1950 (1976).
44. Popov, A.A., and Zaikov, G.E., Dokl.Akad.Nauk SSSR, 244,1178 (1979).
45. Pentin, Yu.A., Tarasevich, B.N., and El′tsefon, B., Zhurn.Fiz. Khim.,46,2116 (1972).
46. Hansen, R.H., Pascale, J.V., De Benedictis, T., Rentzepis, P.M., J.Polym.Sci.,A,3,2205 (1965).
47. MacCallum, J.R., Rankin, C.T., J.Polym.Sci.,9,751 (1971).
48. MacCallum, J.R., Rankin, C.T., Makromol.Chem.,175,2477 (1974).
49. Rabek, J.F., Lucki, J., and Rånby, B., Europ.Polym.J.,15,1089 (1979).
50. Lucki, J., Rabek, J.F., and Rånby, B., Europ.Polym.J.,15,1101 (1979).

RECEIVED September 11,1987

Chapter 15

Photochemical Modifications of Poly(vinyl chloride)

Conducting Polymers and Photostabilization

C. Decker

Laboratoire de Photochimie Générale associé au Centre National de la Recherche Scientifique, Ecole Nationale Supérieure de Chimie, 68200 Mulhouse, France

Polyvinyl chloride has been modified by photochemical reactions in order to either produce a conductive polymer or to improve its light-stability. In the first case, the PVC plate was extensively photochlorinated and then degraded by UV exposure in N_2. Total dehydrochlorination was achieved by a short Ar^+ laser irradiation at 488 nm that leads to a purely carbon polymer which was shown to exhibit an electrical conductivity. In the second case, an epoxy-acrylate resin was coated onto a transparent PVC sheet and crosslinked by UV irradiation in the presence of both a photoinitiator and a UV absorber. This superficial treatment was found to greatly improve the photostability of PVC as well as its surface properties.

UV radiation is known to have deleterious effect on most commercial polymers, thus reducing the service life of these materials for outdoor applications. That is particularly true for PVC, one of the most widely used thermoplastics, whose field of applications still remains restricted by its poor resistance to sunlight (1). In some cases, it is also possible to profit from the energy carried by the photons to induce useful chemical modifications that will generate new materials with improved properties. These photochemical reactions will develop primarily at the surface and in the top layer of the irradiated polymer because of the limited penetration of UV radiation into organic compounds. We describe here two examples of such light-induced surface modifications that were both carried out on a PVC substrate. In the first one, the polymer was exposed successively to UV radiation and to a laser beam in order to produce a purely carbon polymer that was shown to be able to carry electrons. The second example shows how the light stability of PVC can be greatly improved by protecting the surface of PVC-based materials with a UV curable coating that acts as an effective anti-UV filter.

0097-6156/88/0364-0201$06.00/0

CONDUCTING POLYMERS BY LASER CARBONIZATION OF PVC

Synthetic polymers are best known for their insulating dielectric properties which have been exploited for numerous applications in both the electrical and electronic industries. It was found recently that some polymers can also be rendered conductive by an appropriate treatment, thus opening the way to a new field of applications of these materials (2, 3). Usually, electrical conductivity is obtained by doping a neutral polymer, rich in unsaturation, with donor or acceptor molecules. These polymers are rather difficult to synthesize, which makes them very expensive ; besides they are often sensitive to environmental agents, like oxygen or humidity, thus restricting their practical use to oxygen-free systems.

In the present work, a somewhat different approach was chosen in order to produce conducting polymers ; the basic idea was to start with a cheap material, like PVC, and try to remove all the hydrogen and chlorine atoms from the polymer chain. The purely carbon material thus obtained was expected to exhibit the electrical conductivity of a semimetal, while being insensitive to the atmospheric oxygen. In this paper, we report for the first time how PVC can be completely dehydrochlorinated by simple exposure to a powerful laser beam that combines both the photochemical and the thermal effects.

In a previous work (4, 5), we have shown that long conjugated polyene sequences, $-(CH{=}CH{-}CH{=}CH)_n-$, are formed in large amounts during the laser-induced degradation of PVC, leading to a heavily colored polymer film. However, to make this material conductive, doping with an appropriate agent like iodine or boron trifluoride (6) is still necessary, since total dehydrogenation into graphite cannot be worked out under those conditions. The situation is quite different if chlorinated PVC (C-PVC) is used as starting material. This polymer was found to be very sensitive to UV radiation (7), generating chlorinated polyene sequences, $-(CH{=}CCl{-}CH{=}CCl)_n-$, with high quantum yields ; such structures are susceptible to be further dehydrochlorinated into a purely carbon polymer by photochemical or thermal degradation. If a laser is used to perform this reaction, one can expect to thus achieve an extremely fast and extensive carbonization of the polymer and produce conductive patterns in well defined areas. The whole procedure can be represented by the following reaction scheme :

$$-(CH_2-\underset{Cl}{CH})_n- \xrightarrow[\text{laser}]{h\nu} -(CH{=}CH{-}CH{=}CH)_n- + HCl$$

$$\downarrow Cl_2,\ h\nu$$

$$\underset{\text{chlorinated PVC}}{-(\underset{Cl}{CH}-\underset{Cl}{CH})_n-} \xrightarrow{h\nu} \underset{\text{chlorinated polyenes}}{-(CH{=}\underset{Cl}{C}{-}CH{=}\underset{Cl}{C})_n-} \xrightarrow{\text{laser}} \underset{\text{carbon}}{-(C{=}C)_n-} + HCl$$

Basically, it consists of three photochemical processes that are illustrated by figure 1 :

- the light-induced chlorination of PVC,
- the photodegradation of the resulting chlorinated PVC,
- the laser-induced dehydrochlorination of the degraded C-PVC.

Each of these processes will now be described in detail.

1. Photochlorination of PVC

When a PVC film is exposed to the UV-visible radiation of an incandescent lamp in the presence of pure chlorine, at room temperature, the chlorine content of the polymer increases from 56.8 % initially to over 70 % after a few hours of irradiation (8). As the reaction proceeds, the rate of chlorination decreases steadily as shown by the kinetic curves of figure 2, most probably because of the decreasing number of reactive sites on the polymer chain that remain available for the attack by chlorine radicals.

At the same time, large amounts of hydrogen chloride are evolved, at a rate very similar to the rate of chlorine addition to PVC (figure 2). This is in good agreement with the postulated reaction scheme, shown in figure 3, that predicts the formation of one HCl molecule for each chlorine atom fixed to the PVC backbone. Quantum yield measurements have shown that this chain reaction process develops very efficiently in thin PVC films, each chlorine radical generated by photolysis of Cl_2 being able to induce the chlorination of up to 30 methylene sites (9).

In order to evaluate how deep the chlorination can proceed into the polymer film, photochlorination experiments were carried out on PVC samples of various thickness in the 5 to 60 µm range. Figure 4 shows that extensive chlorination only occurs in the top layer, but that chlorine radicals can still penetrate as much as 30 µm deep into the PVC film, thus leading to a gradient of chlorination that leaves essentially unchanged the deep underlying layers. For some specific applications, a thicker layer of conductive polymer, and therefore of C-PVC, might be needed. This can be obtained by using as starting material a PVC powder that has been extensively chlorinated in a fluid bed reactor or in a vibrating photoreactor (10), at conversions above 90 %. C-PVC films of required thicknesses can then be cast on the PVC substrate. The latter method has also to be used when the conducting patterns must appear on a support other than PVC, such as metals, glass or other plastics.

2. Photodegradation of chlorinated PVC

Since chlorinated PVC is totally transparent in the near-UV and visible range, it will not absorb at 488 nm, the emission line of the argon ion laser that we intended to use to perform the carbonization. Therefore C-PVC films were first exposed to the UV radiation of a medium pressure mercury lamp in order to produce the strongly absorbing polyenes. This irradiation was carried out at room temperature in the absence of oxygen, thus preventing the formation of undesirable oxidation products.

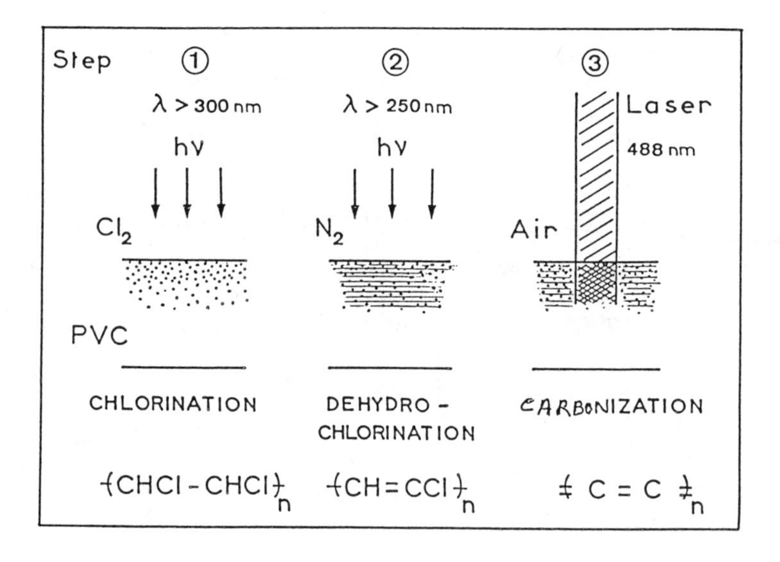

Figure 1. Three step procedure of the carbonization of PVC by UV and laser irradiation

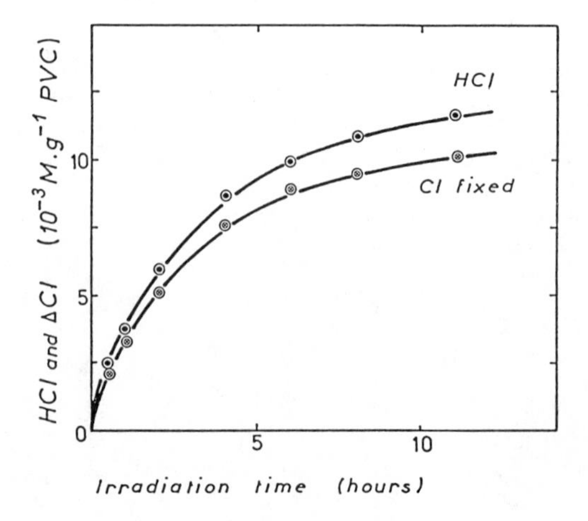

Figure 2. Kinetics of the photochlorination of a PVC film (thickness = 50 µm ; light intensity = 5.10^{-9} E s^{-1} cm^{-2})

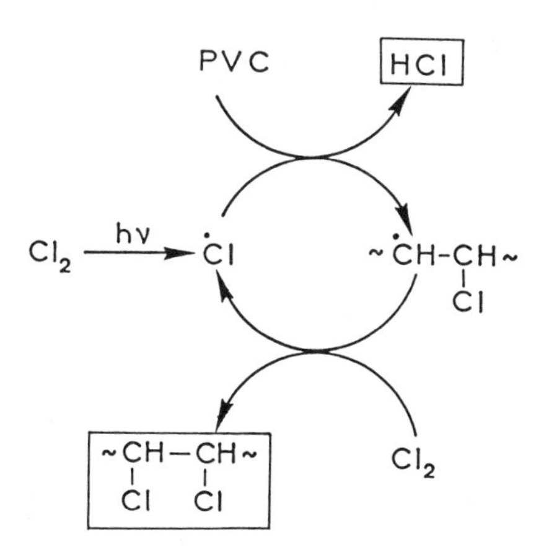

Figure 3. Reaction scheme of the photochlorination of PVC by a chain process

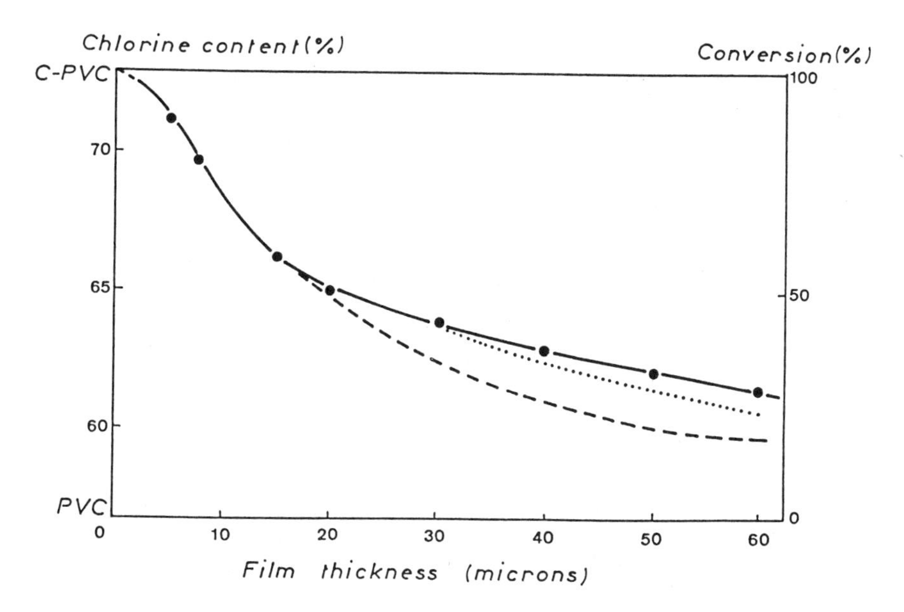

Figure 4. Dependence of the chlorine content of C-PVC on the film thickness, after 7h of UV exposure. Calculated curves if the chlorination were restricted to the 20 μm (---) or 30 μm (...) top layer

The intense discoloration which developed rapidly upon UV exposure reveals the high photosensitivity of C-PVC that is even more pronounced than for PVC itself, as shown by the UV-visible absorption spectra of figure 5. After 15 minutes of irradiation, large amounts of polyenes have already accumulated in C-PVC, with sequence lengths up to 20 conjugated double bonds, while PVC is hardly affected after that short exposure.

The mechanism of the photodegradation of C-PVC has been extensively studied (7,11), and can be summarized by the following reaction scheme :

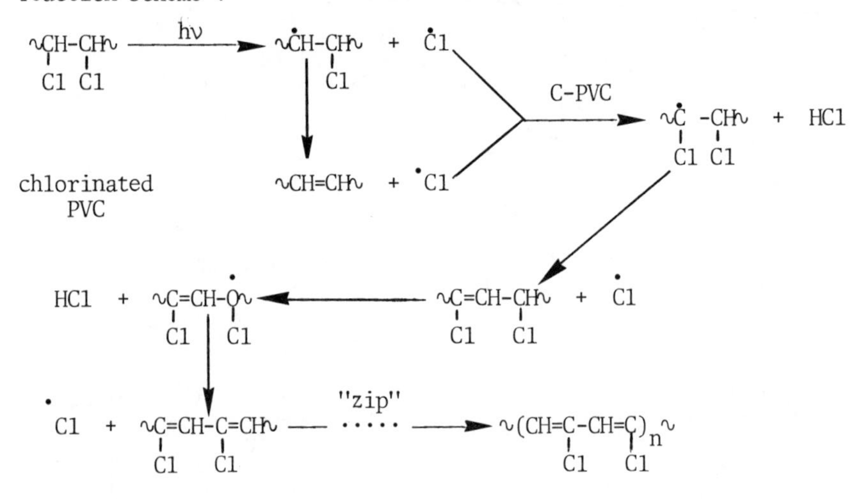

Once UV photons have been absorbed by the polymer, excited states are formed ; they disappear by various routes, one of them leading to the formation of free radicals by cleavage of the C-Cl bonds. The very reactive $\dot{C}l$ radicals evolved are most likely to abstract an hydrogen atom from the surrounding CHCl sites to generate α-β,β' chloro alkyl radicals : $-\underset{Cl}{CH}-\underset{Cl}{\dot{C}}-\underset{Cl}{CH}-$. These radicals are known to stabilize readily by splitting off the β chlorine with formation of a double bond (12). If the $\dot{C}l$ radical evolved reacts with the close by allylic hydrogen, a new unstable radical will be formed so that the dehydrochlorination will develop step by step along the polymer chain, leading to the formation of chlorinated polyene sequences and evolution of hydrogen chloride. It should be mentioned that, besides this efficient chain reaction process, main chain scissions and crosslinking were also found to occur to some extent during the light-induced degradation of C-PVC films (11) ; these reactions are yet not likely to affect the production of a purely carbon polymer in the third and last step of the procedure.

3. Laser carbonization of degraded C-PVC

Since our main objective was to remove all the chlorine and hydrogen atoms from the polymer chain, C-PVC films were further exposed to the UV radiation of the medium pressure mercury-lamp. This led to a dark brown material which was found to be unable to carry an electrical current, even after extended irradiation time. Therefore we turned to a powerful laser source, a 15 W argon ion laser tuned to its continuous emission at 488.1 nm. At that wavelength, the degraded polymer film absorbs about 30 % of the incident laser photons. The sample was placed on a X-Y stage and exposed to the laser beam at scanning rates in the range of 1 to 50 cm s^{-1}, in the presence of air.

A laser power output of 1 W, concentrated on a tiny area of 2 mm^2, proved to be already enough to transform the polymer into a black residue at a scanning speed of 2 cm s^{-1}, which corresponds to an exposure time as short as 0.1 s. The resulting material consists essentially of carbon ; it is strictly insoluble in the organic solvents and was found to totally absorb radiations over the whole wavelength spectrum, from the deep UV to the visible and infra-red regions (13). Upon laser exposure, large amounts of hydrogen chloride were evolved, thus resulting in a substantial weight loss of the laser irradiated sample ; gravimetric measurements have shown that essentially all the chlorine atoms have been removed from the polymer backbone, the apparent density of the laser-irradiated material dropping to about 0.5g cm^{-3}. This result was confirmed by infra-red spectroscopy analysis which clearly revealed that all the functional groups have disappeared, in particular the C-Cl bonds that absorb in the 700 cm^{-1} region (figure 6). All the spectroscopy measurements were carried out on a sodium chloride or quartz plate coated with a 20 μm thick C-PVC film.

If faster scanning rates are required for some specific applications they can easily be reached either by increasing the power output of the laser up to 5 W, or by focusing the beam down into the micron range. Table I compares the results obtained with the unfocused laser beam and with a 100 or 10 μm laser spot. With the most sharply focused beam, carbonization alrealy occured at a scanning speed of 50 cm s^{-1} and the exposure time dropped into the microsecond range.

One of the main characteristics of the laser emission is the huge amount of energy that is concentrated within a narrow beam and can be delivered on a tiny area. In order to take full profit of the high power density available, it is also necessary to use photosensitive systems which obey the reciprocity law, i.e. where the energy required for the reaction is not dependent on the light intensity, which means that the quantum yield remains constant. This condition appears to be almost fullfilled in the present case since the fluence, expressed in J cm^{-2}, was found to increase by only a factor of 4 when the light-intensity was increased by over 4 orders of magnitude (Table I).

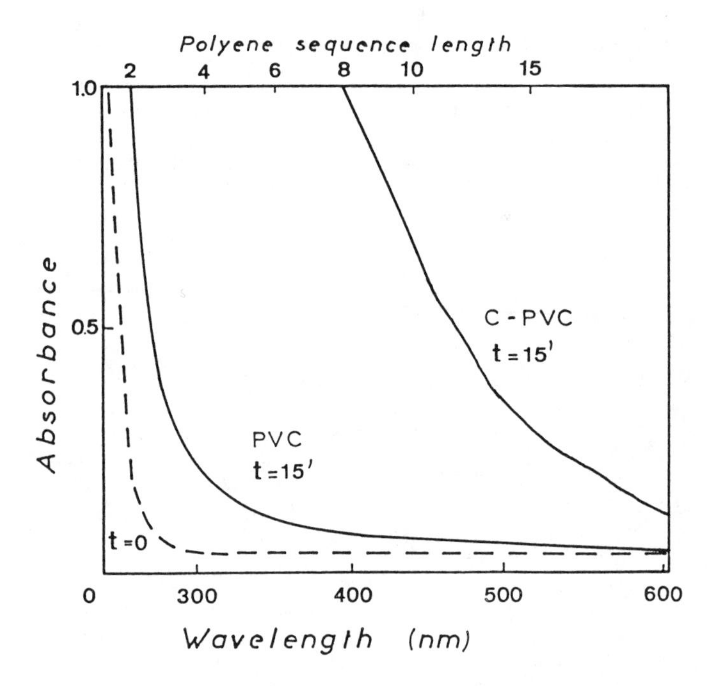

Figure 5. UV-visible absorption spectra of PVC and C-PVC films before and after 15 min of UV irradiation in a N_2 atmosphere

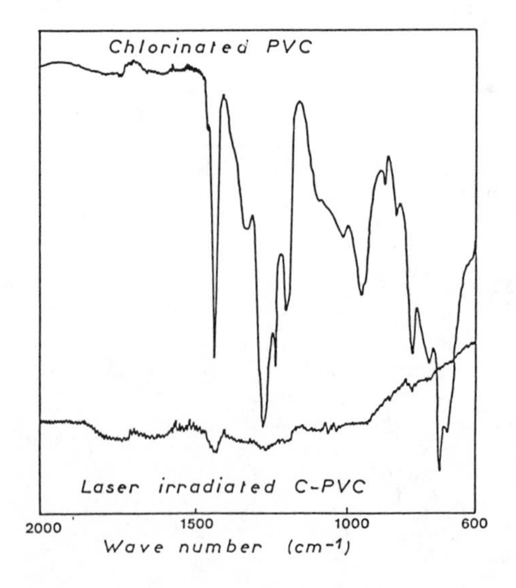

Figure 6. IR absorption spectra of chlorinated PVC before and after laser irradiation at 488 nm for 0.1 s in air

Table I : Influence of the beam focusing in the laser-graphitization of chlorinated PVC (continuous wave mode emission line at 488.1 nm of Ar^+ laser)

λ = 488.1 nm Power = 1 W	Argon-ion laser beam unfocused	focused	
Spot diameter (nm)	1.7	0.1	0.01
Fluence rate (Wcm^{-2})	50	10^4	10^6
Scanning speed ($cm\ s^{-1}$)	2	10	50
Exposure time (s)	0.1	10^{-3}	2×10^{-5}
Fluence (Jcm^{-2})	5	10	20
Quantum yield	12	6	3

The quantum yield of the carbonization process can be evaluated from the amount of HCl evolved during the laser irradiation :

$$\phi_{HCl} = \frac{\text{number of molecules of HCl evolved}}{\text{number of photons absorbed}} = 2\phi_{carbon}$$

taking into account that 2 molecules of HCl are evolved from C-PVC for each =C=C= unit formed.

Since only 30 % of the incident laser photons are absorbed by the polymer film, the amount of energy absorbed in a $1 cm^2$ area illuminated for 0.1 s by the unfocused laser beam will be : $5\ J\ cm^{-2} \times 0.3 = 1.5\ J\ cm^{-2}$ or 6×10^{-6} einstein cm^{-2} (1 einstein associated with the 488 nm emission has an energy of 2.45×10^5 Joule $mole^{-1}$). On the other hand, about 7×10^{-5} mole of HCl are evolved by each square centimeter of a 20 µm C-PVC film transformed into carbon. The quantum yield of HCl evolved can then be calculated from the following ratio :

$$\phi_{HCl} = \frac{7 \times 10^{-5}\ \text{mole cm}^{-2}}{6 \times 10^{-6}\ \text{einstein cm}^{-2}} = 12\ \text{mole einstein}^{-1}$$

and $\phi_{carbon} = 6$ mole $einstein^{-1}$

These quantum yield values appear to be much higher than unity and therefore demonstrate that carbonization occurs by a chain reaction process. The mechanism of the laser-induced dehydrochlorination of photodegraded C-PVC can be schematically represented by the following set of reactions, that leads ultimately to a purely carbon polymer made of either linear sequences or polycondensed double bonds :

$$-(CH=\underset{Cl}{C}-CH=\underset{Cl}{C})_n- \xrightarrow[488\ nm]{laser} -CH=\dot{C}-CH=\underset{Cl}{C}- \quad + \quad \dot{C}l$$

$$\dot{C}l \; + \; -\underset{Cl}{C}=CH-\underset{Cl}{C}=CH- \longrightarrow -\underset{Cl}{C}=\dot{C}-\underset{Cl}{C}=CH- \quad + \quad HCl$$

$$\downarrow$$

$$HCl \; + \; -\underset{Cl}{C}=C=C=\dot{C}-\underset{Cl}{C}- \longleftarrow -\underset{Cl}{C}=C=C=CH- \quad + \quad \dot{C}l$$

$$\downarrow$$

$$\dot{C}l \; + \; -\underset{Cl}{C}=C=C=C=C- \dashrightarrow =(C=C)_n= \quad + \quad HCl$$

This reaction scheme bears some formal analogy with the mechanism previously elaborated for the laser-induced degradation of PVC (3,4), except that the 488 nm laser photons are now absorbed by the chlorinated polyenes, with a sequence length of about 12. One of the important routes of deactivation of the excited states thus formed is by cleavage of the most labile allylic C-Cl bond, with liberation of a very reactive •Cl radical that will readily abstract an hydrogen atom from the surrounding polymer chains. The resulting β,β' chloroallylic radical tends to stabilize by splitting off a chlorine atom, thus generating a =C=C= type structure. As this chain reaction propagates along the polymer backbone, a purely carbon material is finally formed, together with large amounts of HCl.

Since the UV degraded C-PVC still contains substantial amounts of the initial CHCl-CHCl structure, one can expect the chlorine radicals evolved to also initiate the zip-dehydrochlorination of these structures. The resulting chlorinated polyenes will then be further destroyed by the laser irradiation, so that finally all the C-PVC polymer is converted into a purely carbon material within a fraction of a second.

4. Electrical conductivity

The black lines that appear on the C-PVC plate after scanning by the laser beam were found to consist essentially of carbon and were thus expected to exhibit some electrical conductivity. Indeed, when a low voltage was applied to both ends of the laser tracks, the tiny filament turned bright red and even incandescent when a potential over 30 V was applied. This clearly shows that the laser irradiation can transform an insulating polymer like C-PVC into a conductive material. Thus, it becomes possible to write high resolution conductive patterns with this light-pencil which can be easily visualised since they appear as well contrasted black tracks.

By measuring the resistance (R) of these laser tracks, one can evaluate the electrical conductivity (σ) of the polymer formed from the equation :

$$R = \frac{\ell}{\sigma . S}$$

where ℓ is the length of the wire of cross-section S. The conductivity was found to be $\sim 10 \Omega^{-1} cm^{-1}$, a value comparable to the conductivity of pristine graphite ($\sigma_c = 8.3\ \Omega^{-1}\ cm^{-1}$) when it is measured in the direction perpendicular to the planes containing the carbon atoms (14). Since the conductivity of the starting polymer is in the range of $10^{-10}\ \Omega^{-1} cm^{-1}$, we are thus achieving through the laser irradiation a remarkable and instantaneous jump of 11 decades in the electrical conductivity of this material.

A higher conductivity might still be obtained if necessary, either by compacting the porous carbon structure or by inserting acceptor or donor molecules. Thus, in the case of pristine graphite, the perpendicular to the plane conductivity was found to increase to $2.10^3\ \Omega^{-1}\ cm^{-1}$ by insertion of potassium intercalates and as high as $8.10^4\ \Omega^{-1}\ cm^{-1}$ by using lithium (14).

The overall procedure that allows the transformation of PVC into carbon is summarized by the diagram of figure 7 that clearly shows the three successive photochemical processes. It should be mentioned that the laser irradiation of PVC or photodegraded PVC produces no conductive polymer but leads, after prolonged exposure, to a brown material resulting from both thermal and photochemical degradations. One of the main advantages of using chlorinated PVC is that this polymer combines both a high photosensitivity, thus requiring short exposure times, and a good thermal stability (Tg > 150°C) which precludes any phase changes during the laser exposure.

For practical applications of these conducting polymers in the electronic industry, it is still necessary to use a top coat to protect the tiny carbon patterns which are rather fragile due to their porous structure. A photopolymerizable acrylic resin was formulated for that purpose that had a relatively large viscosity and a high reactivity, both factors which prevent any significant diffusion of the resin into the carbon structure during the short time lapse between deposit and final cure. A very resistant coating was thus obtained after UV exposure during a fraction of a second ; it was found to still preserve the electrical conductivity of the laser tracks and allows an easy handling of the plates, without risking to erase the conductive circuits. Besides, such a treatment also ensures a good adhesion of the patterns to the support, in particular if the latter consists of a material other than PVC, where adhesion was found to be poor before treatment by the UV curable coating.

5. Conclusion

In the present study it has been shown for the first time that chlorinated PVC can be readily transformed into a conducting polymer by simple laser irradiation in the presence of air. The resulting material consisted essentially of carbon and proved to be able to carry electrons, without any doping procedure. By focusing the laser beam down into the micron range, it becomes thus possible to directly write highly complex conductive patterns on a chlorinated PVC film coated onto a transparent substrate, at scanning speeds up to 50 cm s^{-1}. For an easy handling of such plates, the fragile image has to be protected by a UV-cured coating that is both resistant to abrasion and scratching and inert toward environmental agents like humidity, acidic pollutants or organic solvents. Potential applications of such organic metals are expected to be found mostly in the microelectronic industry for the production of high-resolution conducting devices.

PHOTOSTABILIZATION OF PVC BY UV CURED COATINGS

Among the most widely used thermoplastic polymers, PVC is known to exhibit the highest sensitivity toward sunlight. Solar radiations were shown (1) to induce a fast dehydrochlorination reaction leading to the production of highly colored polyene structures as well as to the formation of crosslinks and chain scissions, with a subsequent loss in the mechanical performances. The most usual way to improve the durability of a polymer is by introducing light stabilizers or pigments like Ti O_2 or carbon black in the formulation. In the case of transparent PVC, the outdoor service life still does not exceed 5 to 7 years at best, depending on the exposure location.

A somewhat different approach was developed here in order to protect transparent PVC against weathering ; it consists in applying at the surface of the PVC plate a thin coating that will both exhibit a high photostability and be able to screen out the UV portion of the sunlight which has the most harmful effects toward PVC (15). Such coatings can be readily obtained by light-induced polymerization of multifunctional acrylic monomers (16). The highly crosslinked polymer network thus formed was shown (17) to resist UV radiation and chemicals very well, while it exhibits at the same time remarkable mechanical properties. Therefore, an additional advantage that can be expected from this method of stabilization lies in the new surface properties which will be confered to the coated material, in particular a better resistance to scratching, abrasion and environmental attack.

1. UV-curing of acrylic monomers

The basic principle of the light-induced polymerization of multifunctional monomers can be represented schematically as follows:

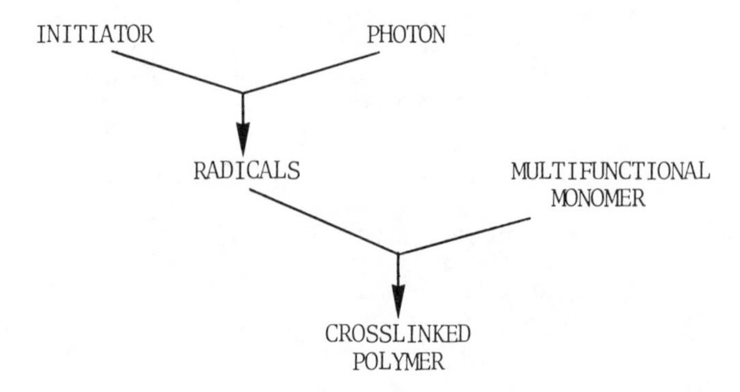

Under UV irradiation, the photoinitiator cleaves into radical fragments that react with the vinyl double bond and thus initiate the polymerization of the monomer. If the latter molecule contains at least two reactive sites, the polymerization will develop in three dimensions to yield a highly crosslinked polymer network.

The photoinitiator selected for this study was 1-benzoyl cyclohexanol (Irgacure 184 from Ciba Geigy), a compound known for its high initiation efficiency and the weak coloration of its photoproducts. The multifunctional monomer was an epoxy-diacrylate derivative of bis-phenol A (Ebecryl 605 from UCB). A reactive diluent, tripropyleneglycol diacrylate, had to be introduced in equal amounts, in order to lower the viscosity of the formulation to about 0.3 Pa.s.

When this resin was exposed as a thin film to the UV radiation of a medium pressure mercury lamp (80 W cm^{-1}), the crosslinking polymerization was found to develop extensively within a fraction of a second (18). The kinetics of this ultra-fast reaction can be followed quantitatively by monitoring the decrease of the IR absorption at 810 cm^{-1} of the acrylic double bond ($CH=CH_2$ twisting). Figure 8 shows a typical kinetic curve obtained for a 20 µm thick film coated onto a NaCl disk and exposed in the presence of air to the UV radiation at a fluence rate of 1.5×10^{-6} einstein s^{-1} cm^{-2}.

The induction period observed at the very beginning of the irradiation is due to the well known inhibition effect of oxygen on these radical-induced reactions. Once it is over, after the ~10 ms needed to consume essentially all of the oxygen dissolved in the liquid film (19), the polymerization starts rapidly to reach 75 % conversion within 0.08 s. Further UV exposure leads only to a slow increase in the cure, mainly because of mobility restrictions in the rigid matrix, so that there still remains about 15 % of acrylic unsaturation in coatings heavily irradiated for 0.4 s.

In order to act as an efficient anti-UV filter, the coating must absorb essentially all the UV radiation of $\lambda < 380$ nm from the solar spectrum. Therefore, an hydroxy-benzotriazole UV absorber (Tinuvin 900 from Ciba Geigy) was introduced in small amounts (0.5%) in the formulation before curing, which allows a good dispersion of this additive in the liquid resin. As expected, the rate of the photopolymerization is then dropping substantially (figure 8), since the UV absorber now competes with the photoinitiator for the absorption of the incident light. Under the experimental conditions used, extensive through cure was still achieved within less than one second of exposure to the UV lamp.

When the photostabilization of a polymer material is to be obtained through such a surface treatment process, it is all important to make sure that the protective effect will last throughout the service life and therefore to ensure a long-term adhesion of the coating onto the substrate. This can be best achieved by promoting a grafting reaction between the two elements (20). For that purpose, the photoinitiator was partly incorporated in the top layer of the PVC plate by a surface treatment with an acetone solution. Upon UV-irradiation of the resin-coated sample, the following reactions are expected to occur :

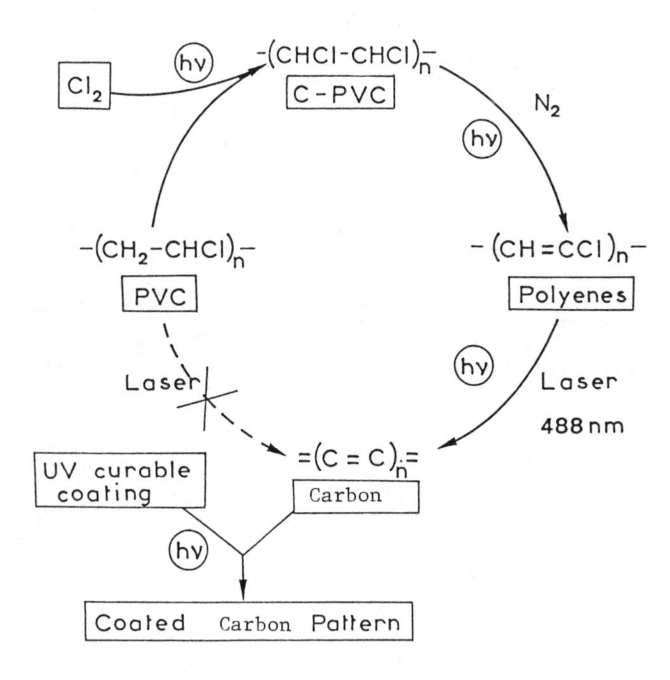

Figure 7. Schematic representation of the overall laser-carbonization process of PVC

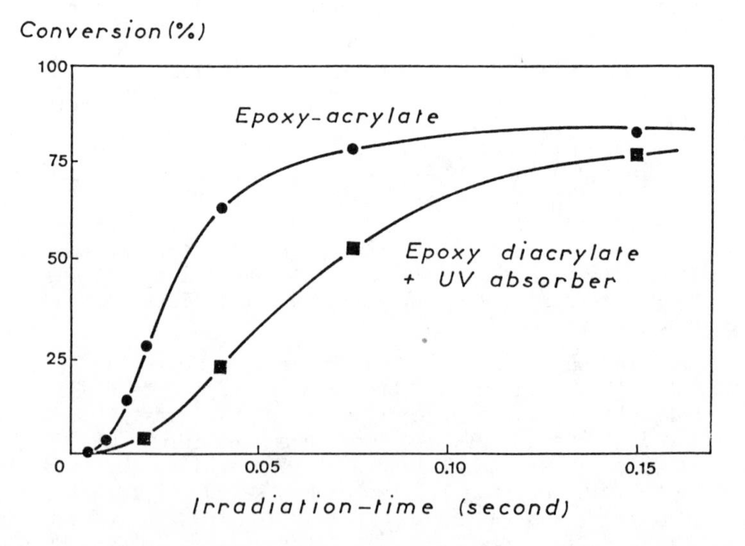

Figure 8. Kinetics of the photopolymerization of an epoxy diacrylate resin with and without UV absorber (0.5 % of Tinuvin 900)

$$C_6H_5-\underset{\underset{O}{\|}}{C}-\underset{\underset{OH}{|}}{C}(C_5H_{10}) \xrightarrow{h\nu} C_6H_5-\underset{\underset{O}{\|}}{C}^{\bullet} + {}^{\bullet}\underset{\underset{OH}{|}}{C}(C_5H_{10})$$

$$C_6H_5-\underset{\underset{O}{\|}}{C}^{\bullet} + \sim CH_2-\underset{\underset{Cl}{|}}{CH}\sim \longrightarrow C_6H_5-\underset{\underset{O}{\|}}{C}-H + \sim CH_2-\underset{\underset{Cl}{|}}{\dot{C}}\sim$$

$$\sim CH_2-\underset{\underset{Cl}{|}}{\dot{C}}\sim + CH_2=CH-\underset{\underset{O}{\|}}{C}-O\sim\sim\sim O-\underset{\underset{O}{\|}}{C}-CH=CH_2$$

$$\downarrow$$

$$Cl-\underset{\underset{CH_2}{|}}{\overset{\overset{CH_2}{|}}{C}}-CH_2-\dot{C}H-\underset{\underset{O}{\|}}{C}-O\sim\sim\sim O-\underset{\underset{O}{\|}}{C}-CH=CH_2 \xrightarrow{\text{Monomer}} \text{grafted network}$$

The initiator radicals formed by photocleavage at the polymer surface are most likely to abstract an hydrogen atom from the surrounding PVC molecules to generate PVC radicals at the resin-polymer interface. By reacting with the acrylic double bonds of the monomer, these radicals will then initiate the crosslinking polymerization leading ultimately to a polymer network that is grafted onto the PVC support. As a consequence of this chemical bonding taking place at the interface, the adhesion of the acrylic coating onto the PVC substrate was much improved and found to remain essentially unchanged after photoaging (20).

2. Light-stability of coated PVC

Previous studies on the photooxidation of UV cured epoxy-acrylate networks have revealed the remarkable resistance of these polymers to UV radiation (17). The quantum yields of the various reactions that occur upon photoaging were found to be considerably lower than in linear polymers of similar chemical structure. This outstanding light-stability results essentially from the high crosslink density of the network which, by decreasing the molecular mobility, is expected to favor cage-recombination of the primary radicals over chain propagation. Even after prolonged UV exposure, no significant changes could be noticed in both the optical properties (color, transparency, glass) and the mechanical characteristics of these crosslinked polymers. It was therefore tempting to use such UV resistant materials as protective coatings in order to improve the light-stability of a photosensitive polymer like PVC.

Transparent PVC plates were coated with a 70 µm thick film of an epoxy-acrylate resin containing 0.5 % of a benzotriazole UV absorber. They were first UV cured for one second and then exposed at 40°C to the low intensity radiations of a QUV accelerated weathering tester. The extent of the degradation was followed by UV-visible spectroscopy, a very sensitive method that permits detection of minor changes in the discoloration which usually precedes the failure in the mechanical properties of photodegraded PVC. Figure 9 shows the absorption spectra of the uncoated PVC samples, before and after QUV aging.

The 2 mm thick PVC transparent plate used here was a commercial material, well stabilized with tin maleate and benzotriazole additives. It nevertheless proved to be quite susceptible to photodegradation since, after 500h of QUV exposure, the PVC plate becames heavily colored, as shown by the strong absorption in the visible range due to the formation of long conjugated polyene sequences. At that time, chain scission and crosslinking have occured to a large extent, which leads to both a loss in transparency and a sharp drop in the impact resistance of the irradiated sample.

By contrast, the coated PVC plate was found to be little affected by QUV aging (figure 9) ; even after 1000h of exposure, it still remained essentially uncolored and perfectly transparent in the visible range. By measuring the light-transmission of the sample at 420, 580 and 680 nm, one can evaluate the yellow index (YI) from the following equation :

$$YI = \frac{(T_\circ - T_t)_{420} - (T_\circ - T_t)_{680}}{(T_\circ)_{580}}$$

where T_o and T_t correspond to the transmission of the sample at the indicated wavelength, before and after irradiation during time t, respectively. In the case of transparent PVC, the yellow index is generally considered as the best parameter to assess the extent of the degradation.

The stabilizing effect of the coating is well demonstrated by fig. 10 which shows the kinetics of the discoloration upon aging in a QUV weatherometer. For the uncoated PVC plate, it takes about 400h of exposure to reach a yellow index of 10, a value which is usually considered as the upper limit acceptable for outdoor applications. For the coated PVC on the contrary, no discoloration was detected after 400h and the yellow index stayed well below 10 after more than 2000h of QUV exposure. Similar results were obtained by photoaging in a weatherometer where it took over 10,000h of exposure for the coated PVC to reach a YI value of 10. If a comparable improvement in the light stability is observed in the natural weathering experiments now in progress, one can expect by this surface treatment to considerably increase the outdoor durability of transparent PVC.

Actually it appears that such an important stabilizing effect is mainly due to the presence in the coating of the UV absorber which is effectively cutting off all the harmful radiations of wavelength below 380 nm. Under the experimental conditions used, it was indeed shown that the benzotriazole light-stabilizer is 20 times more efficient in preventing the absorption of light by the PVC substrate when it is acting as an external filter than if it is dispersed in the bulk of the polymer. Another advantage of introducing the UV absorber in the crosslinked coating is that the loss of stabilizer by exudation during the aging is considerably reduced since

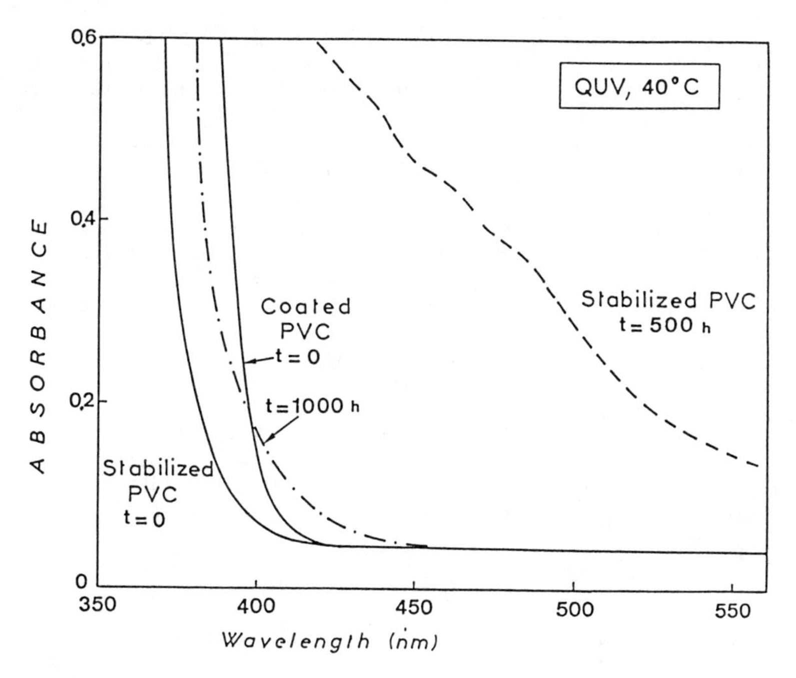

Figure 9. UV-visible absorption spectra of a stabilized PVC plate, with or without a 70 µm UV cured epoxy-acrylate coating, before and after QUV aging at 40°C

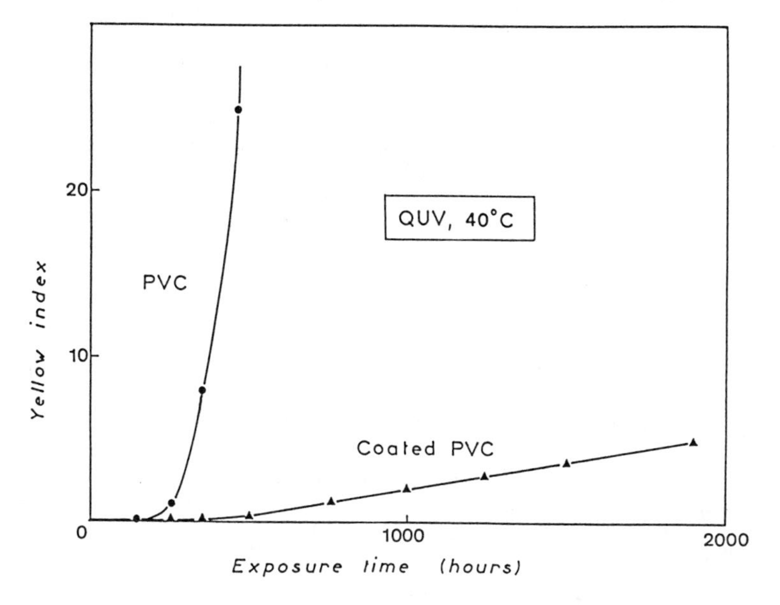

Figure 10. Kinetics of the discoloration of a 2 mm plate of stabilized PVC, with or without a 70 µm epoxy acrylate coating upon QUV aging at 40°C

these molecules are trapped within a tight network and are therefore much less likely to migrate toward the surface of the coating. In this connection, it should be mentioned that, when plasticized PVC was used instead of rigid PVC as a support, the highly-crosslinked coating acted as an efficient superficial barrier that prevents the plasticizer from diffusing out of the PVC substrate.

Finally, in addition to their photoprotective action, such UV-cured coatings also impart new surface properties to the coated polymer. Since the epoxy-acrylate network is strictly insoluble in the organic solvents and even suffers very little swelling, the coated PVC plate becomes insoluble in solvents like dichloroethane or tetrahydrofuran. Furthermore, it was found quite resistant to acidic pollutants, moisture, saline vapors , at temperatures up to 70°C. The coated PVC samples also show a better resistance to abrasion and scratching and exhibit a high gloss, even after extended photoaging. Therefore, such transparent polymer materials are most likely to be used as low cost organic glasses for outdoor applications in the building industry where an outstanding light stability is highly required, together with good mechanical and optical properties.

REFERENCES

1. E. Owen "Degradation and stabilization of PVC" Elsevier Appl. Sci. Publ. London 1984
2. R.H. Baughman, J.L. Bredas, R.R. Chance, R.L. Eisenbaumer, L.W. Shacklette Chem. Rev. 82, 209 (1982)
3. G.L. Baker, Polym. Mat. Sci. Eng. 55, 77 (1986)
4. C. Decker and M. Balandier, J. Photochem. 15, 213 (1981)
5. C. Decker and M. Balandier, J. Photochem. 15, 221 (1981)
6. Kise, H, Sugihara, M and H.E., F.F., J. Appl. Polym. Sci. 30, 1133 (1985)
7. C. Decker and M. Balandier, Makrom. Chem. 183, 1263 (1982)
8. C. Decker, M. Balandier and J. Faure, J. Macromol. Sci. A16, 1463 (1981)
9. C. Decker, M. Balandier and J. Faure, 27 Intern. Symp. on Macromolecules Strasbourg 1981, Preprints vol. 1, p. 445
10. M. Balandier, J. Faure and C. Decker, FR-Patent 2.492.387 (1982)
11. C. Decker and M. Balandier, Polym. Photochem. 5, 267 (1984)
12. R.K. Friedlina "Advances in Free-Radical Chemistry" (Edited by G. Williams) vol. 1, chap. 6, Logos, London (1965)
13. C. Decker J. Polym. Sci., Polym. Letters, 25, 5 (1987)
14. M.S. Dresselhaus and G. Dresslhaus, Advances in Physics, 30, 139 (1981)
15. C. Lorenz, S. Tu and P. Wyman, US Patent 4.129.667 (1978) and 4.135.007 (1979)
16. C.G. Roffey "Photopolymerization of surface coatings" John Wiley New-York 1982
17. T. Bendaikha and C. Decker, J. Radiation curing, 11, 6 (1984)
18. C. Decker and T. Bendaikha, Europ. Polym. J. 20, 753 (1984)
19. C. Decker and A. Jenkins, Macromolecules
20. C. Decker, J. Appl. Polym. Sci. 28, 97 (1983)

RECEIVED August 27, 1987

Chapter 16

Hydrophilization of Polydiene Surfaces by Low-Temperature Ene Reactions

Lee A. Schechtman

Procter & Gamble Company, Miami Valley Laboratories, P.O. Box 39175, Cincinnati, OH 45247

The low temperature ene reactions of 4-substituted-1,2,4-triazoline-3,5-diones (RTD) were used to modify polydiene surfaces. Hydrophilic surfaces (contact angles with water of 30-50°) were obtained on polybutadiene, polyisoprene and styrene-butadiene copolymers by first treating the polymer at room temperature with RTD (R=Ph, Me, NO_2Ph, $ClSO_2$Ph) and then with aqueous base. The use of the difunctional bis(p-1,2,4-triazoline-3,5-dione-4-ylphenyl)methane, $(TD)_2$DPM, also produced hydrophilic surfaces. However, except for the $(TD)_2$DPM treated surfaces, the hydrophilicity of the treated surfaces decreased after 5-10 days. The $(TD)_2$DPM surfaces are presumably crosslinked and remain hydrophilic for at least one year. The degree of reaction, as indicated by contact angle measurements, obtained under these heterogeneous reaction conditions suggest that the relative reactivity of the various RTDs and polydienes parallel the analogous reactions under homogeneous conditions. An ESCA study of $ClSO_2$PhTD treated surfaces confirms that the RTD is a surface species. The transformations of the surface species upon treatment with warm water and then aqueous base, $-SO_2Cl \rightarrow -SO_3H \rightarrow -SO_3^-$, were also substantiated by ESCA.

For many applications the surface properties of a polymer are of equal or greater importance than the polymer's bulk properties. Properties such as wettability (1) adhesion (2,3) permeability (4) biocompatibility (5), or weatherability (6) of polymer surfaces may be key considerations in evaluating a polymer's suitability for a given need. To promote adhesion it is usually desirable to have surfaces of high surface energy (2). Since many synthetic polymers have relatively low surface energies (7), a number of methods have been devised to increase the surface energy of polymers (i.e., to make the surface hydrophilic without making the polymer water soluble or swellable).

Methods currently used for the hydrophilization of polymer surfaces include chemical modification (8), plasma modification (9),

0097-6156/88/0364-0219$06.00/0

and irradiation methods (10). Chemical modification can cause degradation of the polymer molecules at the surface (8), although in some cases polymer degradation is not observed (11,12). Plasma methods can result in surface degradation or surface grafting depending on whether a reactive gas, inert gas, or polymerizable gas is used to generate the plasma as well as other experimental conditions (9). Despite the many unique surface properties obtainable with plasma methods, they are not yet employed widely in large scale commodity processing. While irradiation methods can be used to modify surfaces (e.g., graft polymerization), they may suffer from simultaneous modification of bulk polymer and the difficulty of getting homogeneously covered surfaces (9,10). Thus, surface modification methods that are mild, uniform, and nondestructive are desirable.

Of particular interest to us was to find a method to surface modify elastomers. G. B. Butler and co-workers have demonstrated that 4-substituted-1,2,4-triazoline-3,5-diones, RTDs, readily undergo ene reactions with polydienes at ambient temperatures (13). They found that the solubility and solution properties of the modified

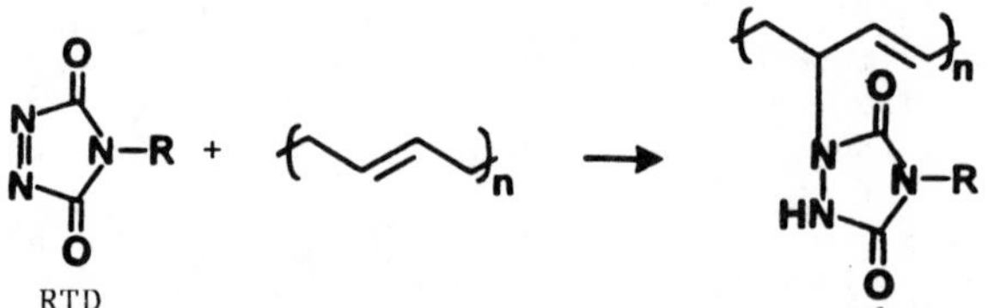

RTD

1

polymers, 1, are quite different from those of the parent polymer. Furthermore, neutralization of the pendant urazole on 1 produced 2 which has still different solution properties. If the degree of substitution is greater than 40%, then 2 is water soluble (14). The degree of substitution depends on the ratio of RTD to monomer units up to degrees of substitution of 1 (15).

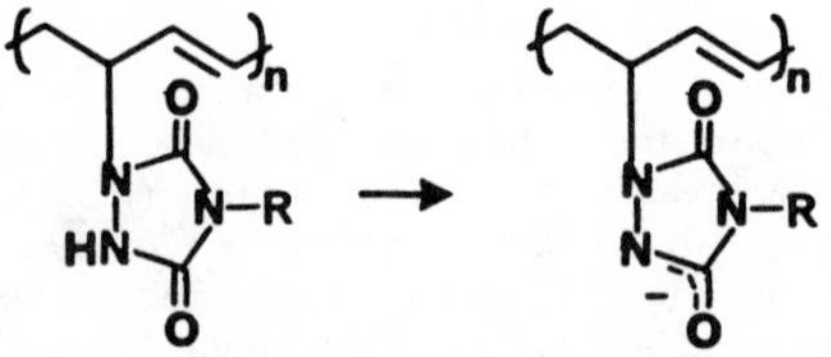

2

Bis-triazolinedione compounds such as bis(p-1,2,4-triazoline-3,5-dione-4-ylphenyl)methane, $(TD)_2DPM$, have been used to crosslink

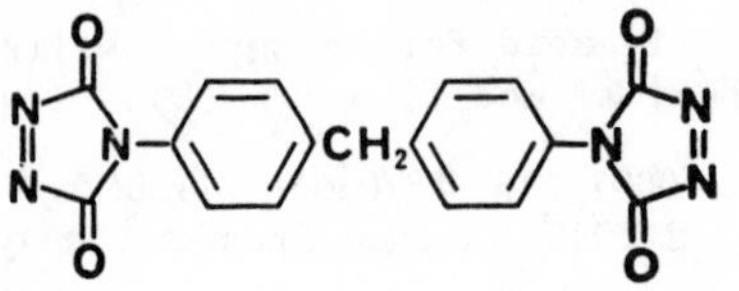

$(TD)_2DPM$

rubbers (15-17). Dipping natural rubber into solutions of $(TD)_2DPM$ causes surface hardening of the rubber (17). Triazolinediones have also been used to treat rubber surfaces to promote adhesion (18). However, these investigators did not subsequently neutralize the surfaces.

This investigation was undertaken to study the important variables in the hydrophilization of polydiene surfaces by ene reaction with triazolinediones (Step 1) followed by neutralization (Step 2) as shown below. These variables included the nature of the

Step 1 → 3 Step 2 → 4

polydiene double bond, the nature of R in RTD, and experimental conditions such as reaction time and RTD concentration.

Experimental

Reagents. Published procedures were used to synthesize PhTD (19), MeTD (19), and $p\text{-}ClSO_2PhTD$ (20) from commercially available urazoles (Aldrich, Alfa), while $(TD)_2DPM$ (17) and $p\text{-}NO_2PhTD$ (19) were prepared from the appropriate isocyanates. The RTDs were purified by sublimation, except for $(TD)_2DPM$ which was purified by precipitation or used as the synthesis product after vacuum drying. The polydienes used were purchased (Aldrich, Polyscience) and used as received except for cis-polyisoprene which was purified by precipitation. These polydienes are listed in Table I. The 3,3-ionene hydroxide used in some of the neutralization experiments was prepared from 3,3-ionene chloride (21) by passage through an ion exchange column in the hydroxide form or treatment with Ag_2O (22). All solvents used were reagent grade and were typically dried over activated molecular sieves.

Table I. Polymers and Abbreviations

Polymer	Abbreviation	$\overline{M}_w$[a]
trans-polybutadiene[b]	t-PB	160,000
cis-polybutadiene	c-PB	300,000
cis-polyisoprene	c-PI	2,000,000
trans-polyisoprene	t-PI	--
styrene/butadiene ABA block copolymer[c]		--

[a] Determined by ultracentrifuge for PB samples and light scattering for PI. [b] t-PB contained 60% trans, 20% cis, and 20% 1,2-addition. [c] 30% styrene.

Polymer Films. The polymer films were prepared by casting 5% toluene solutions onto glass microscope slides. After air drying in a fume hood, the films were vacuum dried. Film thicknesses were typically 0.05 mm.

Surface Modification. A polydiene film (supported on a microscope slide) was immersed in a stirred, room temperature, RTD-acetonitrile solution of known concentration contained in a large glass-stoppered test tube. After a specific reaction time, the film was removed from the solution, washed with acetonitrile, water, and acetonitrile again, and dried under vacuum (Step 1). Films subsequently treated with base were immersed in aqueous solutions for 5-15 min. They were then washed with water and CH_3CN, and vacuum dried (Step 2). Some films were aged in air at room temperature.

Analytical Techniques. Sessile drop contact angles were measured with a NRL C.A. Goniometer (Ramé-Hart, Inc.) using triply distilled water. The contact angles reported are averages of 2-8 identically treated samples with at least three measurements taken on each sample. ESCA spectra were obtained on a Kratos ES-300 X-ray Photoelectron Spectrometer under the control of a DS-300 Data System. Peak area measurements and band resolutions were performed with a DuPont 310 Curve Resolver.

Results and Discussion

Polydiene surfaces treated with PhTD or MeTD and then with base are rendered more hydrophilic than the untreated polymer. Typical values of the contact angle of water with the treated surfaces are given in Table II. The values are given as ranges because there was a wide scatter in the data from one sample to another although great care was exercised to make handling of duplicate samples as uniform as possible. Since contact angle measurements are very sensitive to surface heterogeneity, it is not surprising to see such discrepancies (23,24). The untreated polymers give water contact angles of 90°. A t-PB/PhTD 1:1 polymer prepared as described by Leong and Butler (14) and cast from DMF has a contact angle of 65°. Of course treatment of this polymer with base resulted in complete dissolution of the film.

Table II. Contact Angles (H_2O) of Treated Polydiene Films[a]

Polymer	PhTD		MeTD	
	1	2	1	2
t-PB	75-80	50-75	80-90	60-80
c-PB	75-90	70-80	80-84	65-80
c-PI	60-65	25-60	59-88	30-90
t-PI	50-65	30-45	52-68	20-50

[a] Contact angles are in degrees; values are given as ranges which reflect the sensitivity of contact angle to surface heterogeneity after Step 1 and Step 2 (see experimental) for PhTD and MeTD. Reaction times were 5-60 s in RTD solutions of 0.2-0.5 M.

When the reaction times for Step 1 are 5 min or longer, the samples severely crack, curl, or dissolve. These results suggest that substantial reaction is occurring in the bulk of the polymer. Significant hydrophilization can occur with reaction times as short as 5 s with RTD concentrations of 0.2-0.5 M. However, 0.002-0.02 M solutions of MeTD or PhTD do not allow sufficient reaction rates for surface hydrophilization at the shorter reaction times. Thus, diffusion of MeTD and PhTD into the polymer must occur readily from the acetonitrile solutions. Acetonitrile was used because it does not swell or dissolve the polymer or RTD-polymer adduct, and the RTDs are soluble and stable in it. This solvent is quite polar (dielectric constant, ϵ=38) (25), and this is probably a major factor in the partitioning of the relatively nonpolar RTDs between the polydiene film and the solvent. As noted below, more polar RTDs show less tendency to diffuse into the polymer.

The most hydrophilic surfaces are obtained with the polyisoprenes and PhTD. Since the reaction times are quite short, this increased hydrophilicity is most likely a reflection of the more complete reaction of surface sites in these systems. These results correlate with homogeneous ene reaction reactivities. Two factors favor polyisoprenes being more reactive than polybutadienes in ene reactions. First, primary hydrogens, which PI has but PB does not, are abstracted more readily than secondary hydrogens (26). Second, trialkyl substituted double bonds, such as those in PI, are in general more reactive toward enophiles than 1,2-disubstituted double bonds, such as those in PB (27). Also, PhTD is a more reactive enophile than MeTD by virtue of the greater electron-withdrawing effect of the phenyl ring vs. the methyl group (28).

Significant hydrophilization of c-PI surfaces with NO_2PhTD (0.2 M) could be achieved with reaction times of 1-15 s (contact angle 52°). The nitro group makes this an extremely reactive TD (28). Also, the increased polarity of the molecule was expected to make its diffusion into the polymer less favorable. However, reaction times greater than 1 min do result in films that crack or curl. This is indicative of reaction within the bulk of the sample. Because of its high reactivity, NO_2PhTD was difficult to purify and work with.

In contrast, $ClSO_2$PhTD did not show a tendency to migrate into the bulk of the polymers. When t-PB was treated with a $ClSO_2$PhTD solution (0.001 M) for 20 h, a contact angle of 70° was measured. After being placed in water (50 °C, 1 h) and then dried, the sample had a contact angle of 64°. Neutralization of the sample with aqueous NaOH (1 M) resulted in a film having a contact angle of 36°. Thus, a long reaction time at a low concentration of RTD (conditions that should favor substantial migration of RTD into the polymer) did not result in cracked, curled or soluble polymer film. An ESCA angular depth profile study of similarly treated t-PB showed that sulfur and nitrogen are surface species.

Figure 1 shows the chemical transformations occurring during this sequence of treatments based on the reactions of analogous nonpolymeric model compounds (20). The ESCA spectra of the polymer films taken after each transformation also support this scheme, Figure 2. The high resolution spectra of species 5 and 6 show N(1s) peaks at 401 eV (nonprotonated nitrogen) and 402.7 eV (protonated nitrogen) while samples neutralized with base, 7, do not show protonated nitrogen peaks. Also, the latter sample shows two species

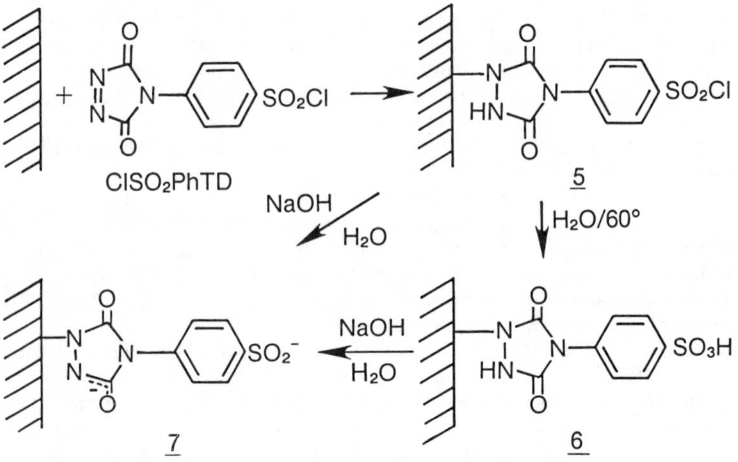

Figure 1. Treatment sequence of samples examined by ESCA.

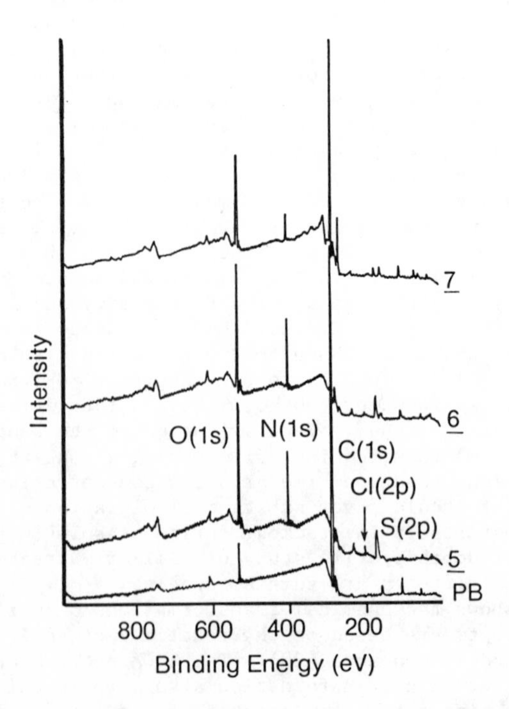

Figure 2. ESCA Spectra of t-PB, 5, 6, and 7.

of sodium. The Cl(2p) peak (199.5 eV) indicates a large abundance of Cl for 5. No chlorine is detected for the base treated samples, 7. Interestingly, the intensity of the S(2p) peak (168 eV) decreases in each sample along the progression of treatment, 5→6→7. This suggests that the some of the surface is removed during each step.

Difunctional $(TD)_2DPM$ was another reagent that could be used at low concentrations and long reaction times while giving only surface modification. Treatment of c-PI and c-PB in 0.01 M $(TD)_2DPM$ solution for 24 h produced films with 30° contact angles after neutralization. Presumably the added size of this reagent and its ability to crosslink the polymer greatly slows its diffusion within the polymer.

Another advantage of this reagent is that the hydrophilic surfaces produced by it show only a small loss of hydrophilicity on aging. Most of the surfaces hydrophilized with mono-TD reagents showed losses in hydrophilicity after 5-10 days; contact angles rose to 70-90°, Figure 3. The increase in contact angle is due to rearrangement of the surface which causes unmodified polymer to migrate to the surface. Low energy surfaces are known to be thermodynamically favored over high energy surfaces (29). Also, entropic factors are expected to favor the mixing of the urazole groups on the surface into the bulk polymer. Such factors have been shown to be important in surface modified polymers that are above or near their Tg (29).

Alternate explanations for the loss of hydrophilicity upon aging include the blooming of hydrophobic impurities in the polymer to the surface (30), or the deposition of ubiquitous airborne contaminants onto the surface. However, since the $(TD)_2DPM$ surfaces can remain relatively hydrophilic for up to 1 year, these are probably not significant factors in the loss of hydrophilicity. It should be noted that oxidized polydiene films (exposed to air 1 year) can display contact angles as low as 65°.

Better surface stability could be achieved if the surface components are less mobile. Some success has been achieved with polyethylene and perhalogenated polymer surfaces (8,11,12,29). Crosslinking the surface should also decrease its mobility, and this is how $(TD)_2DPM$ provides added stability, Figure 3. Some surface rearrangement does appear to occur initially, but samples have retained good hydrophilicity (contact angle 45-55°) for a year.

In order to obtain stable hydrophilic surfaces with $(TD)_2DPM$ it is necessary to use it at the low concentration and long reaction time described above. Polydienes treated with 0.1 M solutions of $(TD)_2DPM$ for 3 h did show good hydrophilicity (contact angle 35°) initially. However, the hydrophilicity of the surfaces decreased similarly to those of PhTD treated surfaces. Apparently, the higher concentration of reagent favors the reaction of the surface double bonds, with different $(TD)_2DPM$ molecules which results in little or no crosslinking. Whereas, the lower concentration of reagent allows time for a $(TD)_2DPM$ molecule to react at one end and then still find double bonds available to react with the remaining TD end. The net result is crosslinking.

Attempts to obtain permanently hydrophilic surfaces by other types of crosslinking chemistry were not successful. The other methods investigated included forming ionic crosslinks, using cured

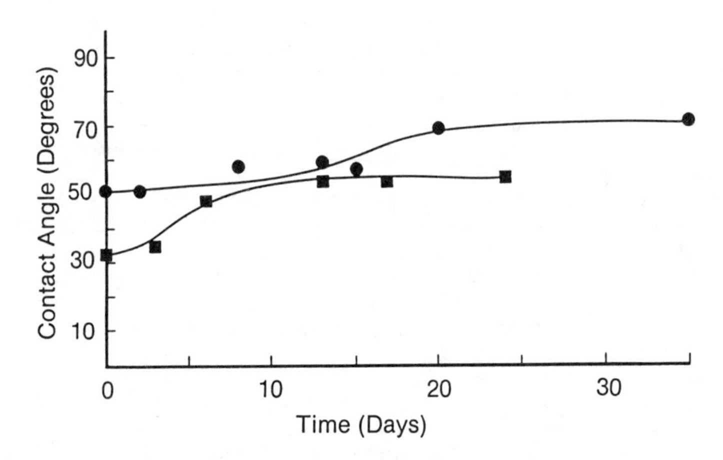

Figure 3. Contact angle vs. time for PB treated with PhTD (●) and PI treated with $(TD)_2DPM$ (■). Both neutralized with NaOH.

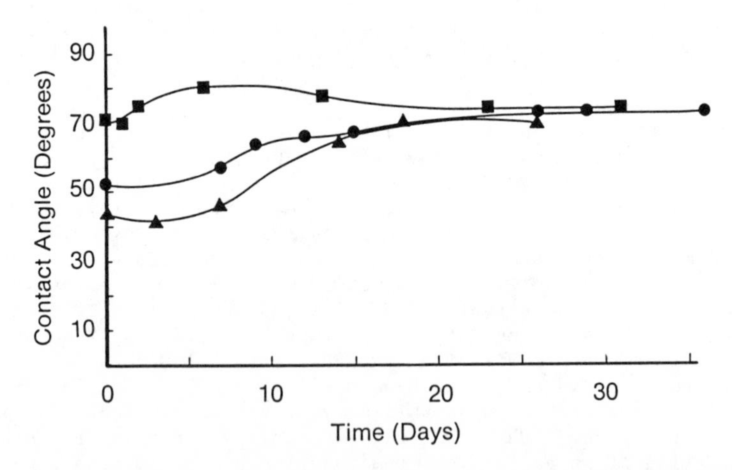

Figure 4. Contact angle vs. time for PB treated with PhTD and neutralized with 3,3-ionene hydroxide (■), styrene-butadiene ABA block copolymer treated with PhTD and neutralized with $NaHCO_3$ (●), and PI treated with PhTD and neutralized with CaO (▲).

polymers as substrates, or using block copolymers. By using a base with a polyvalent cation, one might hope to achieve crosslinking through ionic interactions (31). The use of CaO in place of NaOH did not enhance the permanence of PhTD treated surfaces. The use of 3,3-ionene hydroxide, Figure 4, did not provide good hydrophilicity even initially (contact angle 70°). Currently, the explanation for this behavior is not known. A so called "hydrophobic effect", resulting from the conformation of the ionene on the surface, may be operative.

The use of lightly crosslinked polymers did result in hydrophilic surfaces (contact angle 50°, c-PI, 0.2 M PhTD). However, the surfaces displayed severe cracking after 5 days. Although qualitatively they appeared to remain hydrophilic, reliable contact angle measurements on these surfaces were impossible. Also, the use of a styrene-butadiene-styrene triblock copolymer thermoplastic elastomer did not show improved permanence of the hydrophilicity over other polydienes treated with PhTD. The block copolymer film was cast from toluene, and transmission electron microscopy showed that the continuous phase was the polybutadiene portion of the copolymer. Both polystyrene and polybutadiene domains are present at the surface. This would probably limit the maximum hydrophilicity obtainable since the RTD reagents are not expected to modify the polystyrene domains.

Conclusions

The low temperature ene reaction of triazolinediones with polydienes occur under heterogeneous conditions to yield hydrophilic surfaces, especially after neutralization of the resulting pendant urazole groups. Permanent hydrophilic surfaces can be obtained when $(TD)_2DPM$ is used. The use of the other RTDs tested results in surfaces that lose their hydrophilicity within 5-20 days. In applications such as improving the adhesion of rubber to other substrates, these reagents are probably sufficient (18). However, when more permanent hydrophilic surfaces are desired a bis-triazolinedione such as $(TD)_2DPM$ would be required.

Acknowledgments

Most of the sample preparations and contact angle measurements were made by A. D. Karnas. K. W. Littlepage and G. G. Engerholm did the ESCA spectroscopy. J. Burns did the electron microscopy on the block copolymers. I thank D. F. Hager and G. B. Butler for helpful discussions.

Literature Cited

1. Zisman, W. A. In *Adhesion Science and Technology*; Lee, L.-H., Ed; Plenum: New York, 1975; p. 55.
2. Huntsberger, J. R. In *Contact Angle, Wettability, and Adhesion*; Fowkes, F. M., Ed.; Advances in Chemistry Series No. 43; American Chemical Society: Washington,D. C., 1964; Chapter 7.
3. Andrew, E. H.; King, N. E. In *Polymer Surfaces*; Clark, D. T.; Feast, W. J., Eds.; John Wiley & Sons: New York, 1978; Chapter 3.

4. Bixler, J. J.; Sweeting, O. J. In The Science and Technology of Polymer Films, Vol. II; Sweeting, O. J., Ed.; Wiley-Interscience: New York, 1971; Chapter 1.
5. Lyman, D. J. Angew. Chem. Int. Ed. Engl. 1974, 13, 108.
6. Carlsson, D. J.; Wiles, D. M. J. Macromol Sci., Rev. Macromol. Chem. 1976, 14, 65.
7. Shafrin, E. G. In Polymer Handbook, 2nd ed.; Brandrup, J.; Immergut, E. H., Eds.; Wiley-Interscience: New York, 1975; p. III-221.
8. Rasmussen, J. R.; Stedronsky, E. R.; Whitesides, G. M. J. Am. Chem. Soc. 1977, 99, 4736.
9. Yasuda, H. J. Polym. Sci., Macromol. Rev. 1981, 16, 199.
10. Tazuke, S.; Kimura, H. Makromol. Chem. 1978, 179, 2603.
11. Dias, A. J.; McCarthy, T. J. Macromolecules 1985, 18, 1826.
12. ibid. 1984, 17, 2529.
13. Butler, G. B. Ind. Eng. Chem. Prod. Res. Dev. 1980, 19, 512.
14. Leong, K.-W.; Butler, G. B. J. Macromol. Sci., Chem. 1980, A14, 287.
15. Butler, G. B.; Williams, A. G. J. Polym. Sci., Polym. Chem. Ed. 1979, 17, 1117.
16. Rout, S. P.; Butler, G. B. Polym. Bull. 1980, 2, 513.
17. Saville, B. J. Chem. Soc., Chem. Commun. 1971, 635.
18. Cutts, E.; Knight, G. T. U.S. Patent 3 966 530, 1976.
19. Stickler, J. C.; Pirkle, W. H. J. Org. Chem. 1966, 31, 3444.
20. Keana, J. F. W.; Guzikowski, A. P.; Ward, D. D.; Morat, C.; VanNice, F. L. J. Org. Chem. 1983, 48, 2654.
21. Yen, S. P. S.; Casson, D.; Rembaum, A. In Water Soluble Polymers; Bikales, N. M., Ed.; Plenum: New York, 1973; p. 291.
22. Razvodovskii, E. F.; Nekrasov, A. V.; Enikolopyam, N. S. Vysokomol. Soedin., Ser. B. 1972, 14, 338; Chem. Abstr. 1972, 77, 75533z.
23. Dwight, D. CHEMTECH 1982, 12, 166.
24. Adamson, A. W.; Ling, I. In Contact Angle, Wettability, and Adhesion; Fowkes, F. M., Ed.; Advances in Chemistry Series No. 43; American Chemical Society: Washington, D.C.; 1964, Chapter 3, p. 69.
25. Bates, R. B.; Schaefer, J. P. In Research Techniques in Organic Chemistry; Prentice-Hall: Englewood, NJ 1971; p. 18.
26. Hoffmann, H. M. R. Angew. Chem. Int. Ed. Engl. 1969, 8, 556.
27. Snider, B. B. Acc. Chem. Res. 1980, 13, 426.
28. Ohashi, S.; Butler, G. B. J. Org. Chem. 1980, 45, 3472.
29. Holmes-Farley, S. R.; Whitesides, G. M. Polym. Mater. Sci. Eng. 1985, 53, 127.
30. Brewis, D. M.; Briggs, D. Polymer 1981, 22, 7.
31. Kinsey, R. H. J. Appl. Polym. Chem., Appl. Polym. Symp. 1969, 11, 77.

RECEIVED August 27, 1987

Chapter 17

Surface Heparinization of Poly(ethylene terephthalate) Films Modified with Acrylic Hydrogels

Cristofor I. Simionescu, Monica Leanca, and Ioan I. Negulescu

"Petru Poni" Institute of Macromolecular Chemistry, Iaşi 6600, Romania

A semi-interpenetrated network was obtained by bulk polymerization of 2-hydroxyethyl methacrylate incorporated in DMF treated PET films by solvent-exchange technique, followed by treatment of films in electrical discharges. Heparinization was accomplished by reacting glutaraldehyde with heparin and poly(2-hydroxyethyl methacrylate) present on the surface of modified polyester films. The immobilization of heparin was indirectly evidenced by chromatographying the silylated hydrolyzates of heparinized PET films and heparin, respectively. In vitro experiments demonstrated the enhanced thromboresistance of heparinized films.

More than 20 years have passed from the first report of Gott et al. (1) on the bonding of heparin on colloidal graphite surface and some heparin-containing materials developed since then have found clinical applications (2). Surface modification has long been seen as offering the advantage of very wide variety in the chemical nature of the blood-presenting surface while allowing the choice of the substrate to be on the bases of the mechanical properties needed. The surface modification includes the grafting of polymers, such as macromolecular hydrogels, which are inherently too weak mechanically to be useful as unsupported materials. Polymers of this type have been extensively investigated both because there were early indications that poly(2-hydroxyethyl methacrylate), pHEMA, possessed some degree of thromboresistance and because of the hypothesis that a gel surface would be less recognizable as a foreign surface to the blood (2,3). Polymeric materials with covalently immobilized heparin were shown to display enhanced throm-

0097-6156/88/0364-0229$06.00/0

boresistance in vitro and in vivo experiments since the presence of heparin substantially changes the character of adsorbed proteins on a polymer surface and the number of adhered platelets (4,5).

The aim of this paper is to report the bonding of heparin on poly(ethylene terephthalate) films containing an acrylic hydrogel (pHEMA), the method of preparation of the support material and some of its properties.

Experimental

Strips of unidimensionally oriented PET films (Terom, Romania) of 0.3 mm thickness, washed successively with detergent solution, water and acetone, were treated with dimethylformamide, DMF (Fluka), at various temperatures (50-140°C) for 15 or 30 min. The films were centrifugated and immersed in a mixture of initiator (AIBN, Serva) and monomer (HEMA, Eastman Kodak) for different periods (2-100 hr). Subsequently the films were wiped up with filter paper and introduced in glass ampoules which were degased, filled with nitrogen and sealed. The ampoules were put in an oven and HEMA incorporated in the polyester matrix was allowed to polymerize at 70-90°C for 5-48 hr. The films were then subjected to a nitrogen glow discharge (input power, 40 W; frequency, 2.5 MHz) and immersed thereafter for 35 min in an aqueous adhesive mixture containing (6) polyvinylalcohol (10%), glycerin (4%), magnesium chloride (5%), heparin (Biofarm-Romania, 5000 UI/mL) (1%) and glutaraldehyde (0.25%). The modified PET films designated as control samples were immersed in a similar mixture devoid of heparin. After drying for 5 hr at room temperature the films were heated at 80°C for 100 min and then washed with 3M NaCl solution for removing the unreacted components, heparin included.

The samples for gas-chromatographic determinations were prepared according to Neeser et al. (7) as follows. Known amounts of heparinized films or control samples (0.5g) and heparin (0.01g) were hydrolyzed in separate experiments with 4M CF_3COOH at 125°C for 60 min. The hydrolyzate was dried up by evaporation, pyridine (1 mL) was added and the mixture was transfered in a reaction tube containing methoxyamine hydrochloride (0.003g) which was subsequently kept at 80°C for 2 hr. N,O-bis (trimethylsilyl)trifluoroacetamide (0.2 mL) was introduced after cooling and the mixture was allowed to react at 80°C for 15 min. A determined volume (1 µL) was introduced thereafter in gas-chromatograph (Fractovap-M-2350, Carlo Erba, equipped with FID; column, SE-30).

X-ray measurements were made at room temperature using a TUR M-62 (DDR) spectrometer.

Mechanical properties were determined according to Romanian standards using an Instron 1114.

The thromboresistance of films was determined in vitro as the time of blood-clotting.

Results and Discussion

The in situ bulk polymerization of vinyl monomers in PET and the graft polymerization of vinyl monomers to PET are potential useful tools for the chemical modification of this polymer. The distinction between in situ polymerization and graft polymerization is a relatively minor one, and from a practical point of view may be of no significance. In graft polymerization, the newly formed polymer is covalently bonded to a site on the host polymer (PET), while the in situ bulk polymerization of a vinyl monomer results in a polymer that is physically entraped in the PET. The vinyl polymerization in the PET is usually carried out in the presence of the swelling solvent, thereby maintaining the swollen PET structure during polymerization. The swollen structure allows the monomer to diffuse in sufficient quantities to react at the active centers that have been produced by chemical initiation (with AIBN) before termination takes place. It was shown (8,9) that the pretreatment of PET yarns with certain strongly interacting solvents can lead not only to swelling but also to irreversible modifications of polymers structure. The basis of structural modification during the DMF treatment of PET is solvent-induced crystallization which occurs while the PET structure is swollen by DMF. At low treatment temperatures (i.e., 50-100°C, Table I), only small crystallites are formed and after removal of the solvent the swollen structure cannot be supported by the small crystallites and consequently collapses.

Table I. Structural Parameters of PET Films Treated in DMF at Different Temperatures

Treatment Temp. (°C)	Time (min)	Degree of Crystal. (%)	Dimension of Crystallites (Å)	Amorphous Volume Assoc. (Å^3)
Untreated	-	32.7	106.1	163,200
50	15	35.5	109.3	159,750
50	30	37.1	115.2	146,350
100	15	42.1	123.9	132,470
100	30	44.7	127.5	127,327
120	15	47.5	131.3	114,655
120	30	55.4	144.8	105,979
140	15	54.7	143.7	107,830
140	30	57.3	151.6	97,270

As shown in Figure 1, when the DMF treatment is carried out at higher temperatures (120-140°C), the crystallinity of the polyester increases and formation of larger crystallites occurs (Table I). These larger and more stable crystallites are capable of supporting the solvent-swollen structure to a greater extent so that total collapse of the swollen structure upon removal of the solvent is prevented.

Vinyl monomers can be introduced into the PET by two different procedures. In one procedure, the monomer is introduced into the PET after removal of the DMF. In the other procedure, the monomer is solvent exchanged with the DMF in the PET. However, when DMF was removed from the PET prior to monomer incorporation, lower uptake of monomer was observed as compared with results obtained by the solvent exchange technique (9). Consequently the solvent exchange technique was used in the present work for the incorporation of HEMA in the PET films. The dependence of the HEMA uptake on the temperature of DMF treatment (HEMA-solvent exchange, 72 hr) is shown in Figure 2. It can be seen that only in the case of DMF pretreatments at elevated temperatures, when solvent-induced crystallization in the swollen state takes place, is the monomer incorporated significantly. The time dependence of HEMA incorporation by DMF-treated (140°C, 15 min) PET films is presented in the same figure. It can be seen that the process of monomer incorporation is completed in 60-70 hr. Accordingly, an equilibration time of three days was generally allowed.

Polymerization of HEMA incorporated in the PET films is dependent both on the initiator concentration and reaction temperature. In order to overcome the low initiation efficiency inside the PET, due to the low mobility of the free radicals formed inside the PET structure, as well as that of the monomer itself, high initiator concentrations were used. The results are listed in Table II. However, satisfactory conversions of HEMA in pHEMA were obtained only when the polymerization was carried out at 80-85°C for 20-40 hr, even if the initiator concentration was high (Figure 3).

Table II. Polymerization of HEMA Incorporated in PET Films Treated in DMF (140°C/15 min)

Incorporated HEMA, (%)	AIBN in HEMA (Mol/L)	Incorporated Polymer, (%)	HEMA Conversion, (%)
20.8	0.03	0.36	1.9
20.8	0.06	1.81	8.7
20.8	0.08	7.43	35.7
20.8	0.10	8.33	40.1

The increase of the polymerization temperature to 90°C resulted in a low monomer conversion, despite of a longer reaction time (50 hr), perhaps due to the fact that at this temperature the probability of termination reactions of the newly formed free radicals is vastly increased (9).

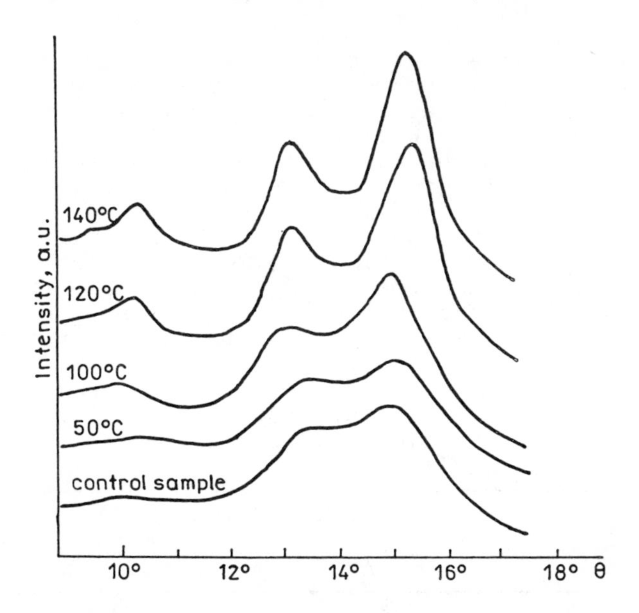

Figure 1. Wide angle X-ray diffraction pattern from PET films exposed to DMF at various temperatures for 15 min.

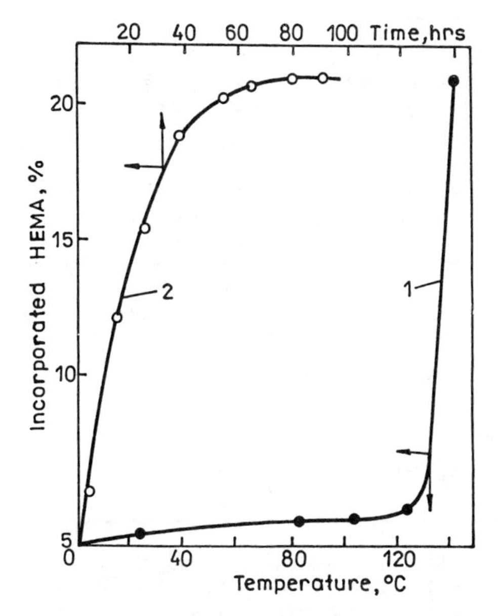

Figure 2. The dependence of HEMA incorporated in PET films on: /1/ the temperature of DMF treatment, and /2/ the time of immersion in HEMA of films treated in DMF at 140°C for 15 min.

When the pHEMA containing PET films were subjected to nitrogen glow discharge, the acrylic polymer was grafted to the host polyester (10,11), giving rise to a semi-interpenetrated network structure which in contact with water assures the formation of HEMA-based hydrogel. Both the solvent treatment and pHEMA grafting processes determined a complete modification of the appearance of PET films, and it can be seen from the scanning microscope imagines shown in Figure 4. On the other hand, according to the data presented in Table III, the mechanical properties of pHEMA containing PET samples subjected to electrical discharges in nitrogen did not change dramatically.

Table III. Mechanical Properties of Parent and Modified PET Films

Sample	Initial Modulus (Kg/cm^2)	Tensile Strength (daN/mm^2)	Break Extension (%)
Parent PET Film	35,000	22	23.3
DMF Treated PET Film (140°C/15 min)	28,000	12	34.0
8.5% pHEMA Containing PET Film	32,000	16	28.7

However, the starting decomposition temperature of films modified with the acrylic polymer decreased with 60-80°C (Table IV), but the thermal resistance remained in practical limits.

Table IV. Thermal Resistance of Parent and Modified PET Films

Sample	Initial Decomposition Temperature (°C)	Temperature of 5% Weight Loss (°C)
Parent PET Film	312	385
DMF Treated PET Film (140°C/15 min)	311	379
8.5% pHEMA Containing PET Film	258	294

The presence of the polar hydrogel on the surface of PET films led, as seen from Table V, to a decrease in their contact angle with water and to a corresponding rise of the critical surface tension through the polar component γ_s^p, γ_s^d being the dispersion component.

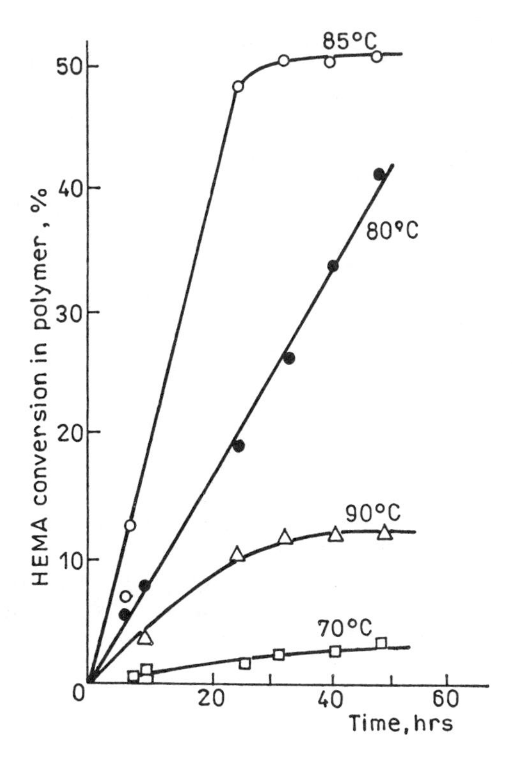

Figure 3. Conversion in polymer of HEMA incorporated in PET films. Dependence on time and temperature of polymerization.

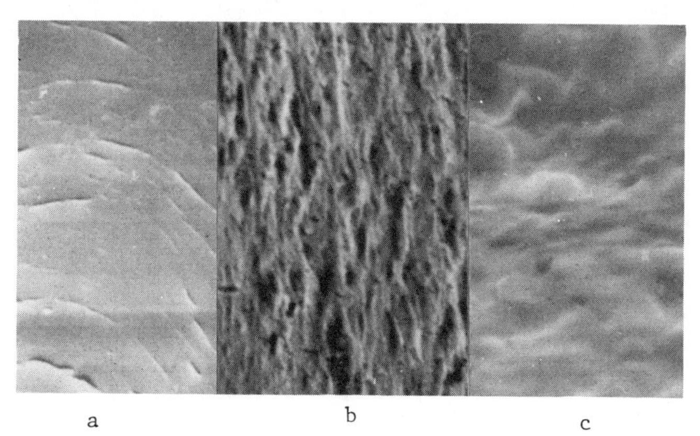

Figure 4. Scanning electronic microscope images (x5000) of (a) parent PET film; (b) PET film treated in DMF at 140°C for 15 min; (c) 10.5% pHEMA containing PET film.

The contact angle with formamide remained almost the same. As shown in Figure 5, the contact angle with water further decreased in time since the absorption of water by pHEMA incorporated in PET is accompanied by swelling of the hydrogel, thereby increasing the surface polarity and ability of the film to be wetted.

Table V. Wetting Angle at the Contact with Water (θ_w) and Formamide (θ_f) and Critical Surface Tension (γ_s) at 20°C

Sample	Wetting Angle (Degree)		Critical Surface Tension (mN/m)		
	θ_w	θ_f	γ_s	γ_s^d	γ_s^p
Parent PET Film	72.5	53.0	38.3	28.8	9.5
10.5% pHEMA Containing PET Film	54.0	51.7	46.3	22.6	23.7

The heparin was bonded to the surface of modified PET films through the agency of hydroxylic groups provided by the acrylic hydrogel. The presence of heparin in these films was evidenced by gas-chromatography. It can be seen from Figure 6 that the "fingerprints" of the heparin-reacted modified PET film and heparin are similar while the chromatogram of control sample is void of peaks common to these two samples.

The thromboresistance of heparinized films, as shown in Table VI, was enhanced by higher hydrogel content since the concentration of hydroxylic groups on modified PET film surface, able to react with glutaraldehyde in order to bind heparin, was higher.

Table VI. Thromboresistance of Heparinized pHEMA Containing PET Films

No.	PET Sample	Incorporated pHEMA (%)	Clotting Time of Blood (min)
1	Control	8.5	8
2	Heparinized PET	2.5	22
3	Heparinized PET	6.6	91
4	Heparinized PET	8.5	108

Blood that has been stored for 60 min in the presence of heparinized PET films (Sample No. 4), when put into another glass vessel was clotted within 5 min, which is the clotting time of blood not exposed to heparinized materials (4). Therefore, any reduction in thrombus formation on the heparinized film surface can only be attributed to the activity of bound heparin, i.e. the free heparin was not eluted into the blood.

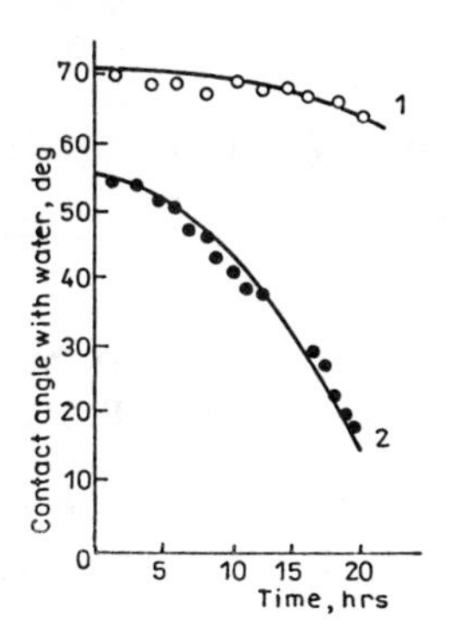

Figure 5. The decrease of contact angle with water for /1/ parent PET film; /2/ 8.5% pHEMA containing PET film.

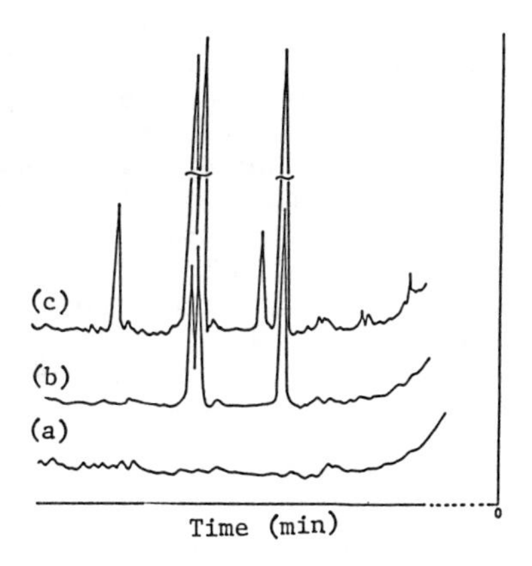

Figure 6. Gas chromatograms of silylated hydrolyzate of (a) control sample; (b) heparinized PET film; (c) heparin.

Literature Cited

1. Gott, V.L.; Whiffen, J.D.; Dutton, K.C., Science 1963, 142, 1297.
2. Leininger, R.I. In Biomedical and Dental Applications of Polymers; Gebelein, C.G.; Koblitz, F.K., Eds.; Plenum: New York, 1980.
3. Andrade, J. Hydrogels for Medical and Related Applications; American Chemical Society: Washington DC, 1976.
4. Plate, N.A.; Valuev, L.I., Biomater. 1983, 4, 14.
5. Heyman, P.W.; Kim, S.W., Makromol. Chem., Suppl. 1985, 9, 119-124.
6. Gooser, M.F.A.; Sefton, M.V., J. Biomed Mater. Res. 1979, 13, 347.
7. Neeser, J.R., Carbohydr. Res. 1985, 138, 189.
8. Weigmann, H.D.; Scott, M.G.; Ribnick, A.S.; Rebenfeld, L., Text Res. J. 1976, 46, 574.
9. Avny, Y.; Rebenfeld, L.; Weigmann, H.D., J. Appl. Polymer Sci. 1978, 22, 125-147.
10. Moshnov, A.; Avny, Y., J. Appl. Polymer Sci. 1980, 25, 89.
11. Simionescu, C.I.; Denes, F.; Macoveanu, M.; Negulescu, I.I., Makromol. Chem., Suppl. 1984, 8, 17-25.

RECEIVED August 27, 1987

Chapter 18

Reactions of Metal Vapors with Polymers

Colin G. Francis[1], Scott Lipera, Pascale D. Morand, Peter R. Morton, John Nash, and Peter P. Radford

Department of Chemistry, University of Southern California, Los Angeles, CA 90089–1062

A novel polysiloxane, containing the isocyanide group pendent to the backbone, has been synthesized. It is observed to react with the metal vapors of chromium, iron and nickel to afford binary metal complexes of the type $M(CN-[P])_n$, where n = 6, 5, 4 respectively, in which the polymer-attached isocyanide group provides the stabilization for the metal center. The product obtained from the reaction with Fe was found to be photosensitive yielding the $Fe_2(CN-[P])_9$ species and extensive cross-linking of the polymer. The Cr and Ni products were able to be oxidized on exposure of thin films to the air, or electrochemically in the presence of an electron relay. The availability of different oxidation states for the metals in these new materials gives hope that novel redox-active polymers may be accessible.

The preparation of elaborate metal species within polymer media represents an important synthetic problem in view of the increasing use of metal-containing polymers in heterogeneous reaction systems (1). Interest in these materials has been prompted in great part by recent advances in photocatalysis and electrocatalysis, particularly as applied to solar energy conversion processes (2). Our previous studies with organometallic polymers (3) have led us to investigate now the synthesis of new polymers which might act as templates for the formation of polymer-encapsulated binary metal complexes at both a monometallic and cluster level.

In this paper, we describe the synthesis and initial reactivity studies of a macromolecule (Scheme 1) containing the isocyanide functional group [Z = $-N{\equiv}C$] tethered to a polysiloxane backbone. The polymer was designed to incorporate the following features:

[1]Current address: Centre Suisse d'Electronique et de Microtechnique S.A., CH–2000 Neuchatel 7, Switzerland

0097–6156/88/0364–0238$06.00/0

(i) simple synthesis - the supporting polymer itself is prepared via a straightforward one-step polymerization procedure (Scheme 1, step 1) and can be fully characterized before inclusion of the metal centers. Direct incorporation of metal atoms by metal vapor routes (4) has been employed in these early studies (Scheme 1, step 2) in order to prevent the formation of side-products whose separation would be difficult.
(ii) versatility - the pendent isocyanide function provides a site of attachment for a wide range of transition metals (5) and is expected to favor formation of diverse metal structures [mono- and multi-metallic]. It is worthwhile to note that the RNC molecule is formally isoelectronic with CO, a ubiquitous ligand in coordination chemistry and one well-known to stabilize a large number of transition metal complexes (6).

$M \leftarrow :C\equiv O$ $M \leftarrow :C\equiv N-R$

A   B

Small-molecule isocyanides possess a ligand chemistry which is only slightly less well-developed (5).
(iii) ease of characterization - the polymer, a polysiloxane, is soluble in organic media, in addition it is capable of completing the coordination sphere around each metal center, thereby facilitating the characterization spectroscopically and structurally, using small molecule compounds.

We shall focus here on the synthesis of the isocyanide-containing polymer. Several reactions of the polymer with the metal vapors of Cr, Fe and Ni using a matrix-scale modeling technique, as well as synthetic-scale metal vapor methods, are then presented in order to demonstrate the reactivity of the isocyanide groups on the polymer. Finally, preliminary studies of the reactivity of the polymer-based metal complexes are described.

RESULTS AND DISCUSSION

Synthesis of Poly(dimethyl-co-isocyanopropylmethylsiloxane)

The isocyano-substituted polysiloxane 2 was synthesized in high yield by the procedure shown in Scheme 2. Formation of 1 was accomplished following known procedures (7) and characterized by its ^{1}H, ^{13}C NMR, infrared and mass spectra. Base-catalyzed polymerization of 1 and $Me_2Si(OEt)_2$ afforded the branched copolymer 2. Although polymerization of siloxanes in the presence of base usually gives rise to low molecular weights for the resulting polymers (8), this was the method of choice due to the acid-sensitivity of the isocyanide function (9). Typically, 1 and $Me_2Si(OEt)_2$ were mixed in an approximately 1:17 molar ratio and stirred for 24 hr. in the presence of a catalytic amount of NaOH and a deficiency of water [i.e. less than 1 mole of water per mole of silane]. Maintaining the reactants in the dark under a nitrogen atmosphere during the polymerization led, after work-up, to polymers which were colorless/pale-yellow viscous fluids. In most preparations a small amount of solid polymer was also obtained. It was possible to separate this

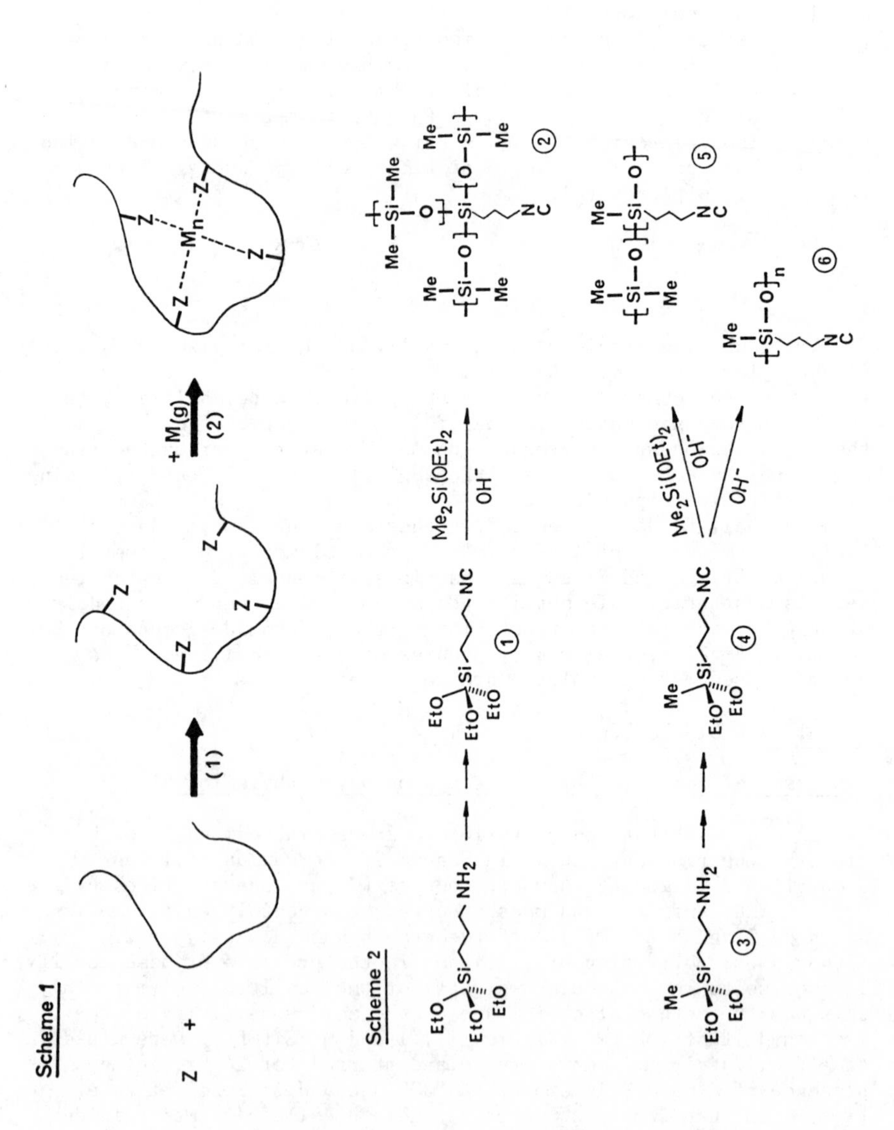
Scheme 1
Z +
(1)
+ M(g)
(2)
Scheme 2
Me2Si(OEt)2
OH-

solid, which is presumed to be the homopolymer of 1, by centrifugation. The nature of the resulting polymer is extremely sensitive to the molar ratio of water to silane, as shown by Table I. On increasing the ratio further to 1:1 an insoluble rubber was obtained. This increase in molecular weight follows a trend which has been observed previously for the homopolymerization of $Me_2Si(OEt)_2$ (10). In addition to molecular weight determinations, which were carried out by gel permeation chromatography, 2 was characterized by ^1H and ^{13}C NMR spectra and its infrared spectrum (Figure 1). The infrared spectrum in the range 2500 - 1400 cm^{-1} shows a characteristic isocyanide stretch at 2148 cm^{-1}. In the region of interest for coordinated isocyanides, the spectrum is clear except for two weak bands associated with the polysiloxane backbone at 1945 and 1580 cm^{-1}. The ^1H NMR spectrum (Figure 1) contains multiplets at δ0.66, 1.72 and 3.34 due to the propyl side-chain, in addition to the resonances of the methyl groups at δ0.08 and 0.05. The polymer chains are terminated by -OEt groups as evidenced by a quartet at δ3.70 and a triplet at δ1.18. The ^{13}C NMR spectrum clearly shows the isocyanide carbon as a broad singlet at 156.0 ppm with the carbon atom of the propyl chain adjacent to the CN- group occuring at 43.8 ppm as three peaks of equal intensity, due to coupling to the nitrogen [J_{NC} = 6.1 Hz].

From the ^1H NMR data the ratio of isocyanide to Me_2SiO- groups was found to vary between 1:4 and 1:13 depending on the water:silane ratio. As shown in Table I the ratio of EtO- end-groups to isocyanide functions was ~2, consistent with approximately 2-3 CN- groups in each polymer chain. In each case the values of M_n determined by GPC indicated a slightly greater number [3-4] of isocyanide groups.

As will be seen later, the existence of residual EtO- groups [or possibly -OH groups] on the polymer 2 can lead to unexpected reactions with some metal atoms. Therefore, 2 was also prepared with the chain-terminating groups being Me_3SiO- by stirring 2 with Me_3SiOEt in the presence of base. The replacement of the EtO- groups was evidenced in the ^1H NMR by the loss of the resonances due to EtO-.

The procedure described here is not limited to the preparation of polymers such as 2. Starting from the difunctional silane 3 we have synthesized a copolymer, poly(dimethyl-co-isocyanopropylmethylsiloxane) 7, as well as a linear homopolymer, poly(isocyanopropylmethylsiloxane) 8 (Scheme 2). Indeed, preparation of a monofunctional analogue of 1 and 4 creates the potential for end-capping with an isocyanide function any polymer containing other functional groups, thereby in principle permitting mixed ligand complexes of polymers to be accessed.

Metal Vapor Studies with 2:

Metal atoms can be incorporated into polymers using two approaches. For probing new reactions between metal atoms and polymers a small-scale spectroscopic approach, sometimes referred to as the Fluid Matrix Technique (11), is used. The coreactant polymer matrix, containing on the order of 0.5 μl of polymer, is preformed on an optical surface. In the case of viscous fluids such as 2 the material is painted on the substrate and held at temperatures ranging typically from 200 to 270 K. The temperature is chosen to maintain low volatility but retain mobility. Under high vacuum [10^{-6} torr]

Table I. Effect of the Water–Silane Molar Ratio

n^a =	0.65	0.77	0.78
Weight average molecular weight (M_w):	1,462	2,463	3,123
Number average molecular weight (M_n):	1,247	1,719	2,468
Polydispersity:	1.172	1.433	1.265
EtO- : -NC[b]	2.0	2.2	2.0
[Me_2SiO] : -NC[b]	4	13	10
Number of -NC per polymer chain:	2.5	2	2

a. moles of water per total mole of silane
b. as determined by end-group analysis

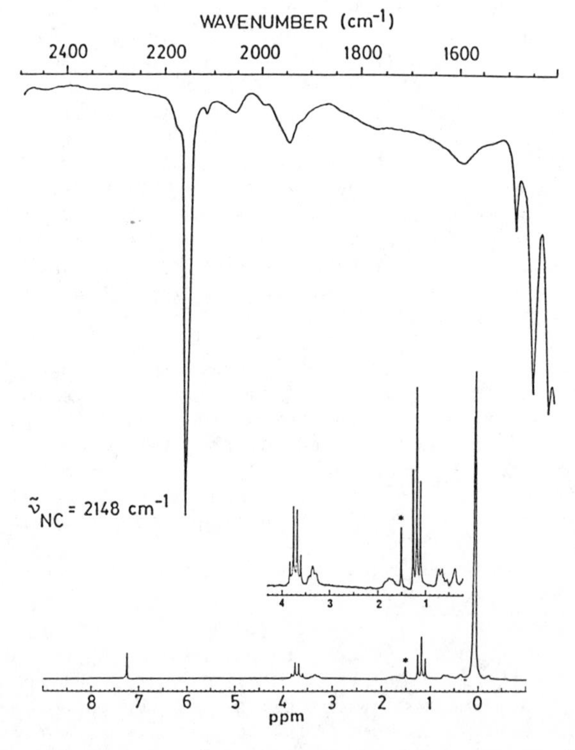

Figure 1. Infrared and ^1H NMR spectra for CN-[P]. (^1H NMR spectrum in $CDCl_3$; peak marked with an asterisk is due to the presence of water in the solvent.)

metal atoms are generated resistively [0.4 μg.min^{-1}] and pass by a line-of-sight path to the substrate where they penetrate and diffuse into the polymer film and react with functional groups present. Product formation can then be monitored using spectroscopic techniques, in this case infrared and ultraviolet-visible spectroscopy.

Clearly this approach is not suitable for preparing large quantities of products, its main purpose being to permit the greatest amount of information to be obtained concerning the reactivity of a new material. If the coreactant is expensive and/or difficult to prepare then this procedure is invaluable. However, it is important to consider that the quantities of derivatized polymer obtained in this approach [$\sim 10^{-6}$ mole based on -N≡C repeat unit] might well represent sufficient material if such a process were to be used for the direct preparation of chemically modified electrodes (12), or incorporated into a planar microfabrication process, with which it would appear to be compatible.

For detailed characterization and extensive studies of reactivity, multi-gram quantities are still needed and large-scale metal vapor synthetic routes are necessary. The equipment required for this is well-documented (4) and so will not be described in detail here. The principles are those of the Fluid Matrix Technique except that in order to accommodate 10-100 gram of polymer, the coreactant is contained within a rotating flask which serves to provide a continuously renewed film as metal atoms are produced under high vacuum.

Focusing on reactions using the Fluid Matrix Technique, we have studied the interaction of chromium vapor with 2 at 200 K (13). The resulting film was found to contain metal complexes encapsulated within the polymer in which the isocyanide group adopts a well-defined octahedral arrangement around the chromium center, i.e. a species of type $Cr(CN\text{-}[P])_6$. Since characterization of this metal complex within the polymer is not trivial we shall develop the analysis in a little detail.

The infrared spectrum of the polymer matrix species shows a characteristic broad band around 1950 cm^{-1}, split into three peaks (Figure 2). The small-molecule model compounds $Cr(CN\text{-}TESP)_6$ [TESP-NC = 1, Scheme 2] or $Cr(CN\text{-}Bu^n)_6$ also show very similar infrared spectra (14). The IR active stretching mode for L in a purely octahedral ML_6 complex is of T_{1u} symmetry. Thus, $\bar{\nu}_{CO}$ for $M(CO)_6$ [M = Cr, Mo, W] appears as a single, sharp peak (15). Therefore, the splitting of this band in the corresponding isocyanide complexes arises from a lowering of symmetry, which may be attributed to a distortion of the $M(CNC\text{-})_6$ fragment and a non-symmetrical orientation of the R groups. The origin of the distortion of $M(CNC\text{-})_6$ can be understood by considering the bonding of [P]-NC to the metal, as shown below. Electron donation from the isocyanide (B) to an electron-rich metal is counterbalanced by back-donation from the metal to vacant orbitals of π-symmetry on the ligand. The effect of populating these orbitals, which are antibonding in nature, is to weaken the C-N bond so that eventually in the limiting case one arrives at a structure analogous to B' in which the extra electron density is localized on the nitrogen and the C-N-R angle approaches 120°. Of course B and B' represent canonical forms, so that in reality a range of structures intermediate between B and B' may be realized.

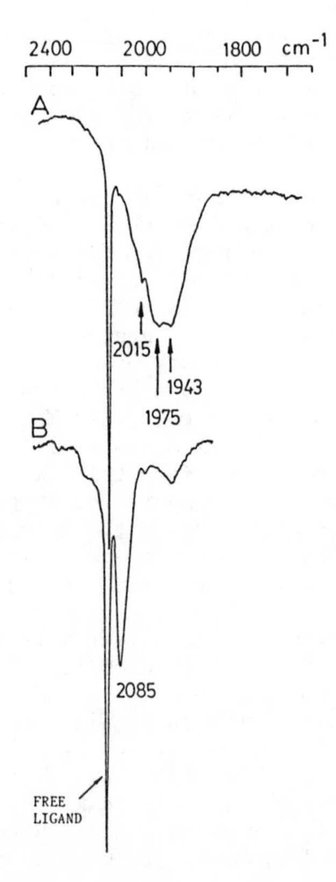

Figure 2. Infrared spectrum obtained after deposition of Cr atoms into CN-[P] at 200 K (A). Exposure to air (B).

M——C≡N-R ⟷ M═C=N̈—R

B B'

A non-symmetrical orientation of the R groups might be anticipated for 2 since, while the propyl chain can be considered an antenna to introduce flexibility for coordination, there are still some steric limitations to how the polymer chain(s) wrap around each metal atom. Interestingly, the contributions of these two factors are revealed when the chromium complex within the polymer is exposed to air. In the IR spectrum the band at 1950 cm^{-1} disappears to be replaced by a comparatively sharp peak at 2085 cm^{-1} (Figure 2) indicative of the formation of the $[Cr(CN\text{-}[P])_6]^+$ species. The sharpness of this band reflects the higher oxidation state of the chromium leading to less back-donation of electron density to the isocyanide, in other words a structure very close to that of B. This observation is consistent with the low symmetry in the original complex being associated with $M(CNC\text{-})_6$ distortion.

The $Cr(CN\text{-}[P])_6$ complex can also be characterized by its uv-visible spectrum (Figure 3) which contains a major absorption band at 298 nm with a high energy shoulder at ~260 nm and two low energy shoulders at ~340 and ~400 nm. While the overall spectrum consists of a rather broad envelope, a Gaussian analysis reveals the individual peaks quite clearly. While the $Cr(CNR)_6$ complexes for TESP-NC and ^{n}BuNC show very similar uv-visible spectra (14), a more useful analogy can be drawn with $Cr(CO)_6$ for which detailed spectroscopic analysis exists (16). Although symmetry factors once again result in the spectrum of $Cr(CN\text{-}[P])_6$ being broader than that of $Cr(CO)_6$ the gross features are essentially the same. In a similar fashion, the spectrum of $Cr(CN\text{-}[P])_6$ can be ascribed to a metal-to-ligand charge transfer band [298 nm] due to formal promotion of an electron from the d-orbitals of chromium to the π^* orbitals of the isocyanide ligand, with the lower intensity shoulders resulting from ligand-field transitions, i.e. reorganization of the d-electrons on Cr.

Deposition of iron atoms into a thin film of 2, with M_W = 3100 (Table I), resulted in a product with an infrared spectrum consisting of bands at 2085 and 1840 cm^{-1}. The former corresponds to an isocyanide ligand possessing a linear structure (B) while the latter is associated with 'bent' isocyanide ligands (B'). Comparison with results for small-molecule analogues shows that the spectrum is consistent with formation of $Fe(CN\text{-}[P])_5$ (17).

After uv-irradiation of the matrix at 180 K the terminal region became more complex - a new peak appeared at 2144 cm^{-1} with a shoulder at 2095 cm^{-1} - and the band at 1840 cm^{-1} disappeared concomitant with growth of a band at 1660 cm^{-1}. This last band is indicative of formation of a complex in which isocyanide ligands bridge two metal centers (5), and we therefore assign the new species as $Fe_2(CN\text{-}[P])_9$. These changes in the IR spectrum were accompanied by a color change of the matrix from yellow to deep orange. In addition to this photochromic behavior, the polymer matrix, fluid and air-sensitive before irradiation, turned into a self-supporting, air-stable film.

The same reaction sequence was monitored by uv-visible spectroscopy and despite the lack of optical data for the zerovalent binary iron-isocyanide complexes, comparison with the spectrum for $Fe(CO)_5$ (18) reveals a close similarity. After the photolysis the original spectrum [an intense absorption at 232 nm with two shoulders at 297 and 340 nm] had decayed to be replaced by a new spectrum consisting of two bands at 220 and 382 nm.

A feature of interest in this system is that the dimeric species $Fe_2(CN\text{-}[P])_9$ may also be generated as a minor product directly on deposition of iron atoms into 2 possessing a lower molecular weight (Table I) of M_w = 1500. The major difference between these two polymers lies in the average spacing between isocyanide groups as shown in Table I. The Me_2SiO- : -NC ratio in the higher molecular weight polymer is 10:1 while that in the lower molecular weight material is only 4:1. The different reactivity toward iron atoms can then be explained in terms of the probability of encounter between two Fe-isocyanide species forming at different points on the polymer backbone. These results are summarized in Scheme 3.

It is also important to mention that deposition of Fe into 2 yielded a side-product in addition to $Fe(CN\text{-}[P])_5$. This species was characterized in the infrared spectrum by a band at 2160 cm^{-1} overlapping with the free ligand stretch. This band grew independently from the others at the beginning of the deposition and reached rapidly a maximum after which its intensity did not vary with increased metal loading. The position of the band to higher energy relative to the free ligand suggests a metal complex containing the metal in a higher oxidation state. Support for the idea that this results from a reaction between an iron-isocyanide species and terminal Si-OH groups, which may be present in small amounts in the polymer, is provided by the observation that this band does not occur when iron atoms are reacted with the Me_3SiO-capped polymer, although the rest of the spectrum appears as normal.

Turning finally to the reaction of Ni atoms with 2, the product shows an infrared spectrum consisting of a single strong band at 2065 cm^{-1} (Figure 4A). This data compares favorably with that for the model compound $Ni(CN\text{-}TESP)_4$ (14), permitting an assignment of the polymer-bound complex to the analogous $Ni(CN\text{-}[P])_4$ (17). In view of recent structural studies with $Ni(CNBu^t)_4$ (19), we can assume that the polymeric species contains a tetrahedral arrangement of ligands around each Ni atom, which is consistent with observations from the IR spectrum. Note the relatively high $\bar{\nu}_{NC}$ for this complex, particularly when compared with that for $Cr(CN\text{-}[P])_6$, which is a consequence of the less favorable overlap between metal d and ligand π^* orbitals in a tetrahedral geometry. The electronic spectrum of $Ni(CN\text{-}[P])_4$ is also consistent with this picture, containing three intense bands at 207, 244 and 279 nm, in good agreement with the results for $Ni(CN\text{-}TESP)_4$ and $Ni(CO)_4$ (14).

Exposure to air of the matrix species [at 200 K] was accompanied in the IR spectrum by growth of a double band at 2193/2220 cm^{-1} (Figure 4B), while in the electronic spectrum bands appear at 228 and 294 nm.

In order to interpret these observations, results from model studies need to be examined. When $Ni(CNBu^t)_4$ is mulled in oxygenated Nujol, the infrared spectrum of the product shows two sharp bands at 2200 and 2180 cm^{-1} and an additional peak at 900 cm^{-1} (19). This

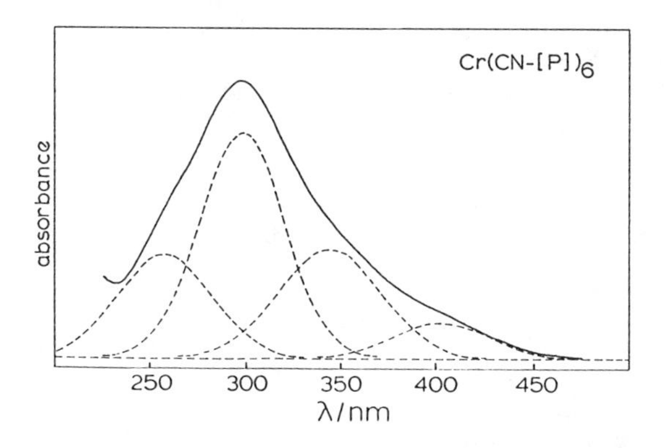

Figure 3. Ultraviolet-visible spectrum of $Cr(CN\text{-}[P])_6$, with Gaussian resolution.

Scheme 3

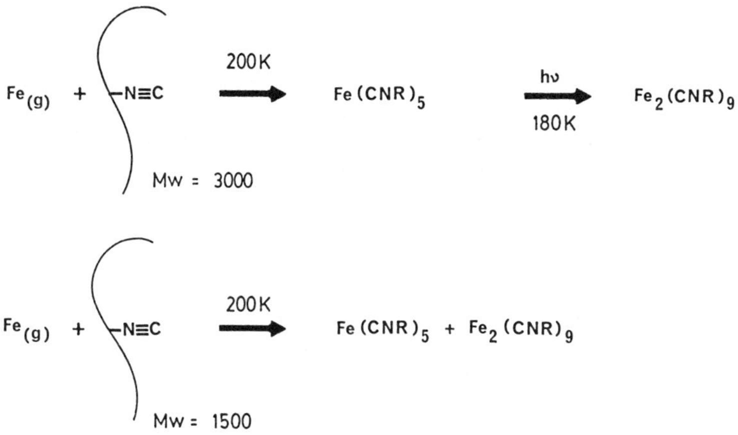

$$Fe_{(g)} + \text{—N}{\equiv}\text{C} \xrightarrow{200K} Fe(CNR)_5 \xrightarrow[180K]{h\nu} Fe_2(CNR)_9$$

Mw = 3000

$$Fe_{(g)} + \text{—N}{\equiv}\text{C} \xrightarrow{200K} Fe(CNR)_5 + Fe_2(CNR)_9$$

Mw = 1500

spectrum can be attributed to a dioxygen adduct $Ni(CNBu^t)_2(O_2)$ (20). While it is tempting to associate Figure 4B with an analogous dioxygen complex, the electronic spectrum of the air-exposed polymer matrix is not consistent with its formation. In a separate series of experiments (21) it was shown that $[Ni(CN-[P])_4][ClO_4]_2$ exhibits a $\bar{\nu}_{NC}$ at 2228 cm^{-1}. It is also known (22) that conproportionation of tetracoordinate Ni(0) and Ni(II) complexes of MeNC affords a dimeric complex, i.e. $[Ni_2(CNMe)_8]^{2+}$, which possesses a $\bar{\nu}_{NC}$ at 2220 cm^{-1} with an electronic spectrum containing a band at 304 nm. Consideration of these observations in the light of our Ni/polymer studies suggests that the initial phase of the oxidation is to the $[Ni(CN-[P]_4]^{2+}$ as shown in Scheme 4. Some of the Ni(II) species then reacts with excess $Ni(CN-[P])_4$ in the polymer matrix to give the $Ni_2(CN-[P])_8]^{2+}$ dimer. Independent electrochemical measurements (21) support the hypothesis that Ni(0) and Ni(II) present in the same polymer matrix can react to give a dimeric species.

An interesting question, in view of the demonstrated existence of $Ni(CNBu^t)_2(O_2)$ (20), is why oxidation of the polymer does not lead to an analogous species. A possible explanation is that when water is present [via exposure of a cold (200 K) substrate to air] the overall reaction becomes that in which oxidation of Ni(0) to Ni(II) is accompanied by reduction of oxygen to OH^-. Further studies are anticipated to clarify this point.

The $Cr(CN-[P])_6$ and $Ni(CN-[P])_4$ complexes of 2 have been prepared as well under macroscale conditions. 100 - 200 mg of metal were deposited typically into a blend of 2 in a poly(dimethylsiloxane) at 270 K [normally ~10 ml of 2 in 40 ml of diluent]. The products for each metal were dark orange, viscous oils which gave IR and uv-visible spectra in good agreement with those obtained from the matrix experiments.

We have discussed previously that the $Cr(CN-[P])_6$ and $Ni(CN-[P])_4$ species can be oxidized by air to the mono- and dication respectively. Further studies with the materials obtained from the macroscale experiments have shown that oxidation of the metal centers can be achieved electrochemically in the presence of an electron transfer mediator or 'relay' such as ferrocene. The experiment for the Cr/polymer combination is summarized in Figure 5, although the principles apply equally for the Ni analogue. Figure 5A shows the cyclic voltammetric response for ferrocene at a Pt electrode - the expected chemically and electrochemically reversible couple due to $Fe(Cp)_2/Fe(Cp)_2^+$. Addition of 2 was found to lead to an increase in the peak-to-peak separation $[\Delta E_p]$, attributed to the increased uncompensated cell resistance or so-called iR drop, although the chemical reversibility remained intact (Figure 5B). However, replacement of 2 by the derivatized polymer $Cr(CN-[P])_6$ led to a loss of the chemical reversibility, i.e. i_c/i_a less than 1, although there was no further change in ΔE_p (Figure 5C).

These observations for $Fe(Cp)_2$ in the presence of the polymer-bound Cr complex are consistent with $Fe(Cp)_2^+$, generated electrochemically, undergoing a reaction with $Cr(CN-[P])_6$ resulting in the chemical reduction of $Fe(Cp)_2^+$ and oxidation of the Cr species. Therefore, when the cathodic part of the $Fe(Cp)_2/Fe(Cp)_2^+$ wave is scanned, little ferricenium ion remains to be reduced electrochemically. As a result, the ferrocene molecule has effected the transfer of electrons from the polymer to the electrode.

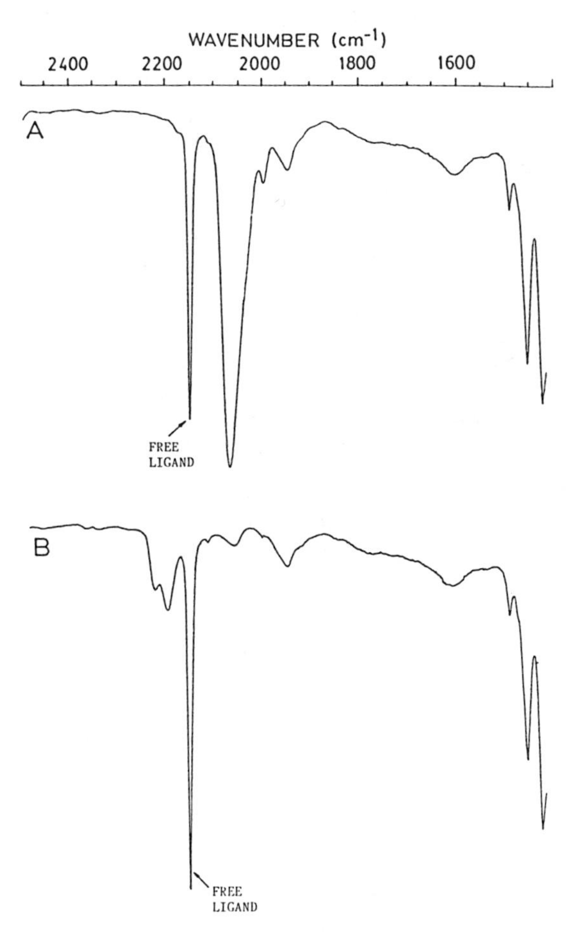

Figure 4. Infrared (FTIR) spectrum obtained after deposition of Ni atoms into CN–[P] at 200 K (A). Spectrum after exposure to air (B).

Scheme 4

$$\mathrm{Ni(CN\text{-}[P])_4} \xrightarrow{O_2} \mathrm{Ni(CN\text{-}[P])_4}^{2+}$$

$$\mathrm{Ni(CN\text{-}[P])_4}^{2+} \xrightarrow{\mathrm{Ni(CN\text{-}[P])_4}} \mathrm{Ni_2(CN\text{-}[P])_8}^{2+}$$

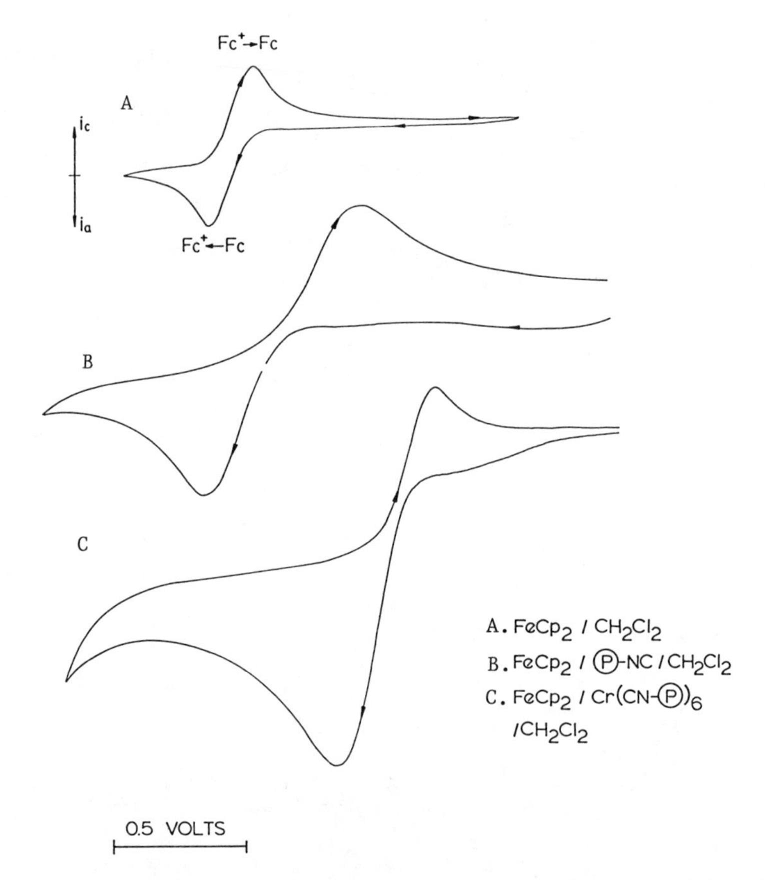

Figure 5. Cyclic voltammetric response of Ferrocene - (A) in CH_2Cl_2, (B) with added CN-[P], (C) after addition of $Cr(CN-[P])_6$.

We can see that ferrocene is ideally suited to this application from model compounds. Using $Cr(CNBu^n)_6$ as our model for the polymer-based Cr complex (21), we can estimate that the K_{eq} for reaction between ferricenium and $Cr(CN-[P])_6$, leading to removal of one electron, is approximately 10^{24}. It is also thermodynamically favorable $[K_{eq} \approx 10^{12}]$ for a second electron to be removed to give the bound Cr(II) species.

In neither case - $Cr(CN-[P])_6$ or $Ni(CN-[P])_4$ - was a cyclic voltammetric response observed directly for the metal center at a Pt electrode, for either solution phase or surface-confined materials. The solution phase behavior in this respect is very similar to the situation encountered with many biological macromolecules (23). Since model compounds reveal well-defined cyclic voltammograms for the $Cr(CNR)_6$ and $Ni(CNR)_6$ complexes (21) the origin of the electro-inactivity of the polymers is not obvious. A possible explanation (12) is that the ohmic resistance across the interface between the electrode and polymer, due to the absence of ions within the polymer, renders the potentially electroactive groups electrochemically inert, assuming the absence of an electronic conduction path. It is also important to consider that the nature of the electrode surface may influence the type of polymer film obtained. A recent observation which bears on these points is that when one starts with the chromium polymer in the $[Cr(CN-[P])_6]^{2+}$ state, an electroactive polymer film may be obtained on a glassy carbon electrode. This will constitute the subject of a future paper.

EXPERIMENTAL

NMR spectra were recorded on a JEOL FX90Q (90 MHz) spectrometer. ^1H chemical shifts are reported relative to an internal standard of $CHCl_3$ in $CDCl_3$, while ^{13}C chemical shifts are reported relative to the $CDCl_3$ used as solvent. Infrared spectra were recorded on an IBM IR32 FTIR or a Perkin-Elmer 1330 IR spectrophotometer. Uv-visible spectra were obtained using a Varian DMS90 spectrophotometer.

Molecular weights were measured by gel permeation chromatography on a Perkin-Elmer Series 10 Liquid Chromatograph using tetrahydrofuran as solvent and refractive index as the detection mode. Standards were polystyrene, and reported molecular weights for the polysiloxanes do not include a correction.

Triethoxysilapropylamine was purchased from Petrarch Systems. Dimethyldiethoxysilane and methyltriethoxysilane were purchased from Aldrich.

Synthesis of 2:

The triethoxysilapropylisocyanide 1 was prepared as outlined in reference 7, and characterized by its ^1H, ^{13}C NMR, IR and mass spectra. For details see reference 14. Aqueous NaOH [0.09 g in 3 ml H_2O] was added dropwise to a mixture of 30 ml [0.17 mole] of $(EtO)_2SiMe_2$ and 2.5 ml [0.01 mole] of $(EtO)_3Si(CH_2)_3NC$. The solution was stirred under an inert atmosphere for 24 hr. in the dark at room temperature. After completion of the reaction, the residue was washed thoroughly with water until neutral. It was then resuspended in hexane and dried over $MgSO_4$ for several hours. The solution was decanted, the solvent removed in vacuo and the polymer purified by sublimation, affording a clear, viscous liquid.

In the case where end-capping of the chains was necessary, this was achieved by stirring 2 with a five-fold excess of $(EtO)_3SiMe$ in aqueous base for 24 hr. The resulting polymer was worked-up as before.

2 - ^1H nmr ($CDCl_3$) : δ3.70 (q, $CH_3\underline{CH_2}O$), 3.34 (t, $\underline{CH_2}NC$), 1.72 (m, $\underline{CH_2}CH_2NC$), 1.18 (t, $\underline{CH_3}CH_2O$), 0.66 (m, $\underline{CH_2}Si$), 0.08, 0.05 ($\underline{CH_3}Si$).

^{13}C nmr ($CDCl_3$): 156.0 ($\underline{NC}$), 57.6 ($CH_3\underline{CH_2}O$), 43.8 (t, J_{NC} = 6.1 Hz, $\underline{CH_2}NC$), 23.5 ($\underline{CH_2}CH_2NC$), 18.2 ($\underline{CH_3}CH_2O$), 10.8, 9.7, 8.6 ($\underline{CH_2}Si$), 0.89, -1.17 ($\underline{CH_3}Si$) /ppm

IR (neat) : 2965s, 2905m, 2148s ($\bar{\nu}_{NC}$), 1392m, 1261vs, 1090vs, 954s, 802vs cm^{-1}.

Metal Vapor Polymer Matrix Depositions:

The metal vapors were generated by resistive heating of metal filaments [Fe, Ni] or Knudsen cells [Cr] containing the desired powder. The polymer matrix was prepared by applying 2 to the surface of the optical support [quartz for uv-visible; NaCl, KBr or CsI for IR monitoring]. The matrix was allowed to degas thoroughly under vacuum and was then maintained at 200 K under dynamic vacuum [10^{-6}-10^{-7} torr]. The metal vapor was deposited into the cold film at deposition rates on the order of 0.4 $\mu g.min^{-1}$. The ensuing reactions were monitored by uv-visible and infrared spectroscopy.

Macroscale Metal Vapor Reactions:

In a typical reaction 100 - 200 mg of metal [Cr or Ni] was evaporated from a preformed alumina crucible over a period of 60 - 90 min and deposited into a mixture of 2 in poly(dimethylsiloxane) [Petrarch Systems; 0.1 P.] within a rotary solution metal vapor reactor operating at 10^{-4} torr. The reaction flask was cooled to approximately 270 K by an iced-water bath. For a description of the apparatus see Chapter 3 of reference 4. The product in each case was a dark orange viscous liquid and was characterized as obtained from the reaction vessel.

Electrochemical Measurements:

All electrochemical measurements were performed employing a three-electrode arrangement with Pt or glassy carbon working and auxiliary electrodes and a Ag wire pseudo-reference electrode separated from the bulk solution by a Vicor membrane. The wave generator/potentiostat was a BAS Instruments Model CV-1B connected to a Houston Instruments 2000 recorder. All materials were loaded into the cell under anaerobic conditions and dried, degassed solvents [CH_2Cl_2, CH_3CN, thf] and degassed supporting electrolyte [tetra-^{n}butylammonium-hexafluorophosphate] were used. In general, potentials are reported relative to ferrocene as internal standard.

ACKNOWLEDGMENTS

We wish to thank the 3M Company for financial support of this research and the Van't Hoff Fund for the purchase of the electrochemical equipment. PPR acknowledges the University of Southern California for the award of a Moulton Fellowship.

LITERATURE CITED

1. Kaneko, M.; Yamada, A. Adv. Polymer Sci. 1984, 55, 2.
2. Grätzel, M. Energy Resources through Photochemistry and Catalisis; Academic Press: New York, 1983.
3. Francis, C.G.; Morand, P.D.; Spare, N.J. Organometallics 1985, 4, 1958. [and references therein]
4. Moskovits, M.; Ozin, G.A. Cryochemistry; Wiley: New York, 1976.
5. Singleton, E.; Oosthuizen, H.E. Adv. Organometal. Chem. 1983, 22, 209.
6. Cotton, F.A.; Wilkinson, G. Advanced Inorganic Chemistry; Wiley: New York, 1980, pp 1049 - 1094.
7. Howell, J.A.S.; Berry, M. J. Chem. Soc., Chem. Commun. 1980, 1039.
8. Eaborn, C. Organosilicon Compounds; Academic Press: London, 1960.
9. Ugi, I. Isonitrile Chemistry; Academic Press: New York, 1971.
10. Fletcher, H.J.; Hunter, M.J. J. Am. Chem. Soc. 1949, 71, 2918.
11. Francis, C.G., Ozin, G.A. J. Macromol. Sci.-Chem. 1981, A16, 167.
12. Murray, R.W. In Electroanalytical Chemistry; Bard, A.J., Ed.; Marcel Dekker: New York, 1983; Vol. 13.
13. Francis, C.G.; Klein, D.; Morand, P.D. J. Chem. Soc., Chem. Commun. 1985, 1142.
14. Morand, P.D. Ph.D. Thesis, Univ. Southern California, 1986.
15. Amster, R.L.; Hannan, R.B.; Tobin, M.C. Spectrochim. Acta 1963, 19, 1489. [and references therein]
16. Beach, N.A.; Gray, H.B. J. Am. Chem. Soc. 1968, 90, 5713.
17. Francis, C.G.; Morand, P.D.; Radford, P.P. J. Chem. Soc., Chem. Commun. 1986, 211.
18. Dartiguenave, M.; Dartiguenave, Y.; Gray, H.B. Bull. Soc. Chim. France 1969, 4223.
19. Morton, P.R. Ph.D. Thesis, Univ. Southern California, 1987.
20. Otsuka, S.; Nakamura, A.; Tatsuno, Y. J. Am. Chem. Soc. 1969, 91, 6994.
21. Radford, P.P. Ph.D. Thesis, Univ. Southern California, 1987.
22. DeLaet, D.L.; Powell, D.R.; Kubiak, C.P. Organometallics 1985, 4, 954.
23. Margoliash, E.; Schejter, A. In Advances in Protein Chemistry; Anfinsen, C.B.; Anson, M.L.; Edsall, J.T.; Richards, F.M., Eds.; Academic Press: New York, 1966; Vol. 21.

RECEIVED August 27, 1987

SPECIALTY POLYMERS WITH POLAR/IONIC GROUPS

Introduction to Specialty Polymers with Polar/Ionic Groups

Chemical reactions on polymers to introduce polar or ionic functional groups constitute a specialized area in the field of chemical modification. In recent years, the area has experienced high levels of research activity. This is because the polar/ionic substituents impart many useful and desireable properties to the polymers. Among those properties are solubility, wettability, adhesion, and conductivity. In addition, the polar/ionic groups can form aggregates or clusters which are analogous to the microphase separated hard segments in conventional block polymers, and which provide improved mechanical properties in the polymer. Hydrogen bonding of the polar groups can also have a positive effect on the thermal stability of the polymer. The polar or ionic modified polymers have seen increased use in coating applications, where solubility, wettability, and adhesion are important, as well as in electroactive application areas.

Large variety is available in both the polymer and the functional groups. Addition polymers, e.g. polydienes, polyolefins, polystyrenes, acrylics, as well as condensation polymers, e.g. polyesters, polyamides, polyamines and polyimines, epoxies, polyurethanes, have been used. The substituent groups have included sulfonate, sulfone, phosphonate, carboxylate, nitro, and hydroxyl. The substituents can be introduced directly on the polymer backbone or, alternatively, can be introduced on a side chain. Therefore, the number of resulting polar/ionic polymers is very large.

A portion of the research in the field has related to the preparation of high carboxylate functionality in polymers in order to achieve high water solubility or high water absorption. Although direct polymerization of carboxylate containing monomers, e.g. acrylic or methacrylic acid, is possible, the resulting polymer has a random distribution of carboxylate functionality. This makes correlation of the polymer structure with polymer properties difficult. Therefore, a major route to high carboxylate functionality has been the hydrolysis of pendant ester functionality in block polymers. Block polymers have been prepared from acrylate-acrylate or styrene-acrylate monomers, and the resulting polymers hydrolyzed to produce the free carboxylic acid or carboxylate salt. It has been shown that t-butyl methacrylate is a particularly attractive monomer since the t-butyl group can be selectively hdrolyzed in the presence of other ester groups. The ability to prepare polymers of controlled structure and functionality via chemical modification is providing a foundation for understanding the structure property relationships for ion-containing polymers.

Other polar functional groups which are suited to hydrogen bonding interactions can be introduced by addition reactions. Examples include the addition of carbethoxy carbene to polydienes, and the free radical addition of hexafluoroacetone to polyolefins. The hydrogen bonding groups render the polymer more soluble in conventional solvents, and it has been suggested in some studies that the polar groups may contribute to improved compatibility of the polymer in polymer blends.

The variety of polar/ionic modified polymers is high, and increasing, as demonstrated by growing numbers of publications in this area. Fundamental studies of chemical, morphological, rheological, and mechanical properties are currently being pursued on polymers of controlled composition and structure. New and expanded applications for these polymers are also being identified. The papers in this section of the book are representative of recent developments in this evolving area of chemical reactions on polymers.

Chapter 19

Synthesis and Characterization of Block Copolymers Containing Acid and Ionomeric Functionalities

T. E. Long [1], R. D. Allen [2], and J. E. McGrath [3]

Department of Chemistry and Polymer Materials and Interfaces Laboratory, Virginia Polytechnic Institute and State University, Blacksburg, VA 24061-0699

By employing anionic techniques, alkyl methacrylate containing block copolymer systems have been synthesized with controlled compositions, predictable molecular weights and narrow molecular weight distributions. Subsequent hydrolysis of the ester functionality to the metal carboxylate or carboxylic acid can be achieved either by potassium superoxide or the acid catalyzed hydrolysis of t-butyl methacrylate blocks. The presence of acid and ion groups has a profound effect on the solution and bulk mechanical behavior of the derived systems. The synthesis and characterization of various substituted styrene and all-acrylic block copolymer precursors with alkyl methacrylates will be discussed. In addition, the polymer modification reactions leading to acidic and ionomeric functionalities are described in detail. The derived ion-containing block copolymers may aid in the correlation of chemical architecture with ionomer morphology and properties.

The necessity of novel polymeric systems for specific electroactive applications has motivated significant research efforts in polymer synthesis and characterization (1-4). In addition to the preparation of conducting, charge transfer, and non-linear optic polymeric materials, an abundance of industrial and academic attention has been devoted to the introduction of ions to a hydrocarbon backbone (5-7). The interest in ionomers has been encouraged by the realization that small amounts of ions (10 mole percent or less) can drastically modify polymer properties. However, a detailed understanding of the origin of ionomer properties has been difficult to achieve. This is presumably due to the complex aggregate morphology that results from

[1]Current address: Eastman Kodak Company, 1999 Lake Avenue, Rochester, NY 14650
[2]Current address: IBM Almaden Research Center, 650 Harry Road, San Jose, CA 95120-6099
[3]Correspondence should be addressed to this author.

0097-6156/88/0364-0258$06.00/0

electrostatic interactions. In most investigations, the role of polymer architecture has not been addressed and it is believed that this variable will have a profound influence on the development of the morphology. For instance, the controlled placement of an ionic functionality in block-like sequences versus random incorporation may significantly alter the extent and mechanisms of coulombic interaction. Several complimentary techniques have been developed in our laboratories which allow for the synthesis of charged polymers with controlled architecture. The synthetic capabilities described herein will facilitate the correlation of ionomer structure with various thermal, mechanical, rheological, and scattering properties. Although the focus of this work is on the synthesis of model ion-containing block copolymers, the copolymer precursors may also demonstrate potential utility.

Carboxylic acid containing polymers have been conventionally prepared by the direct free radical polymerization of acrylic or methacrylic acid with various vinyl comonomers (8,9). The corresponding carboxylate ionomers are then obtained by partially or completely neutralizing the acid groups with a variety of bases. The selection of the base dictates the valency of the counterion and has been shown to be important in the determination of rheological and mechanical properties (10,11). These synthetic routes result in the random placement of ionic groups along polymer backbones. The heterogeneous composition of these materials has made characterization and interpretation of structure-property correlation quite difficult.

Anionic techniques, in favorable cases, enable one to synthesize block copolymers with controlled composition and architecture, predictable molecular weights, and narrow molecular weight distributions. In particular, we have devoted significant attention to the synthesis of acrylic-acrylic and styrenic-acrylic block copolymers (12). Present research efforts are focused on the synthesis of diene-acrylic block copolymers and hydrogenated derivatives (13). The dielectric constant of the first block may play a significant role in determining the final aggregate morphology. One can systematically vary the polarity of the first block from the very nonpolar diene phase to the moderately polar alkyl methacrylate.

The introduction of ionomeric functionalities can be accomplished in a controlled fashion by the subsequent hydrolysis of the polymeric ester functionality of various alkyl methacrylates. Our earlier investigations have introduced the use of potassium superoxide (KO_2) for the hydrolysis of a wide range of poly(alkyl methacrylates) (14). Recently, we have developed the proper techniques for the introduction of t-butyl methacrylate (TBMA) into block copolymer systems. It is well-established that the hydrolysis of esters containing alkyl groups which can stabilize carbenium ions, e.g. t-butyl, undergo mild acid catalyzed alkyl oxygen cleavage (A_{AL1}) (15). In addition to the ability to form stable carbenium ions, the presence of β-hydrogen in the ester alkyl group provides for a mechanism of elimination, e.g.

isobutylene for TBMA hydrolysis. This attribute drives the reaction to very high conversions (>95%). We have repeatedly employed this second hydrolysis approach for the synthesis of tactic poly(methacrylic acid), e.g. 100% isotactic poly(methacrylic acid). Due to the limited conversion (60-70%) of the KO_2 route, the carboxylate block would contain unreacted ester functionality. On the other hand, a block comprised solely of metal carboxylate can be prepared by the acid catalyzed hydrolysis.

It is well known that inter- and intramolecular forces play a very important role in determining polymer morphologies and properties. For the case of acid containing polymers, it has been reported that the occurrence of hydrogen bonding drastically affects such properties as the glass transition temperature (16,17). In addition, mechanical and dielectric relaxations may shift with increasing acid content. In a similar fashion, polymer properties drastically change with the incorporation of ionic groups due to preferential aggregation of these moieties (8,18). Eisenberg has proposed a very useful scheme involving two types of aggregates. Multiplets consist of small groups of ion pairs with no polymeric content which serve as simple multifunctional crosslink points. Clusters arise from further aggregation of multiplets to form a larger, more loosely associated, structure which may have appreciable hydrocarbon (polymer) content. Clusters are thus analogous to a microphase separated "hard segmented" in a conventional block or segmented copolymer. A discussion of the various molecular parameters affecting cluster formation in ionomers enables one to better understand the driving forces for aggregation. Thus, one can control the morphology and physical properties of these systems.

The molecular parameters affecting aggregation in ionomers include: ion content, ion type, counterion, percent neutralization, backbone dielectric constant and relative position of ionic groups in a polymer chain. The first five factors are easily controlled and have been shown to affect the morphology and physical properties of ionomers. Less attention has been focused on ion placement due primarily to the synthetic difficulties involved. Preliminary investigations indicate that this factor may exert a strong influence on ionomer properties. Anionic techniques in combination with various subsequent polymer modification reactions yield a wide range of ion-containing materials and provide extensive insight into their unique properties.

Experimental

Materials. Styrene (Aldrich) and *t*-butylstyrene (Dow Chemical) were purified by distillation from dibutyl magnesium (DBM). This reagent has been shown by Fetters et al. (19) to successfully remove water and air from various hydrocarbon monomers. In some cases, the styrenic monomers can be passed through columns of silica and activated alumina followed by degassing to obtain anionic grade monomers.

Diphenylethylene (DPE) was obtained from Eastman Kodak Co. and was purified by vacuum distillation from s-butyllithium. The yellowish color associated with the crude DPE disappears after purification.

The alkyl methacrylate monomers were available from various sources. Isobutyl methacrylate (IBMA) (Rohm and Haas) and t-butyl methacrylate (TBMA) (Rohm Tech) may be purified first by distillation from CaH_2, followed by distillation from trialkyl aluminum reagents as described in detail earlier (20,21). In particular, t-butyl methacrylate (b.pt. ~150°C) was successfully purified by distillation, from triethyl aluminum containing small amounts of diisobutyl aluminum hydride. The trialkyl aluminum and dialkyl aluminum hydride reagents were obtained from the Ethyl Corporation as 25 weight percent solutions in hexane. The initiator, s-butyllithium, was obtained from the Lithco Division of FMC, and analyzed by the Gilman "double titration" (22).

Most polymerizations were carried out in tetrahydrofuran (THF) (Fisher, Certified Grade) which had been previously distilled from the purple sodium/benzophenone ketyl.

Polymerization. All glassware was rigorously cleaned and dried immediately prior to use. For small scale polymerizations, the reactor consisted of a 250 mL, 1 neck, round bottom flask equipped with a magnetic stirrer and a rubber septum. The septum was secured in place with copper wire in order that a positive pressure of prepurified nitrogen could be maintained. The reactor was assembled while hot and subsequently flamed under a nitrogen purge. After the flask had cooled, the polymerization solvent was added to the reactor via a double-ended needle. The reactor was submerged into a -78°C bath and allowed to reach thermal equilibrium. Purified styrene monomer was charged into the reactor with a syringe. The calculated charge of initiator was quickly syringed into the reactor and immediately one could see the formation of the orange styryl anion. The first block was allowed to polymerize for over 20 minutes to ensure complete conversion. A slight excess of DPE (2-3 molar excess compared to lithium) was syringed into the reactor in order to cap the front block. This capping procedure was essential in most cases to prevent carbonyl attack (1,2 addition) of the alkyl methacrylate monomer. The successful conversion of the highly delocalized diphenyl ethylene derived anion was witnessed by the rapid formation of a deep red color. After several minutes, highly purified alkyl methacrylate monomer was slowly added to the living capped first block. Initiation of the second block was rapid and was characterized by the formation of a colorless poly(alkyl methacrylate) enolate anion. After 15-20 minutes, the polymerization could be terminated by the addition of a few drops of degassed methanol. The polymer can finally be obtained by precipitating in a large excess (10X) of nonsolvent such as methanol or methanol/isopropanol depending on the solubility characteristics of the polymer.

Hydrolysis. Two methods of hydrolysis have been employed. t-Butyl methacrylate homopolymers and block copolymers were easily hydrolyzed using catalytic amounts of p-toluene sulfonic acid monohydrate (PTSA) in solution at mild temperatures (60-80°C). Other acid catalysts have also been studied. The choice of solvent was critical since maintaining solubility leads to higher degrees of hydrolysis. Toluene (Fisher, Certified Grade) is an excellent solvent for the block copolymers since the amount of incorporated t-butyl methacrylate was generally less than 10% by weight. For homopolymers, the addition of methanol will maintain solubility throughout the reaction. Although this technique is facile, it is only applicable to certain esters as described in the introduction. Scheme I depicts the hydrolysis of poly(t-butylstyrene)-b-poly(t-butyl methacrylate). Conversion to the metal carboxylate is accomplished by simple titration of the acid groups with an appropriate base, e.g. KOH (Normal/10 in methanol), using phenolphthalein as an indicator. The techniques for superoxide cleavage have been described in detail earlier (14).

Scheme I.

Hydrolysis of Poly(t-Butyl Methacrylate) Containing Block Copolymers

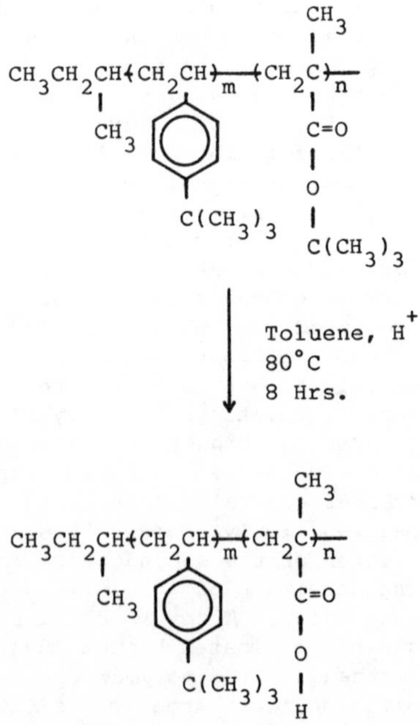

$$CH_3CH_2CH(CH_3){-}\!\left(CH_2CH(C_6H_4C(CH_3)_3)\right)_m\!{-}\!\left(CH_2C(CH_3)(C{=}O\text{-}O\text{-}C(CH_3)_3)\right)_n{-}$$

Toluene, H^+
80°C
8 Hrs.

$$CH_3CH_2CH(CH_3){-}\!\left(CH_2CH(C_6H_4C(CH_3)_3)\right)_m\!{-}\!\left(CH_2C(CH_3)(C{=}O\text{-}O\text{-}H)\right)_n{-}$$

Characterization. Molecular weight and molecular weight distributions of the diblock precursors and acid containing polymers were determined primarily by size exclusion chromatography (SEC). A variable temperature Waters 590 GPC equipped with ultrastyragel columns of 500Å, 10^3Å, 10^4Å and 10^5Å was employed. A Waters 490 programmable wavelength detector and a Waters R401 differential refractive index detector were utilized. Both PMMA and PS standards (Polymer Laboratories) were used for the construction of calibration curves.

Fourier transform infrared (FTIR) spectroscopy was performed on a Nicolet 10DX spectrometer. Nuclear magnetic resonance (^1H) characterization was accomplished using an IBM 270 SL. Both techniques can successfully be utilized to analyze both the diblock precursors as well as the derived acid containing polymers.

Thermal analysis (DSC, TMA, TGA) was performed with either a Perkin-Elmer Model 2 or Model 4.

Results and Discussion

A very critical, but not always appreciated, aspect of block copolymer synthesis by living polymerization techniques is monomer purity. This consideration is particularly important when high molecular weight polymers ($>5.0x10^4$ g/mole) are desired, and the corresponding anion concentration is quite low (<0.20 mmoles). Regarding the synthesis of poly(styrene)-*b*-poly(alkyl methacrylate) block copolymers, the deleterious effects of styrene impurities are avoided since any protic impurities are scavenged prior to initiation. On the other hand, any impurities present in the alkyl methacrylate monomer will quickly terminate the living polystyryl or capped polystyryl anion. Our earlier investigations have outlined a novel purification methodology for alkyl methacrylate monomers (23). This technique employs trialkyl aluminum reagents which quickly react with any acid, alcohol or water impurities present in many methacrylate monomers.

Recent developments in this purification methodology involve the purification of branched alkyl methacrylate monomers, e.g. *t*-butyl methacrylate (TBMA). Special attention must be given to their purification, since the branched alcoholic impurities are more difficult to remove, due to the lower reactivity of branched alcohols with R_3Al (24). In particular, we have utilized two approaches to obtain anionic polymerization grade branched alkyl methacrylates. First, longer contact times between the alkyl methacrylate monomer and triethylaluminum are necessary for complete conversion of the alcohol to the aluminum alkoxide. Although this approach seems quite feasible, it is complicated by the fact that the possibility of premature polymerization prior to distillation increases. This undesirable complication arises due to the inherent instability of the ester carbonyl-trialkyl aluminum complex. Despite attempts using longer reaction times, molecular weight distributions range from 1.3 to 1.5. The second approach involves the addition of small amounts (50 wt.% or less) of diisobutyl aluminum hydride (DIBAH) to the TEA solution. It is known that the aluminum hydrogen bond is significantly more labile to nucleophilic substitution than the aluminum alkyl bond (25). This mixed reagent approach is necessary for two reasons. First, the aluminum hydride, unlike the trialkyl aluminum reagent, does not form a colored complex with the α,β-unsaturated ester. Second, the mixed system is believed to tone down the reactivity of the aluminum hydride. A TEA/DIBAH mixture can be added to cold (-78°C) monomer until the stable colored complex forms. The purification reaction is then allowed to proceed for ~60 minutes at room temperature. This procedure allows for removal of impurities without reduction of the ester. Significantly narrower gel permeation chromatograms ($\bar{M}w/\bar{M}n$ <1.25) of poly(t-butyl methacrylate) are obtained when the samples are prepared from TEA/DIBAH purified monomer.

Various substituted styrene-alkyl methacrylate block copolymers and all-acrylic block copolymers have been synthesized in a controlled fashion demonstrating predictable molecular weight and narrow molecular weight distributions. Table I depicts various poly (t-butylstyrene)-b-poly(t-butyl methacrylate) (PTBS-PTBMA) and poly(methyl methacrylate)-b-poly(t-butyl methacrylate) (PMMA-PTBMA) samples. In addition, all-acrylic block copolymers based on poly(2-ethylhexyl methacrylate)-b-poly(t-butyl methacrylate) have been recently synthesized and offer many unique possibilities due to the low glass transition temperature of PEHMA. In most cases, a range of 5-25 wt.% of alkyl methacrylate was incorporated into the block copolymer. This composition not only facilitated solubility during subsequent hydrolysis but also limited the maximum level of derived ionic functionality.

Table I. PTBMA Containing Block Copolymers: Molecular Weight Control

Sample	Wt.% TBMA	$\bar{M}n$(th)	$\bar{M}n$(gpc)	$\bar{M}w/\bar{M}n$
PTBS-PTBMA	5	52,300	57,900	1.03
PTBS-PTBMA	10	55,000	54,700	1.03
PMMA-PTBMA	17	27,000	29,300	1.12
PMMA-PTBMA	23	65,000	51,100	1.20

In the past, ionomers have generally consisted of 10-12 mole percent of ions and it is our intention to be consistent with the corresponding random ionomers previously discussed in the literature. In addition to gel permeation chromatography (GPC), ^{1}H and ^{13}C NMR can readily be utilized to verify the relative amount of monomer successfully incorporated into the block copolymer. For example, the composition of a PMMA-PTBMA diblock can be verified by ^{1}H NMR ratioing the methyl ester integration (3.5 ppm) to the t-butyl ester integration (1.36 ppm). Figure 1 depicts the t-butyl ester chemical shift which appears reproducibly at 1.36 ppm. ^{13}C or FTIR can be utilized in certain instances when ^{1}H NMR chemical shifts overlap. For example, one can witness the infrared carbonyl absorbance increase with an increase in TBMA monomer charge. Figure 2 illustrates an FTIR of a PTBS-PTBMA block copolymer. In particular, one should note the carbonyl absorbance (1723 cm^{-1}) for PTBMA and the characteristic styrene band about 3000 cm^{-1} associated with the aromatic C-H stretch. These well-defined precursors can be subsequently hydrolyzed to yield metal carboxylate containing block copolymers. The percent of incorporated ion groups is dictated by the amount of alkyl methacrylate monomer charged and the extent conversion of hydrolysis.

Two criteria must be met for the polymer hydrolysis reactions to be advantageous. First, the hydrolysis reagents must not affect the hydrocarbon block or result in any chain cleavage. The absence of degradation can be verified by gel permeation chromatography (GPC) of the modified precursor coupled with explainable spectroscopic changes. For example, the appropriate shift to higher elution volumes (lower molecular weight) is evident while maintaining a narrow molecular weight distribution. The exact magnitude of the shift cannot be directly compared to the loss of the ester alkyl group presumably due to simultaneous hydrodynamic volume changes. Second, the extent of hydrolysis should be controlled. It would be most desirable for the hydrolysis reaction to be quantitative since the final acid or ion content could be dictated by the weight percent of alkyl methacrylate in the copolymer. The first method of hydrolysis involves the reaction of potassium superoxide (KO_2) with various poly(alkyl methacrylates) to directly generate carboxylate ionomers. This method of hydrolysis has been reported earlier for small molecules (26). In addition, this reagent has been utilized for modifications of PVC (27) and PAN (28). A variety of alkyl methacrylate containing polymers can successfully be hydrolyzed using KO_2. The size of the ester alkyl group drastically affects the ease of hydrolysis. In particular, as the ester group size increases, e.g. isobutyl, the required reaction time increases. It is interesting to note at this point that it has also been shown (21,29) that the size of the ester alkyl group also controls the propagating enolate stability in alkyl methacrylate polymerizations. This is attributed in a similar fashion to the deactivation of the carbonyl by the bulky ester alkyl group. The hydrolysis of poly(styrene)-*b*-poly(isobutyl methacrylate) diblocks was allowed to continue for 12 hours at 90°C in a solvent combination of DMSO and THF. GPC indicated that no chain degradation occurred despite the long reaction times. Figure 3 depicts GPC behavior as a function of hydrolysis time. Solvent combinations of toluene and DMSO do permit higher reaction temperatures and corresponding shorter reaction times. Unless the resulting materials were acidified with small amounts of acid, they remained insoluble in most solvents. Partial conversions (~70%) do lead to an ionomer that is not truly blocky due to the presence of residual alkyl methacrylate between the carboxylates. It is probable that in a polymer modification reaction when an ionic reactant such as

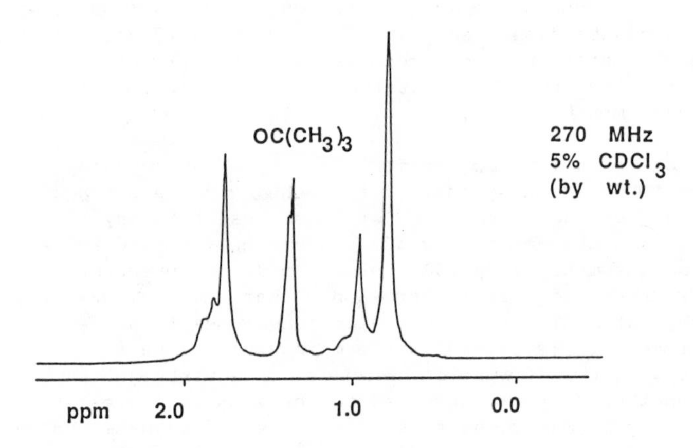

Figure 1. 270 MHz ^{1}H NMR spectrum of PMMA-*b*-PTBMA precursor.

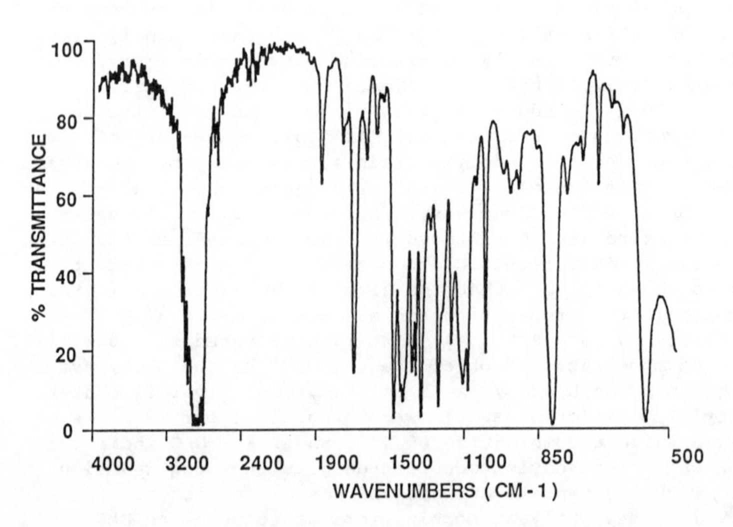

Figure 2. FTIR spectrum of poly(*t*-butyl styrene)-*b*-poly(*t*-butyl methacrylate) (10% TBMA by weight).

KO_2 approaches a charged backbone that yields are limited to 60-70%. This is presumably due to electrostatic repulsion between the reactant molecule and polymer substrate.

The second approach involves the acid catalyzed (~5-10% by weight compared to weight % ester) hydrolysis of t-butyl methacrylate (TBMA) containing block copolymers as illustrated in Scheme I. In most cases, the amount of TBMA in the copolymer ranged from 2-20% by weight. In addition to remaining consistent with previous investigations, the selection of this compositional range also eliminated the possibility of precipitation during hydrolysis. The propensity of the polymer to precipitate is a result of the enhanced polarity that develops as the acid units are formed. Consequently, the selection of hydrolysis solvent is also critical since maintaining solubility throughout the hydrolysis reaction leads to higher yields of the corresponding carboxylic acid. However, for this compositional range, toluene was an excellent solvent for all block copolymer hydrolysis reactions. On the other hand, the hydrolysis of PTBMA homopolymers at 80°C in toluene lead to premature precipitation at approximately 50% conversion in only 8 minutes and an alcohol such as methanol was required to restore solubility.

A variety of substituted polystyrene and poly(alkyl methacrylate) first blocks have been utilized. Poly(t-Butylstyrene) (PTBS) is particularly interesting due to its high glass transition temperature (~145°C) and its low solubility parameter (8.0 $(cal/cm^3)^{1/2}$) (30). This second attribute provides for excellent solubility of the acid-containing polymer in hydrocarbon solvents and for a unique surfactant-like molecule. Figure 4 is a FTIR spectrum (2000-1600 cm^{-1}) of a poly(t-butylstyrene)-b-poly(methacrylic acid) film obtained after only 90 minutes of hydrolysis at 80°C. The carbonyl band associated with the TBMA ester (1723 cm^{-1}) has nearly disappeared and the corresponding dimeric methacrylic acid carbonyl band has appeared at 1707 cm^{-1}. Although not shown in Figure 4, the broad COOH band is also quite evident in the 3200-2700 cm^{-1} range. Despite this very rapid conversion, the reaction is allowed to proceed for 6-8 hours to ensure complete conversion. The structural integrity of the block copolymers has also been verified using ^1H and ^{13}C NMR. In fact, ^1H NMR can successfully be used to follow the transformation of poly(methyl methacrylate)-b-poly(t-butyl methacrylate) into the corresponding acid containing polymer by following the disappearance of the t-butyl chemical shift at 1.36 ppm (Figure 5). Also, the simultaneous appearance of the acid proton chemical shift (~12 ppm) can be used to verify conversion. Figure 7 shows the ^1H NMR spectrum of a isotactic polymethacrylic acid sample obtained by hydrolysis of isotactic poly(t-butyl methacrylate). In all cases, conversions are greater than 95% providing for a very distinct hydrophobic-hydrophilic molecule. GPC indicates that the reaction proceeds without degradation of the polymer backbone and simultaneous increase in the molecular weight distribution. Figure 6 illustrates the chromatogram of a PMMA-PTBMA precursor and the

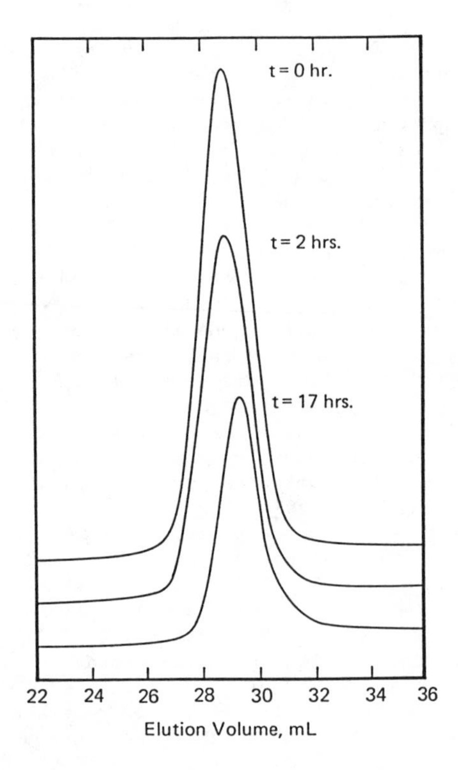

Figure 3. GPC behavior versus time for KO_2 hydrolysis route.

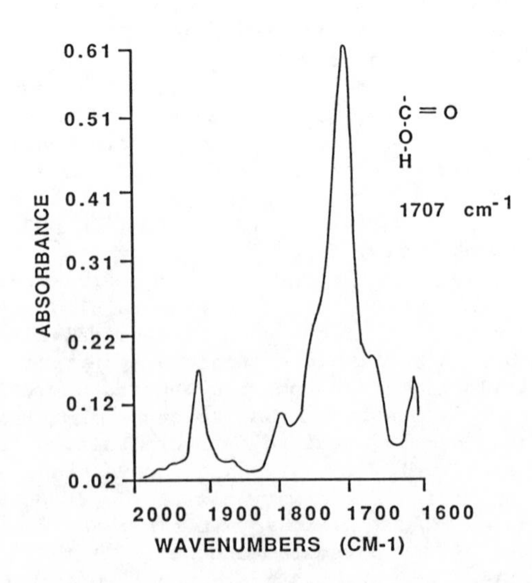

Figure 4. Carbonyl region of poly(t-butyl styrene)-b-poly(methacrylic acid) after 90 min. of acid hydrolysis.

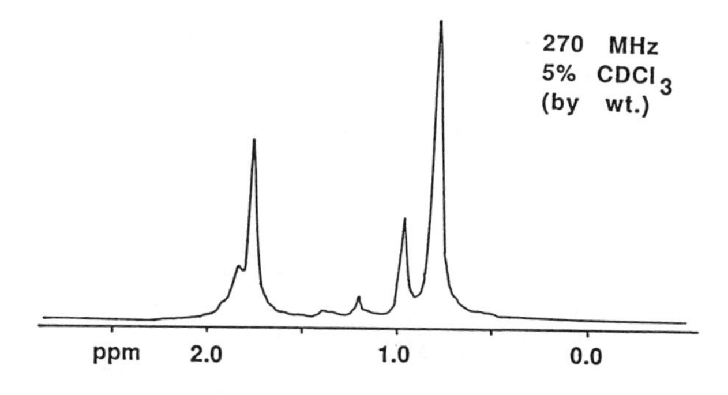

Figure 5. 270 MHz ^{1}H NMR of PMMA-b-PMAA: disappearance of t-butyl chemical shift.

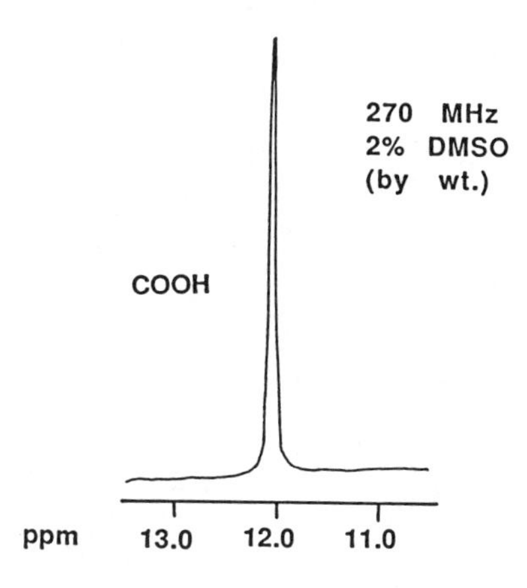

Figure 6. GPC behavior of PMMA-b-PTBMA diblock precursor and acid catalyzed hydrolysis product.

final hydrolyzed product. The appropriate shift to lower $\bar{M}n$ is witnessed while the molecular weight distribution remains narrow. This reaction has also been monitored in the solid state by utilizing thermogravimetric analysis (TGA) and noting the weight loss associated with the t-butyl group.

Although the potassium superoxide route can be universally applied to various alkyl methacrylates, it is experimentally more difficult than simple acid hydrolysis. In addition, limited yields do not permit well-defined hydrophobic-hydrophilic blocks. On the other hand, acid catalyzed hydrolysis is limited to only a few esters such as TBMA, but yields of carboxylate are quantitative. Hydrolysis attempts of poly(methyl methacrylate) (PMMA) and poly(isopropyl methacrylate) (PIPMA) do not yield an observable amount of conversion to the carboxylic acid under the established conditions for poly(t-butyl methacrylate) (PTBMA). This allows for selective hydrolysis of all-acrylic block copolymers. Other workers have also hydrolyzed trimethylsilyl esters in a controlled fashion (31). We believe there is a very significant amount of synthetic utility for t-butyl methacrylate in the area of model carboxylate-containing block copolymers. In addition to its hydrolyzability, the bulky t-butyl group decreases the probability of deleterious side reactions associated with the carbonyl during anionic polymerization. In fact, the TBMA lithium enolate is stable at room temperature, e.g. as high as 37°C in polar solvents such as THF. This is in sharp contrast to the anions of methyl and ethyl methacrylate which undergo significant termination with the ester carbonyl at ~ -48°C, or higher temperatures. The TBMA enolate stability in hydrocarbon solvents has been investigated as well (32,33) and living polymerizations are possible. In fact, ^{13}C NMR reveals that these homopolymers are essentially 100% isotactic. Thus, it appears that the syntheses of isotactic all-acrylic block copolymers and isotactic poly(methacrylic acid) are feasible. The enhanced enolate stability greatly facilitates the controlled synthesis of the diblock precursors. The "protected" nature of the carbonyl also facilitates the crossover reaction from the poly(styryl)lithium anion. In fact, unlike MMA, the capping of the first block with DPE is not necessary and eliminates the possibility of any termination due to impurities in DPE.

The physical properties of the acid- and ion-containing polymers are quite interesting. The storage moduli vs. temperature behavior (**Figure** 8) was determined by dynamic mechanical thermal analysis (DMTA) for the PS-PIBMA diblock precursor, the polystyrene diblock ionomer and the poly(styrene)-b-poly(isobutyl methacrylate-co-methacrylic acid) diblock. The last two samples were obtained by the KO_2 hydrolysis approach. It is important to note that these three curves are offset for clarity, i.e. the modulus of the precursor is not necessarily higher than the ionomer. In particular, one should note the same Tg of the polystyrene block before and after ionomer formation, and the extension of the rubbery plateau past 200°C. In contrast, flow occurred in

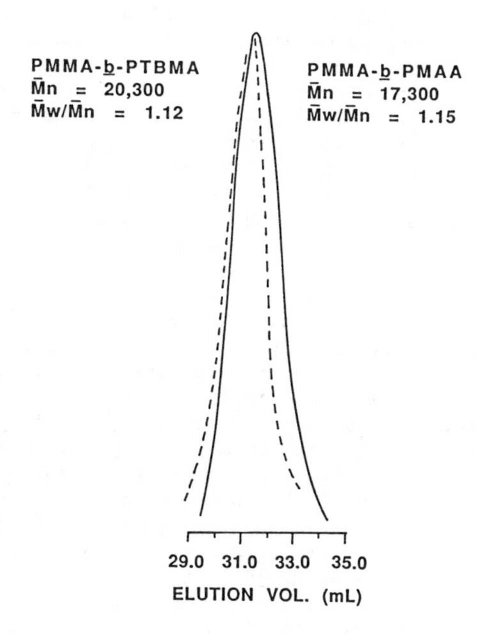

Figure 7. 270 MHz ^{1}H NMR spectrum of isotactic poly(methacrylic acid) in DMSO-d_6.

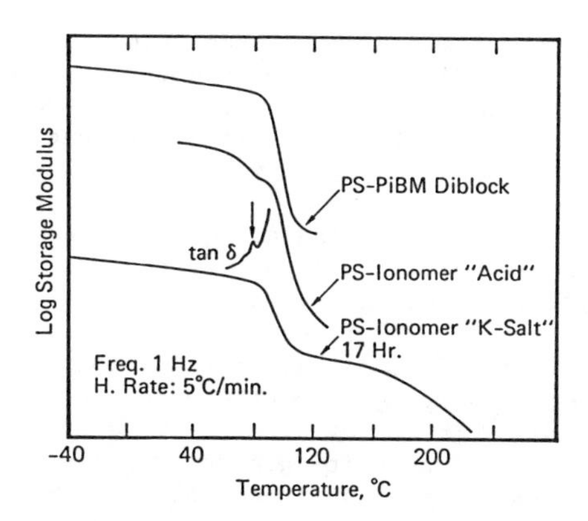

Figure 8. DMTA behavior for poly(styrene)-b-poly(isobutyl methacrylate-co-methacrylic acid) potassium salt.

the precursor and acidified version immediately after Tg. Similar formations of a rubbery plateau are evident in the derived ionomers based on hydrolysis of PTBMA.

The acid containing diblocks, especially those based upon poly(t-butylstyrene)-b-poly(methacrylic acid), demonstrate a potential for emulsifier applications. Thermal analysis of acid-containing block copolymers has shown that a large endotherm occurs near 200°C. The magnitude of this endotherm depends on the mole percentage of poly(methacrylic acid) in the system. A significant endotherm is quite evident even at poly(methacrylic acid) compositions less than 10 mole percent. Figure 9 depicts differential scanning colorimetry (DSC) analysis of a poly(t-butylstyrene)-b-poly(methacrylic acid) block copolymer containing ca. 10 mole percent acid. FTIR and mass spectroscopic analysis have attributed this phenomena to the formation of anhydride units. Other workers have investigated this phenomenon in relation to the sensitivity of lithographic materials (34). This observation prompted the TGA analysis of predominately syndiotactic poly(t-butyl methacrylate). Figure 10 shows that PTBMA reproducibly loses ~46% of its weight beginning at 250°C. This percentage corresponds to the loss of the t-butyl group and subsequent anhydride formation. Figure 10 also demonstrates the relative thermal stability of various anionic poly(alkyl methacrylates). Although not shown in Figure 10, isotactic (100%) PTBMA demonstrates exactly the same thermal stability as syndiotactic (52%) PTBMA. This can be attributed to the fact that TBMA undergoes a living polymerization in hydrocarbon solvents; consequently, thermally weak linkages due to side reactions do not form resulting in enhanced thermal stability.

CONCLUSIONS

Anionic techniques provide for the synthesis of well-defined block copolymers. In addition to the interesting properties of the block copolymers, these materials can also serve as precursors for subsequent polymer modification reactions. We have demonstrated two possible hydrolysis routes leading to acid- and ion-containing polymers. Although both techniques have their inherent limitations and advantages, they are very complimentary in their utility and applicability. The modified diblocks may have significant potential for such applications as surfactants, membranes, elastomers or adhesives. This investigation has also provided insight into an ionomer molecular parameter largely ignored, the effect of ion location on ionomer properties. It is believed that the synthesis of these model ion- and acid-containing block copolymers will surely compliment random systems already described in detail. Future efforts involve the scattering and rheological analysis of various block ionomers. In addition, the preparation of block ion-containing polydienes is currently being pursued.

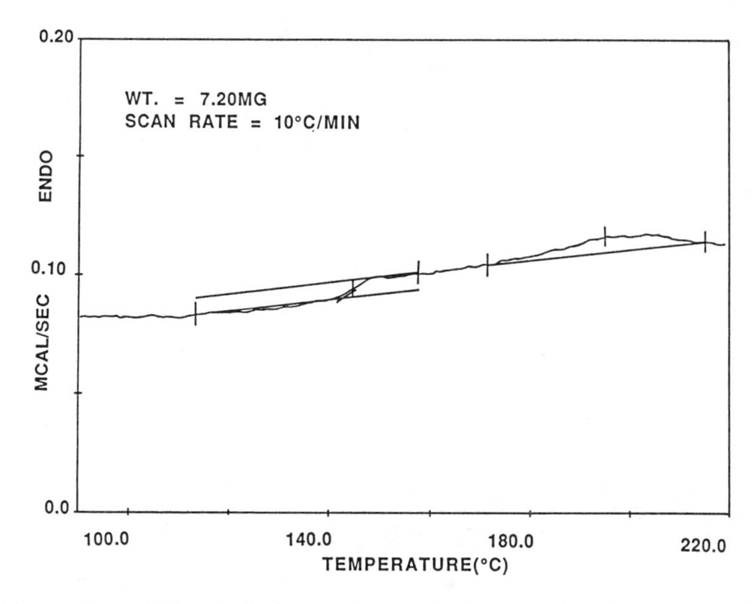

Figure 9. Differential scanning calorimetry (DSC) analysis of PTBS-b-PMMA (10 wt.% acid).

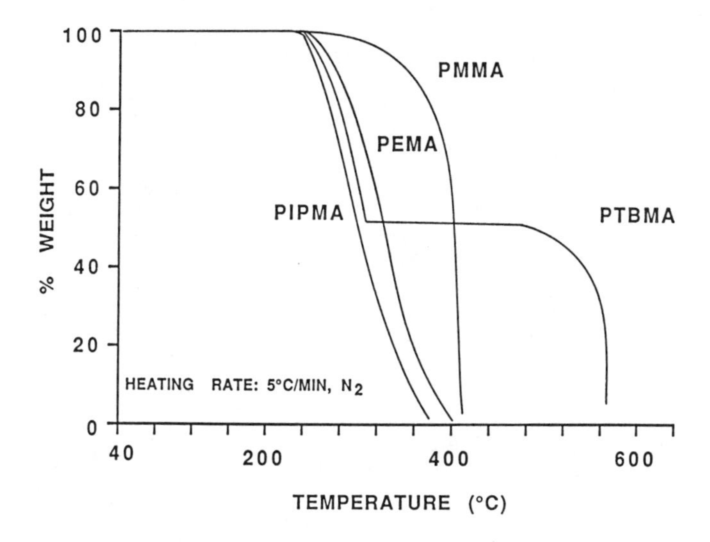

Figure 10. Thermogravimetric analysis (TGA) of various syndiotactic poly(alkyl methacrylates): PMMA, methyl; PEMA, ethyl; PIPMA, isopropyl; PTBMA, t-butyl.

ACKNOWLEDGMENTS

The authors would like to thank Mr. R. H. Bott and Ms. B. E. McGrath for thermal analysis measurements and Dr. Wunderlich (Rohm Tech, West Germany) for the generous donation of various acrylic monomers. We also appreciate the support of the Exxon Foundation, Dow USA, the Army Research Office (ARO) and the Defense Advanced Research Projects Agency (DARPA).

Literature Cited

1. Allen, R. D.; Yilgor, I.; McGrath, J. E. In Coulombic Interactions in Macromolecular Systems; Eisenberg, A.; Bailey, F. E., Eds.; ACS Symposium Series No. 302; American Chemical Society: Washington, DC, 1986; p 79.
2. Lundberg, R. D.; Makowski, H. S. In Ions in Polymers; Eisenberg, A., Ed.; Advances in Chemistry Series No. 187; American Chemical Society: Washington, DC, 1980; Chapter 2.
3. Williams, C. E.; Russell, T. P.; Jérôme, R.; Horrion, J. Macromolecules 1986, 19, 2877.
4. Major, M. D.; Torkelson, J. M. Macromolecules 1986, 19, 2801.
5. MacKnight, W. J.; Lundberg, R. D. Rubber Chem. and Tech. 1984, 57, 652.
6. Eisenberg, A.; King, M. In Ion-Containing Polymers: Physical Properties and Structures; Academic: New York, 1977; Vol. 2, p 169.
7. Muggee, J.; Vogl, O. J. Poly. Sci.: Part A: Polymer Chemistry 1986, 24, 2327.
8. Eisenberg, A. Macromolecules 1970, 3, 147.
9. Fitzgerald, J. J.; Weiss, R. A. Polymer Preprints 1986, 27(2), 163.
10. Murayamo, T. In Dynamic Mechanical Analysis of Polymeric Materials; Elsevier, 1978.
11. Brenner, D.; Oswald, A. In Ions in Polymers; Eisenberg, A., Ed.; Advances in Chem. Series No. 187; American Chemical Society: Washington, DC, 1980; p 53.
12. Allen, R. D.; Smith, S. D.; Long, T. E.; McGrath, J. E. Polymer Preprints 1985, 26(1), 247.
13. Long, T. E.; Broske, A. D.; Bradley, D. J.; McGrath, J. E. Polymer Preprints 1987, 28(1), 384.
14. Allen, R. D.; Huang, T. L.; Mohanty, D. K.; Huang, S. S.; Qin, H. D.; McGrath, J. E. Polymer Preprints 1983, 24(2), 41.
15. March, J. In Advanced Organic Chemistry: Reactions, Mechanisms, and Structure; McGraw-Hill, 2nd ed., 1977; p 352.
16. Otoka, E. P.; Kwei, T. K. Macromolecules 1968, 1, 244.
17. Newmann, R. M.; MacKnight, W. J. J. Polym. Sci.; Polym. Symp. No. 53, 1978; p 281.
18. Bagrodia, S.; Mohajer, Y.; Wilkes, G. L.; Storey, R. F.; Kennedy, J. P. Polym. Bull. 1983, 9, 174.

19. Morton, M.; Fetters, L. J. Rubber Chem. and Tech. 1975, 48, 359.
20. Allen, R. D.; Long, T. E.; McGrath, J. E. In Advances in Polymer Synthesis; Culbertson, B. M.; McGrath, J. E., Eds.; Plenum, 1985; p 347.
21. Long, T. E.; Subramanian, R.; Ward, T. C.; McGrath, J. E. Polymer Preprints 1986, 27(2), 258.
22. Gilman, H.; Cartledge, F. K. J. Organometal. Chem. 1964, 2, 447.
23. Allen, R. D.; Long, T. E.; McGrath, J. E. Polym. Bull. 1986, 15(2), 127.
24. Mole, T.; Jeffrey, E. A. In Organoaluminum Compounds; Elsevier, 1972; Chapter 12.
25. Mole, T.; Jeffrey, E. A. In Organoaluminum Compounds; Elsevier, 1972; Chapter 2.
26. Kornblum, N.; Singaram, S. J. Org. Chem. 1979, 44, 4727.
27. Osawa, Z.; Nakamo, H.; Nitsui, E. J. Polym. Sci., Polym. Chem. Ed. 1979, 17, 139.
28. Han, Y. K.; Jin, M. Y.; Choi, S. K. J. Polym. Sci., Polym. Chem. Ed. 1983, 21, 73.
29. Müller, A. H. E. Makromol. Chem. 1981, 182, 2863.
30. Hoover, J. M.; Ward, T. C.; McGrath, J. E. Polymer Preprints 1985, 26(1), 253.
31. Chapman, A.; Jenkins, A. D. J. Polym. Sci., Polym. Chem. Ed. 1977, 15, 3075.
32. McGrath, J. E.; Allen, R. D.; Hoover, J. M.; Long, T. E.; Broske, A. D.; Smith, S. D.; Mohanty, D. K. Polymer Preprints 1986, 27(1), 183.
33. Müller, A. H. E.; Jeuck, H.; Johann, C.; Kilz, P. Polymer Preprints 1986, 27(1), 153.
34. Thompson, L. F.; Willson, C. G.; Bowden, M. J. In Introduction to Microlithography; ACS Symposium Series No. 219; American Chemical Society: Washington, DC, 1983; p 125.

RECEIVED October 7, 1987

Chapter 20

AB Block Copolymers Containing Methacrylic Acid and/or Metal Methacrylate Blocks

Preparation by Selective Cleavage of Methacrylic Esters

Douglas E. Bugner

Eastman Kodak Company, 1999 Lake Avenue, Rochester, NY 14650

We have successfully prepared block copolymers containing blocks of methacrylic acid and/or alkali metal methacrylate from the corresponding methacrylic ester block copolymers by judicious choice of both the ester and the reaction conditions (1). Such copolymers are not readily prepared by conventional anionic techniques. Although poly(methyl methacrylate) can be hydrolyzed with either acid or base, block copolymers containing less than 10 mol% methyl methacrylate, e.g., poly(styrene-b-methyl methacrylate) (S-b-MM), are very difficult to hydrolyze. Other, nonhydrolytic methods for the cleavage of esters are known in the synthetic literature, however. For example, trimethylsilyl iodide is known to cleave alkyl esters under ambient conditions to yield the corresponding acid in high yield upon workup. Interestingly, S-b-MM is again unreactive, but the t-butyl methacrylate copolymer, S-b-tBM, is cleanly converted to the methacrylic acid copolymer, S-b-MA, under the same conditions. This remarkable selectivity has been exploited to synthesize the previously unknown poly(methyl methacrylate-b-methacrylic acid) (MM-b-MA). These and other reactions are described.

Block copolymers comprising oleophilic and hydrophilic blocks are of interest on both practical and theoretical grounds (2). The preparation of such copolymers, however, is not an easy task. The technique of sequential anionic polymerization is generally not amenable to polar and/or ionic monomers (3). Heterogeneous, free radical copolymerizations have been shown to give "blocky" copolymers, but the microstructure is usually ill-defined, and the polymerization conditions are often specific to a given pair of monomers (4,5). An alternate approach to polymers containing well-defined ionomeric blocks is to first prepare a precursor block copolymer by the standard anionic method followed by conversion of one of the blocks into an ion-bearing group.

0097-6156/88/0364-0276$06.00/0

$$-\!\!\left(CH_2\underset{R}{C}H\right)_n\!\!\left(CH_2\overset{CH_3}{\underset{CO_2R'}{C}}\right)_m\!\!- \quad \nearrow \quad -\!\!\left(CH_2\underset{R}{C}H\right)_n\!\!\left(CH_2\overset{CH_3}{\underset{CO_2H}{C}}\right)_m\!\!-$$

$$HX \uparrow \; \downarrow MOH$$

$$\searrow \quad -\!\!\left(CH_2\underset{R}{C}H\right)_n\!\!\left(CH_2\overset{CH_3}{\underset{CO_2^{\ominus}\,M^{\oplus}}{C}}\right)_m\!\!-$$

Brown and White employed this approach to prepare block copolymers of styrene and methacrylic acid (6). They were able to hydrolyze poly(styrene-b-methyl methacrylate) (S-b-MM) with p-toluenesulfonic acid (TsOH). Allen, et al., have recently reported acidic hydrolysis of poly(styrene-b-t-butyl methacrylate) (S-b-tBM) (7-10). These same workers have also prepared potassium methacrylate blocks directly by treating blocks of alkyl methacrylates with potassium superoxide (7-10).

Our requirements for certain applications called for the preparation of block copolymers of styrene and alkali metal methacrylates with molecular weights of about 20,000 and methacrylate contents of about 10 mol%. In this report we describe the preparation and reactions of S-b-MM and S-b-tBM. In the course of our investigation, we have found several new methods for the conversion of alkyl methacrylate blocks into methacrylic acid and/or metal methacrylate blocks. Of particular interest is the reaction with trimethylsilyl iodide. Under the same mild conditions, MM blocks are completely unreactive, while tBM blocks are cleanly converted to either methacrylic acid or metal methacrylate blocks. As a consequence of this unexpected selectivity, we also report the preparation of the new block copolymers, poly(methyl methacrylate-b-potassium methacrylate) (MM-b-MA.K) and poly(methyl methacrylate-b-methacrylic acid) (MM-b-MA).

Experimental

Materials. Methyl methacrylate was a product of Rohm and Haas, and t-butyl methacrylate was obtained from Polysciences, Inc. Potassium trimethylsilanolate (PTMS) was obtained from Petrarch Systems, Inc. Anhydrous lithium iodide, trimethylsilyl iodide (TMSI), and n-butyllitium (in hexanes) were purchased from Aldrich Chemical Co. All other reagents and solvents were obtained from Kodak Laboratory Chemicals. All monomers were purified by distillation under an inert atmosphere. They were distilled from and collected over 3 Å molecular sieves just prior to the polymerizations (11). TMSI was distilled under an inert atmosphere and stored over copper powder. It remained colorless and retained its reactivity for well over a year under these conditions. Fluorene was recrystallized from absolute ethanol and stored in vacuum dessicator. Tetrahydrofuran (THF) was distilled directly into the reaction vessel from a solution of sodium benzophenone ketyl. Toluene and dichloromethane were dried with 3 Å molecular sieves. All transfers of monomers and TMSI were made with syringes or stainless steel cannulas. All polymerizations were carried out under a slight positive pressure of dry, oxygen-free argon.

Analytical Methods. Proton magnetic resonance spectra (^{1}H NMR) were recorded on a General Electric QE-300 instrument. Infrared spectra (IR) were obtained with Perkin-Elmer 297 or 298 spectrophotometers. Microanalyses and inductively coupled plasma spectoscopy (ICP) were performed at Kodak Research Laboratories. Copolymer compositions for S-b-MM and S-b-tBM were determined by taking the average of two or more measurments for C, H, and O, and then normalizing these averages to 100%. Molecular weight determinations were made by using gel permeation chromatography (GPC) in THF on a 5-column set of µStyragel columns which were calibrated with polystyrene standards. The molecular weights reported are thus expressed as polystyrene equivalents (pse). Carboxylate contents were measured by non-aqueous titration (NAT) in THF using 0.1 N perchloric acid in 1:25 methanol-THF titrant. Titrations for carboxylic acid were done in 1:9 water-THF using 0.1 N aqueous NaOH titrant.

Preparation of Block Copolymers. Poly(styrene-b-methyl methacrylate) and poly(styrene-b-t-butyl methacrylate) were prepared by procedures similar to those reported for poly(styrene-b-methyl methacrylate (12,13). Poly(methyl methacrylate-b-t-butyl methacrylate) was synthesized by adaptation of the method published (14) for syndiotactic poly(methyl methacrylate): polymerization of methyl methacrylate was initiated with fluorenyllithium, and prior to termination, t-butyl methacrylate was added to give the block copolymer. Pertinent analytical data are as follows.

S-b-MM: ^{1}H NMR ($CDCl_3$): δ 0.85, 1.0, and 3.59 ppm, plus the usual polystyrene signals, integration of the MM signal at 3.59 ppm against the styrene aromatic region reveals 10.5 mol% (10.1 wt%) MM; IR (film): ν 1730, 1273, 1242, 1193, and 1151 cm^{-1}, plus polystyrene bands; GPC (THF): $\overline{Mw}$ = 60600 pse, $\overline{Mw}/\overline{Mn}$ = 1.46; Anal. found: 88.8% C, 7.75% H, and 3.48% O; S-b-MM (89:11) requires: 88.8% C, 7.78% H, and 3.39% O.

S-b-tBM: ^{1}H NMR: ($CDCl_3$): δ 1.41 ppm, plus polystyrene signals, integration reveals 9.7 mol% (13 wt%) tBM (see also Figure 1a); IR (film): ν 2977, 1728, 1397, 1374, 1247, 1130, and 839 cm^{-1}, plus polystyrene bands (see also Figure 2a); GPC (THF): $\overline{Mw}$ = 31400 pse, $\overline{Mw}/\overline{Mn}$ = 1.39; Anal. found: 88.8% C, 8.13% H, and 2.96% O; S-b-tBM (90:10) requires: 89.0% C, 8.03% H, and 2.96% O.

MM-b-tBM: ^{1}H NMR ($CDCl_3$): δ 0.85, 1.02, 1.42, 1.81, and 3.60 ppm, integration reveals 5.5 mol% (7.6 wt%) tBM and 72% syndiotacticity (see also Figure 3a); IR (film): ν 2995, 2960, 2850, 1726, 1484, 1445, 1390, 1367, 1270, 1240, 1192, 1146, 1068, 990, 972, 921, 852, 833, 816, 756, and 738 cm^{-1} (see also Figure 4a); GPC (THF): $\overline{Mw}$ = 16000 pse, $\overline{Mw}/\overline{Mn}$ = 1.17 (see also Figure 5a).

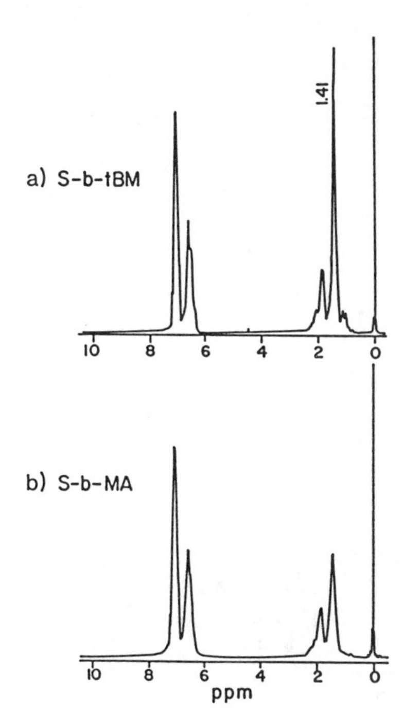

Figure 1. Comparison of ¹H NMR spectra for S-b-tBM and S-b-MA.

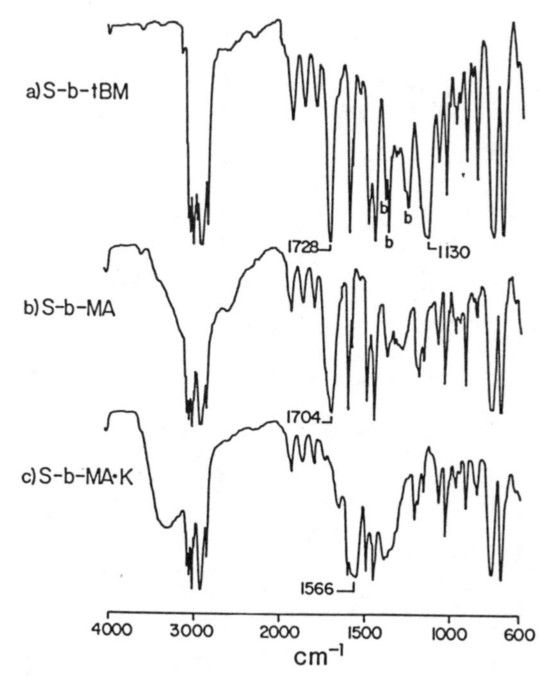

Figure 2. Comparison of IR spectra for S-b-tBM, S-b-MA, and S-b-MA.K.

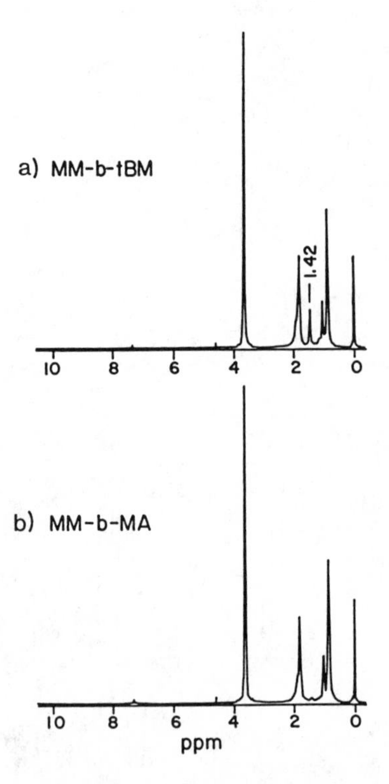

Figure 3. Comparison of ^{1}H NMR spectra for MM-b-tBM and MM-b-MA.

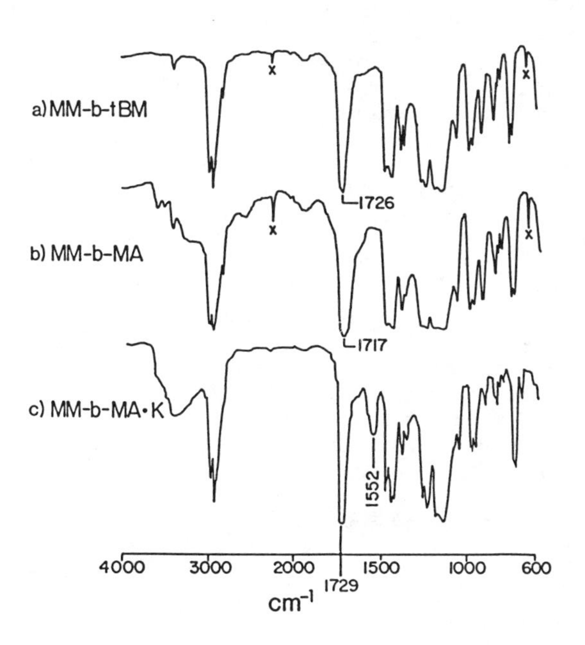

Figure 4. Comparison of IR spectra for MM-b-tBM, MM-b-MA, and MM-b-MA.K.

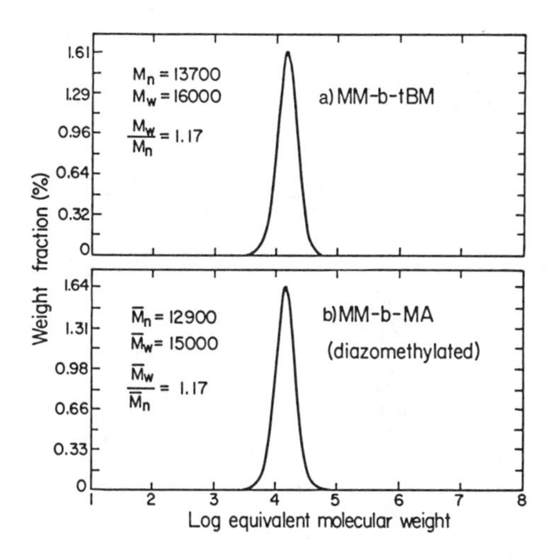

Figure 5. Comparison of GPC traces for MM-b-tBM and MM-b-MA (diazomethylated).

Attempted Hydrolysis of S-b-MM with TsOH. Following the procedure of Brown and White (6), a mixture of S-b-MM-90/10-wt (1.52 g, 1.5 meq MM), $TsOH \cdot H_2O$ (3.02 g, 15.9 mmol), acetic acid (100 g), dimethylformamide (20 mL), and water (0.5 mL) was refluxed for 47 hr. The resulting suspension was poured into methanol (500 mL). The solids were collected, washed with water (2 x 250 mL) and methanol (500 mL), and dried in a vacuum oven. The dried polymer was dissolved in dichloromethane and reprecipitated from methanol. The precipitate was washed with methanol and dried in a vacuum oven, yielding 1.25 g of product. Both ^1H NMR and IR indicate essentially unreacted S-b-MM.

Attempted Hydrolysis of S-b-MM with KOH. A solution of S-b-MM-90/10-wt (1.50 g, 1.5 meq MM) in 1:9 water-THF (50 mL) was treated with KOH (1.57 g, 28.0 mmol), and the mixture was refluxed for 47 hr. The solvents were evaporated, and the residue was shredded in a blender containing 200 mL of water. The solids were filtered, washed with water and methanol, and dried. The dried polymer was dissolved in dichloromethane and precipitated from methanol. The precipitate was washed with methanol and dried in a vacuum oven, yielding 1.23 g of product. Both ^1H NMR and IR indicate essentially unreacted S-b-MM.

Reaction of S-b-MM with Lithium Iodide. The reaction was carried out under nitrogen in a 250-mL, round-bottom flask equipped with a magnetic stirrer and a reflux condenser. To a solution of S-b-MM-90/10-wt (3.00 g, 3.0 meq MM) in toluene (100 mL) was added LiI (0.45 g, 3.4 mmol), and the mixture was refluxed for 20 hr. The solvent was partially evaporated, and the polymer was then precipitated from methanol. The precipitate was washed with methanol, water, and again with methanol. The dried polymer was reprecipitated, washed, and dried as described above, yielding 2.58 g of product. Analysis by ^1H NMR shows only a trace of MM signal at 3.59 ppm; IR (film): ν 1723 and 1570 cm^{-1} (shoulder); ICP: 0.106, 0.110 wt% Li (0.156 meq Li/g); expect 1.07 meq Li/g.

Reaction of S-b-MM with Potassium Trimethylsilanolate. The reaction was carried out in a 250-mL, round-bottom flask equipped with a magnetic stirrer and a reflux condenser. To a solution of S-b-MM (6.02 g, 6.0 meq MM) in toluene (100 mL) was added PTMS (0.87 g, 6.8 mmol). The mixture was refluxed for 1 hr, and was then isolated as described above. Attempts to redissolve the product in dichloromethane resulted in a thick suspension. This was precipitated from methanol, washed, and dried as before, yielding 4.55 g of product. Analysis by ^1H NMR shows no signal at 3.59 ppm; IR (film): 1727, 1569 (shoulder), 1256. 1206, and 1089 cm^{-1}, bands for MM no longer present; ICP: 1.44, 1.38 wt% K (0.361 meq K/g); expect 1.03 meq K/g.

A 2.09 g portion of this product was dissolved in THF (50 mL) and treated with conc. HCl (3.33 mL, 40 meq) in water (5 mL). The mixture was refluxed for 4 hr. It was cooled, filtered to remove insoluble residue, and precipitated from 1:1 water-methanol. The precipitate was resuspended in dichloromethane, reprecipitated from methanol, washed, and dried as before, yielding 1.27 g: IR (film): ν 1732, 1701, 1199, and 1083 cm^{-1}.

A 2.02 g portion of the initial product was dissolved in THF (50 mL) and treated with a solution of KOH (2.02 g, 36.1 mmol) in water (5 mL). The mixture was refluxed for 4 hr, and was then isolated and reprecipitated as before, yielding 1.66 g: IR (film): ν 1725, 1664, 1572 (shoulder), 1274, 1204, and 1081 cm^{-1}; ICP: 1.83, 1.87 wt% K (0.473 meq K/g); expect 1.03 meq K/g.

Reaction of S-b-MM with Trimethylsilyl Iodide. The reaction was carried out under nitrogen in a 250-mL, round-bottom flask equipped with a magnetic stirrer. To a solution of S-b-MM-94/6-wt (5.01 g, 2.9 meq MM) in dichloromethane (50 mL) was added TMSI (1.3 g, 6.3 mmol) via syringe. The solution was refluxed for 22 hr. It was cooled, precipitated from methanol, washed with methanol, and dried, yielding 4.39 g. NMR, IR, and GPC analyses were virtually identical to those of the starting material.

Hydrolysis of S-b-tBM with TsOH. Using the same procedure as that described above for the reaction of S-b-MM with TsOH, a 1.51 g portion of S-b-tBM-87/13-wt (1.4 meq tBM) yielded 1.20 g of S-b-MA: ^1H NMR ($CDCl_3$): signal at 1.41 ppm substantially reduced in intensity (see also Figure 1b); IR (film): ν 1704 and 1280 cm^{-1}, bands for tBM no longer present (see also Figure 2b); GPC (THF, diazomethylated): $\overline{M}w$ = 30800 pse, $\overline{M}w/\overline{M}n$ = 1.35; titration: 0.94 meq COOH/g; expect 0.98 meq/g.

A 0.49 g portion of S-b-MA as obtained above was treated with KOH (0.51 g, 9.1 mmol) in 1:10 water-THF (33 mL). The mixture was refluxed for 4 hr, and then was isolated and reprecipitated as described above for the reaction of S-b-MM with KOH, yielding 0.42 g of S-b-MA.K: IR (film): ν 1664, 1566, and 1206 cm^{-1} (see also Figure 2c); ICP: 3.84, 3.82 wt% K (0.980 meq K/g); NAT: 0.91 meq CO_2^-/g; expect 0.94 meq/g.

Attempted Hydrolysis of S-b-tBM with KOH. Using the same procedure as that described above for the attempted hydrolysis of S-b-MM, 1.57 g of S-b-tBM-87/13-wt (1.4 meq tBM) yielded 1.35 g of product. IR analysis indicated that essentially no reaction had occurred.

Reaction of S-b-tBM with Potassium Trimethylsilanolate. Using the same procedure as that employed for the reaction of S-b-MM with PTMS, 3.0 g of S-b-tBM-87/13-wt (2.8 meq tBM) yielded 2.5 g of product. The IR spectrum was identical to that of the starting polymer. ^1H NMR showed a slight reduction in the integral for the aliphatic region of the spectrum. ICP indicated 0.36, 0.37 wt% K (0.093 meq K/g); expect 0.94 meq K/g.

Reaction of S-b-tBM with Trimethylsilyl Iodide. The reaction was carried out under conditions similar to that employed for the reaction of S-b-MM with TMSI. A mixture of S-b-tBM-87/13-wt (10.0 g, 9.3 meq tBM) in dichloromethane (100 mL, dried over 3 Å sieves) was treated with TMSI (5.0 g, 25 mmol) and was stirred for 4 hr at room temperature, resulting in a dark red solution. The solvent was partially evaporated, and the residue was precipitated from methanol. The precipitate was washed with several portions of methanol and was dried. It was redissolved in 1:9 water-THF (300 mL), 3 mL of conc. HCl was added, and the mixture was refluxed for 2 hr. The solvents were evaporated, and the residue was redissolved in THF (30 mL) and reprecipitated from methanol. The precipitate was collected, washed with methanol, and dried, yielding 8.50 g of S-b-MA. Analyses: ^1H NMR and IR spectra identical to those observed for S-b-MA prepared by hydrolysis of S-b-MM with TsOH; GPC (THF, diazomethylated): $\overline{Mw}$ = 30400 pse, $\overline{Mw}/\overline{Mn}$ = 1.36; titration: 0.91 meq COOH/g; expect 0.98 meq/g.

A 0.55 g portion of S-b-MA as obtained above was converted to the potassium salt in the same manner as that described above following the hydrolysis of S-b-tBM with TsOH, yielding 0.43 g of S-b-MA.K: IR spectrum identical to that described above for S-b-MA.K; NAT: 0.91 meq CO_2^-/g; ICP: 4.16, 4.13 wt% K (1.06 meq K/g); expect 0.94 meq K/g.

Reaction of MM-b-tBM with Trimethylsilyl Iodide. The reaction was carried out exactly as described above for the reaction of S-b-tBM with TMSI. Thus MM-b-tBM-92/8-wt (10.0 g, 5.5 meq tBM) was treated with TMSI (2.8 g, 14 mmol) in dichloromethane (100 mL) for 4 hr at room temperature. This time the product was isolated by precipitation from cyclohexane, yielding 9.41 g of product.

A 3.12 g portion of this product was stirred in 1 N HCL (240 mL, 1:1 methanol-water) for 2 hr. It was filtered, washed with methanol, and dried, yielding 2.82 g of MM-b-MA. Analyses: ^1H NMR ($CDCl_3$): signal at 1.41 ppm now gone (see also Figure 3b); IR (film): ν band at 1726 cm broadened and shifted to 1717 cm^{-1}(see also Figure 4b); GPC (THF, diazomethylated): $\overline{Mw}$ = 15000 pse, $\overline{Mw}/\overline{Mn}$ = 1.17 (see also Figure 5b); titration: 0.583 meq COOH/g; expect 0.56 meq/g.

A 0.42 g portion of the initial product from the reaction of MM-b-tBM with TMSI was converted to its potassium salt in the same manner as that previously described above for S-b-MA, yielding 0.20 g of MM-b-MA.K. Analyses: IR (film): ν 1729 and 1552 cm^{-1}; ICP: 2.29, 2.24 wt% K (0.68 meq K/g); expect 0.54 meq K/g.

Results and Discussion

Preparation and Reactions of S-b-MM. As mentioned in the introduction, we were interested in block copolymers of styrene and alkali metal methacrylates with overall molecular weights of about 20,000 and methacrylate contents on the order of 10 mol%. The preparation of such copolymers by the usual anionic techniques is not feasible. An alternative is to prepare block copolymers of styrene and methacrylic esters by sequential anionic polymerization, followed by a post-polymerization reaction to produce the desired block copolymers. The obvious first choice of methacrylic esters is methyl methacrylate. It is inexpensive, readily available, and its block copolymers with styrene are well-known. In fact, Brown and White have reported the preparation and hydrolyses of a series of S-b-MM copolymers of varying MM content using p-toluenesulfonic acid (TsOH) (6). The resulting methacrylic acid copolymers were easily converted to their sodium carboxylates by neutralization with sodium hydroxide.

$$-(CH_2CH)_n-(CH_2C(CH_3)(CO_2CH_3))_m- \xrightarrow[AcOH]{TsOH} -(CH_2CH)_n-(CH_2C(CH_3)(CO_2H))_m-$$

S-b-MM S-b-MA

$$\xrightarrow{NaOH} -(CH_2CH)_n-(CH_2C(CH_3)(CO_2^{\ominus}\ Na^{\oplus}))_m-$$

S-b-MA.Na

In their study, molecular weights were kept quite low (ca. 1200), and the lowest MM content used was about 13 mol%. There was also some question as to the effect of the tacticity of the MM block on the extent of hydrolysis. MM blocks have reportedly been directly converted into the potassium methacrylate (MA.K) blocks by reaction with potassium superoxide (7-10). The synthetic details of this transformation, however, were sketchy. We decided to reinvestigate these and other methods for the preparation of the block copolymers as defined above.

S-b-MM was prepared according to the published procedures (4-6). Molecular weights in the desired range and with narrow, unimodal distibutions were obtained without resorting to extensive monomer purification (11) or capping of the styrene block with diphenylethylene (4,5,7-10). The S-b-MM contained about 10 mol% MM, and was conveniently characterized by ^1H NMR and IR spectroscopy. The methyl ester gives rise to a fairly sharp singlet at 3.59 ppm, and the ester carbonyl exhibits an infrared band at 1730 cm^{-1}. The MM content of the copolymer is easily ascertained by integration of the ^1H NMR spectrum and may be corroborated by elemental analysis.

Attempts to hydrolyze S-b-MM under either acidic or basic conditions were unsuccessful. Reaction of S-b-MM with TsOH under the same conditions described by Brown and White (6) resulted in no reaction as evidenced by comparison of the ^1H NMR and IR spectra of the product with that of the starting material. At the molecular weights and MM contents that we were working with, the S-b-MM was insoluble in the reaction medium, which may explain its lack of reactivity. Although the hydrolysis of methyl methacrylate under basic conditions is known to be quite sluggish, we nevertheless subjected S-b-MM to two days of reflux in aqueous THF in the presence of excess KOH. As expected, the product was spectroscopically identical to the starting copolymer.

Two alternatives to conventional acid/base hydrolyses for cleaving esters are Sn2 displacement of the carboxylate group by reactive nucleophiles and nucleophilic attack at the carbonyl carbon. In this latter context we investigated the reaction of S-b-MM with potassium trimethylsilanolate, a so-called potassium superoxide equivalent (15). One advantage that this reagent has over potassium superoxide is its greater solubility in organic solvents. Its reaction with S-b-MM in refluxing toluene was quite rapid. In less than an hour, the ^{1}H NMR spectrum of the product showed complete loss of the methyl ester. The IR spectrum, however, still exhibits a strong band at 1727 cm^{-1} and only a weak band for the carboxylate stretch near 1560 cm^{-1}. Analysis for potassium by inductively coupled plasma spectroscopy (ICP) indicated only a fraction of the expected amount. The polymer was also poorly soluble in most organic solvents. Further treatment of the polymer by either HCl or KOH in refluxing aqueous THF failed to eliminate the IR band at 1727 cm^{-1}. In the IR spectrum of the acidified polymer, the weak carboxylate band near 1560 cm^{-1} was replaced with a new, weak carbonyl band at 1701 cm^{-1}, attributable to methacrylic acid residues. Apparently a side-reaction, probably involving some crosslinking of the copolymer and/or formation of a new ester, was occurring in addition to cleavage of the methyl ester.

At this point a comparison of these observations with those reported by Allen, et al. (7-10), for the reaction of poly(styrene-b-isobutyl methacrylate) with potassium superoxide should be made. They too obtained a product which was poorly soluble in THF and many other solvent combinations. Acidification with HCl in THF improved the solubility, but the product displayed an IR band at 1730 cm^{-1} in addition to a methacrylic acid at 1700 cm^{-1}. They attributed the band at 1730 cm^{-1} to unreacted isobutyl methacrylate groups, but in light of our observations, it may be due to the same type of side-reaction.

Another attractive reagent is lithium iodide, which is known to produce the lithium carboxylate directly in the case of non-polymeric esters by an Sn2 mechanism (16). Although pyridine is usually the solvent of choice for this reaction, we found only partial cleavage (by ^{1}H NMR) of the methyl ester for S-b-MM after extended periods of reaction with LiI in refluxing pyridine. A better solvent appeared to be refluxing toluene. After 20 hr, the signal for the methyl ester in the NMR spectrum was eliminated. Analogous to the reaction with potassium trimethylsilanolate, however, the IR spectrum still displayed a carbonyl band at 1723 cm^{-1} along with a shoulder for the carboxylate ion at 1570 cm^{-1}. Analysis for Li (ICP) showed only about 15% of the expected amount. It thus appeared that quite similar chemistry was occurring with these two reagents, the details of which are currently under study.

Another nonsaponicative method for cleaving esters employs trimethylsilyl iodide (TMSI) (17,18). Unfortunately, reaction of S-b-MM with TMSI in refluxing dichloromethane for 22 hr gave back unchanged starting material. Interestingly, the literature reports that t-butyl and benzyl esters are significantly more reactive toward TMSI than methyl esters (17). This reactivity pattern is similar to that observed for acid-catalyzed ester hydrolysis (7-10). Therefore, confounded by our inability to efficiently cleave methyl methacrylate copolymers, it was decided to investigate the preparation and reactivity of the analogous t-butyl methacrylate block copolymers.

Preparation and Reactions of S-b-tBM. S-b-tBM copolymers were prepared in exactly the same manner as that described above for S-b-MM. Characterization by ^{1}H NMR confirmed the presence of the t-butyl protons at 1.41 ppm, coincident with the signal for the backbone methylene protons (Figure 1a). The IR spectrum displayed the ester carbonyl stretch at 1728 cm^{-1}, as well as t-butyl C-H vibrations at 2977, 1397, 1374, and 1247 cm^{-1}, and the C-O-C stretching band at 1130 cm^{-1} (Figure 2a). The molecular weight distribution was again narrow, unimodal, and in the expected range.

We first attempted to hydrolyze S-b-tBM with TsOH under the same conditions which were unsuccessful for S-b-MM. This time, although the polymer was again incompletely soluble in the reaction milieu, the t-butyl methacrylate block appeared to be quantiatively hydrolyzed. The t-butyl bands listed above are no longer observed in the IR spectrum (Figure 2b). The carbonyl band is broadened and shifted to 1704 cm^{-1}, and a C-O-H stretch is observed at 1280 cm^{-1}. A weak, broad band at 2625 cm^{-1} and a shoulder at 1735 cm^{-1} can be attributed to hydrogen-bonded O-H and C=O stretches, respectively. Such bands are not observed in random copolymers of styrene and methacrylic acid (Guistina, R. A., Eastman Kodak, personal communication, 1981). The signal at 1.41 ppm in the NMR spectrum is considerably reduced, also consistent with the loss of the t-butyl group (Figure 1b). A GPC trace of the diazomethylated product indicates that no significant chain degradation has occurred. The slight decrease in the observed $\overline{M}w$ is expected since the t-butyl group of the starting copolymer has been replaced with a much smaller methyl group. Titration with 0.1 N sodium hydroxide in aqueous THF indicates 0.94 meq COOH/g, in good agreement with the expected 0.98 meq/g of methacrylic acid, based on the tBM content of the precursor.

$$-(CH_2CH(C_6H_5))_n-(CH_2C(CH_3)(CO_2\,t\text{-}Bu))_m- \xrightarrow[\text{Ac OH}]{\text{Ts OH}} -(CH_2CH(C_6H_5))_n-(CH_2C(CH_3)(CO_2H))_m-$$

S-b-tBM S-b-MA

$$\xrightarrow[\text{aq. THF}]{\text{KOH}} -(CH_2CH(C_6H_5))_n-(CH_2C(CH_3)(CO_2^{\ominus}K^{\oplus}))_m-$$

S-b-MA.K

Neutralization with KOH in aqueous THF gave the desired poly(styrene-b-potassium methacrylate) (S-b-MA.K). The carbonyl band in the IR spectrum is replaced with a strong, broad carboxylate absorption centered near 1566 cm^{-1} (Figure 2c). The carboxylate and potassium contents were assayed by non-aqueous titration and ICP, respectively. The resulting values of 0.91 meq CO_2^-/g and 0.98 meq K/g indicate essentially quantitative conversion to the potassium methacrylate. S-b-MA.K obtained in this manner is easily dissolved in solvents such as THF and dichloromethane, in contrast to the product of of the reaction of S-b-MM with potassium trimethylsilanolate.

We next investigated the dealkylation of S-b-tBM with TMSI. Unlike the reaction with S-b-MM, it required only 4 hr at room temperature to completely cleave the t-butyl ester. Work-up under acidic conditions gave S-b-MA which was virtually identical by NMR, IR, GPC, and titration with that just described above. Likewise, neutralization with KOH resulted in quantitative conversion to S-b-MA.K. Although the initially formed product of the reaction of alkyl esters with TMSI is presumably the trimethylsilyl ester (17), we were not able to isolate or characterize this copolymer. It is known that trimethysilyl methacrylate and its polymers spontaneously hydrolyze even in moist air (19). Any traces of water in the methanol used to precipitate the reaction mixture would thus preclude isolation of the intermediate trimethylsislyl ester.

We also explored the direct conversion of S-b-tBM to S-b-MA.K. Hydrolysis under basic conditions (KOH in refluxing aqueous THF) was again resulted in unchanged S-b-tBM. The reaction with potassium trimethylsilanolate for 1 hr in refluxing toluene gave very little reaction. Only 10% of the expected amount of potassium was found by ICP, and the NMR and IR spectra were little changed from those of the starting copolymer. This difference in reactivity between S-b-MM and S-b-tBM parallels that observed for the reaction of alkyl methacrylate blocks with potassium superoxide (7-10).

The Preparation of MM-b-MA and MM-b-MA.K. Inspired by the unexpected selectivity of the reaction of TMSI with S-b-MM and S-b-tBM, we decided to attempt the preparation of poly(methyl methacrylate-b-t-butyl methacrylate) (MM-b-tBM) and its unprecedented conversion to MM-b-MA.

$$\underset{\textbf{MM-b-tBM}}{-\!\left(CH_2\overset{CH_3}{\underset{CO_2Me}{C}}\right)_n\!\left(CH_2\overset{CH_3}{\underset{CO_2t\text{-}Bu}{C}}\right)_m\!-} \xrightarrow[\text{HCl}]{\text{TMSI}} \underset{\textbf{MM-b-MA}}{-\!\left(CH_2\overset{CH_3}{\underset{CO_2Me}{C}}\right)_n\!\left(CH_2\overset{CH_3}{\underset{CO_2H}{C}}\right)_m\!-}$$

$$\xrightarrow[\text{aq. THF}]{\text{KOH}} \underset{\textbf{MM-b-MA.K}}{-\!\left(CH_2\overset{CH_3}{\underset{CO_2Me}{C}}\right)_n\!\left(CH_2\overset{CH_3}{\underset{CO_2^{\ominus}K^{\oplus}}{C}}\right)_m\!-}$$

The anionic polymerization of the MM block was initiated with fluorenyllithium (14) in THF at -78°C. After several hours, t-butyl methacrylate was introduced, and the polymerization was allowed to slowly rise to room temperature. The reaction was quenched with a few drops of methanol and precipitated from ligroin. The dried polymer was analyzed by NMR, IR, and GPC. The ^{1}H NMR spectrum displays signals at 0.85, 1.02, and 1.13 (shoulder) ppm for the

α-methyl group, 1.42 ppm for the t-butyl ester protons, 1.81 ppm for the backbone methine protons, and 3.60 ppm for the methyl ester (Figure 3a). Integration of the α-methyl region indicates a predominantly syndiotactic (72%) polymer, and comparison of the integrals for the t-butyl and methyl esters reveals a molar ratio of 94.6/5.4, consistent with the feed ratio of the two monomers. The IR spectrum features a single, slightly broadened carbonyl band centered at 1726 cm^{-1} (Figure 4a), and the GPC trace exhibits a narrow ($\overline{M}w/\overline{M}n$ = 1.17) molecular weight distribution with $\overline{M}w$ = 16000 pse Figure 5a).

Treatment of this polymer with TMSI under the same conditions employed for the reaction with S-b-tBM resulted in a quantitative production of MM-b-MA. The t-butyl signal in the NMR spectrum is now gone (Figure 3b), and the carbonyl band in the IR spectrum is further broadened and shifted to 1717 cm^{-1} (Figure 4b). Titration for MA resulted in 0.583 meq COOH/g, in accord with the value of 0.56 meq/g calculated based on the amount of tBM present in the NMR spectrum. Conversion to the potassium methacrylate copolymer was straightforward. IR analysis of the product shows the carboxylate band at 1552 cm^{-1}, and the ester band at 1729 cm^{-1} (Figure 4c). Assay for potassium (ICP) confirmed that the neutralization was quantitative.

In summary, we have examined several new methods for cleaving ester groups in poly(styrene-b-alkyl methacrylates). Short blocks of methyl methacrylate are very difficult to hydrolyze, but can be cleaved with reagents such as lithium iodide and potassium trimethylsilanolate. These latter reagents, however, result in side-reactions which appear to crosslink the polymer.

A more interesting reagent is TMSI. Although MM blocks are unreactive toward TMSI, tBM blocks are cleanly converted to MA blocks under very mild conditions. This chemoselectivity of TMSI has been exploited to prepare the novel block copolymers MM-b-MA and MM-b-MA.K. The tBM blocks can also be readily hydrolyzed with TsOH, corroborating previous reports of this transformation.

The results of this work are not limited to just S-b-MM and S-b-tBM, but may be extended to include styrene derivatives such as p-methylstyrene and p-t-butylstyrene (1). In addition to t-butyl methacrylate, other alkyl esters capable of stabilizing a carbonium ion, such as benzyl methacrylate and allyl methacrylate, should exhibit similar reactivity toward acidic hydrolysis and TMSI. In contrasting the hydrolysis of tBM blocks with TsOH and their reaction with TMSI, it should be noted that the hydrolysis is reportedly catalytic in nature (7-10), whereas the reaction with TMSI is stoichimetric. Therefore the latter approach may allow one to more easily "dial in" a desired level of methacrylic acid or metal methacrylate.

Acknowledgments

The author would like to acknowledge the following analytical assistance: Mr. R. A. Guistina, titrations and many helpful discussions; Mr. J. Reiff, GPC; Mr. Ed McLean, ICP; and the friendly crew in the combustion analysis laboratory at Kodak Research Labs. Without their expert help this work would not have been possible.

Literature Cited

1. Preliminary results of this work have been published: Bugner, D. E. Polym. Prep. 1986, 27(2), 57.
2. Riess, G.; Hurtrez, G.; and Bahdur, P. In Encyclopedia of Polymer Science and Engineering; Mark, H.; Bikales, N.; Overberger C.; Menges G., Eds.; Wiley-Interscience: New York, 1985; Vol. 2, pp 324-434.
3. Noshay, A.; McGrath, J. E. Block Copolymers: Overview and Critical Survey; Academic Press: New York, 1977.
4. Dunn, A. S.; Melville, H. W. Nature 1952, 169, 699.
5. Brown, C. W.; Taylor, B. A. J. Appl. Polym. Sci. 1969, 13, 629.
6. Brown, C. W.; White, I. F. J. Appl. Polym. Sci. 1972, 16, 2671.
7. Allen, R. D.; Huang, T. L.; Mohanty, D. K; Huang, S. S.; Qin, H. D.; McGrath, J. E. Polym. Prep. 1983, 24(2), 41.
8. Allen, R. D.; Yilgor, I.; McGrath, J. E. In Coulombic Interactions in Macromolecular Systems; Eisenberg, A.; Bailey, F. E., Eds.; ACS Symposium Series No. 302; American Chemical Society: Washington, DC, 1986; p 79.
9. McGrath, J. E.; Allen, R. D.; Hoover, J. M.; Long, T. E.; Broske, A. D.; Smith, S. D.; Mohanty, D. K. Polym. Prep. 1986, 27(1), 183.
10. Long, T. E.; Allen, R. D.; McGrath, J. E. Polym. Prep. 1986, 27(2), 54.
11. Allen, R. D.; Long, T. E.; McGrath, J. E. In Advances in Polymer Synthesis; Culbertson B. M.; McGrath, J. E., Eds.; Plenum: New York, 1985; Vol. 31, p 347.
12. Freyss, D.; Rempp, P.; Benoit, H. Polym. Lett. 1964, 2, 217.
13. Kotaka, T.; Tanaka, T.; Inagaki, H. Polym. J. 1972, 3, 327.
14. Sorenson W.; Campbell, T. In Preparative Methods of Polymer Chemistry; Interscience: New York, 1968; 2nd ed., pp 285-6.
15. Laganis E. D.; Chenard, B. L. Tetrahedron Letts. 1984, 25, 5831.
16. McMurray, J. Org. React. 1976, 24, 187.
17. Jung M. E.; Lyster, M. A. J. Am. Chem. Soc. 1977, 99, 968.
18. Ho T.; Olah, G. Angew. Chem., Int. Ed. Eng. 1976, 15, 774.
19. Chapman A.; Jenkins, A. D. J. Polym. Sci., Polym. Chem. Ed. 1977, 15, 3075.

RECEIVED August 27, 1987

Chapter 21

Imide Hydrolytic Stability of *N*-Substituted Dimethacrylamide Cyclopolymers: Poly[*N*-(4-sulfophenyl)dimethacrylamide]

Joseph J. Kozakiewics, Sun-Yi Huang, Daniel R. Draney, and JoAnn L. Villamizar

American Cyanamid Company, Stamford, CT 06904-0060

The hydrolytic stability of water soluble poly[N-(4-sulfophenyl)dimethacrylamide] (PSPDM) was studied at 90°C in aqueous solutions at pH 7, pH 1.2 (0.1M HCl), and pH 12.3 (0.1M NaOH). PSPDM, which possesses predominantly 5-membered ring imides, was prepared by the cyclopolymerization and subsequent sulfonation of N-phenyldimethacrylamide. No detectable PSPDM imide hydrolysis occurred after 30 days at pH 7 or pH 1.2. Under basic conditions, however, complete hydrolysis to amic acid occurred after one day. The resulting Nsubstituted amide was extremely stable to further basic hydrolysis.

Cyclopolymerization of substituted dimethacrylamides is well known and has recently been employed as a possible route for the synthesis of head-to-head poly(methacrylic esters).[1,2] Unexpectedly, the resulting backbone 5-membered ring imides were found to be nearly impossible to hydrolyze. Xi and Vogl found that poly(N-phenyldimethacrylamide) could not be hydrolyzed in very concentrated sodium hydroxide or potassium hydroxide solutions.[1] Additionally, hydrazinolysis with both anhydrous hydrazine and hydrazine monohydrate was unsuccessful. All attempts to hydrolyze poly(N-phenyldimethacrylamide) were unsuccessful. Otsu and Ohya attempted to hydrolyze poly(Nmethyldimethacrylamide) and poly(N-propyldimethacrylamide) in both very strong acidic and very strong basic aqueous solutions. No hydrolysis was observed with the possible exception of the reaction of poly(N-propyldimethacrylamide) in 60% aqueous KOH at 100°C for 14 days. The hydrolyzed polymer, however, was not isolated. Otsu and Ohya concluded that their imide-containing polymers were not hydrolyzed under acidic or basic conditions.

The extreme resistance of these imide groups to hydrolysis is surprising. Imides are well known to be readily hydrolyzed under both

0097-6156/88/0364-0291$06.00/0

acidic and basic conditions.[3] 5-membered ring succinimides and 6-membered ring glutarimides are hydrolyzed under mildly alkaline conditions at room temperature.[4-6] Herd et al. have shown, however, that alkyl substitution at the second and third position of N-phenyl succinimides can significantly effect the hydrolysis rate.[7] N-phenyl succinimide hydrolyzed 83 times faster than N-phenyl-2,2,3,3-tetramethylsuccinimide. Imide groups in polymers have also been shown to be readily hydrolyzed. Hydrolysis of the 5-membered ring imides of anhydropolyaspartic acid has been studied extensively.[8-10] There is no reason to expect the imide groups of N-substituted dimethacrylamide copolymers to be unusually hydrolytically stable based on these studies.

To study the factors leading to the unusual hydrolytic stability of polydimethacrylamides, we synthesized poly[N-(4-sulfophenyl)dimethacrylamide] (PSPDM) (IV). PSPDM is water soluble, in contrast to the previously studied polymers.[1,2] The hydrolytic stability of PSPDM was studied in acidic, neutral, and basic aqueous solutions at 90°C and above. The results of these experiments will be described.

EXPERIMENTAL

N-phenyldimethacrylamide (PDMA) was synthesized by the method of Butler and Meyers.[14] Poly(N-phenyldimethacrylamide) (PPDMA) was prepared by free radical cyclopolymerization of PDMA. AIBN (0.050g., 0.3 mmoles) and PDMA (5.00 g., 21.8 mmoles) were dissolved in 8.0 g. dry DMF. The solution was purged with nitrogen for 30 minutes and then heated to 50°C. After 19 hours, additional AIBN (0.050 g., 0.3 mmoles) was added and the heating continued for a total of four days. The solution was then poured into 400 ml. methanol. The white precipitated polymer was filtered, washed with methanol, and dried under vacuum at 50°C to give 3.78 g. (76% yield) of polymer with inh 0.23 (0.5% polymer in DMF, 30°C.). Poly[N-(4-sulfophenyl)dimethacrylamide] (PSPDM) was prepared by sulfonation of PPDMA. PPDMA (1.50 g.) was slowly dissolved in fuming sulfuric acid (40 g., 18-24% SO_3) at room temperature. The solution was then poured into two liters saturated sodium chloride. The precipitate was filtered, dissolved in 200 ml deionized water, dialyzed against deionized water for two days and freeze dried. The resulting polymer (1.40 g.) was soluble in water and DMSO.

Hydrolytic stability studies were performed by preparing one wt.-% aqueous solutions in culture tubes, adjusting the solutions to the desired pH and placing the sealed tubes in an oven at the desired temperature. The tubes were removed at various times, cooled to room temperature, dialyzed and freeze-dried.
The ^{13}C NMR spectra were obtained at 50.3 MHz on a Varian XL200 spectrometer. IR spectra were obtained on a Digilab F75-15 FTIR and Perkin-Elmer 1310 infra red spectrophotometer.

RESULTS AND DISCUSSION

Free radical cyclopolymerization of N-phenyldimethacrylamide (PDMA) (I) yielded PPDMA possessing predominantly 5-membered ring imides (II) (Figure 1) as determined by ^{13}C NMR and IR (Figure 2). The ^{13}C NMR peak at 50.0 indicates that PPDMA also contains a small amount of 6-membered ring imide (III). No pendant vinyl groups were found. This is in agreement with PPDMA prepared by Xi and Vogel.[1] PPDMA was soluble in DMF, DMSO and chloroform, but was insoluble in water.

Sulfonation of PPDMA proceeded quickly at room temperature to give PSPDM (IV) (Figure 3). Sulfonation of the aromatic rings went to nearly 100% completion in the para position as determined by ^{13}C NMR and IR (Figure 4a and b). The small peaks between 128 and 133 ppm are attributable to a small fraction of PPDMA rings which were not sulfonated. PSPDM, in contrast to PPDMA, is very soluble in water.

The hydrolytic stability of PSPDM was studied initially at 90°C in aqueous solution at pH 1.2 (0.1*M* HCl), pH 7 and pH 12.3 (0.1M NaOH). At both pH 1.2 and pH 7, no detectable hydrolysis occurred after thirty days. The ^{13}C NMR spectra of these samples were unchanged. Under acidic and neutral conditions, the hydrolytic stability of the imide was excellent.

After only one day at pH 12.3, however, the imide rings were nearly quantitatively opened to amic acid (V) (Figure 5). The aromatic carbon region of the ^{13}C NMR spectrum shows three new major aromatic carbon peaks from the N-sulfophenylamide group at 127.9, 134.1 and 144.6 ppm and the disappearance of the corresponding imide aromatic peaks (Figure 6). New carbonyl peaks also appear at 178 and 182 ppm attributable to the amide carbonyl carbon and the carboxylate carbonyl carbon, respectively. Under basic conditions, the imide rings are readily hydrolyzed to the respective carboxylate and amide. Surprisingly, however, no further hydrolysis of the amide occurred on heating for an additional 29 days. The amide, unlike the starting imide, is hydrolytically stable at pH 12.3 and 90°C.

In an attempt to hydrolyze PSPDM completely to head-to-head poly(methacrylic acid), higher temperature, and more strongly basic conditions were employed. PSPDM was heated at 125°C for four days in 5M NaOH. Like the 0.1M NaOH experiments, the imide was hydrolyzed to the amic acid and no further. The amide of the ring opened imide (V) is extremely resistant to basic hydrolysis.

The amide hydrolytic stability may be attributable to several factors. The amide is located right next to the carboxylate group which is formed during imide hydrolysis. Under basic conditions, this carboxylate group may screen the amide group from incoming hydroxide ions. The carboxyl group of the amide is also quite crowded by the polymer chain and neighboring pendant groups. This

(I) (II) (III)

Figure 1. Free Radical Polymerization of PDMA to PPDMA.

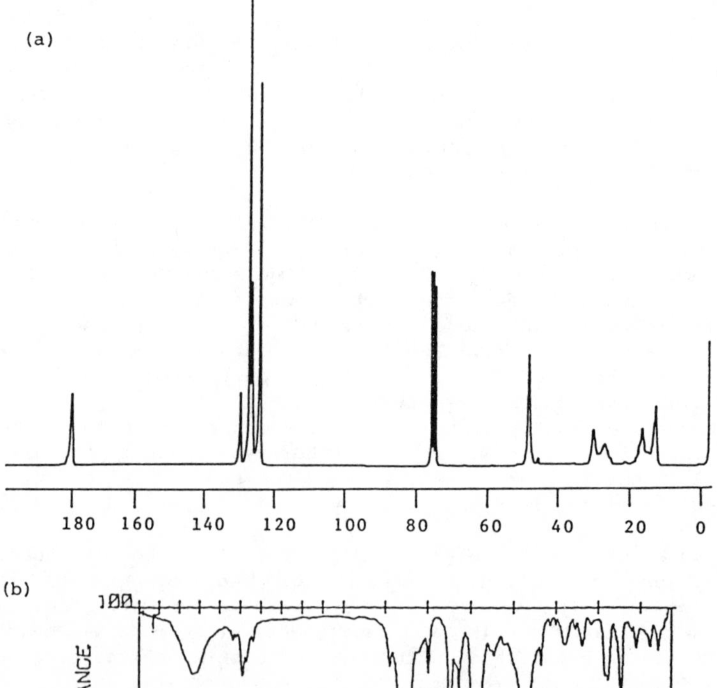

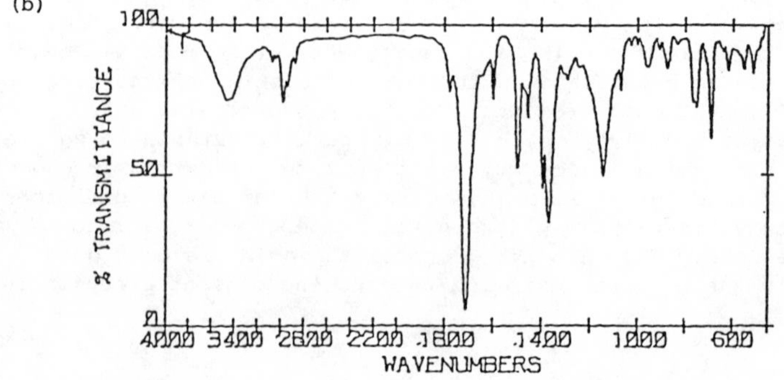

Figure 2. (a) ^{13}C NMR (50.3 MHz, $CDCl_3$, TMS); (b) IR of PPDMA.

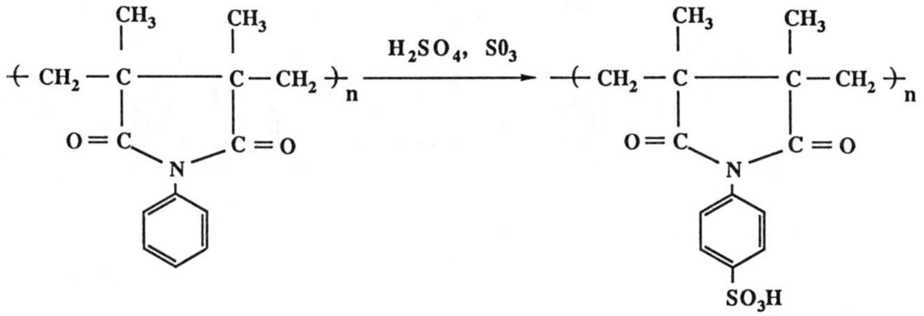

$$\left(CH_2-C(CH_3)-C(CH_3)-CH_2\right)_n \xrightarrow{H_2SO_4,\ SO_3} \left(CH_2-C(CH_3)-C(CH_3)-CH_2\right)_n$$

(II) (IV)

Figure 3. Sulfonation of PPDMA to PSPDM.

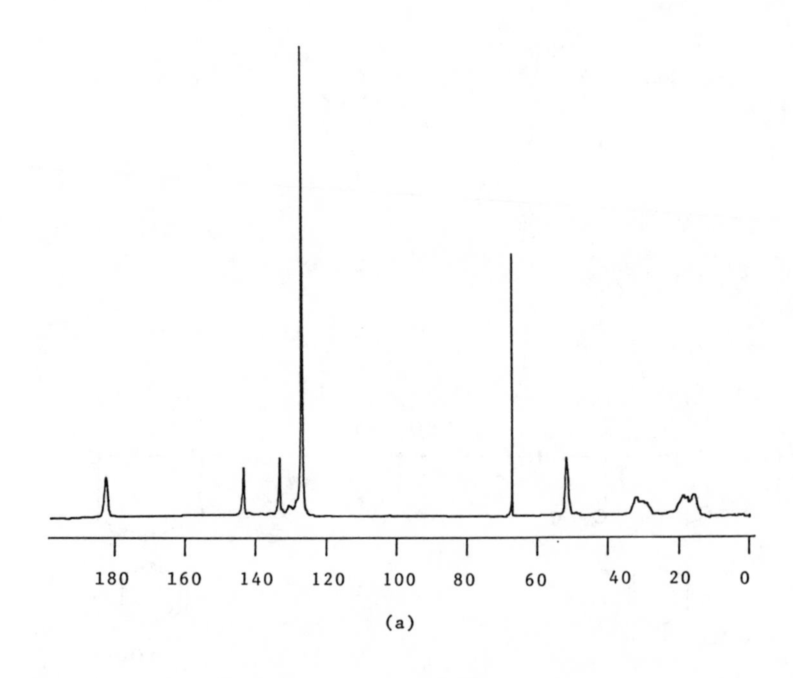

(a)

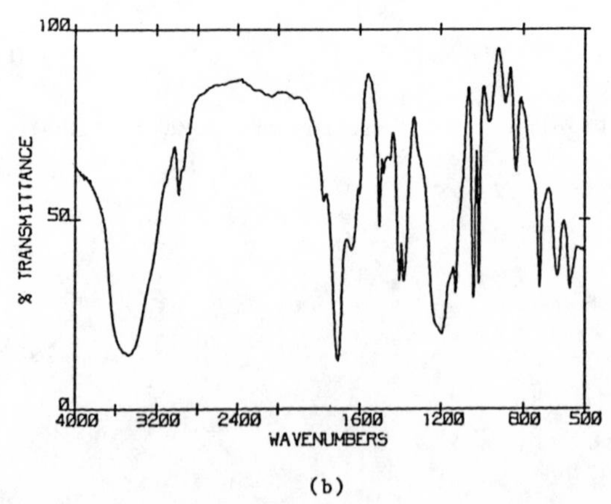

(b)

Figure 4. (a) ^{13}C NMR (50.3MHz, D_2O) of PSPDM; (b) IR of PSPDM.

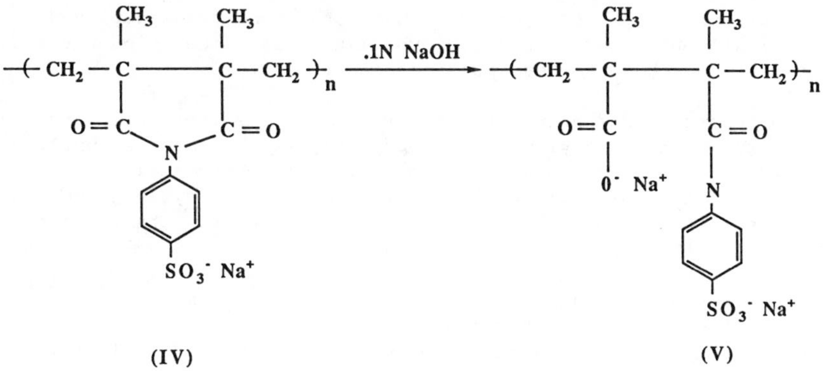

$$-\!\!\left(CH_2-\underset{O=C\diagdown N(C_6H_4SO_3^-Na^+)\diagup C=O}{\overset{CH_3}{C}-\!\!-\overset{CH_3}{C}}-CH_2\right)_n \xrightarrow{.1N\ NaOH} -\!\!\left(CH_2-\underset{O=C-O^-\ Na^+}{\overset{CH_3}{C}}-\underset{C=O-N(C_6H_4SO_3^-Na^+)}{\overset{CH_3}{C}}-CH_2\right)_n$$

(IV) (V)

Figure 5. Alkaline hydrolysis of PSPDM after one day at 90°C and pH 12.3 (0.1M NaOH).

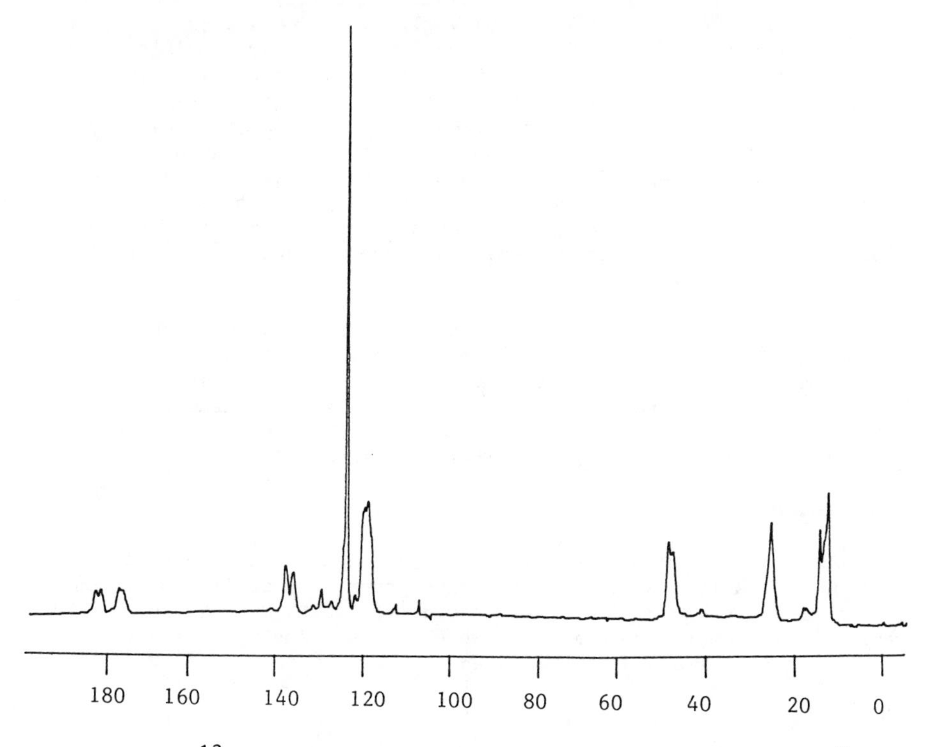

Figure 6. ^{13}C NMR (50.3MHz, D_2O) of PSPDM after heating at 90°C for one day at pH 12.3 (0.1M NaOH).

crowding may restrict the accessibility of the amide group to incoming hydroxide ions. It is also possible that the polymeric amide is inherently resistant to hydrolysis under the conditions employed. Further attempts will be made to hydrolyze this amide to enable the preparation of head-to-head poly(methacrylic acid).

An important difference between PSPDM and the previously studied Nsubstituted dimethacrylamide cyclopolymers is the water solubility of PSPDM. Hydroxide ion probably had limited access to a large fraction of the imide groups in these other dimethacrylamide cyclopolymers owing to their limited solubility in water. Their hydrolytic stability under basic conditions may have been governed solely by solubility. With the aqueous soluble PSPDM, however, hydroxide ion has greater accessibility to the imide groups. It is also possible that any imide rings hydrolyzed in the previous studies were inadvertently ring closed during polymer isolation and analysis. Hsieh et al. have recently reported that acidification of hydrolyzed polyacrylamides can cause imidization between neighboring amides and acids under relatively mild conditions.[13]

The hydrolysis of PSPDM under the basic conditions studied was unanticipated in light of the hydrolytic stability of N-substituted poly(dimethacrylamides) studied by Vogl,[1] Otsu,[2] and co-workers. The mild electron-withdrawing sulfonate group (= 0.09[11]) is expected to make the imide slightly less resistant to hydrolysis. Electron withdrawing substituents on the phenyl ring of N-phenyl succinimide have been shown to increase the rate of imide hydrolysis.[12] Electron donating groups decrease the rate of imide hydrolysis. Consequently, PSPDM would be predicted to hydrolyze marginally faster than PPDMA. The effect is not expected to be large enough to be the sole factor determining whether the imide groups of poly(N-substituted dimethacrylamides) hydrolyze under the conditions studied.

Five-membered ring imides in cyclopolymers of N-substituted dimethacrylamides such as PSPDM can be hydrolyzed to amic acids under moderately basic conditions. The resulting N-substituted amide is extremely resistant to basic hydrolysis. Consequently, this basic hydrolysis approach can still not be employed for the preparation of head-to-head poly(methacrylic acid).

CONCLUSION

Sulfonation of PPDMA occurs readily to yield the water soluble PSPDM. PSPDM is stable at neutral and moderately acidic conditions, but is readily hydrolyzed to amic acid (V)in 0.1N NaOH at 90°C. The amide of the hydrolyzed imide, however, is extremely resistant to further hydrolysis under basic conditions.

REFERENCES

1. Xi, F. and Vogl, O., J. Macromol. Sci.-Chem., A20 (3), 321 (1983).
2. Otsu, T. and Ohya, T., J. Macromol. Sci.-Chem., A21 (1), 1 (1984).

3. Hargreaves, M.K., Pritchard, J.G., and Dave, H.R., Chem. Rev., 70 (40), 439 (1970).
4. Sircar, S.S.G., J. Chem. Soc., 600 (1927).
5. Sircar, S.S.G., J. Chem. Soc., 1252 (1927).
6. Edward, J.T. and Terry, K.A., J. Chem. Soc., 3527 (1957).
7. Herd, A.K., Eberson, L., and Higuchi, T., J. Pharm. Sci., 55, 162 (1966).
8. Vegotsky, A., Harada, K., and Fox, S.W., J. Am. Chem. Soc., 80, 3361 (1958).
9. Harada, K., J. Org. Chem., 24, 1662 (1959).
10. Kovacs, J., Nagy Kovacs, H., Konyves, I., Csaszar, J., Vajda, T., and Mix, H., J. Org. Chem., 26, 1084 (1961).
11. Gordon, A.J., and Ford, R.A., "The Chemists Companion", J. Wiley & Sons, NY, NY, p. 147 (1972).
12. Tirouflet, J. and le Trouit, E., C.R. Acad. Sci., Paris, 241, 1053 (1955).
13. Hsieh, E.T., Westerman, I.J., and Moradi-Araghi, A., Poly. Mat. Sci. Eng. Prepr., 55, 700 (1986).
14. Butler, G.B., and Meyers, G.R., J. Macromol. Sci.-Chem., A5, 105 (1971).

RECEIVED August 27, 1987

Chapter 22

Radical-Promoted Functionalization of Polyethylene

Controlled Incorporation of Hydrogen-Bonding Groups

M. F. Schlecht, E. M. Pearce, T. K. Kwei, and W. Cheung

Department of Chemistry, Polytechnic University, 333 Jay Street, Brooklyn, NY 11201

The radical-promoted addition of hexafluoroacetone to polyethylene has been examined with the aim of maximizing the extent of incorporation and characterizing the functionalized polymer. Reaction conditions were found which produce samples modified to the extent of from 0.4 to 5.6 residues per hundred methylenes, as determined by elemental combustion analysis. The functionalization reaction was performed on small molecule substrates, and for comparison model compounds containing the fluoroalkanol and fluoroalkyl ether groups were prepared by independent synthetic routes. Analysis of the modified polyethylene samples by ^{19}F-NMR showed that of the pendent groups introduced, four in five are fluoroalkanol groups, and one in five is a fluoroalkyl ether. The functionalized polymers are soluble in tetrahydrofuran and in aromatic solvents, and show the expected thermal properties in proportion to the increasing percent modification, e.g. decreasing degradation temperature, increasing glass transition temperature, and decreasing melting point. Some preliminary results indicate that the functionalized polymers form miscible blends with poly(methyl methacrylate).

The chemical modification of polymers is a powerful preparative method for obtaining functionalized macromolecules from simple precursor polymers, without the necessity of synthesizing specialty monomers. Among the methods available, the radical-promoted addition of functional groups is perhaps the most widely used, yet the least understood. The patent literature contains a large number of such preparations, and in addition several systematic studies have been reported. Hydrocarbon polymers have been modified by addition of maleic anhydride to introduce the succinoyl group (1-5), and by reaction with phosphorus trichloride and oxygen to introduce the phosphoryl group (6). The radical-promoted addition of hexafluoroacetone (7) to various organic substrates, including polyethylene, was

0097-6156/88/0364-0300$06.00/0

reported by a Dupont group to yield the substituted bis(trifluoromethyl)carbinols derived by insertion of the carbonyl group of hexafluoroacetone into the C-H bond of the substrate (8). The modified polymers which result from this type of reaction are of interest to us because of the pronounced changes in physical and chemical properties which arise due to the incorporated fluoroalkanol groups. Fluoroalkanols are reported to have greatly enhanced acidity (pKa = 9.3 for hexafluoro-2-propanol vs. pKa = 18 for 2-propanol)(9), which suggests that polymers bearing such pendent groups would exhibit useful hydrogen bond donor properties. In fact, we have found that polyethylene, which is normally quite insoluble, becomes readily soluble in tetrahydrofuran or benzene with even very low levels of incorporation. A polymer bearing fluoroalkanol pendent groups would exhibit hydrogen-bonding power intermediate between simple alcohols (e.g. polyvinyl alcohol), and a carboxylic acid (e.g. polyacrylic acid). By virtue of the shielding influence of the bulky trifluoromethyl substituents in the alcohol pendent group, these functionalized macromolecules would likely not suffer the problem of competing self-association to the same extent found with the usual hydrogen-bonding moieties. All of these factors suggest that these modified polymers should provide unique materials for further study.

With the ultimate objective of using these materials as components of blends, and in applications as polymer reagents, we have made a thorough investigation of the addition of hexafluoroacetone to polyethylene, and to two cycloalkane model substrates, with the goal of optimizing the reaction conditions. We have prepared, for comparison, authentic samples of of the small molecule model compounds by alternative synthetic procedures. The samples of modified polyethylene with varying degrees of incorporation of the fluoroalkanol group have been characterized by combustion analysis, thermal analytic techniques, molecular weight determinations, and spectroscopically with comparisons to small molecule model compounds.

Experimental Section

Materials. Low density polyethylene (LDPE) was obtained from Dow Chemical Co., PE 510. High density polyethylene (HDPE) was obtained from Phillips Chemical, Marlex 6001. Polypropylene (PP) was obtained from Exxon Corp., Exxon PP #8216. Di-t-butylperoxide, 99%, (DtBP) was obtained from Polysciences. Gaseous hexafluoroacetone was obtained from Nippon Mektron Ltd., Japan.

Methods. Elemental combustion analysis was performed by Galbraith Laboratories. Infrared spectra were determined using a Shimadzu 435 infrared spectrophotometer: samples were prepared as hot hydraulic pressed films for LDPE, as solvent-cast films on NaCl windows for the modified polymer samples, and as a thin film on NaCl windows for the model compounds. All values are in reciprocal centimeters, cm^{-1}; (w) = weak, (m) = medium, (s) = strong, (br) = broad. ^{1}H-NMR spectra were determined at 90 MHz on a Varian EM390 NMR spectrometer, and are reported in ppm downfield from internal tetramethylsilane standard as follows: chemical shift (multiplicity, integration, coupling constant). ^{19}F-NMR spectra were determined at 85 MHz on a JEOL FX90Q FT NMR spectrometer and are reported in ppm upfield from internal fluorotrichloromethane in the above format with the integration given in percentage where appropriate. The gel permeation chromatography (GPC) was performed on a Waters 590 HPLC in THF (except that 1,2,4-trichlorobenzene was used for LDPE itself) and calibrated with polystyrene standards.

Purification. The commercial LDPE was dissolved in hot toluene and precipitated in methanol. The resulting small chips were dried in a vacuum oven overnight at 80° C. HDPE was purified in the same manner, except that in this case much of the commercial material was insoluble in toluene, and so significant fractionation apparently occurs during the purification. The model compounds were purified by the flash chromatographic technique of Still et al. (10).

Modification Reactions. The procedure of Howard et al. (8) was taken as a standard. A Parr pressure reactor (500 mL capacity) was charged with 20 g of polymer (1.43 mol of "CH_2" for LDPE or HDPE, or 0.974 mol of the equivalent for PP), 200 mL of dry benzene, and 0.2 mL (1.09 mmol), or other measured amount, of di-t-butylperoxide. The resulting mixture was degassed through three cycles of freeze-evacuate-thaw-nitrogen purge. Gaseous hexafluoroacetone was passed into a calibrated vessel cooled to -78° C until 20 mL (0.159 mol) had collected, and this portion was then distilled into the cooled (-78° C) reactor. The reactor was sealed and heated to 135° C with vigorous mechanical stirring for 16 h (or other measured time). After cooling, the reaction mixture was filtered, and the solids obtained were dissolved in tetrahydrofuran (THF), precipitated in methanol, and dried under vacuum. The reaction conditions are detailed in Table I for modified LDPE (samples **1** - **8**), for modified HDPE (sample **9**) and for modified PP (sample **10**). In modification trials employing 1.2 mL (6.53 mmol) of di-t-butylperoxide, the product was insoluble in THF and was assumed to be crosslinked. The modification reaction was also applied to two cycloalkanes as model substrates, and these were carried out in much the same fashion (reaction time 3 h, 20 g of substrate, 30 mL (0.239 mol) of hexafluoroacetone, 0.6 mL (3.27 mmol) of di-t-butylperoxide) although the crude products were not purified: cyclohexane produced sample **11** (37% mass recovery) and cyclododecane produced sample **12** (220% mass recovery).

Selected IR data:

For LDPE:

[80° C]: 2980 (s), 2840 (s), 2700 (w), 1480 (s), 1380(s) cm^{-1};

[120° C, difference]: 1700 (m) cm^{-1}.

For Sample **1**: [room temp.]: 3600-3300 (b), 2920 (s), 2840 (s), 1440 (s), 1180 (s) cm^{-1};

[80° C, difference]: 3630 (w), 2700 (w), 1700 (s) cm^{-1};

[120 degrees C, difference]: 2700 (m) cm^{-1}.

For Sample **5**:

[room temp.]: 3600 (w), 3300-3550 (w br), 2930 (s), 2860 (s), 1710 (w), 1620 (ww) cm^{-1};

[330° C, difference]: 3600 (ww), 3300-3550 (ww br), 1620 (lost) cm^{-1};

[410° C, difference]: 3600 (lost), 3300-3550 (lost) cm^{-1}.

For Sample **6**: 3640 (w), 3600-3300 (w br), 2940 (s), 2880 (s), 1705 (w), 1480 (s), 1370 (s), 1290 (s), 1220 (s), 1130 (s), 1100 (s) cm^{-1}.

For Sample **7**: 3630 (w), 3600-3300 (s br), 2940 (s), 2880 (s), 1710 (ww), 1480 (s), 1370 (s), 1290 (s), 1220 (s), 1130 (s), 1100 (s) cm^{-1}.

For Sample **10**: 3640 (w), 3600-3300 (s br), 2990 (s), 2960

(s), 2940 (s), 2900 (s), 1680 (m), 1600 (s), 1470 (s), 1380 (s), 1360 (s), 1270 (s), 1220 (s), 1160 (s), 1100 (s) cm^{-1}.
Selected ^{19}F-NMR spectra (d_8-toluene): Sample **4**: 73.86 (80%), 74.18 (20%). Sample **5**: 74.00 (80%), 74.28 (20%). Sample **6**: 73.97, 74.1 (shoulder). Sample **11**: 73.52 (67%), 74.83 (33%). Sample **12**: 74.61 (88%), 75.32 (12%).

Synthesis of Model Fluoroalkanols **13** and **14**. Although the addition of secondary Grignard reagents to hexafluoroacetone is known to proceed in low yield (11-12), this was deemed the most expedient synthesis of **13** and **14**. A three-neck 250 mL flask equipped with a dewar condenser, two rubber septa and a magnetic stirbar, was flame-dried under nitrogen flush, and was chilled to -78° C in a dry ice/acetone bath. The flask was charged with 30 mL of a 3 M solution of cyclohexylmagnesium chloride in ether (90 mmol), and this was stirred while gaseous hexafluoroacetone was condensed into the solution. After 15 m, the originally translucent solution turned clear and the system developed positive pressure, so the flow of hexafluoroacetone was stopped. The apparatus was purged with nitrogen, and warmed to room temperature, and the reaction mixture was poured into 1 L of 5% aqueous hydrochloric acid. This was extracted with three 100 mL portions of ether, and the combined ether portions were dried over sodium sulfate and concentrated to give 15 g of light yellow oil. Purification by chromatography (75 g of silica/petroleum ether with increasing methylene chloride) yielded 1.923 g (9%) of 2-cyclohexyl-1,1,1,3,3,3-hexafluoropropan-2-ol, **13**, as a clear oil: IR (thin film) 3560 (m), 3450 (m br), 2920 (s), 2850 (s), 1450 (s), 1390 (w), 1345 (m), 1280 (s br), 1220 (s br), 1185 (s), 1130 (s), 1110 (s) cm^{-1}. ^{1}H-NMR ($CDCl_3$) δ 1.35 (m, 5H); 2.0 (m, 6H); 3.03 (s, 1H). ^{19}F-NMR ($CDCl_3$) 73.48 (s), (d_8-toluene) 73.09 (s). In a similar fashion 2-cyclododecyl-1,1,1,3,3,3-hexafluoropropan-2-ol, **14**, was prepared from cyclododecylmagnesium bromide in 4% yield: IR (thin film) 3560 (m), 3460 (w br), 2920 (s), 2850 (s), 1470 (s), 1445 (s), 1370 (w), 1350 (m), 1150-1310 (s br), 1130 (s), 1100 (s), 1045 (m), 1010 (m) cm^{-1}. ^{1}H-NMR ($CDCl_3$) δ 1.35 (m, 18H); 1.9 (m, 5H); 2.97 (s, 1H). ^{19}F-NMR ($CDCl_3$) 73.70 (s), (d_8-toluene) 73.40 (s).

Synthesis of Model Fluoroalkyl Ethers 15 and 16. A modification was used of the procedure of Schneider and Busch (13), who employed chlorosulfonic acid as catalyst. Chlorosulfonic acid was ineffective in producing ether **16** in our hands, and perchloric acid performed better. A mixture of 16 g of 1,1,1,3,3,3-hexafluoropropan-2-ol (95.2 mmol) and 4.1 g of cyclohexene (50.0 mmol) was charged with four drops of 70% perchloric acid, and stirred at room temperature under nitrogen for 24 h. The reaction mixture was poured into 300 mL of saturated aqueous sodium bicarbonate, and the aqueous solution was extracted with three 100 mL portions of dichloromethane. The organic phase was dried over sodium sulfate and concentrated to give 6.892 g of light yellow oil. The crude product was purified by chromatography (170 g of silica, petroleum ether) to yield 2 g (16%) of cyclohexyl 1,1,1,3,3,3-hexafluoroprop-2-yl ether, **15**, as a clear oil: IR (thin film) 2920 (s), 2850 (s), 1450 (m), 1410 (m), 1365 (s), 1285 (s), 1265 (s), 1220 (s), 1190 (s), 1155 (s), 1130 (s), 1100 (s), 1050 (w), 1040 (m), 1020 (m) cm^{-1}. ^{1}H-NMR ($CDCl_3$) δ 1.3 (m, 8H); 1.8 (m, 3H); 3.65 (m, 1H); 4.12 (quintet, 1H, J=7 Hz). ^{19}F-NMR ($CDCl_3$) 74.88

(d, J=6.9 Hz); ^{19}F-NMR (d_8-toluene) 74.74 (d, J=7.3 Hz). Cyclododecyl 1,1,1,3,3,3-hexafluoroprop-2-yl ether, **16**, was prepared analogously from cyclododecene and hexafluoroisopropanol in 17% yield: IR (thin film) 2920 (s), 2850 (s), 1470 (s), 1445 (s), 1395 (m), 1365 (s), 1280 (s br), 1260 (s), 1220 (s br), 1190 (s br), 1150 (s), 1130 (s), 1095 (s), 1040 (w) cm^{-1}. ^{1}H-NMR ($CDCl_3$) δ 1.3 (m, 20H); 1.6 (m, 3H); 3.85 (m, 1H); 4.12 (m, 1H, J=7 Hz). ^{19}F-NMR (d-chloroform) 74.61 (d, J=7.3 Hz); ^{19}F-NMR (d_8-toluene) 74.39 (d, J=7.3 Hz).

Results and Discussion

The radical-promoted reaction between polyethylene and hexafluoroacetone is shown in Equation 1. It had been demonstrated previously in the case of simple hydrocarbons (8) that the addition of a carbon radical to the carbonyl group of hexafluoroacetone can take place in two modes, to yield the product of substitution with either a fluoro-

$$-(CH_2-CH_2)_x- \xrightarrow[C_6H_6,\ 135^\circ C]{CF_3C(=O)CF_3,\ (tBuO)_2} -(CH(C(CF_3)_2OH))_m-(CH_2)_n-(CH(OCH(CF_3)_2))_o- \quad (1)$$

$x/2 = m+n+o$

1 — 9

alkanol group or a fluoroalkyl ether group. In the case of the reaction with a macromolecule, it is likely that both groups are present in varying proportions. We have also carried out the modification reaction on two cycloalkanes as model substrates for polyethylene. Cyclohexane was used by the Dupont group (8), and we have examined this substrate as well as cyclododecane. We felt that the singular conformational properties of cyclohexane might make it an imperfect model for polyethylene, and that a larger ring would more closely approximate polymer behavior (14). Selected results from our survey of reaction conditions for the preparation of modified polymer samples **1-10** are found in Table I.

The assumption is made at present that elemental combustion analysis for carbon, hydrogen, and fluorine provides a good approximation to the extent of incorporation of fluoroalkyl residues, i.e. alcohols and ethers. We have ruled out trifluoromethylcarbonyl groups since no evidence is seen for their presence in either the infrared spectra or the ^{19}F-NMR spectra. Thus, our values for percent modification reflect the best fit of the combustion data to an idealized stoichiometry for the product in Equation 1, where (m+n+o) = 100, and the percent modification (% mod.) is given by the expression [100 x (m+o)/(m+n+o)], equivalent to the number of fluoroalkyl residues per one hundred methylenes. An appropriately normalized formula was used to fit the data for polypropylene (sample 10).

Table I. Reaction Conditions and Elemental Combustion Analysis Data

Sample	1	2	3	4	5	6	7	8	9	10
REACTION CONDITIONS										
S.M.	LDPE	LDPE	LDPE	LDPE	LDPE	LDPE	**6**	**7**	HDPE	PP
time hrs.	17.5	24	3.5	2.5	6	15	19	3	16	16
mol s.m.	1.43	1.43	1.43	1.43	1.43	1.43	1.43	0.72	1.43	0.73
mol HFA	0.159	0.159	0.239	0.239	0.239	0.159	0.159	0.159	0.159	0.159
mmol DtBP	1.09	3.27	3.27	3.27	3.27	4.35	3.27	1.63	3.27	3.27
% mass recov.	95	163	130	165	140	124	69	97	43	99
ELEMENTAL COMBUSTION ANALYSIS DATA										
%C	82.59	70.52	64.16	63.48	59.02	67.20	60.03	60.69	84.20	79.97
%H	13.74	10.62	9.72	9.38	8.52	9.98	8.26	8.59	13.78	12.32
%F	3.01	16.47	19.73	22.41	26.77	19.19	26.99	25.47	0.44	6.24
% mod.	0.40	2.70	3.59	4.19	5.58	3.90	5.50	5.10	0.05	1.00

Not surprisingly, The amount of di-t-butylperoxide (DtBP) is an important factor affecting the outcome of the reaction. The level of incorporation increases in proportion to the amount of DtBP (compare samples **1**, **2**, **6**). Too much of the radical promoter is deleterious, since sample **6** was partially crosslinked, and an attempted modification reaction using 6.53 mmol of DtBP produced a completely insoluble product, which apparently was highly crosslinked.

A higher level of incorporation can be achieved by resubjecting the modified polymer to the reaction conditions, as seen in the conversion of sample **6** -> **7**. A third exposure to the conditions, which involved the conversion of sample **7** -> **8**, does not give higher incorporation, and in fact appears to result in a decrease. Thus the reaction appears to be self-limiting, and the existing fluoroalkanol or fluoroalkyl ether groups may act to quench the radical promoter, and inhibit further incorporation.

Another important factor is the amount of hexafluoroacetone available, giving a proportional increase in the level of incorporation (compare samples **2** and **6** with **3**, **4**, and **5**). We have yet to determine the upper limit of the effect of this factor on the percent incorporation of fluoroalkanol residues.

When the modification reaction was attempted on high density polyethylene (HDPE), a very low level of incorporation was obtained (sample **9**). The diminished reactivity probably results from a low solubility of HDPE in the reaction medium, limiting the reaction to the surface of a solid substrate.

Commercial samples of polyethylene contain varying amounts of defects at the molecular level, depending upon the method of preparation. These include methyl or higher alkyl group branches, double bonds, crosslinks, etc. Such defects strongly influence the outcome of reactions on polymers, particularly when the defect presents a site more reactive to the radical promoter, and if the resulting backbone radical intermediate is less or unreactive to the group being added. The modification reaction was carried out on polypropylene (PP, sample 10) in order to test the behavior of a branched backbone. Under conditions comparable to those used for LDPE (sample 2), the percent incorporation was roughly one-third as much with PP.

The effect of the modification reaction on the molecular weight distribution was examined, and the results are given in Table II. Generally, the molecular weight distribution is narrower after modification, and the molecular weight decreases (although these could be artifacts due to the fact that the GPC for LDPE was run in 1,2,4-trichlorobenzene at 135°C. All others run in tetrahydrofuran at room temperature.). It is evident from Table II that a rough correlation exists between molecular weight and degree of modification: the least highly functionalized samples (i.e. 1 and 2) have the lowest molecular weights, and the most highly functionalized sample (i.e. 5 and 7) have the highest molecular weights. The molecular weights are comparatively lower for the products of functionalization reactions which utilized a lesser quantity of hexafluoroacetone (samples 1, 2 and 6 vs. 3, 4 and 5). Under such conditions, fewer radical sites on the backbone will result in product formation, and are thus more available for fragmentation or recombination reactions. Of particular interest is the sequence of successive modifications, 6 -> 7 -> 8. The first stage of modification is accompanied by a decrease in the molecular weight by half. This preparation involved the largest quantity of the radical promoter. In going from sample 6 to sample 7, the molecular weight doubles, accompanied by an increase in the percent modification. With the third cycle of functionalization, the molecular weight drops again, and the level of modification also decreases.

The decrease in molecular weight can be explained by degradation of the polymer samples under the reaction conditions, or by crosslinking and/or selective precipitation of the higher molecular weight

Table II. Molecular Weight Characteristics

Sample	LDPE	1	2	3	4	5	6	7	8	10
% mod.	0.0	0.4	2.7	3.6	4.2	5.6	3.9	5.5	5.1	1.0
M_N	1.08 x10^4	7.42 x10^2	4.2 x10^3	1.84 x10^4	2.26 x10^3	2.29 x10^4	6.66 x10^3	1.89 x10^4	1.66 x10^4	1.02 x10^4
M_W	1.26 x10^5	2.71 x10^3	1.72 x10^4	7.41 x10^4	6.38 x10^4	1.52 x10^5	4.75 x10^4	1.55 x10^5	9.3 x10^4	3.26 x10^4
Molec. Weight Dist.	11.7	3.66	4.09	4.03	2.83	6.62	7.14	8.22	5.64	3.18

fractions during the purification process. An alternative explanation is that the solubility is related to the degree of modification, and so it is logical that with a higher percent modification, more of the higher molecular weight fractions will be soluble. A distinction between these possibilities awaits a series of calibration GPC determinations in which both the starting polymer and modified samples are run in 1,2,4-trichlorobenzene at higher temperature. In most of the cases, the molecular weight decreases by less than a factor of ten, and under optimal conditions (i.e. samples **5** and **7**) is fairly stable. The practicality of the successive modification procedure seems to be limited to two stages, since the percent functionalization and the molecular weight decrease after the third stage. While we have no data in this series for the starting polypropylene, preliminary results from viscosity measurements indicate that degradation takes place during the modification of this polymer.

Evidence is seen in the infrared spectrum for the presence of hydroxyl functionality in all of the products. In some cases, samples which were subjected to a cycle of heating and cooling back to room temperature gave spectra in which the O-H stretch absorption is resolved into free (3640-3630 cm^{-1}) and associated (3600-3300 cm^{-1}) portions, probably by the driving off of adsorbed THF. Some samples also showed carbonyl absorption (1710-1700 cm^{-1}), presumably arising via autoxidation of the polymer. The alternative explanation of conversion of the hexafluoro-2-propanol group to a trifluoromethylcarbonyl group (fluoroform reaction), is unlikely since our study of the change in infrared spectrum of sample **5** with heating does not show an increase in carbonyl absorption proportional to the loss of hydroxyl absorption. In this experiment, upon heating from 100° C to 330° C the hydroxyl absorptions at 3600 and 3300-3550 cm^{-1} decrease roughly by half. From 330° C to 410° C the hydroxyl absorptions disappear, with no increase in the magnitude of the weak carbonyl absorption at 1710 cm^{-1}. These results indicate the the fluoroalkanol group is split off from the polymer backbone between 100° and 330° C, and fits together nicely with the thermal analytic data (vide infra). The infrared for sample **10** is interesting, for the absorptions at 1680 and 1600 cm^{-1} indicate the presence of unsaturated carbonyl and unsymmetrical double bond, resp. These would arise from radical-promoted chain scission, as discussed above, and this result can be rationalized as follows: tertiary hydrogens are more easily abstracted than secondary ones (relative reactivity 9.8:1,15), but the resulting radical is more stable and more hindered than its secondary counterpart. The addition to hexafluoroacetone would be expected to be more sluggish, as is indeed the case with the addition of Grignard reagents (11-12), resulting in a lower level of incorporation, and the radical scission reaction would be more favorable.

To aid in the characterization of the modified polymer samples **1-10**, and of the mixtures **11** and **12** obtained from functionalizing the two cycloalkanes, we prepared the model fluoroalkanols **13** and **14** as shown in Equation 2 (11-12), and the model fluoroalkyl ethers **15** and **16** as shown in Equation 3 (13). Compounds **13** and **15** were previously isolated by the Dupont group (8) from their mixture **11**. Products **13-16** proved to be good comparison compounds for the infrared spectra, and particularly for the ^{19}F-NMR spectra. The ^{19}F-NMR signal for the fluoroalkanol is approximately 1 ppm downfield from that of the fluoroalkyl ether, which is enough to get semiquantitative ratios of

$$RBr \xrightarrow[2)\ (CF_3)_2C{=}O]{1)\ Mg/ether} R{-}C(OH)(CF_3)_2 \quad \mathbf{13, 14} \qquad (2)$$

R = cyclo-C_6H_{11}, cyclo-$C_{12}H_{23}$

$$\text{cyclo-}[(CH_2)_n{-}CH{=}CH] \xrightarrow[HClO_4]{(CF_3)_2CHOH} \text{cyclo-}[(CH_2)_{n+1}{-}CH]{-}O{-}CH(CF_3)_2 \quad \mathbf{15, 16} \qquad (3)$$

n = 4,10

these residues in the functionalized products. The results for the polymer samples were that a fairly consistent 4:1 ratio of alcohol to ether was found. For the model substrates the product mixtures obtained were complex, but through ^{19}F-NMR analysis a ratio of 2:1 alcohol/ether was determined for these constituents in the modified cyclohexane mixture **11**, and a 7:1 ratio of alcohol/ether for the corresponding components in the cyclododecane mixture **12**. Our results are roughly comparable with the value obtained by the Dupont group (8), who found a 5:1 ratio of alcohol/ether for cyclohexane.

Some thermogravimetric analytic data for our samples are presented in Table III. Good trends are seen, excluding samples 7 and 8 from the successive modification series. It is evident that there is a decrease in the degradation onset temperature with increasing percent modification (i.e. from 463° C at 0% mod. to 393° C at 5.6% mod.). This would be expected with the introduction of reactive functional groups which would act as foci for degradative processes. The temperature of maximum degradation likewise decreases over this range, but less precipitously (i.e. from 480° C at 0% mod. to 463° C at 5.6% mod.). Of particular interest is the early weight loss that takes place at approximately 250° C, with a loss of roughly 4-7% of the

Table III. Thermogravimetric Analytic Data

Sample	LDPE	2	3	5	6	7	8	9	PP	10
% mod	0.00	2.70	3.59	5.58	3.90	5.50	5.10	0.05	0.00	1.00
% early wt. loss	-	5.0	3.9	6.8	4.7	7.3	6.7	neg	-	neg
Temperatures (in °C):										
early wt. loss	-	261.5	345.2	265	256	250.8	245	-	-	-
onset degrad.	463.0	433.3	425.8	393.6	428.0	417.8	393.6	448.4	424.1	401.7
max. degrad.	480.7	472.5	471.1	463.2	473.0	450.7	444.5	475.6	452.8	444.8

mass. This weight loss corresponds roughly to the weight percent of the pendent groups, and together with the infrared results for the film of sample **5** heated in stages up to 410° C is suggestive of a specific thermal degradation in which the functional group is split from the backbone. We plan to investigate this reaction further under controlled conditions using our model compounds.

Some representative differential scanning calorimetry results are shown in Table IV. These data fall into two groups: the samples in the first series, **3**, **4**, and **5**, which were prepared with a larger quantity of hexafluoroacetone and are assumed to be less crosslinked; and the second series, samples **6**, **7**, **8**, which are the products of successive modification reactions, and which we believe are relatively more crosslinked. In both series we see a decrease in the melting temperature T_m with increasing percent modification. The glass transition temperature T_g rises in proportion to increasing percent modification, an indication of increasing interactions between the pendent functional groups. In the second series, sample **6** gives a T_g of the expected magnitude, but after the second stage of modification the T_g rises profoundly, with a slight decrease after the third stage of modification. This effect is likely due to crosslinking processes which we propose take place in a cumulative fashion over successive modifications. For the sample of modified polypropylene, the T_m decreases as expected.

Table IV. Differential Scanning Calorimetry Data

Sample	LDPE	3	4	5	6	7	8	9	PP	10
% mod	0.00	3.59	4.19	5.58	3.90	5.50	5.10	0.05	0.00	1.00
T_g (°C)	-129.	-116.	-112.	-90.0	-111.	22.4	-7.0	-87.4		
T_m (°C)	110.2	93.0	90.6	85.9	103.6	84.9	72.8	130.6	157.8	103.6
onset T_m (°C)	98.2	73.6	68.2	66.2	85.4	69.8	54.0	119.1	143.5	85.4

We have recently initiated our investigation of blends by examining the compatibility between our modified polymer sample **4** and poly(methyl methacrylate). Mixtures with a composition of between 10% and 30% of sample **4** yield compatible blends which are transparent under a polarized light microscope, and are characterized by a single T_g. Mixtures richer than 60% of **4** undergo complete phase separation.

Conclusions

Low density polyethylene can be modified bv the incorporation of hexafluoroacetone to the extent of from 0.4 to 5.6 residues per hundred methylenes (by elemental combustion analysis) by the judicious choice of reaction conditions. Our results can be rationalized within the context of the three competing radical-promoted processes: functionalization (addition of hexafluoroacetone), crosslinking (radical site recombination), and degradation (radical site fragmentation or destruction of introduced crosslinks). Functionalization of polyethylene predominates with higher concentrations of hexafluoroacetone, with the quantity of radical promoter kept low to prevent crosslinking and degradation, and with shorter reaction times in order to prevent the buildup in concentration of radical species which facilitates crosslinking. Indications are that polypropylene undergoes a more competitive degradation due to the higher stability of tertiary radicals, and their lower reactivity toward addition to hexafluoroacetone. High density polyethylene is not modified appreciably under these conditions, most likely as a result of its poor solubility in the reaction medium. Four out of five of the introduced residues are fluoroalkanol groups, with the remainder as fluoroalkyl ether groups (by ^{19}F-NMR). The analysis of structure is facilitated by comparison with functionalized cycloalkane model compounds which were prepared by independent synthesis, and these small molecule models should also prove to be useful systems for exporing the reactivity of the functionalized polymers. A decrease in molecular weight seems to occur during the modification, at least under the less optimal reaction conditions, but since molecular weight also correlates with percent modification this may be a result of selective solubility. The thermal analytic properties show the expected trends: with increasing percent modification, the degradation temperature decreases, the glass transition temperature increases, and the melting temperature decreases. Some aspects of the chemistry require further study; specifically, chain scission, and crosslink formation will be investigated further, and the proposed thermal splitting of the fluoroalkanol groups at higher temperature (TGA, IR) must be quantified. Preliminary results indicate that modification of polyethylene by the attachment of fluoroalkanol hydrogen bond donor groups renders it able to form compatible blends with hydrogen-bond accepting polymers such as poly(methyl methacrylate).

Acknowledgments

We would like to thank Ms. J. P. Huang for providing the ^{19}F-NMR spectra. We acknowledge support of this work from the National Science Foundation through Materials Research Group Grant DMR-8508084, and through Materials Research Instrumentation Grant DMR-8411022.

Literature Cited

1. Braun, D.; Eisenlohr, U. Angew. Makromol. Chem. **1976** 55 43.
2. Ostroverkhov, V. G.; Glavati, O. L.; Chermenin, A. P.; Rabinovich, I. L.; Dets, M. M. Khim. Tekhnol. Topl. Masel. **1980** 20; Chem. Abstr. **1980** 93 98086a.
3. Gabara, W.; Porejko, S. J. Polym. Sci. A-1 **1967** 5 1547.
4. Porejko, S.; Gabara, W.; Kulesza, J. J. Polym. Sci. A-1 **1967** 5 1563.
5. Gaylord, N. G.; Mehta, M.; Kumar, V. In Polymer Science and Technology. Carraher, C. E., Jr.; Moore, J. A. Ed.; Plenum Press: New York, 1983; Vol. 21, p171.
6. Weiss, R. A.; Lenz, R. W.; MacKnight, W. J. J. Polym. Sci. Polym. Phys. Ed. **1977** 15 1409.
7. For a review of the chemistry and properties of hexafluoroacetone, see: Krespan, C. G.; Middleton, W. J. Fluorine Chem. Rev. **1967** 1 145.
8. Howard, E. G.; Sargeant, P. B.; Krespan, C. G. J. Am. Chem. Soc. **1967** 89 1422.
9. Middleton, W. J.; Lindsey, R. V. Jr. ibid. **1966** 86 4948.
10. Still, W. C.; Kahn, M.; Mitra, A. J. Org. Chem. **1978** 43 2923.
11. Knunyants, I. L.; Ch'en, C.-y.; Gambaryan, N. P. Bull. Acad. Sci. U.S.S.R., Div. Chem. Sci. (Engl. Trans.) **1960** 647; Chem. Abstr. **1960** 54 22484h.
12. Knunyants, I. L.; Gambaryan, N. P.; Ch'en, C.-y.; Rokhlin, E. M. Bull. Acad. Sci. U.S.S.R., Div. Chem. Sci. (Engl. Trans.) **1962** 633; Chem. Abstr. **1962** 57 12305i.
13. Schneider, H.-J.; Busch, R. J. Org. Chem. **1982** 47 1766.
14. See for example: Deneke, M.; Broeker, H. C. Makromol. Chem. **1975** 176 1471.
15. Cheng, H. N.; Schilling, F. C.; Bovey, F. A. Macrolmol. **1976** 9 365.

RECEIVED August 27, 1987

CHEMICAL MODIFICATION FOR ANALYTICAL CHARACTERIZATION

Introduction to Chemical Modification for Analytical Characterization

Chemical reactions on polymers can be used for analytical purposes in a number of different ways. For example, a functional group on a polymer may be used as a coreactant in an analysis scheme. The functionalized polymer may act as a species donor (for example, donating an H+ from a carboxylic acid or a metal ion from a carboxylate salt) or as an acceptor for some externally available species (for example, amine groups that form salts or quarternize, etc.). The polymer bound functional group may also be a coreactant in an analytical scheme that does not involve ionization, but instead relies on a net non-ionic coupling reaction. Examples include polymer bound -NCO reacting with small molecule OH groups or vice versa, reactions of various species like -SH with maleate units incorporated onto vinyl aromatic polymers, etc. A polymer bound functional group may even be utilized as a catalyst in an analytically pertinent reaction, for example a polymeric tertiary amine catalyst for the -NCO plus -OH reaction. In most of these cases, the polymer bound reactive species is used as a direct substitute for the analogous small molecule, with the polymeric version being preferred because of ease of separation, analysis, or sometimes reuse. In a few cases, the polymeric version exhibits additional benefits, like moderation of reactions because of accessibility considerations, or even a different degree of completion because of different pK's between macro- and small molecules. Ion exchange resins, available with a variety of groups in both acidic and basic forms, can be used in many of the above applications. Monomeric or polymeric precursors to the ion exchange resins allow ready synthesis of specialized materials for many additional analytical uses. In yet others, a range of highly individualized polymeric materials have been synthesized for specific analytical characterization applications. Some of the materials in all three of the above categories are synthesized directly by copolymerization techniques, while others are prepared by post-polymerization chemical modification. Note that in the latter case, the overall scheme of preparation and utilization involves two, usually very different, chemical modification reactions on the polymer.

Polymeric materials are also widely used in analytical characterization in the form of chromatography substrates. Gel permeation chromatography (GPC) is principally based on pore size/size exclusion mechanisms (a sort of mechanical as opposed to chemical process), so is not considered among these chemical reactions of polymers. Liquid chromatography (LC), including high pressure/performance liquid chromatography (HPLC), gas liquid chromatography (GLC), and classical column chromatography depend to a much greater degree on molecular level chemical interactions. The interactions may be of the so-called non-specific type, usually meaning generalized hydrophobic interactions like those between an alkane and a polyolefin. Alternatively, the interactions may be more group specific, as in ionization that can involve species dislocation/migration through a fairly large sphere of interaction/reaction, but require stoichiometric pair-wise (or greater) interaction. In other cases, interactions may occur with a specific stoichiometry in a tighter locus, like those involving H-bonding, where an H atom bonded to an electronegative atom must find, interact with, and (at least temporarily) become coupled with another electronegative atom. Geometric implications become important here, too, since interactions depend on relative spatial arrangements of interacting groups. This is especially true in cases where chelation is used for separation, since successful interactions require several molecular units to arrange in a very specific geometry. So the polymeric substrate used in these applications must contain the "correct" number of specific functional groups arranged in specific geometric fashion. Obviously, the polymer modification reactions that situate these functional groups properly on the substrate must be very well understood and carefully conducted.

Much ingenuity has been used in design of materials for the applications mentioned in these two paragraphs. One important consideration involves the timing and homogeneity of the modification reaction. If the functional group is introduced into the polymer in homogeneous solution, and the polymer is then used in homogeneous solution - or is lightly crosslinked and used in highly swollen condition - the accessibility of the groups and reaction efficiency can be quite high. If the groups are introduced in homogeneous solution, and the polymer is then isolated and used under heterogeneous conditions, a significant fraction of the functional groups is not accessible. If the polymer substrate is in its final form and size, then the introduction is conducted under heterogeneous conditions, the groups are introduced onto surface regions that are also accessible to coreactants or cointeractors, again allowing high efficiency.

Chemical reactions on a polymer can assist in characterizing the polymer itself. This is useful well beyond the usual determination of functional group content. Especially valuable are reactions that are sensitive to stereochemical or compositional sequence distribution, or to relative spatial arrangement/independence/cooperativity of sites. Certain elimination and cyclization reactions have proven useful, as have selected chain degradation reactions. (Each of these is illustrated by works in this chapter.) Analysis of the rates and extents of monitored chemical modification reactions, and of the derived products, can often provide a detailed understanding of the polymer's microstructure and (ideally) of the molecular level mechanisms by which it was formed.

When it is well enough defined, a chemical modification reaction on a polymer can even be used as an analytical probe of process or environmental conditions. For example, the severity and consequences of conditions involving high temperature, high shear, oxidative or photooxidative environments, etc., can often be characterized and monitored by the degree of chemical modification induced on the "probe" polymer.

The individual studies in this section illustrate very nicely the scope (both breadth and depth) of chemical reactions on polymers for analytical applications.

Chapter 23

Recent Studies of Polymer Reactivity

Herbert Morawetz

Department of Chemistry, Polytechnic University, 333 Jay Street, Brooklyn, NY 11201

Three studies are reviewed: (a) The kinetics of the basic hydrolysis of polyacrylamide cannot be fully accounted for by nearest neighbor interaction effects, probably because longer-range Coulombic interactions are also involved. A small fraction of monomer residues added head-to-head are characterized by imide formation much faster than the amide hydrolysis. (b) The kinetics of a reaction involving a polymer-bound reactive group and a small molecule in a solvent by which the polymer is swollen were studied on model systems in which the reaction led to the quenching of fluorescence. When the polymer was not crosslinked, simple exponential decay of fluorescence was observed. When the fluorescent label was carried by a crosslinked network or by linear chains entrapped in a network, a dispersion of the rate constant was observed. (c) Reaction kinetics of a poly(vinylbenzyl chloride) latex with low molecular weight amines, linear chain molecules terminated by an amino group and with proteins are described.

In principle, the reactivity of a functional group should not be altered when it is attached to a polymer (1). However, special effects may be encountered when a reagent is attracted to a polymer or repelled from it, when the polymer-bound reactive group is activated or inhibited by a neighboring group or when the local polarity of the polymer domain differs from that of the bulk solvent. A review of studies of such effects

0097-6156/88/0364-0317$06.00/0

up to 1975 (2) has been supplemented by more recent publications (3-5). Here I shall discuss three more recent investigations. The first was designed to compare the theoretical prediction of kinetics of the basic hydrolysis of polyacrylamide based on a model in which the reactivity of an amide group depends on whether neighboring amide groups have reacted, with experimental data. As an unexpected dividend, this study also yielded information on the polymer microstructure. In the second study a sensitive fluorescence method was etablished by which it could be shown whether all reactive groups appended to a crosslinked network are equally reactive. The third study was concerned with interfacial reactions of groups at the surface of a polymer phase with small and macromolecular reagents in a solution phase.

Degradation of Polyacrylamide in Basic Solution

The reactions of polymers in which each monomer residue carries a reactive substituent are frequently subject to neighboring group activation or inhibition. Some such neighboring group effects depend on the stereoisomerism of the chain backbone (6,7), but if we neglect this complication and if interactions with groups attached further on the chain backbone can be assumed to be negligible, the reaction of such a polymer should be characterized by three rate constants, k_0, k_1, k_2, which describe the reactivity of groups with 0,1 and 2 reacted neighbors. Given these rate constants, Keller (8) has derived expressions for the course of the reaction.

Unfortunately, the number of systems in which it can be established whether Keller's model is realistic for a particular case is severely limited since the original polymer is usually not soluble in the same medium as the ultimate reaction product. In cases where the entire course of the reaction can be followed, as in the basic hydrolysis of polyacrylamide, investigators have analyzed their results by a computer search for the k_0, k_1, k_2 values which fit best their kinetic data (9). This, of course, does not answer the question whether the model using these three rate constants provides a full description of a particular case.

We have, therefore, taken a different approach (10). In a copolymer of acrylamide with acrylic acid, the initial rate constant for amide hydrolysis is given by

$$k_{in} = (1 - \beta)^2 k_0 + 2\beta(1 - \beta) k_1 + \beta^2 k_2 \quad (1)$$

where β is the probability that the site next to an

acrylamide is occupied by an acrylic acid residue. This probability is given by

$$\beta = (1 + r_{Aam}Q)^{-1} \qquad (2)$$

Here r_{Aam} is the reactivity ratio of an acrylamide radical with acrylamide and acrylic acid and Q is the ratio of acrylamide and acrylic acid in the monomer mixture from which the copolymer was derived. Thus, the determination of k_{in} for at least three copolymers allows a derivation of k_0, k_1 and k_2. With this approach we found $k_1/k_0 = 0.11$ and $k_2/k_0 = 0.013$.

Figure 1 compares the conversion predicted for any reduced time $\tau = k_0 t$ with the use of Keller's theory and the above values of k_1/k_0 and k_2/k_0 with experimental results obtained when polyacrylamide was exposed to 0.2N NaOH at 53°C. It may be seen that the reaction slows down at large τ much more than predicted by Keller's model. In fact, this decrease of the reaction rate is even more pronounced than predicted by Keller's equations for the case where a single reacted nearest neighbor completely inhibits amide hydrolysis. We believe that this discrepancy is due to the repulsion of the catalyzing hydroxyl ions from amide residues by non-neighboring carboxylate groups.

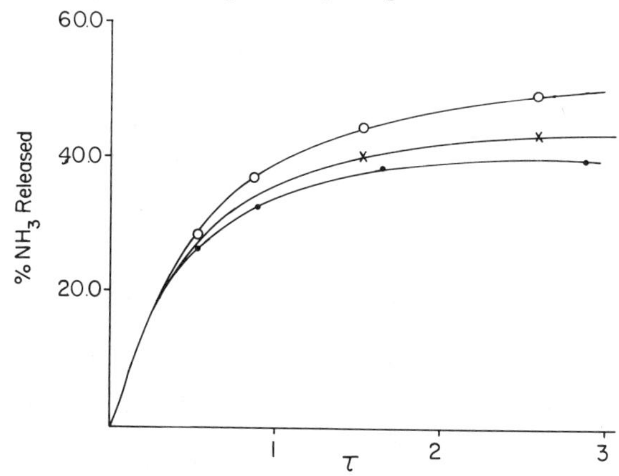

Figure 1: Hydrolysis of polyacrylamide in 0.2 N NaOH at 53°C. (●) Experimental data. Prediction of the Keller theory with $k_1/k_0 = 0.11$; $k_2/k_0 = 0.013$ (o) and with $k_1 = k_2 = 0$ (x).

Unexpectedly, it was found that a small amount of ammonia was released at 25°C in a reaction much faster than the amide hydrolysis. It was suspected that this was due to imide formation, in analogy with the reported behavior of succinamide (11):

$$\begin{array}{l}CH_2CONH_2\\ |\\ CH_2CONH_2\end{array} + OH^- \longrightarrow \begin{array}{l}CH_2CO\\ |\\ CH_2CO\end{array}\!\!>N^- + NH_3 + H_2O$$

$$\begin{array}{l}CH_2CO\\ |\\ CH_2CO\end{array}\!\!>N^- + OH^- + H_2O \longrightarrow \begin{array}{l}CH_2COO^-\\ |\\ CH_2COO^-\end{array} \quad (3)$$

Fortunately, the imide has a characteristic peak in the UV which facilitates the monitoring of its formation and decomposition. Figure 2 shows that this takes place with polyacrylamide in Ø.2N NaOH solution at 25°C. No such imide intermediate was observed in

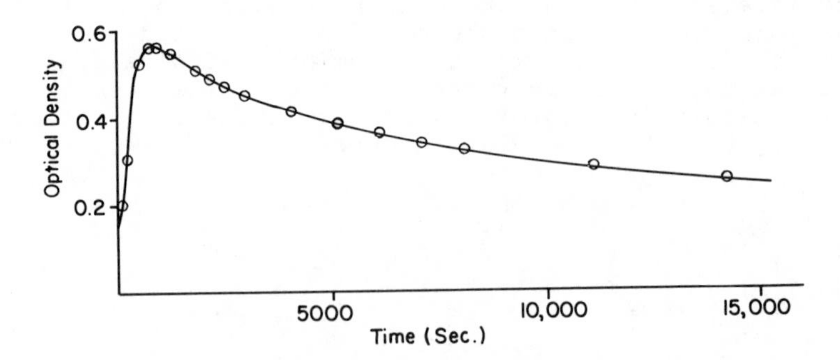

Figure 2. Imide formation and hydrolysis from polyacrylamide (2.2 mg/mL) in Ø.22 N NaOH at 25°C as monitored by UV absorption at 235 nm.

the conversion of glutaramide to glutarate since the glutarimide hydrolysis is abbout 3ØØ times as rapid as its formation. This suggested that the imide intermediate observed with polyacrylamide must be formed from amide groups attached to neighboring atoms of the chain backbone, i.e., from monomer residues added head-to-head during the polymerization. In fact, the rate constants for imide formation and hydrolysis from racemic α,α'-dimethylsuccinamide were found to be extremely close to the rate constants of the corresponding reactions of polyacrylamide. The data were interpreted as showing that 4.5% of the monomer residues added head-to-head and that two thirds of such pairs had the racemic configuration.

Reaction Kinetics of Functional Groups Attached to a Swollen Polymer Gel

In the synthesis of polypeptides with biological activity on a crosslinked polymer support as pioneered by Merrifield (12) a strict control of the amino acid sequence requires that each of the consecutive reactions should go virtually to completion. Thus, for the preparation of a polypeptide with 60 amino acid residues, even an average conversion of 99% would contaminate the product with an unacceptable amount of "defect chains". Yet, it has been observed (13) that with a large excess of an amino acid reagent in the solution reacting with a polymer-bound polypeptide, the reaction kinetics deviate significantly from the expected exponential approach to quantitative conversion, indicating that the reactive sites on the polymer are not equally reactive.

Deviations from a kinetic equivalence of chemically similar groups attached to a polymer in a model system can be studied with very high sensitivity if the polymer is exposed to a large excess of a reagent in the solvent with which the polymer is swollen and if the progress of the reaction is followed by a signal which characterizes the polymer-bound reagent and disappears in the reaction product. In that case, any deviation from linearity in a plot of the logarithm of the signal intensity against time will indicate a kinetic non-equivalence of the reactive groups.

We have applied this principle to the reaction of copolymers containing a small proportion of p-aminostyrene residues with acetic anhydride, which results in the quenching of fluorescence (14). As shown in Figure 3, with the reactive copolymer in solution, the linearity of the semilogarithmic plot of the fluorescence intensity against time can be followed over 7-8 half-lives, demonstrating the equal reactivity of all fluorescent residues. A linear plot was also obtained when uncrosslinked chains were swollen with a solvent containing the acylating reagent. However, even a low degree of crosslinking led to a dispersion of the rate constant as revealed in a curvature of the plot. This seemed to exclude the possibility that the dispersion of the rate constant is the result of the relative inaccessibility of reactive groups attached to the polymer network in the vicinity of a crosslink. On the other hand, we found that the acylation of p-toluidine is seventy times as fast in cyclohexane as in a number of polar solvents. We have tentatively proposed that the kinetic behavior of p-aminostyrene copolymers reflects the polarity of the micro-environment of the reactive groups. In the absence of crosslinks, the averaging of the nature of this environment will be much faster, because of microbrownian motion, than the acylation reaction and the process will follow first order kinetics. However, in the presence of crosslinks conformational transitions are hindered and some aminostyrene residues may be trapped in highly polar regions where the reaction is slow.

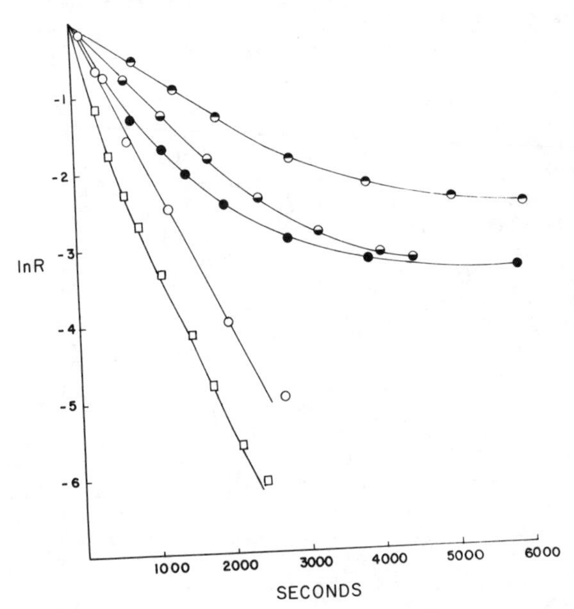

Figure 3. Time dependence of the fraction R of unreacted aminostyrene residues during acetylation by Ø.14 M acetic anhydride at 3Ø°C. Methyl methacrylate copolymer in acetonitrile solution (O); linear poly-(methyl methacrylate-co-butyl methacrylate) swollen with acetonitrile (□); methyl methacrylate copolymer crosslinked with 1 mole% (◒) and with 15 mole% (◓) ethylene dimethacrylate; poly(methacrylate crosslinked with 3 mole% ethylene dimethacrylate containing entrapped poly(methyl acrylate-co-aminostyrene) (●).

Kinetics of Polymer Surface Reactions

Few studies of reactions involving a polymer surface and a reagent in a solution phase have been reported. A study of this type was carried out by Rasmussen et al. (15) on reactions of acid chloride groups on the surface of a polyethylene sheet with nucleophiles in aqueous solution, but because of the small number of reacting groups, highly sensitive spectroscopic techniques had to be used to follow the process.

The experimental problems was greatly simplified by the use of a reactive polymer latex with its enormous surface area. Vinylbenzyl chloride can be polymerized in emulsion to yield a latex with uniform particle size with virtually no loss of the reactive chloromethyl groups during the polymerization (16). The nature of the reaction of these groups with amines in the aqueous phase depends on whether the nucleophile is soluble in the polymer. The reaction with trimethylamine (Figure 4a) exhibits an accelerating phase, reflecting the diffusion of the nucleophile into the latex particle. The reaction goes to completion resulting in a homogeneous solution of poly(vinylbenzyl trimethylammonium chloride). By contrast, the polymer-insoluble glycinate ion attacks only the surface of the latex particle (Figure 4b).

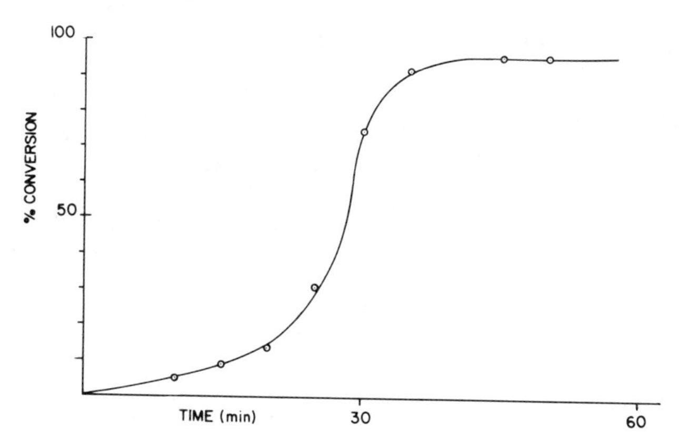

Figure 4a. Reaction of poly(vinylbenzyl chloride) latex with 0.85 M trimethylamine at 37°C.

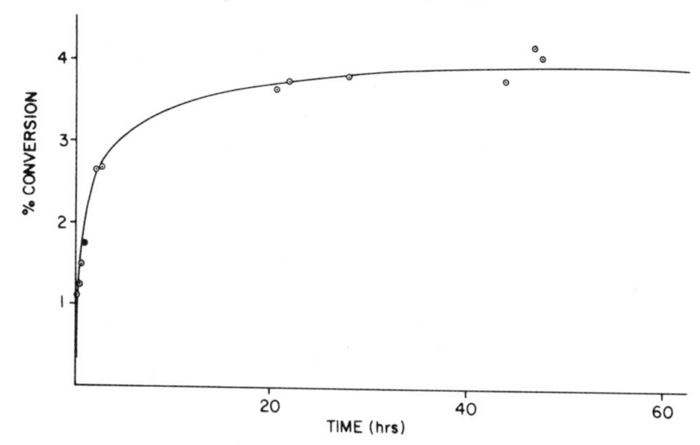

Figure 4b. Reaction of poly(vinylbenzyl chloride) latex with 0.35 M glycinate at 52.5°C. Diameter of latex particles 160 nm.

Proteins may be covalently attached to the latex particle by a reaction of the chloromethyl group with α-amino groups of lysine residues. We studied this process (17) using bovine serum albumin as a model protein - the reaction is of considerable interest because latex-bound antigens or antibodies may be used for highly sensitive immunoassays. The temperature dependence of the rate of protein attachment to the latex particle was unusually small - this rate increased only by 27% when the temperature was raised from 25°C to 35°C. This suggests that non-covalent protein adsorption on the polymer is rate determining. On the other hand, the rate of chloride release increases in this temperature interval by a factor of 17 and while the protein is bound to the latex particle by only 2 bonds at 25°C, 22 bonds are formed at 35°C. This change reflects the increasing deformability of the globular protein molecule as the temperature is raised.

Finally, we have studied the reaction of amino-terminated polyoxyethylenes with the poly(vinylbenzyl chloride) latex (18). The rate of this reaction was found to be independent of the length of the chain carrying the terminal amine. Attachment of these chains stabilized the latex against coagulation, in analogy with the "steric stabilization" produced by adsorbed polymer chains (19).

Acknowledgment. The author is grateful to the National Science Foundation, Polymers Program, for financial support of these studies by Grants DMR 77-07210 and DMR 85-00712.

References

1. P.J. Flory, J. Am. Chem. Soc., 61, 3334(1939).
2. H. Morawetz, "Macromolecules in Solution", Second Edition, Wiley, New York, 1975, Chapter IX.
3. H. Morawetz, Israel J. Chem., 17, 287(1979).
4. H. Morawetz, J. Macromol. Sci. Chem., A13, 311(1979).
5. H. Morawetz, Pure Appl. Chem., 51, 2307(1979).
6. E. Gaetjens and H. Morawetz, J. Am. Chem. Soc., 83, 1738(1961).
7. J. J. Harwood and T. K. Chen, Polym. Prepr., Am. Chem. Soc., Div. Polym. Chem., 21, (1), 2(1980).
8. J. P. Keller, J. Chem. Phys., 37, 2584(1962); 38, 325(1963).
9. E.A. Boucher, Progr. Polym. Sci., 6, 63(1978).

10. S. Sawant and H. Morawetz, Macromolecules, 17, 2427(1984).
11. B. Vigneron, P. Crooy and A. Bruylants, Bull. Soc. Chim. Belg., 69, 616(1960).
12. R. B. Merrifield, Adv. Enzymol., 52, 221(1969).
13. W. Geising and S. Hornle, in "Proceedings of the 11th European Polypeptide Symposium", H. Nesvatba, Ed., North Holland, Amsterdam, 1973, p. 146.
14. S. S. Pan and H. Morawetz, Macromolecules, 13, 1157(1980).
15. J. R. Rasmussen, E. R. Stedronsky and G. M. Whitesides, J. Am. Chem. Soc., 99, 4736(1977).
16. C.-H. Suen and H. Morawetz, Macromolecules, 17, 1800(1984).
17. C.-H. Suen and H. Morawetz, Makromol. Chem., 186, 255(1985).
18. Y.-H. Huang, Z.-M. Li and H. Morawetz, J. Polym. Sci., Chem. Ed., 23, 795(1985).
19. D. H. Napper, J. Colloid Interface Sci., 58, 390(1977).

RECEIVED August 27, 1987

Chapter 24

Analytical Study of Photodegraded *p*-Aramid in an $^{18}O_2$ Atmosphere

Madeline S. Toy and Roger S. Stringham

Science Applications International Corporation, 5150 El Camino Real, Los Altos, CA 94022

The photodegradation of para-aramid in an $^{18}O_2$ atmosphere allows the differentiation between the accelerated experimental photooxidative conditions from its usual daylight exposure effects. This study illustrated an estimation of the rates of photooxidation of a commercial para-aramid product (i.e., DuPont's Kevlar-29 woven fabric) based on the oxygen-18-labelled carbon dioxide ($^{46}CO_2$ and $^{48}CO_2$) decarboxylated from the sample. The oxygen-18-labelled atoms, which are inserted in the macromolecules, were analyzed for the photodegradation processes. This technique also allows the radial ^{18}O-distribution measurement from the fiber surface toward the fiber center.

The 'aramids' is a series of isomeric fully aromatic polyamides, which can withstand service-life stress at high temperature without deformation and degradation(1,2). The 'aramids' inherent flame resistance, high thermal and chemical stability, and high modulus fulfill a new source for engineering materials. Para-aramid shows one of the greatest high modulus fibers developed to date(3). The two members of the aramid family, which have achieved the commercial importance, are developed by DuPont: 1) Nomex, poly(m-phenylene isophthalamide), and 2) Kevlar, poly(p-benzamide) or poly(p-phenylene terephthalamide).

Kevlar fibers are supplied by DuPont as Kevlar-29 and 49. The former is characterized by high tensile strength and the latter by high initial modulus(4). Some Kevlar-29 end uses are in ropes and cables, which are as strong as steel at one-fifth the weight, and in ballistic vests. Some Kevlar-49 end uses are in reinforcing resins and composites for aerospace structures, boat hulls, and sport equipments.

The molecular structure of Kevlar is a regular alternating p-directed benzene ring and amide group. The amide linkage shows a degree of double bond character and is regarded as a rigid planar unit(5). The intermolecular hydrogen-bonding, which the amide link facilitates, imposes a very ordered arrangement on the polymer chains generating a laminar structure(6). The fibers have a Tg (glass

0097-6156/88/0364-0326$06.00/0

transition temperature) above 300°C, and can be heated without decomposition or melting to temperatures exceeding 500°C(7).

Many applications of this para-aramid fabric result in the sunlight exposure between 300 and 450 nm, where the light absorption of para-aramid and the solar spectrum overlap(8). In this region the sunlight deteriorates the aramids and has been reported to occur in the absence of oxygen and in air(9,10). The photodegraded para-aramid in air generates carboxylic acid groups accompanied by a rapid loss of molecular weight and mechanical properties (10). This paper reports an analytical study of the photodegraded Kelvar-29 fabric in 0.2 atm $^{18}O_2$ (99%).

Experimental

Reagents. Oxygen-18-labelled gas (99%) was purchased from ICON, a gaseous mixture of 100 ppm ethane and helium from Alltech, helium from Matheson, concentrated sulfuric acid (Ultrex) from J. T. Baker, and chloroform and dimethylacetamide (DMAc) from Aldrich. The reagents were used as received unless otherwise specified.

Fabric Cleaning. The Kevlar-29 woven fabric was obtained through the courtesy of Naval Weapons Center. A special pair of serrated shears was purchased from Technology Associates for cutting the fabric. The fabric (2.5 cm x 18 cm) was placed in a Soxhlet thimble and extracted by 100 ml of chloroform for 24 hours to remove its surface lubricants (about 3% by weight). The fabric was then removed from the thimble and agitated in a 20 ml of hot distilled DMAc for 15 minutes, before it was placed back into the thimble and extracted for another 8 hours using fresh chloroform solvent. The solvent-cleaned fabric was dried in a vacuo at room temperature.

Photolysis Chamber. A high pressure ozone-free mercury-xenon arc lamp (200 watts) was purchased from UVP (San Gabriel, CA). Its irradiance was estimated at 125 times of the sun(11). The photolysis chamber consisted of a Pyrex conical pipe outside and a quartz sleeve inside. The top and bottom aluminum end plates with the Pyrex (7.6 cm id) and the quartz (4.5 cm od) tubes enclosed an annular chamber of 21.5 cm height with gas inlet and outlet valves attached. The photolysis chamber's ozone free Hg-Xe arc lamp, which was vertically suspended at the center of the quartz sleeve, using two Lucite high voltage connectors, passed its irradiance through the quartz wall onto the fabric in a confinded 99% oxygen-18-labelled atmosphere at a pressure of 0.2 atm (the partial pressure of O_2 in the atmosphere) and at a specified temperature and time.

Photolysis Procedure. The solvent-cleaned Kevlar-29 fabric swatch (2.5 cm x 18 cm) was placed around the outside quartz tube inside the photolysis chamber, which was subsequently evacuated, before $^{18}O_2$ (99%) was introduced to 0.2 atm. The photolysis chamber was preheated to the specified photooxidation temperature, before the Hg-Xe lamp was turned on. The temperature, which was held constant in the chamber by adjusting the air flow around the lamp, was monitored by a thermocouple placed next to the fabric sample inside the chamber. After the photooxidation had continued for the specified time period, the lamp was turned off and the photolysis chamber was evacuated and cooled to ambient temperature. Room air was let in and the photolysis chamber was opened to remove the oxidatively photodegraded fabric swatch. This procedure was repeated on different swatches at several temperatures versus photolysis times.

Preparation of Soluble Sample for Decarboxylation. The photolyzed fabric swatch was placed in 20 ml of distilled DMAc at 100°C for 5 minutes. The DMAc solution was filtered to remove any insoluble fibrous material from the decarboxylation flask (a round 30 ml two-neck Pyrex flask with a protruded bottom well) and dried in vacuo with the temperature kept below 50°C leaving a residual film on one side of the flask.

The purified sulfuric acid (i.e., preheated the commercial ultrapure grade H_2SO_4 at 200°C for 24 hr under vacuo to remove its decarboxylating contaminants and subsequently cooled to ambient temperature), 0.5 ml, was added to the bottom well of the decarboxylation flask and subsequently placed under vacuo. A known amount of a GC-standard (640 mm of 100 ppm C_2H_6 in He) was introduced into the flask for the total CO_2 determination. The acid was then allowed to dissolve the photodegraded residual film on the side of the flask by tilting. The acid solution was left standing for 20 minutes at room temperature.

Analytical Instruments for Carbon Dioxide Analyses. A gas chromatography (Carle Series 100) and a GC-mass spectrometer (LKB9000) were used with identical GC-columns (2 mm id and 6 m length), which were purchased from Alltech and packed with 80% Porapak Q 80/100 and 20% Porapak N 80/100. The six-port mini-switching valve was purchased from Hach and used to trap the total carbon dioxide evolved and the C_2H_6 standard from the decarboxylation flask into its collection/injection loop at liquid nitrogen temperature. This cold loop was warmed to ambient temperature and its trapped contents were vaporized. The collected vapor mixutre in the loop was swept into the GC's helium stream (2 atm pressure) and into the column and then into the thermistor detector, when the valve was turned to injection mode. The sample was analyzed for the total CO_2 content and the C_2H_6 standard. After the CO_2 had passed through the GC's detector, the vapor mixture was then collected in a portable gas cell, which consisted of a six-port mini-switching valve and a collection/injection loop, again chilled to liquid nitrogen temperature.

The collected sample at -196°C was isolated from the flow of the GC's helium gas stream and then the loop was warmed to ambient temperature for GC-mass spectroscopic analyses. The gas cell, which contained the isotopic CO_2 and the C_2H_6 standard in helium at one atmosphere, was placed in the injection helium flow of the GC-mass spectrometer for ten minutes, before the mini-switching valve was turned to inject the vapor contents into the instrument. After three minutes, the CO_2 peak eluted. The superimposed peaks were sampled ten times during their elution and the relative isotopic quantities of $^{45}CO_2$, $^{46}CO_2$, $^{47}CO_2$ and $^{48}CO_2$ were determined.

Results and Discussion

The oxidation under daylight exposures is a continuous ongoing process, that alters the para-aramid sample. The use of $^{18}O_2$ atmosphere allows the differentiation between the accelerated experimental photooxidation conditions and its usual daylight exposure effects. This paper determines the rates of photo-oxidations of Kevlar-29 fabric based on the oxygen-18-labelled carbon dioxide (i.e., $^{46}CO_2$ and $^{48}CO_2$) decarboxylated from the sample. In other words, the oxygen-18-labelled atoms, which are inserted in the Kevlar macromolecules, are being analyzed to determine the photodegradation processes.

At 25°C the Kevlar macromolecule retains its polymeric structure in concentrated sulfuric acid solvent. The rapid decarboxylation observed at 25°C in Table I, which was exposed for 7 min at 100°C in 0.2 atm $^{18}O_2$ in the photolysis chamber, is suggested to originate from the oxidized terminal groups of the macromolecules (10,12). In contrast, the acid decarboxylation at 196°C appears to break down the Kevlar macromolecules completely. A thermal degradation pattern has been recognized for the decarboxylation of Kevlar in sulfuric acid at 196°C. This thermal decomposition pattern is used as a decarboxylation model, which constitutes the same two types of decarboxylation reactions at 196°C (e.g., R1 and R2 in Table I): one yields one mole of CO_2 per $\leftarrow C_7H_5NO \rightarrow$ moiety and the other gives two moles of CO_2 per $\leftarrow C_7H_5NO \rightarrow$ moiety. The half life of the former (R2) is 4 to 12 hr and is suggested to originate from the amide linkages' carbonyl groups. The latter's half life (R1) is 660 hr and is from the two of the six carbons of its aromatic rings.

Figure 1 shows the types of decarboxylation (R1 and R2) as the pseudo first order reactions. They are first order reactions, because the bimolecular second order process (decarboxylation of Kevlar in concentrated H_2SO_4) contains one of the reactants (concentrated H_2SO_4) in a great excess to the other reactant (the solvated surface layer). Figure 1 also illustrates the changes in concentrations with times of R1 and R2. The R2 intercept is 32 x 10^{-6} moles, i.e., the initial concentration of R2 at about 11 x 10^{-6} moles; whereas the initial concentration of R1 is about 21 x 10^{-6} moles. The ratio of R1 to R2 reactions is about 2:1, which conforms with all the decarboxylation reactions (e.g., Table I) at 196°C, whether the sample was or was not exposed to photolysis chamber conditions. There are consistently three moles of CO_2 evolved per $\leftarrow C_7H_5NO \rightarrow$ moiety of seven carbons eliminating as one mole CO_2 at $t_{1/2}$ of 4 to 12 hr (R2) and two moles CO_2 at $t_{1/2}$ of 660 hr (R1).

Table I also shows that the main component of the total CO_2 evolved, which is determined by GC, is the $^{44}CO_2$, and the total of the other CO_2 isotopes (e.g., $^{46}CO_2$ and $^{48}CO_2$) is about 10^{-4} of CO_2. The low concentrations of the isotopic $^{46}CO_2$ and $^{48}CO_2$ are still, however, an easily measurable quantity of GC-mass spectrometer. The $^{45}CO_2$ is used as the standard, because the measurement of $^{45}CO_2$ relates directly to the quantity of the abundant $^{44}CO_2$. Its low concentration levels help to improve the measuring accuracy of the other CO_2 isotopes, which are also in low concentration. The moles

Table 1. Decarboxylation Data of Photodegraded Kevlar

CO_2 Isotope	Decarboxylation[a] at			
	25°C		196°C	
	Moles of Total CO_2 Evolved	Rate of CO_2 Evolution, $t_{1/2}$	Moles of Total CO_2 Evolved[b]	Rate of CO_2 Evolution, $t_{1/2}$
$^{44}CO_2$	1.6×10^{-4}	1 to 10 min		Two types of reactions[c]:
			6.98×10^{-3}	(R1) 660 hr
			3.49×10^{-3}	(R2) 4 to 12 hr
$^{46}CO_2$	1.9×10^{-9}	1 to 10 min	4.0×10^{-7}	42 hr[d]
$^{48}CO_2$	7.0×10^{-11}	1 to 10 min	4.3×10^{-9}	8 min

[a]The total carbon available for decarboxylation from 0.415 g Kevlar fabric sample [expressed as $\text{-(-}C_7H_5NO\text{-)-}$ moieties in moles, i.e., 3.49×10^{-3} moles] = $3 \times 3.49 \times 10^{-3}$ = 1.05×10^{-4} moles, where 3 designates three carbon atoms from a $\text{-(-}C_7H_5NO\text{-)-}$ moiety of 7 carbons eliminating as CO_2.

[b]Includes extrapolated data.

[c]There are consistently two types of decarboxylation reactions that occur at 196°C (R1 and R2) of a given Kevlar sample [expressed as $\text{-(-}C_7H_5NO\text{-)-}$ moieties in moles], whether it was exposed or not exposed to photolysis: one yields one mole of CO_2 per $\text{-(-}C_7H_5NO\text{-)-}$ moiety and the other gives two moles of CO_2 per $\text{-(-}C_7H_5NO\text{-)-}$ moiety (e.g., Fig. 1).

[d]Two rates (a fast and a slow) may be involved.

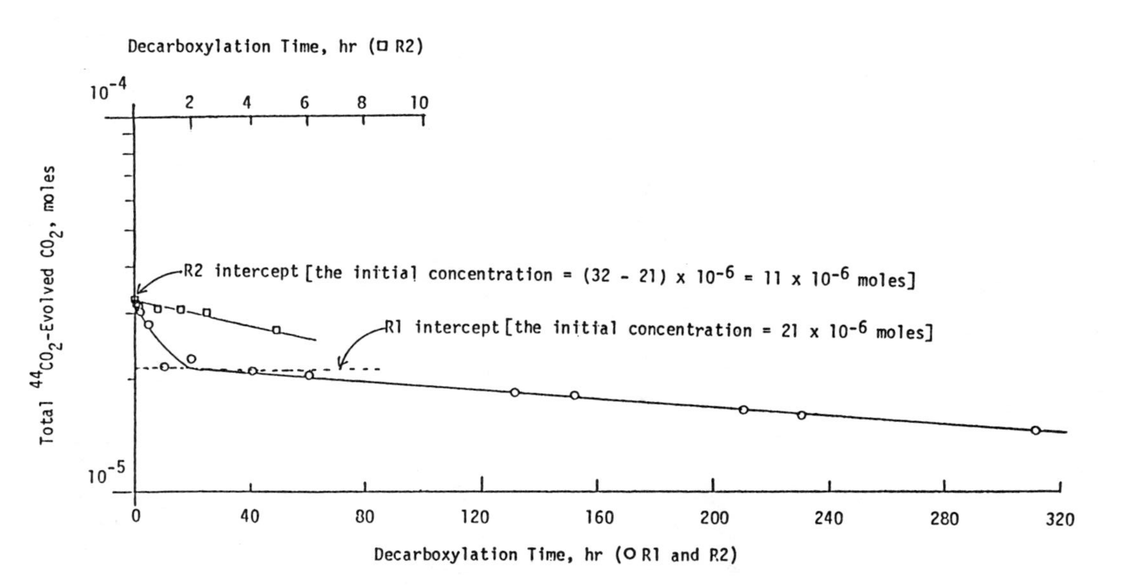

Figure 1. A representative plot of $^{44}CO_2$ concentration versus decarboxylation time for the two types of pseudo first order decarboxylation reactions at 196°C. The ratio R2 intercept over R1 intercept is 1:2.

of the total CO_2 from the GC determination is used for the calculation of the moles of $^{45}CO_2$ as follows:

$$\text{Moles of the total } CO_2 \times 0.01108 = \text{moles of } {}^{45}CO_2$$

where 0.01108 is the fraction of ^{13}C in carbon at mass 12. The presence of $^{46}CO_2$ and $^{48}CO_2$ are measured from their respective GC/ms peak areas compared to that of $^{45}CO_2$. The moles of $^{48}CO_2$ are then corrected by subtracting the $^{46}CO_2$ and $^{48}CO_2$ occurring naturally in the CO_2 sample.

Some of the possible photooxidized sites of the Kevlar macromolecules before decarboxylation are illustrated below (Equations 1 and 2):

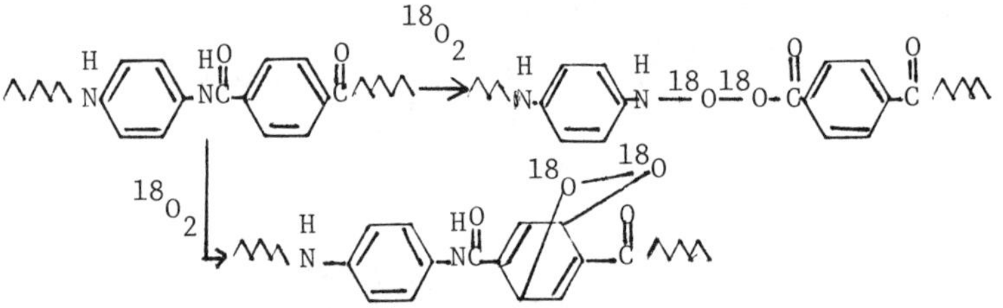

(1)

(2)

These $^{18}O_2$-inserted macromolecules can be differentiated from the ongoing $^{16}O_2$-oxidized macromolecules by GC-mass spectrometer.

When a sample of the same Kevlar-29 fabric was identically treated in $^{18}O_2$-atmosphere except in the absence of photolysis (in the dark) for 24 hours, the results of CO_2 evolved at 25° and 196°C from the GC-analysis were the same as the samples under photolysis (e.g., Table I). However, the gross difference was in the analyses of the GC-mass spectroscopic data, where no oxygen-18-labelled CO_2 isotopes evolved (i.e., there was no $^{46}CO_2$ or $^{48}CO_2$ above the natural background levels).

Figures 2 and 3 illustrate the pseudo first order decarboxylation reactions of $^{46}CO_2$ (the slow reaction with $t_{1/2}$ = 42 hr) and $^{48}CO_2$ (the fast reaction with $t_{1/2}$ = 8 min). The photodegraded sample was exposed for 7 minutes at 100°C in 0.2 atm $^{18}O_2$. The concentration of ^{18}O-Kevlar is expressed as $\text{+}C_7H_5NO\text{+}$ in moles and is equal to the total $^{46}CO_2$ minus evolved $^{46}CO_2$ in moles for Figure 2 and the total $^{48}CO_2$ minus evolved $^{48}CO_2$ in moles for Figure 3.

Figure 4 shows the radial ^{18}O-distribution from the fiber surface toward the fiber center (0.00065 cm). The area under the curves are the total mole ratios $^{46}CO_2/\text{+}C_7H_5NO\text{+}$ at 1.15×10^{-4} for curve a and $^{48}CO_2/\text{+}C_7H_5NO\text{+}$ at 1.23×10^{-5} for curve b. Most of the $^{48}CO_2$ (95%) was evolved during the first 30 minutes. Successive surface layers, which were then dissolved separately in distilled DMAc solvent and decarboxylated in concentrated sulfuric acid, where each was analyzed for $^{46}CO_2$ and $^{48}CO_2$. Although the most severely photodegraded part of the fiber occurred at the light exposed surface, it appeared that the ^{18}O-distributions were throughout the fiber. Table II summarizes the decarboxylation data of Kevlar and its radial distribution of oxygen as a constituent in $^{44}CO_2$, $^{46}CO_2$ and $^{48}CO_2$. The data were obtained from 0.415g Kevlar-29 fabric,

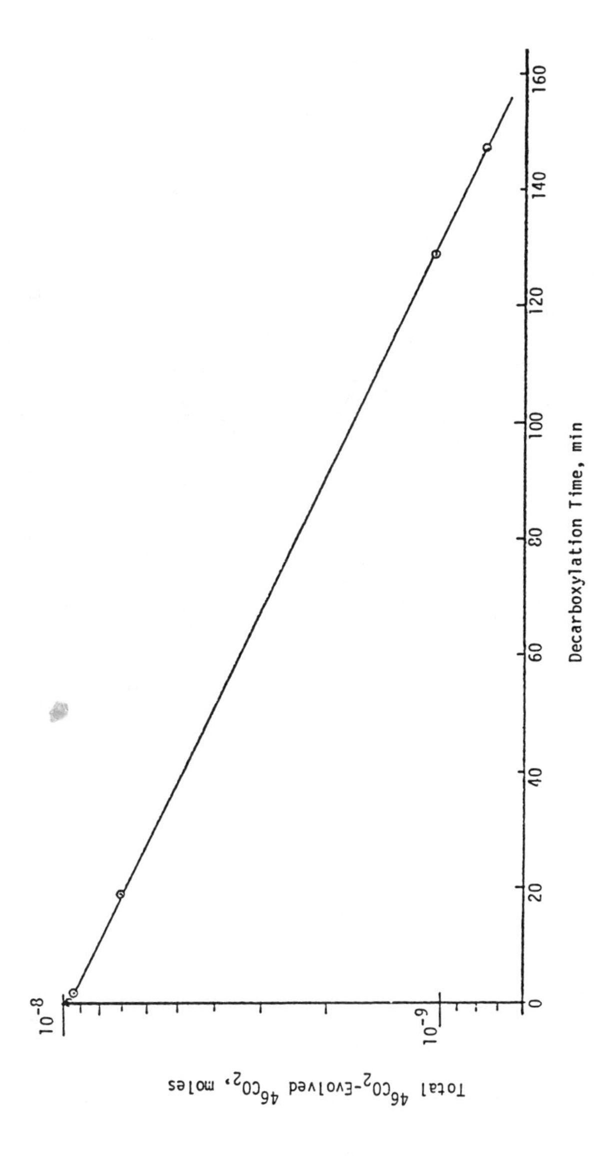

Figure 2. An example of $^{46}CO_2$ concentration versus decarboxylation time for pseudo first order reaction at 196°C.

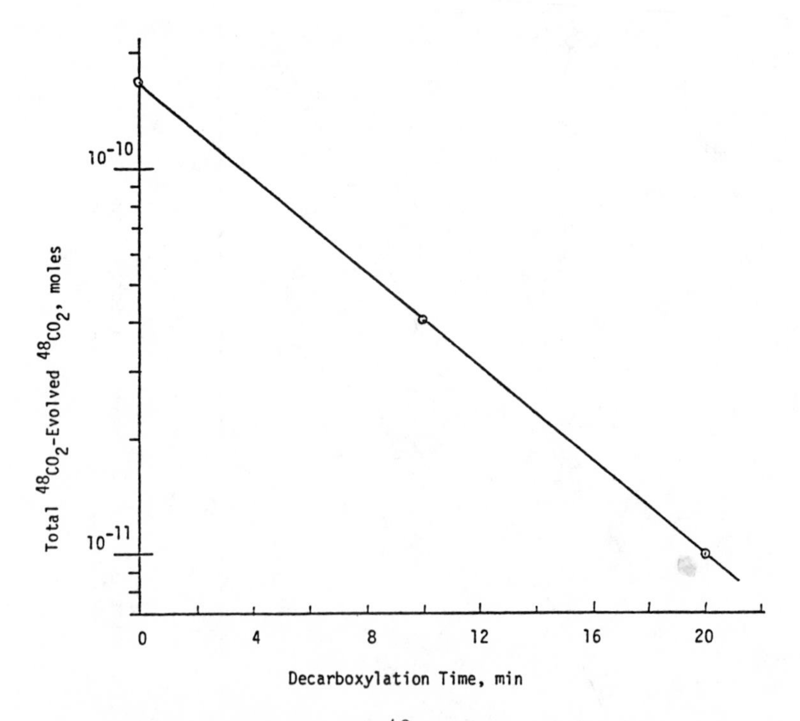

Figure 3. An example plot of $^{48}CO_2$ concentration versus decarboxylation time for pseudo first order reaction at 196°C.

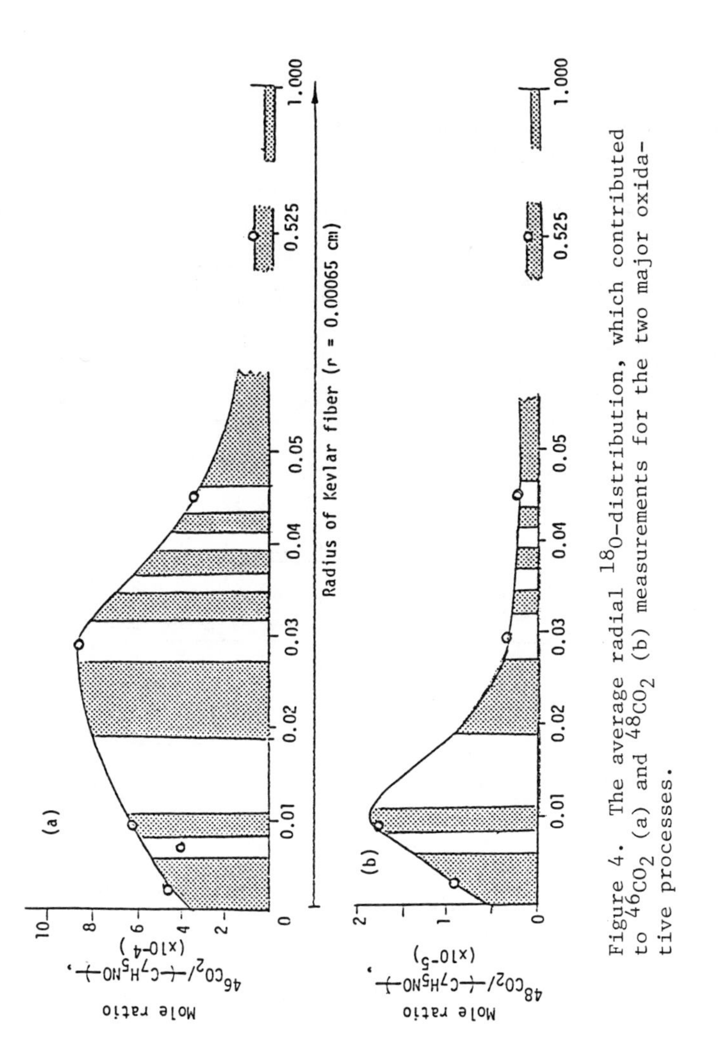

Figure 4. The average radial ^{18}O-distribution, which contributed to $^{46}CO_2$ (a) and $^{48}CO_2$ (b) measurements for the two major oxidative processes.

Table II. Summary Data on Decarboxylation[a] of Kevlar and its Radial Distribution of Oxygen (as a Constituent in $^{44}CO_2$, $^{46}CO_2$ and $^{48}CO_2$)

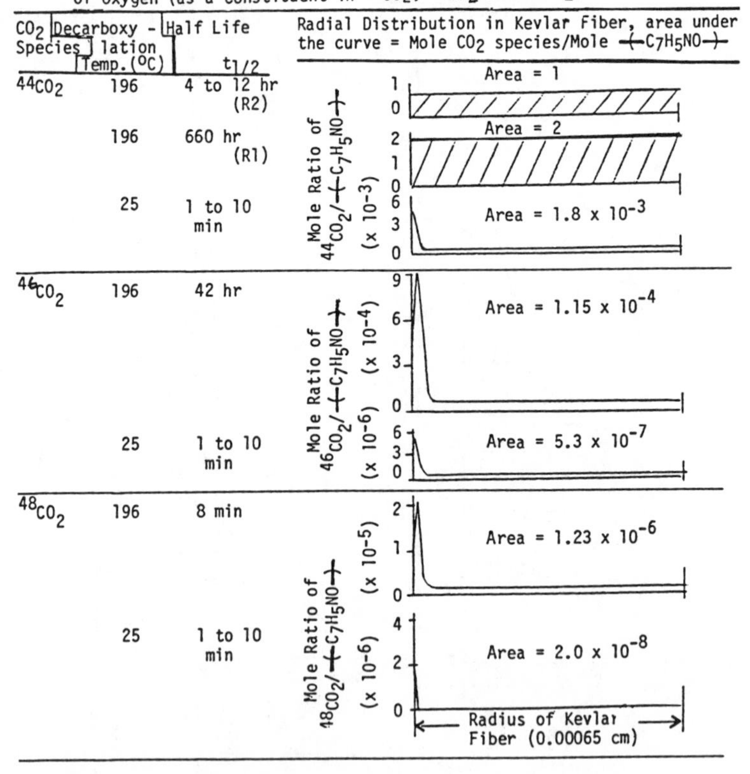

CO_2 Species	Decarboxy-lation Temp. (°C)	Half Life $t_{1/2}$	Radial Distribution in Kevlar Fiber, area under the curve = Mole CO_2 species/Mole $\text{-(-}C_7H_5NO\text{-)-}$
$^{44}CO_2$	196	4 to 12 hr (R2)	Area = 1
	196	660 hr (R1)	Area = 2
	25	1 to 10 min	Area = 1.8 x 10^{-3}
$^{46}CO_2$	196	42 hr	Area = 1.15 x 10^{-4}
	25	1 to 10 min	Area = 5.3 x 10^{-7}
$^{48}CO_2$	196	8 min	Area = 1.23 x 10^{-6}
	25	1 to 10 min	Area = 2.0 x 10^{-8}

which was expressed as $\leftarrow C_7H_5NO \rightarrow$ in moles (3.49×10^{-3} moles) and was exposed for 7 minutes at 100°C in 0.2 atm $^{18}O_2$ in the photolysis chamber. The isotopic $^{46}CO_2$ decarboxylation reactions at 196° and 25°C diminished toward the fiber center. The concentration of $^{46}CO_2$ per $\leftarrow C_7H_5NO \rightarrow$ moiety was highest near the outer fiber surface. The isotopic $^{48}CO_2$ was found only in the outermost surface for the rapid decarboxylation reaction at 25°C, but at 196°C the concentration of $^{48}CO_2$ per $\leftarrow C_7H_5NO \rightarrow$ moiety was also present throughout the fiber. By using the radial ^{18}O-distribution data in Table II, the other photodegraded samples at different photolysis times and temperatures can be estimated from the surface analyses alone by assuming similar ^{18}O-distributions. Further work and verification in this area is recommended.

Table III lists the photooxidation rates, which were deduced from four pseudo first order decarboxylations of $^{46}CO_2$ at 196° and 25°C and $^{48}CO_2$ at 196° and 25°C. The initial concentrations of $\leftarrow C_7H_5NO \rightarrow_n$ to produce ½ $^{18}O_2$ (i.e., to produce $^{46}CO_2$ product) is 6.1 mole l^{-1} and to produce $^{18}O_2$ (i.e., to produce $^{48}CO_2$ product) is 12.2 mole l^{-1} using the density of Kevlar at 1.45 g/cm^3(2). The $^{18}O_2$ is assumed as an ideal gas at 100°C and 0.2 atm. The 0.415 g Kevlar-29 fabric expressed as $\leftarrow C_7H_5NO \rightarrow$ moieties in moles is 3.49×10^{-3} moles.

Figure 5 shows the plots of the two major photooxidative processes versus photolysis time and temperature. Table IV summarizes the two major photooxidative degradation rate constants and activation energies.

Conclusions

This new and novel method to study the photochemical degradation of Kevlar-29 fabric in air divides into four steps: (1) fabric cleaning, (2) photolysis at specified temperature and time in 0.2 atm $^{18}O_2$, (3) preparation of the degraded (DMAc-soluble) sample surface for decarboxylation at 25° and 196°C in the concentrated sulfuric acid, and (4) the total carbon dioxide analyses by gas chromatography and the isotopic carbon dioxide ($^{46}CO_2$ and $^{48}CO_2$) ratios by GC-mass spectrometer.

This new analytical method determines the rate constant and activation energy of Kevlar's photooxidative processes. The 0.2 atm of oxygen-18-labelled environment in a solar chamber simulates the air-exposure under sunlight conditions. The technique also allows the radial ^{18}O-distribution measurement from the fiber surface toward the fiber center. The data from the accelerated experimental conditions in the solar chamber in an $^{18}O_2$-atmosphere are differentiated from the usual daylight exposure effects.

A thermal decomposition pattern of Kevlar in concentrated sulfuric acid at 196°C is recognized to give two types of decarboxylations: one yields one mole of CO_2 per $\leftarrow C_7H_5NO \rightarrow$ moiety and the other gives two moles of CO_2 per $\leftarrow C_7H_5NO \rightarrow$ moiety. The half life of the former is 4 to 12 hr and is suggested to originate from the amide linkages' carbonyl groups. The latter's half life is 660 hr and is from the two of the six carbons of its aromatic rings.

Table III Photooxidation Rates at 100°C of Pseudo First Order Decarboxylations

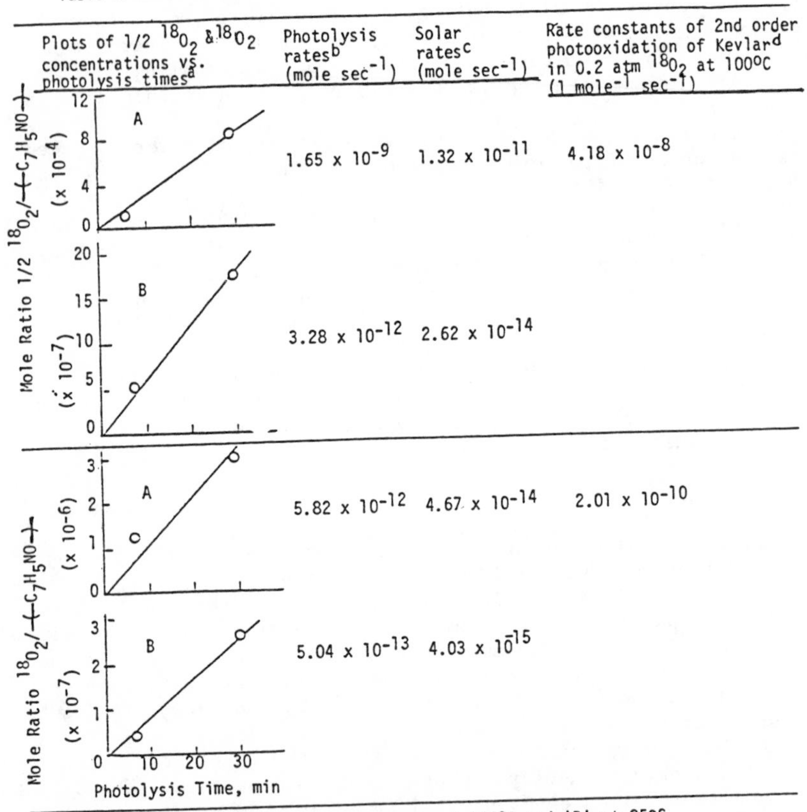

Plots of 1/2 $^{18}O_2$ & $^{18}O_2$ concentrations vs. photolysis times[a]	Photolysis rates[b] (mole sec^{-1})	Solar rates[c] (mole sec^{-1})	Rate constants of 2nd order photooxidation of Kevlar[d] in 0.2 atm $^{18}O_2$ at 100°C (l $mole^{-1}$ sec^{-1})
A	1.65×10^{-9}	1.32×10^{-11}	4.18×10^{-8}
B	3.28×10^{-12}	2.62×10^{-14}	
A	5.82×10^{-12}	4.67×10^{-14}	2.01×10^{-10}
B	5.04×10^{-13}	4.03×10^{-15}	

[a] 'A' designates decarboxylation temperature at 196°C and 'B' at 25°C.

[b] Photolysis rates are expressed as moles of $-(C_7H_5NO)-$/photolysis times (sec).

[c] Solar rates are photolysis rates/125.

[d] The initial concentrations of $-(C_7H_5NO)_n-$ to produce 1/2 $^{18}O_2$ (i.e., to produce $^{46}O_2$ product) is 6.1 mole l^{-1} and to produce $^{18}O_2$ (i.e., to produce $^{48}CO_2$ product) is 12.2 mole l^{-1} using the density of Kevlar at 1.45 g/cm^3. The $^{18}O_2$ is assumed as an ideal gas at 100°C and 0.2 atm. The 0.415 g Kevlar -29 fabric expressed as $-(C_7H_5NO)-$ moieties in moles is 3.49×10^{-3} moles.

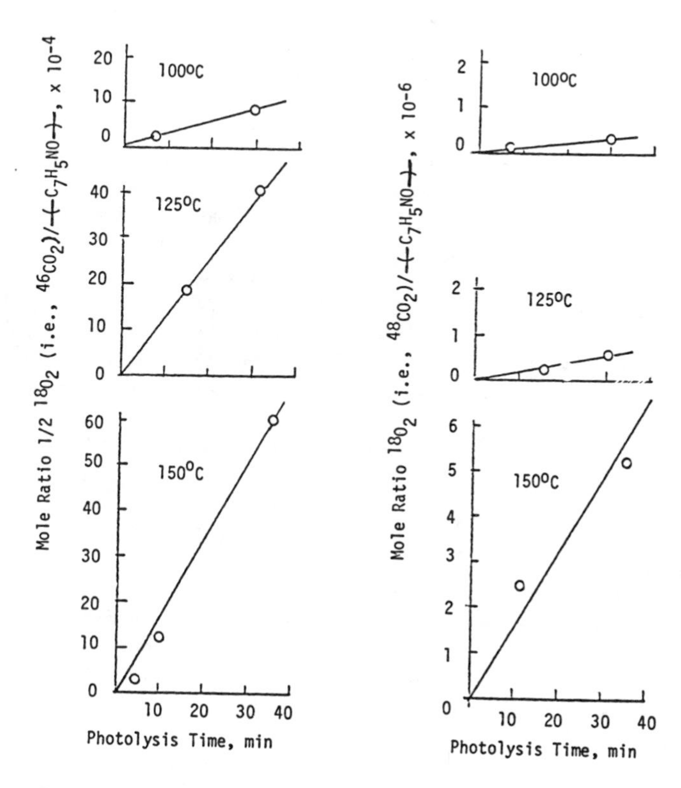

Figure 5. Plots of oxygen-18 concentration versus photolysis time of pseudo first order decarboxylations at 196°C for 20 min from the top surface layer.

Table IV. Summary Data on the Two Major Photooxidative Degradation Rate Constants and Activation Energies[a]

Temperatures °C	°K	1000/°K	Photolysis Rates[b] (mole sec^{-1})	Solar Rates[c] (mole sec^{-1})	Rate Constants[d] (l $mole^{-1}$ sec^{-1})	Activation Energies[e] (kcal $mole^{-1}$)
I. From plots of 1/2 $^{18}O_2$ (i.e., $^{46}CO_2$) concentrations vs. photolysis times (Fig. 5 LEFT)						
25	398	3.356			1.10×10^{-8}	
100	373	2.681	1.65×10^{-9}	1.32×10^{-11}	4.18×10^{-8}	
125	398	2.513	7.76×10^{-9}	6.21×10^{-11}	1.96×10^{-7}	
150	423	2.364	9.69×10^{-9}	7.75×10^{-11}	2.36×10^{-7}	10.8
II. From plots of $^{18}O_2$ (i.e., $^{48}CO_2$) concentrations vs. photolysis times (Fig. 5 RIGHT)						
25	398	3.356			1.03×10^{-12}	
100	373	2.681	5.82×10^{-12}	4.66×10^{-14}	2.0×10^{-10}	
125	398	2.513	9.69×10^{-12}	7.75×10^{-14}	2.37×10^{-10}	
150	423	2.364	9.69×10^{-11}	7.75×10^{-13}	2.43×10^{-9}	15.7

[a]Data are deduced from pseudo first order decarboxylations at 196°C for 20 min from the top surface layer of the photolyzed Kevlar.

[b,c,d]Same as footnotes b, c, d of Table III.

[e]Activation energies are calculated from photolysis runs at 100° and 150°C using calculated rate constants (column 4) at the two temperatures by $\Delta E = (RT_1T_2/T_2 - T_1) \ln (k_2/k_1)$.

This analytical methodology deduces the four photooxidative processes. The data on the total CO_2 evolved from the samples were measured by gas chromatography and the isotopic CO_2 ($^{46}CO_2$ and $^{48}CO_2$) data by GC/mass spectroscopy. The rate constants of the two major photooxidative degradation processes at 25°C were deduced from ½ $^{18}O_2$ per $\leftarrow C_7H_5NO \rightarrow$ (i.e., to produce $^{46}CO_2$ product, $t_{1/2}$ = 42 hr) at 196°C and the other from $^{18}O_2$ per $\leftarrow C_7H_5NO \rightarrow$ (i.e., to produce $^{48}CO_2$ product, $t_{1/2}$ = 8 min) at 196°C. The rate constants of the former process was estimated as 1.10×10^{-8} l $mole^{-1}$ sec^{-1} and the latter as 1.03×10^{-12} l $mole^{-1}$ sec^{-1}. The activation energies of these two processes were deduced as 10.8 kcal/mole for the former and 15.7 kcal/mole for the latter.

Acknowledgments

The authors wish to acknowledge the Naval Weapons Center for support of this work under Contract N60530-83-C-0112, helpful discussion with Mr. R. Tubis and the assistance of Dr. D. Thomas for GC-mass spectra.

Literature Cited

1. Frazer, A. H., High Temperature Resistant Polymers; Wiley, New York, 1968.
2. Jones, J. I., Chem. Brit, 1970, 6, 251.
3. Kwolek, S. L. (DuPont), British Patent 1 283 064, 1972.
4. Chaio, T. T.; Moore, D. L., Composits, 1973, 4, 31.
5. Ward, I. M., ed., Structure and Properties of Oriented Polymers, Ch. 13, 1973, London: Applied Science.
6. Hodd, K. A.; Turley, D. C., Chem. Brit., 1978, 14, 545.
7. Magat, E. E.; Morrison, R. E., Chem. Tech., Nov. 1976, 703.
8. DuPont Information Memo, No. 375, Sept. 28, 1976.
9. Carlsson, D. J.; Gan, L. H.; Wiles, D. M., J. Polym. Sci., 1978, 16, 2353.
10. Carlsson, D. J.; Gan, L. H.; Wiles, D. M., J. Polym. Sci., 1978, 16, 2365.
11. Toy, M. S.; Stringham, R. S., Am. Chem. Soc. Polym. Matls. Sci. Engr. Preprints, 1984, 51(12), 146.
12. Toy, M. S.; Stringham, R. S., Am. Chem. Soc. Polym. Preprints, 1986, 27(2), 83.

RECEIVED August 27, 1987

Chapter 25

Reaction of Atomic Oxygen [O(^3P)] with Polybutadienes and Related Polymers

Morton A. Golub, Narcinda R. Lerner, and Theodore Wydeven

Ames Research Center, National Aeronautics and Space Administration, Moffett Field, CA 94035

Thin films of closely related polymers were exposed at ambient temperature to ground-state oxygen atoms [O(^{3}P)], generated by a radio-frequency glow discharge in O_2. The polymers were cis- and trans-1,4-polybutadienes (CB and TB), atactic 1,2-polybutadiene (VB), polybutadienes with different 1,4/1,2 contents, trans polypentenamer (TP), cis and trans polyoctenamers (CO and TO), and ethylene-propylene rubber (EPM). Transmission infrared spectra of CB and TB films exposed to O(^{3}P) revealed extensive surface recession (etching), unaccompanied by any microstructural changes within the films; this demonstrated that the reactions were confined to the surface layers. There was no O(^{3}P)-induced cis-trans isomerization in CB or TB. From weight-loss measurements, etch rates for elastomeric polybutadienes were found to be very sensitive to the vinyl content, decreasing by two orders of magnitude from CB (2% 1,2) to structures with ≥20-40% 1,2 double bonds, thereafter remaining substantially constant up to VB (97% 1,2). Relative etch rates for EPM and the elastomeric polyalkenamers were in the order: EPM > CO > TP > CB. The highly crystalline TB had an etch rate about six times that of CB, ascribable to a morphology difference, while the partially crystalline TO had an etch rate somewhat higher than that of amorphous CO. Cis/trans content had little or no effect on the etch rate of the polyalkenamers. A mechanism involving crosslinking through vinyl units is proposed to explain the unexpected protection imparted to vinylene-rich polybutadienes by the presence of 1,2 double bonds.

Recently, much interest has been shown in the effects of atomic oxygen on various polymeric materials subjected to low Earth orbital environment (1-4). This interest has focused on significant weight loss or surface recession observed in polymers exposed to ground-state oxygen atoms [$O(^3P)$], with scarcely any attention given to the detailed chemical changes in the polymers themselves. Although there is a substantial literature on the chemical reactions of $O(^3P)$ with organic compounds (5), there have been no mechanistic studies of analogous reactions involving polymers. What is known is mainly the result of weight-loss measurements on a variety of polymers (6-9). With the possible exception of the work of MacCallum and Rankin (9), the other studies on atomic oxygen reactions with polymer films have involved their exposure in oxygen plasmas which contain (10) O_2^+, O_2^-, O^+, O^-, $O(^1D)$, free electrons, and ultraviolet radiation, in addition to singlet oxygen [$O_2(^1\Delta_g)$, or $^1O_2$] and $O(^3P)$. By suitable design and operation of an oxygen discharge apparatus, it is possible to eliminate all but the latter two species and ground-state molecular oxygen (11). Reactions of $^1O_2$ with unsaturated polymers are well-known (12) (leading to allylic hydroperoxides and shifted double bonds but with no chain scission—processes which, in the case of polymer films, are confined to the surface (13,14)), while $^1O_2$ is unreactive toward saturated polymers (15). Thus, the role of $O(^3P)$ in promoting chemical changes in unsaturated or saturated polymers should be distinguishable from that of singlet oxygen.

There is a need to study the mechanisms of $O(^3P)$ reactions with a family of closely related polymers in order to identify the various steps commencing with the initial oxygen atom attack and culminating in degradation of those polymers. As a start in this direction, this paper examines reactions of $O(^3P)$ with polybutadienes having different 1,4/1,2 contents and with their polyalkenamer homologues. Special interest in studying unsaturated hydrocarbon polymers derives from the fact that oxygen atom-olefin reactions constitute the most extensively studied class of $O(^3P)$ reactions (5,16), characterized by the formation of epoxides and carbonyl compounds as well as fragmentation products. Thus, the macromolecular counterparts of these reactions might be more readily observed and interpreted than those produced in saturated polymers which, by analogy to simple alkanes (5), should involve mainly processes ensuant on abstraction of hydrogen. There is added interest in examining the reaction of $O(^3P)$ with *cis*-1,4-polybutadiene since Rabek and co-workers (17,18) reported that this polymer underwent cis-trans isomerization on exposure to atomic oxygen, a process not observed in the corresponding reaction with its low-molecular weight analogue, *cis*-2-butene (19,20).

Experimental

Polymers. The polymers used in this study comprised *cis*- and *trans*-1,4-polybutadienes (CB and TB), amorphous 1,2-polybutadiene (VB), a number of polybutadienes with different 1,4/1,2 contents, trans polypentenamer (TP), cis and trans polyoctenamers (CO and TO), and ethylene-propylene rubber (EPM), the structures and sources of which are indicated in Tables I and II. All of these polymers were elastomers except for the highly crystalline TB and the partially (33%) crystalline TO. The polymers were purified by reprecipitation from benzene solution using methanol as precipitant. Films of CB, VB, and TB, cast from benzene stock solutions onto 0.5-inch diameter NaCl or KBr disks to a thickness of ~5-15 μm, were exposed at ambient temperature to $O(^3P)$ for various periods of time. Transmission and ATR infrared spectra of these films, before and after reaction with $O(^3P)$, were obtained with a Perkin-Elmer Model 621 spectrophotometer and an IBM IR/85 Spectrometer with ATR attachment (zinc selenide crystal), respectively. Thin films of all the polymers mentioned above were also cast onto glass cover slips, dried to constant weight, and subjected to $O(^3P)$ for weight-loss measurements. The weight of each cover slip was ~100 mg; the initial weight of each film was ~10 mg, and its area was 2.5 cm^2. The weight of the cover slip plus film before and after exposure to $O(^3P)$ was measured with a Cahn electrobalance to a precision of ±0.01 mg.

$O(^3P)$ Reactor. Oxygen atoms were produced from Matheson Gas Products ultra-high purity oxygen (99.99% O_2) in a parallel-plate plasma flow reactor (illustrated schematically in Figure 1) driven by a 13.56-MHz radio-frequency power supply. Power, supplied through a matching network to copper electrodes (3.97 cm × 1.59 cm; 1.59 cm apart) attached to the exterior wall of the glass reactor, was measured by a Bird Model 43 rf-wattmeter. To prevent in-line exposure of the samples to ultraviolet radiation from the plasma, a right-angle bend was located in the plasma reactor between the origin of the discharge and the sample. Additionally, the samples were located 12.7 cm beyond the tail of any visible glow, which ended before the bend.

The O atom flow rate was measured by NO_2 titration as described elsewhere (21). At an O_2 flow rate of 6.5×10^{-2} cm^3 (STP)/s, reactor pressure of 73 Pa (0.55 torr) with the discharge off, and a power level of 15 W, the flow rate of O atoms was found to be 2.4×10^{-2} cm^3 (STP)/s; the latter figure represented an 18% conversion of O_2 to O atoms. Assuming complete $O(^3P)$-induced oxidation of the polymer samples to CO_2 and H_2O, the flow rate of O atoms was at least eight times that required to maintain the highest etch rate observed (0.8 mg/cm^2-h, in TB).

Polymer sample temperature, measured with a thermocouple situated beneath a thin glass platform supporting the sample, depended on etch rate of the polymer; the maximum temperature rise encountered in any run was 9 K, exhibited by TB after 20 min exposure to $O(^3P)$. The temperature rise for most samples was insignificant even after prolonged exposure.

Results and Discussion

Infrared Indications of Etching. As may be seen in Figure 2, CB film undergoes substantial thinning, or etching, on prolonged exposure to $O(^3P)$, with no indications of any microstructural changes and certainly no cis-trans isomerization. Thus, the transmission IR spectra show greatly decreased intensities (or increased transmittance) of all bands, while the relative intensities of the cis-1,4 and trans-1,4 bands at 13.6 and 10.3 μm, respectively, are unchanged. Figure 2 also implies what had been noted by previous workers (6,9), namely, that the reaction of $O(^3P)$ with polymer films at ambient temperature is confined to the surface layers. This view is reinforced by the corresponding ATR IR spectra which revealed no microstructural changes in the etched CB film to a depth of ~100 nm; since the ATR spectra are essentially the same as the transmission spectra, there is no need to present them here. Neither of these spectra show any new absorption at 2.8 μm (-OOH), which would indicate involvement of singlet oxygen in the $O(^3P)$-induced etching, although ATR spectra of $^1O_2$-reacted CB films have displayed such absorption (13,14).

Figure 2 certainly differs from Figure 2c in Reference 17, reported to be CB film exposed to atomic oxygen generated by mercury photosensitized decomposition of N_2O. Rabek and co-workers (17) recognized that their observed cis-trans isomerization might have been due instead to NO_2, a byproduct of the N_2O decomposition, since the cited figure showed various bands attributable to NO_2 groups attached to the CB backbone. They supported this view by exhibiting the IR spectra of CB exposed 15 min to NO_2 at 1 atm (Fig. 4 in Ref. 17) which indeed showed NO_2 bands as well as spectral changes associated with cis-trans isomerization. That view is confirmed by Figure 3 which presents our IR spectra of CB film before and after exposure to NO_2 for 15 s (!) at reduced pressure (0.42 atm); Figure 3 is very similar to Rabek and co-workers' Figure 4 in that it shows NO_2-related bands at 6.12, 6.45, 7.40, 7.80, and 11.8 μm, together with evidence of isomerization (strong 10.3- and weak 13.6-μm bands in he treated CB). Actually, the NO_2-induced cis-trans isomerization of CB had been reported 25 years ago by Soviet workers (22). However, the present work clearly demonstrates that $O(^3P)$ causes surface etching of CB, with no observable microstructural changes in the bulk polymer nor any cis-trans isomerization. For completeness, Figure 4 shows that TB likewise undergoes extensive etching on exposure to $O(^3P)$, with no trans → cis isomerization; in the very thin, etched TB film, there is a new, weak band at 12.6 μm, the assignment of which is as yet unknown, indicating minor microstructural modification of this polymer, presumably at or near the surface. Since the transmission IR spectra of the other polymers studied show only the effects of film-thinning or etching, they were omitted from this paper.

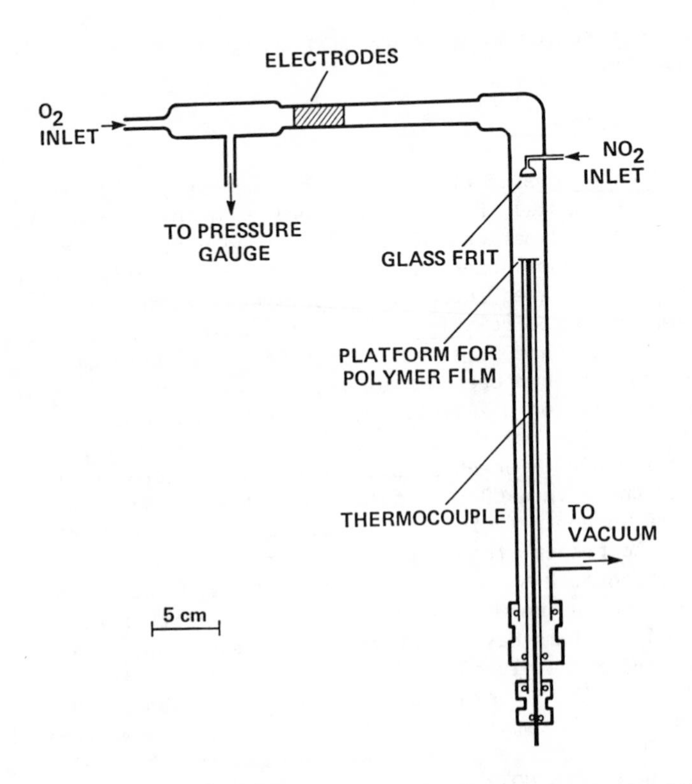

Figure 1. Apparatus for exposure of polymer films to O(^{3}P).

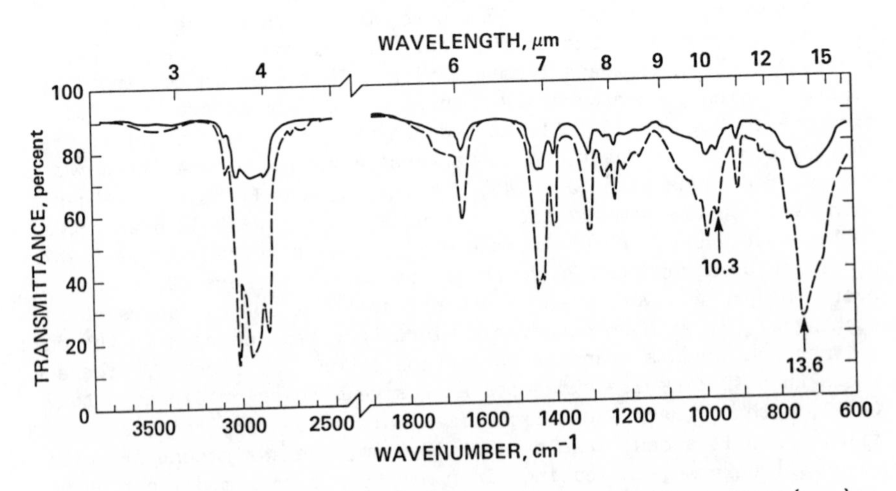

Figure 2. Transmission IR spectra of CB film on KBr before (---) and after (—) exposure to O(^{3}P) for 16 h.

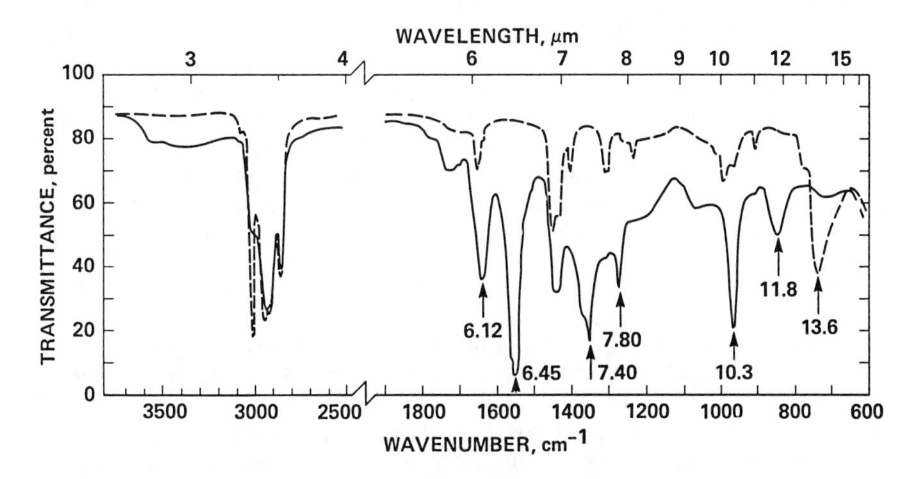

Figure 3. Transmission IR spectra of CB film on NaCl before (---) and after (—) exposure to NO_2 at reduced pressure for 15 s.

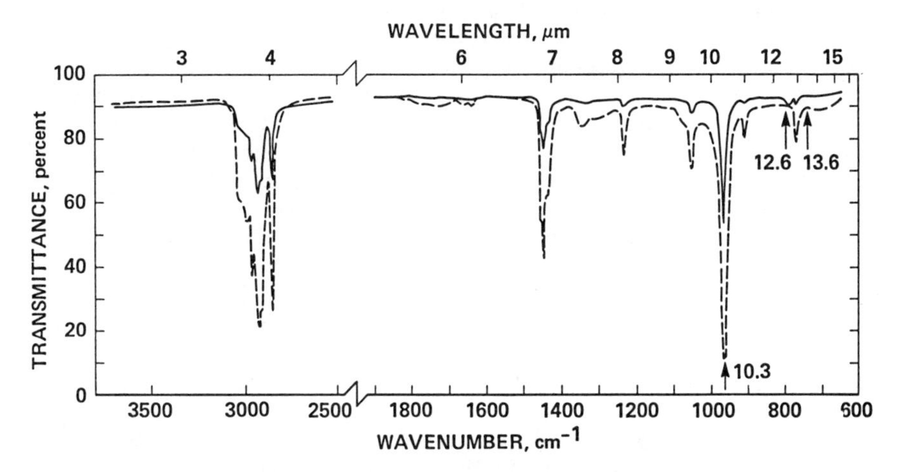

Figure 4. Transmission IR spectra of TB film on KBr before (---) and after (—) exposure to $O(^3P)$ for 4.5 h.

Weight-Loss Indications of Etching. Figure 5 shows typical zero-order kinetic plots for $O(^3P)$-induced weight loss in various polybutadienes and polyalkenamers. From such plots, etch rate data were obtained for a family of closely related polymers and are summarized in Tables I and II. For unknown reasons, the kinetic plots for several other polymers (not shown in the figure) exhibited induction periods, but the plots were almost always linear once the surface erosion commenced. An example of a plot with an induction period may be seen in Figure 4 of the preprint (23) of this paper. Another example may be found in Figure 1 of Reference 9, which shows zero-order plots through the origin for four different polymers, and a straight-line plot with an intercept on the time-axis for a fifth polymer.

Although etch rate data for a particular polymer film yielded straight-line kinetic plots, the data from one film to another for any given polymer exhibited considerable scatter; this is indicated by the large standard deviations in Tables I and II. Because of the scatter, the cause of which is under investigation, the etch rate data reported here have only semiquantitative significance.

As may be seen from the data of Table I, the presence of vinyl units has a marked effect on the etch rates for polybutadienes, decreasing by about two orders of magnitude from CB (with 2% vinyl units) to V22-V40, thereafter remaining substantially constant up to VB (with 97% vinyl units). This demonstrates at once that the effects of $O(^3P)$ reaction with the 1,4 and 1,2 double bonds are not additive and that the vinyl groups impart a special protection to polybutadienes. Unfortunately, a sample of CB with no vinyl double bonds could not be obtained for comparison with the vinyl-free polyalkenamers (Table II), but such a polymer would be expected to have an etch rate somewhat higher than the value indicated for CB in Table I. That the etch rate for TB is about six times that of its cis isomer, CB, is believed to be due to a morphology difference between the respective polymers: scanning electron micrographs showed the highly crystalline TB film to have a much greater surface roughness than the amorphous or elastomeric CB film.

The data of Table II indicate that the etch rates for CB and its "homologues"—TP, CO (or TO), and EPM—tend to increase monotonically with a decrease in vinylene (-CH=CH-) unsaturation. The elastomeric EPM was chosen instead of crystalline polyethylene as a model for the fully 'saturated' CB to avoid a morphology factor in etch rates, as was observed with crystalline TB. The difference in etch rates for the partially crystalline TO and the elastomeric CO (ratio of about 1.2:1.0) is attributable more to a morphology difference between these polyoctenamers than to the difference in their cis/trans content. Cis/trans content had likewise no perceptible effect on etch rates in the vinyl-containing polybutadienes (see Table I); if there was a small effect, it was certainly masked by the dominant effect of the vinyl groups.

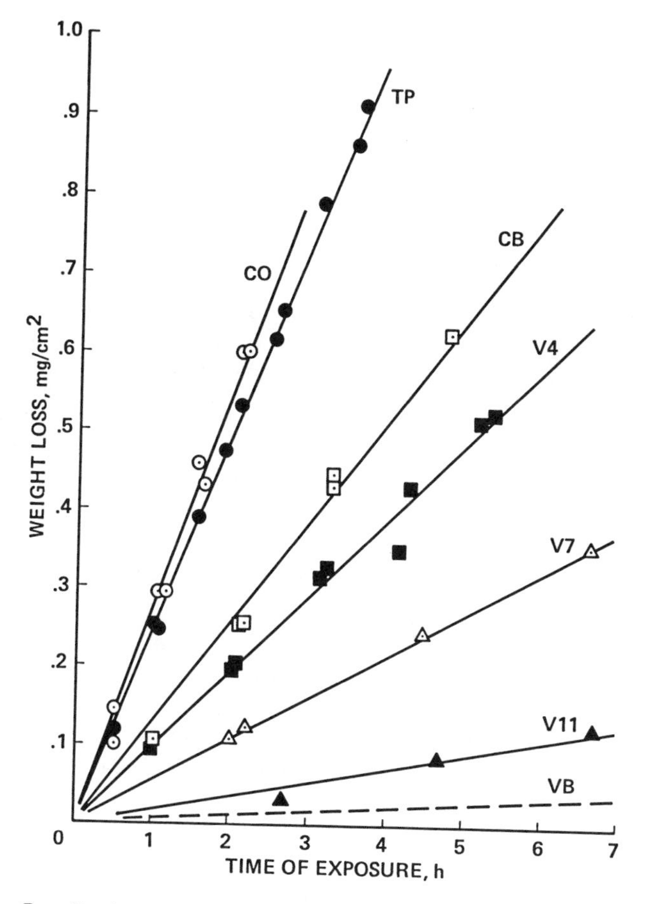

Figure 5. Typical kinetic plots for O(^{3}P)-induced weight loss in various polybutadienes and polyalkenamers.

Table I. Etch Rate Data for Various Polybutadienes[a]

Polymer[b]	% Double bonds			Etch rate mg/cm²-h	Source
	cis-1,4	trans-1,4	Vinyl		
TB	2	96	2	0.80 ± 0.40	Goodrich[c]
CB	96	2	2	0.132 ± 0.011	Polysar[d]
V4	20	76	4	0.093 ± 0.011	GenCorp[e]
V7	48	45	7	0.077 ± 0.031	GenCorp
V10	47	43	10	0.026 ± 0.020	Aldrich
V11	47	42	11	0.092 ± 0.050	Firestone[f]
V22	40	38	22	0.0016 ± 0.0012	GenCorp
V33	31	36	33	0.0046 ± 0.0053	Firestone
V40	28	32	40	0.0028 ± 0.0016	GenCorp
V70	15	15	70	0.0015 ± 0.0006	Goodyear[g]
V82	9	9	82	0.0020 ± 0.0012	Firestone
VB	2	1	97	0.0022 ± 0.0007	Firestone

[a]$-CH_2CH{=}CHCH_2-...-CH_2CH(CH{=}CH_2)-...$
[b]V4 through V82 denote polybutadienes with indicated vinyl contents.
[c]Mr. J. J. Shipman (dec.), BFGoodrich Research Center, Brecksville, OH.
[d]Dr. S. E. Horne, Polysar Inc., Stow, OH.
[e]Dr. I. G. Hargis, GenCorp, Akron, OH.
[f]Dr. T. A. Antkowiak, The Firestone Tire & Rubber Co., Akron, OH.
[g]Dr. A. F. Halasa, Goodyear Tire & Rubber Co., Akron, OH.

A sample of partially hydrogenated V70 (same source as V70; see Table I), hydrogenated with a proprietary catalyst, was also examined. This polymer, designated HV70, had 13% cis-1,4, 12% trans-1,4 and 10% 1,2 double bonds, the remaining saturated monomer units comprising 5% hydrogenated 1,4 and 60% hydrogenated 1,2 double bonds. The etch rate for HV70, with its greatly reduced vinyl content, was found to be 0.010 mg/cm²-h, or about seven times that of V70. This result is consistent with the above-mentioned observation that 1,2 double bonds protect polybutadiene against $O(^3P)$-induced etching.

Table II. Etch Rate Data for cis-1,4-Polybutadiene and its Homologues[a]

Polymer	Double bonds/ carbon atom	Distribution of double bonds % cis	% trans	% Vinyl	Etch rate mg/cm^2-h	Source
CB	0.25	96	2	2	0.132 ± 0.011	Polysar[b]
TP	0.20	17	83	-	0.240 ± 0.016	Goodyear[c]
CO	0.125	81	19	-	0.293 ± 0.009	Hüls[d]
TO	0.125	20	80	-	0.355 ± 0.049	Hüls
EPM	0	-	-	-	0.347 ± 0.014	Polysar

[a] $-[CH=CH-(CH_2)_x]_n-$; $\underline{x}$ = 2 (CB), 3 (TP), 6 (CO, TO), ∞ (EPM, with 2.4:1.0 ethylene-propylene molar ratio—a quasi model for the fully saturated polyalkenamer).
[b] Dr. S. E. Horne, Polysar Corp., Akron, OH.
[c] Dr. E. A. Ofstead, Goodyear Tire & Rubber Co., Akron, OH.
[d] Dr. E. O. E. Siebert, Hüls Corp., Piscataway, NJ.

Mechanistic Considerations

To account for the major findings in this work, namely, the protective effect of vinyl units in 1,4-/1,2-polybutadienes and the increased etch rate with decrease in vinylene unsaturation in the polyalkenamers, we invoke the generally accepted mechanism for the reaction of $O(^3P)$ with simple olefins. This mechanism, elaborated by Cvetanović some 25 years ago (20) and recently updated (16), involves addition of $O(^3P)$ to the double bond to form a transitory biradical adduct which either rearranges to a vibrationally hot epoxide or carbonyl product, or undergoes pressure-independent fragmentation (PIF). The "hot" products in turn are either collisionally deactivated to the corresponding stable compounds or undergo pressure-dependent fragmentation (PDF). PDF, which is completely suppressed at high pressure or in condensed media, would be unimportant in the case of polymer films. PIF, however, which is completely suppressed in condensed media at cryogenic temperatures (<100K), may be important in $O(^3P)$ reactions with unsaturated polymers at ambient temperature, if PIF occurs in the solid state under non-cryogenic conditions. Hydrogen abstraction from simple olefins to form OH· and the corresponding radical is considered to be at most a minor process at ambient temperature, so that addition of $O(^3P)$ to the double bond may well be the dominant initial reaction in polybutadienes. On the other hand, in the fully saturated polymer EPM, where double bonds are absent, hydrogen abstraction by $O(^3P)$ must be the sole initial reaction, as it is in alkanes (5). For the polyalkenamers TP, CO and TO, where the double bonds are farther apart than they are in CB, we visualize increasing competition between hydrogen abstraction and addition to the double bond.

By analogy to simple olefins, we propose that $O(^3P)$ initially adds to the 1,4 or 1,2 double bonds in polybutadienes at ambient temperature. Since the rate constants for $O(^3P)$ addition to cis-2-butene and 1-butene (as models for 1,4 and 1,2 double bonds, respectively) are in the ratio 4.2:1 at 298 K (16), preferential addition to the 1,4 double bonds is assumed to persist to very high vinyl contents (~80%). The biradical adducts then rearrange to epoxides and carbonyl compounds or give rise to chain rupture and/or crosslinking as a consequence of PIF, according to the scheme:

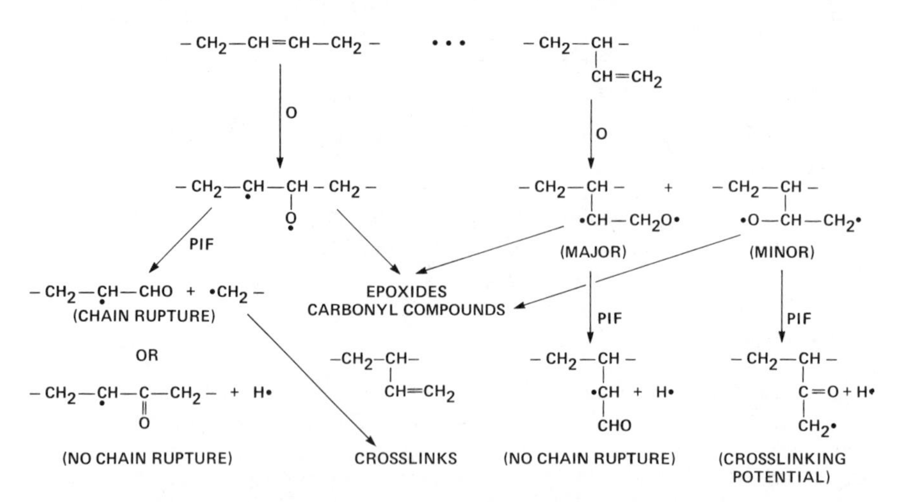

Whereas $O(^3P)$ addition to the 1,2 double bond has no direct route to chain rupture, addition to the 1,4 double bond does. Such chain rupture would be a precursor to polymer fragmentation and weight loss, while crosslinking—a potential result from $O(^3P)$ addition to the 1,2 double bond—would counteract fragmentation and thereby "protect" the polymer against surface erosion. This scheme can thus account for the finding that the 1,2-polybutadiene VB had a much lower etch rate than the 1,4-polybutadiene CB. For polybutadienes with both 1,4 and 1,2 double bonds, additional crosslinking can occur through attack of the macromolecular radical $\cdot CH_2$- (generated in PIF of the biradical formed from the 1,4 double bond) onto a nearby vinyl unit, forming a new polymeric radical that can propagate the crosslinking process. This can account for the sharp drop in etch rates for polybutadienes as the 1,2 double bond content is increased from 2 to ~20-40%. The leveling off of the etch rate for vinyl contents in excess of ~20% suggests that the "protective capacity" of the vinyl groups by capturing the nascent $\cdot CH_2$ radicals is reached at that vinyl content.

Although hydrogen abstraction is not expected to be an important factor in $O(^3P)$ reactions with polybutadienes, additional crosslinking could result from abstraction of the tertiary hydrogen in the vinyl unit. The resulting, resonating radical:

$$-CH_2-\dot{C}(|)-CH=CH_2 \rightleftarrows -CH_2-C(|)=CH-CH_2\cdot$$

could readily attack a vinyl double bond in another polymer molecule to produce a crosslink and a new propagating polymer radical. This process would have the effect of protecting the polymer against fragmentation, since crosslinked polymers are more resistant to etching than are their uncrosslinked counterparts (6).

To explain the progressive increase in etch rate with decrease in unsaturation in the polyalkenamers, we postulate that the fragmentation processes subsequent to hydrogen abstraction are much more efficient than fragmentation processes ensuant on $O(^3P)$ addition to double bonds. This view follows from the fact that the etch rate for the fully saturated EPM is ~2.6 times that for CB, even though the rate constant for $O(^3P)$ addition to the vinylene double bond (as in cis-2-butene) at 298 K is ≈135-175 times the rate constant for hydrogen abstraction from a tertiary C-H bond (in alkanes) (5,16). Thus, as the -CH=CH- double bonds become farther apart in the polyalkenamers, the increasing likelihood for hydrogen abstraction translates into a progressively increasing etch rate in the homologous series from $x = 2$ to ∞. The hydrogen abstraction reaction:

$$O + RH \rightarrow \cdot OH + R\cdot$$

must be rapidly followed by:

$$R\cdot + O \rightarrow RO\cdot^*$$

Since alkoxy radicals are known precursors to chain scission in autoxidation (24), the "hot" alkoxy radicals formed as shown should undergo facile chain scission or fragmentation. The chain scission is illustrated for the alkoxy radical derived from either the ethylene or propylene monomer unit in EPM:

$$-CH_2-C(R)(O\cdot)-CH_2-^* \rightarrow -CH_2-C(=O)-R + \cdot CH_2-$$

where R is H or CH_3

Conclusions

The major findings of this study are that vinyl groups exert a strong protective effect in polybutadienes against $O(^3P)$-induced etching, that the etch rates for the polyalkenamers increase with decrease in -CH=CH- unsaturation, that the reactions are confined to the polymer surfaces, and that there is no cis-trans isomerization of polybutadiene films in the bulk on exposure to atomic oxygen. The Cvetanović mechanism for reactions of $O(^3P)$ with simple olefins is applied to the unsaturated polymers. The protective effect of the vinyls in polybutadienes is explained in part by crosslinking through these double bonds, and in part by crosslinking initiated through abstraction of tertiary hydrogen atoms in the vinyl monomer units. The etch rate data for the polyalkenamers are accounted for on the basis of increasing competition between hydrogen abstraction and addition to the double bond, the former process giving rise to fragmentation with a high efficiency.

Acknowledgments

The authors are grateful to the individuals mentioned in footnotes of Tables I and II for their gifts of the various polymer samples used in this study.

Literature Cited

1. Visentein, J. T.; Leger, L. J.; Kuminecz, J. F.; Spiker, I. K. Paper 85-0415, AIAA 23rd Aerospace Sciences Meeting, January 1985, and references cited therein.
2. Zimcik, D. G.; Tennyson, R. C.; Kok, L. J.; Maag, C. R. Eur. Space Agency Spec. Publ., 1985, ESA SP-232; Chem. Abstr. 1986, 104, 149894, and references cited therein.
3. Banks, B. A.; Mirtich, M. J.; Rutledge, S. K.; Swec, D. M. Thin Solid Films 1985, 127, 107.
4. Arnold, G. S.; Peplinski, D. R. AIAA J. 1985, 23, 1621.
5. Huie, R. E.; Herron, J. T. Prog. Reaction Kinetics 1975, 8, 1.
6. Hansen, R. H.; Pascale, J. V.; De Benedictis, T.; Rentzepis, P. M. J. Polym. Sci. 1965, A3, 2205.
7. Taylor, G. N.; Wolf, T. M. Polym. Eng. Sci. 1980, 20, 1087.
8. Ueno, T.; Shiraishi, H.; Iwayanagi, T.; Nonogaki, S. J. Electrochem. Soc. 1985, 132, 1168.
9. MacCallum, J. R.; Rankin, C. T. Makromol. Chem. 1974, 175, 2477.
10. Lawton, E. J. J. Polym. Sci., A-1, 1972, 10, 1857.
11. Westenberg, A. A.; de Haas, N. J. Chem. Phys. 1964, 40, 3087.
12. Golub, M. A. Pure Appl. Chem. 1980, 52, 305.
13. Kaplan, M. L.; Kelleher, P. G. J. Polym. Sci., A-1, 1970, 8, 3163; Rubber Chem. Technol. 1972, 45, 423.
14. Rabek, J. F.; Rånby, B. J. Polym. Sci., Polym. Chem. Ed. 1976, 14, 1463.
15. Carlsson, D. J.; Wiles, D. M. J. Polym. Sci., Polym. Chem. Ed. 1974, 12, 2217.

16. Cvetanović, R. J.; Singleton, D. L. Rev. Chem. Intermed. 1984, 5, 183.
17. Rabek, J. F.; Lucki, J.; Rånby, B. Eur. Polym. J. 1979, 15, 1089.
18. Rabek, J. F.; Rånby, B. Photochem. Photobiol. 1979, 30, 133.
19. Klein, R.; Scheer, M. D. J. Phys. Chem., 1966, 72, 616.
20. Cvetanović, R. J. Advan. Photochem., 1963, 1, 115.
21. Kaufman, F.; Kelso, J. R. In 8th Symposium (International) on Combustion; The Combustion Institute, Williams and Wilkins: Baltimore, 1960, p. 230.
22. Ermakova, I. I.; Dolgoplosk, B. A.; Kropacheva, E. N. Dokl. Akad. Nauk. USSR 1961, 141, 1363; Rubber Chem. Technol. 1962, 35, 618.
23. Golub, M. A.; Lerner, N. R.; Wydeven, T. Polym. Prepr. 1986, 27 (2), 87.
24. Shelton, J. R. Rubber Chem. Technol. 1983, 56, G71.

RECEIVED August 27, 1987

Chapter 26

Tri-n-butyltin Hydride Reduction of Poly(vinyl chloride)

Kinetics of Dechlorination for 2,4-Dichloropentane and 2,4,6-Trichloroheptane

Fabian A. Jameison [1], Frederic C. Schilling, and Alan E. Tonelli

AT&T Bell Laboratories, Murray Hill, NJ 07974

2,4-Dichloropentane (DCP) and 2,4,6-trichloroheptane (TCH) were reductively dechlorinated with tri-n-butyltin hydride ($(n\text{-Bu})_3SnH$) directly in the NMR sample tube. ^{13}C NMR spectra were recorded periodically to monitor the progress of DCP and TCH dechlorination. From these observations the following kinetic conclusions were drawn: i. meso (m) DCP was reduced 30% faster than racemic (r) DCP; ii. the Cl from DCP was removed 4 times faster than the Cl in 2-chloropentane or 2-chlorooctane; iii. the 4-Cl in mm-TCH is removed faster than the 4-Cl in mr-TCH which in turn is more reactive than the 4-Cl in the rr isomer; and iv. the 4-Cl in TCH is removed 1.5 times faster than the 2- or 6-Cl's. Conclusions i. and ii. were previously observed at the diad level, at least qualitatively, in the $(n\text{-Bu})_3SnH$ reduction of poly(vinyl chloride) (PVC) to ethylene-vinyl chloride (E-V) copolymers. Using the kinetic information obtained from the reduction of DCP and TCH, an attempt was made to simulate the $(n\text{-Bu})_3SnH$ reduction of PVC to E-V copolymers. Comparison of the structures of the E-V copolymers simulated on the computer with those determined for $(n\text{-Bu})_3SnH$ reduced PVC by ^{13}C NMR permits us to conclude that DCP and TCH are model compounds appropriate for studying the reductive dechlorination of PVC.

Starnes and Bovey (**1**) pioneered the method of ^{13}C NMR analysis of reduced poly(vinyl chloride) (PVC) to study the microstructure of PVC. Tri-n-butyltin hydride ($(n\text{-Bu})_3SnH$) was found to completely dechlorinate PVC resulting in polyethylene (PE) whose microstructure (branching, end-groups, etc.) could be sensitively studied by ^{13}C NMR.

The present authors (**2**) subsequently produced a series of ethylene (E) - vinyl chloride (V) copolymers (E-V) by using less than the

[1] Current address: Department of Chemistry, State University of New York at Stony Brook, Stony Brook, NY 11794

0097-6156/88/0364-0356$06.00/0

stoichiometric amount of (n-Bu)$_3$SnH during the reductive dechlorination of PVC. Traditional means of obtaining E-V copolymers suffer from several shortcomings. Chlorination of PE (**3**) results in head-to-head (vicinal) and multiple (geminal) chlorination leading to structures which are not characteristic of E-V copolymers. Direct copolymerization of E and V monomers does not usually lead to random E-V copolymers covering the entire range of comonomer composition. Free-radical copolymerization (**4,5**) at low pressure yields E-V copolymers with V contents from 60 to 100 mol %. The γ-ray induced copolymerization (**6**) under high pressure yields E-V copolymers with increased amounts of E, but it appears difficult to achieve degrees of E incorporation greater than 60 mol % without producing blocky samples.

The series of E-V copolymers obtained by partial reduction of PVC with (n-Bu)$_3$SnH were found (**2**) to have the same chain length as the starting PVC ($\sim$ 1000 repeat units). Their microstructures were determined by ^{13}C NMR analysis (**2**) as indicated in Figure 1. The results of this analysis are presented in Table I in terms of comonomer diad and triad probabilities.

A close examination of the data in Figure 1 and Table I leads to two interesting observations. First, as the amount of chlorine (Cl) removed was increased, the ratio of racemic (r) to meso (m) VV diads increased, and second the disappearance of VV diads was greater than anticipated for the random removal of Cl's. Consequently, we concluded from the (n-Bu)$_3$SnH reduction of PVC to E-V copolymers and eventually to PE, that Cl's belonging to VV diads are preferentially removed relative to isolated Cl's (EVE) and that m-VV diads are reduced faster than r-VV diads.

Subsequent studies of the physical properties of this series of E-V copolymers obtained via the (n-Bu)$_3$SnH reduction of PVC have revealed that their properties, both in the solid state and in solution, are sensitive to their detailed microstructure (**7-10**). These observations prompted the present study concerning the mechanisms of the reductive dechlorination of PVC with (n-Bu)$_3$SnH.

We have chosen the PVC diad and triad compounds 2,4-dichloropentane (DCP) and 2,4,6-trichloroheptane(TCH) as subjects for our attempt to obtain quantitative kinetic data characterizing their (n-Bu)$_3$SnH reduction in the hope that they will serve as useful models for the reduction of PVC to E-V copolymers. Unlike the polymers (PVC and E-V), DCP and TCH are low molecular weight liquids whose high resolution ^{13}C NMR spectra can be recorded from their concentrated solutions in a matter of minutes. Thus, it is possible to monitor their (n-Bu)$_3$SnH reduction directly in the NMR tube and follow the kinetics of their dechlorination.

Finally the kinetic data are compared to the microstructures of the E-V copolymers obtained by (n-Bu)$_3$SnH reduction of PVC to test the suitability of DCP and TCH as model compounds for PVC reduction. This is achieved by computer modeling the reduction of PVC to E-V copolymers with the aid of the kinetic parameters obtained from the study of DCP and TCH reduction, and then comparing the observed and modeled E-V microstructures.

EXPERIMENTAL

MATERIALS. The 2-chloro-4-methylpentane, 2-chlorooctane, and 4-

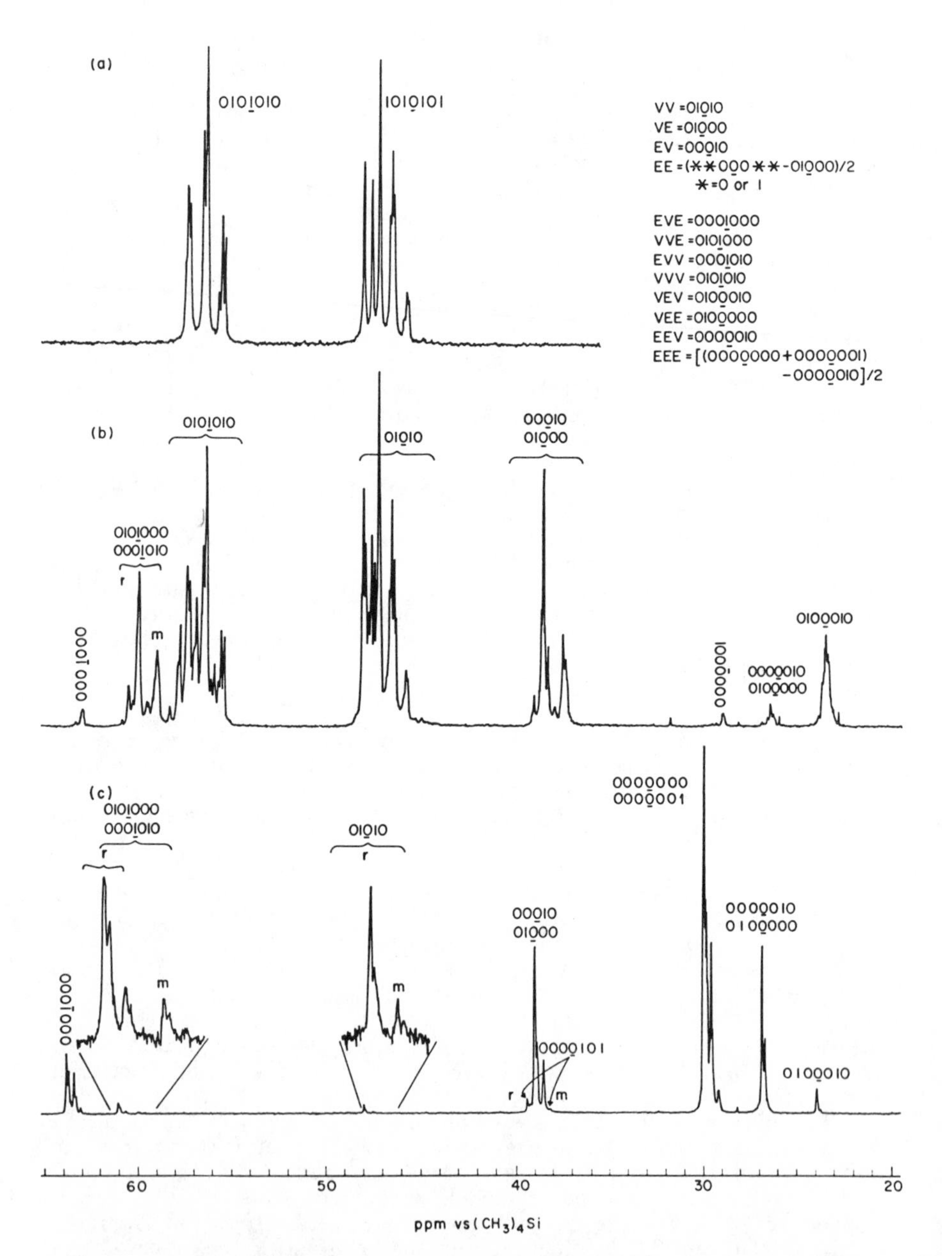

Figure 1. 50.31 MHz ^{13}C NMR spectra of PVC (a) and two partially reduced PVC's, E-V-84 (b) and E-V-21 (c). Please note the table of E-V microstructural designations in the upper right-hand corner of the Figure, where 0,1 ≡ CH_2, CHCl carbons. Resonances correspond to underlined carbons. The assignment of different stereosequences is given in reference 2.

Table I

Diad and Triad Probabilities for E-V Copolymers

Copolymer	P_{VV}	$P_{VE} = P_{EV}$	P_{EE}	P_{EVE}	$P_{VVE} = P_{EVV}$	P_{VVV}	P_{VEV}	$P_{VEE} = P_{EEV}$	P_{EEE}
E-V-85	.742	.124	.011	.015	.115	.619	.114	.011	0.0
E-V-84	.709	.134	.023	.025	.108	.615	.101	.019	.004
E-V-71	.470	.239	.052	.063	.175	.310	.175	.048	.008
E-V-62	.344	.278	.099	.116	.177	.177	.177	.075	.027
E-V-61	.343	.275	.107	.121	.173	.198	.141	.083	.029
E-V-60	.316	.285	.114	.141	.167	.154	.179	.077	.038
E-V-50	.200	.297	.205	.192	.133	.073	.166	.129	.045
E-V-46	.147	.309	.235	.205	.116	.037	.149	.140	.098
E-V-37	.087	.286	.342	.219	.078	.012	.115	.158	.183
E-V-35	.061	.278	.383	.224	.064	.015	.090	.168	.208
E-V-21	.014	.197	.593	.190	.016	0.0	.035	.153	.436
E-V-14	0.0	.127	.746	.104	0.0	0.0	.051	.123	.599
E-V-2	0.0	.025	.950	.021	0.0	0.0	0.0	.026	.926

chlorooctane were purchased from Wiley Organics and used as received. The 2,4-dichloropentane was obtained from Pfaltz & Bauer and also used as received. Tri-*n*-butyltin hydride (Alfa Division, Ventron Corp.) was vacuum distilled and stored under argon before use. The free radical initiator azobis(isobutyronitrile) (AIBN) used in the reduction was also purchased from Alfa Division. The 2,4,6-trichloroheptane was obtained from a new synthesis which involves the hydrohalogenation of 1,6-heptene-4-diol and the chlorination of the resulting alcohol. A detailed description of this method can be found elsewhere (**11**).

SAMPLE PREPARATION. In a small vial 22.5 mg of AIBN was mixed with 1.7 ml of perdeuterobenzene and the mixture held at 0 °C to permit dissolution of the AIBN. The chloroalkane and 0.2 ml of the NMR reference material hexamethyldisiloxane (HMDS) were placed in a 10 mm NMR tube. The AIBN/benzene solution was added to the NMR tube and placed under an argon atmosphere in a glove bag. The freshly distilled $(n\text{-Bu})_3SnH$ (1.0-1.7 ml) was transferred by syringe into the solution. Following a thorough mixing, the sample was degassed with argon for several minutes and sealed with paraffin film. The amount of chlorinated alkane varied between 0.2 and 0.4 ml (7.0-12.5% v/v) and the $(n\text{-Bu})_3SnH$ added was equal to the molar concentration of chlorine atoms present. The sample was placed in the NMR spectrometer at 50 °C and the ^{13}C NMR spectra were recorded as the reduction proceeded.

NMR MEASUREMENTS. Initially, the reduction of 2-chloro-4-methylpentane was carried out in order to ascertain the ideal temperature which would lead to complete reduction in about six hours. The progress of this reduction was followed by ^1H NMR, recording a single scan every thirty minutes. It was found that at 50 °C the reaction reaches 80% of completion after 5 hr. All subsequent reductions were carried out at this temperature.

The 50.31 MHz ^{13}C NMR spectra of the chlorinated alkanes were recorded on a Varian XL-200 NMR spectrometer. The temperature for all measurements was 50 °C. It was necessary to record 10 scans at each sampling point as the reduction proceeded. A delay of 30 s was employed between each scan. In order to verify the quantitative nature of the NMR data, carbon-13 T_1 data were recorded for all materials using the standard 180° - τ -90° inversion-recovery sequence. Relaxation data were obtained on $(n\text{-Bu})_3SnH$, $(n\text{-Bu})_3SnCl$, DCP, TCH, pentane, and heptane under the same solvent and temperature conditions used in the reduction experiments. In addition, relaxation measurements were carried out on partially reduced (70%) samples of DCP and TCH in order to obtain T_1 data on 2-chloropentane, 2,4-dichloroheptane, 2,6-dichloroheptane, 4-chloroheptane, and 2-chloroheptane. The results of these measurements are presented in Table II. In the NMR analysis of the chloroalkane reductions, we measured the intensity of carbon nuclei with T_1 values such that a delay time of 30 s represents at least 3 T_1. The only exception to this is heptane where the shortest T_1 is 12.3 s (delay = $2.5T_1$). However, the error generated would be less than 10%, and, in addition, heptane concentration can also be obtained by product difference measurements in the TCH reduction. Measurements of the nuclear Overhauser enhancement (NOE) for carbon nuclei in the model compounds indicate uniform and full enhancements for those nuclei used in the quantitative measurements. Table II also contains the chemical

Table II

^{13}C NMR Spin Lattice Relaxation Times (T_1) and Chemical Shifts (δ)

Compound	Carbon	δ		T_1 (s)
H−Sn(C−C−C−C)$_3$ (carbons a b c d)	a	8.30		6.0
	b	27.41		8.5
	c	30.24		7.3
	d	13.82		8.1
Cl−Sn(C−C−C−C)$_3$ (carbons a b c d)	a	17.53		4.5
	b	27.04		7.0
	c	28.19		5.9
	d	13.64		7.3
C−C(Cl)−C−C(Cl)−C, m,r (carbons a b c)	a	25.50	(r)	6.7
		24.49	(m)	6.4
	b	55.38	(r)	15.7
		54.24	(m)	15.8
	c	50.86	(r)	8.9
		50.51	(m)	8.8
C−C(Cl)−C−C−C (carbons a b c d e)	a	25.41		9.0
	b	57.58		21.8
	c	42.82		13.3
	d	20.06		14.8
	e	13.56		-
C−C−C−C−C (carbons a b c)	a	14.12		10.8
	b	22.62		24.6
	c	34.46		24.4
C−C(Cl)−C−C(Cl)−C−C(Cl)−C, m,r m,r (carbons a b c d)	a	25.34	(rr)	3.5→ 3.7
		24.21	(mr)	
		24.11	(rm)	
			(mm)	
	b	55.19	(rr)	8.1→ 8.4
		55.07	(mr)	
		54.14	(rm)	
		53.93	(mm)	
	c	49.02	(rr)	4.4→ 4.6
		48.84	(mr)	
		48.33	(rm)	
		47.81	(mm)	
	d	58.32	(rr)	8.2
		57.44	(mr) (rm)	8.1
		56.39	(mm)	8.0

Continued on next page

Table II (*continued*)

^{13}C NMR Spin Lattice Relaxation Times (T_1) and Chemical Shifts (δ)

Structure	Carbon	δ		T_1 (s)
C–C(Cl)–C–C(Cl)–C–C–C (m,r); a b c d e f g	a	25.59	(r)	-
		23.50	(m)	-
	b	55.56	(r)	-
		54.41	(m)	-
	c	49.30	(r)	-
		48.55	(m)	-
	d	60.51	(r)	-
		59.58	(m)	-
	e	41.10	(r)	-
		40.07	(m)	-
	f	19.76	(r)	-
		19.61	(m)	-
	g	14.07	(r) (m)	12.4
C–C(Cl)–C–C–C–C(Cl)–C (m,r); a b c d	a	25.33	(r)	~5.7
		25.30	(m)	
	b	57.80	(r) (m)	~11.0
	c	40.07	(r)	6.4
		40.02	(m)	
	d	24.10	(r)	~3.6
		24.02	(m)	
C–C–C–C(Cl)–C–C–C; a b c d	a	14.07		12.4
	b	19.96		8.9
	c	41.01		6.3
	d	63.08		-
C–C(Cl)–C–C–C–C–C; a b c d e f g	a	25.40		5.7
	b	58.25		14.6
	c	40.71		7.2
	d	26.60		8.5
	e	31.64		9.3
	f	22.80		10.5
	g	14.20		12.3
C–C–C–C–C–C–C; a b c d	a	14.20		12.3
	b	23.01		13.2
	c	32.24		12.3
	d	29.35		12.4

shift data for all compounds studied. The chemical shift data for the TCH sample agrees well with that of an earlier report where the shift assignment of each stereoisomer was established (**12**). The percent reduction was determined by comparing the amounts of $(n\text{-Bu})_3SnH$ and $(n\text{-Bu})_3SnCl$ at each measurement point.

KINETICS OF $(n\text{-BU})_3SnH$ REDUCTION OF DCP AND TCH.

DCP REDUCTION. As illustrated below DCP (D) is sequentially transformed into 2-chloropentane (M) and then to pentane (P) during its reduction with $(n\text{-Bu})_3SnH$.

$$D \xrightarrow{k_D} M \xrightarrow{k_M} P$$

The ratio of rate constants $K=(k_M/k_D)$ can be obtained (**13**) from the concentrations of D and M measured at various degrees of reduction x according to

$$\frac{M_x}{D_x} = \frac{1-\left\{\frac{D_x}{D_0}\right\}^{K-1}}{K-1} \quad (1)$$

where the subscripts o and x indicate concentrations initially and after % reduction x.

An alternative means to determine the relative rates of reduction of M and D, ie. $K = k_M/k_D$, is afforded by comparing the simultaneous $(n\text{-Bu})_3SnH$ reductions of DCP and 2-chlorooctane (M′) to pentane and octane (O), respectively.

$$D \xrightarrow{k_D} M \xrightarrow{k_M} P$$

$$M' \xrightarrow{k_{M'}} O$$

In this case $K' = k_{M'}/k_D$ is given by (**13**)

$$K' = \frac{\ln\left(\frac{M_x'}{M_0'}\right)}{\ln\left(\frac{D_x}{D_0}\right)} \quad (2)$$

Equation (2) can also be used to determine the relative rates of reduction of meso (m) and racemic (r) DCP (D_m, D_r) where M′ and D are replaced by D_m and D_r.

$$D_m \xrightarrow{k_{D_m}} M \xrightarrow{k_M} P$$

$$D_r \xrightarrow{k_{D_r}} M \xrightarrow{k_M} P$$

TCH REDUCTION. In the early stages of the reduction of TCH (T) with $(n\text{-Bu})_3SnH$ it is possible to compare the relative reactivities of the central (4) and terminal (2,6) chlorines. At these levels of reduction only 2,6 and 2,4-dichloroheptanes (2,6-D and 2,4-D) are produced, as shown below.

$$T \xrightarrow{k_C} 2{,}6\text{-}D$$

$$T \xrightarrow{k_E} 2{,}4\text{-}D$$

We can establish the relative reactivities, k_C/k_E, of the central (C) and terminal (E) chlorines directly from the relative concentrations of the resulting dichloroheptanes.

$$\frac{k_C}{k_E} = \frac{2{,}6\text{-}D}{2{,}4\text{-}D} \qquad (3)$$

RESULTS AND DISCUSSION

KINETIC RESULTS FOR DCP AND TCH. The portion of the 50.13 MHz ^{13}C NMR spectra containing the methylene and methine carbon resonances of DCP and the resultant products of its $(n\text{-Bu})_3SnH$ reduction are presented in Figure 2 at several degrees of reduction. Comparison of the intensities of resonances possessing similar T_1 relaxation times (see above) permits a quantitative accounting of the amounts of each species (D,M,P) present at any degree of reduction.

In Figure 3 the percentages of D (DCP), M (2-chloropentane), and P (pentane) observed during the $(n\text{-Bu})_3SnH$ reduction of DCP are plotted against the degree of reduction x. Equation (1) is solved for $K=k_M/k_D$ by least-squares fitting the calculated and observed values of the ratio (M_x/D_x). The observed ratios (D_x/D_0) are substituted into Eq. 1 to obtain the calculated ratios (M_x/D_x) corresponding to the assumed $K = k_M/k_D$ and these are compared with the observed ratios (M_x/D_x). This procedure yields $K = k_M/k_D = 0.26$, which means that DCP is ~4 times more easily reduced than 2-chloropentane. Comparison of the simultaneous reduction of DCP and 2-chlorooctane gave according to Equation (2) $K' = k_{M'}/k_D = 0.24$, lending further support to the observation that chlorines belonging to a VV diad are removed 4 times faster than an isolated chlorine in say an EVE triad. Furthermore, the observed rates of $(n\text{-Bu})_3SnH$ reductions of 2- and 4-chlorooctanes were

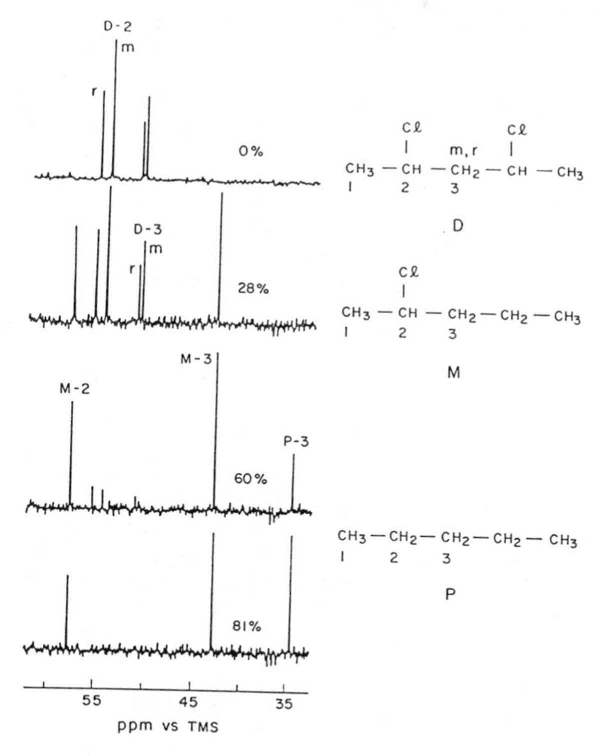

Figure 2. 50.31 MHz ^{13}C NMR spectra of DCP (D) and its products (M and P) resulting from 0, 28, 60, and 81% reduction with $(n\text{-}Bu)_3SnH$.

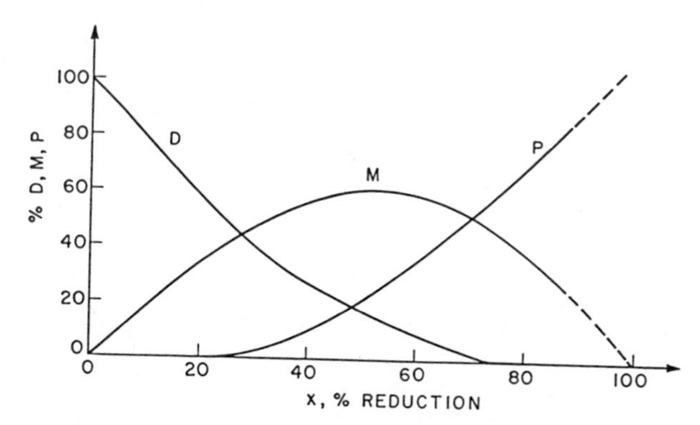

Figure 3. Distribution of reactants (D, M) and products (M, P) observed in the $(n\text{-}Bu)_3SnH$ reduction of DCP. D = DCP, M = 2-chloropentane and P = pentane (see Figure 2).

identical within experimental error. This means that the reactivity of an isolated chlorine is independent of structural position or chain end effects.

The observed ratio of m to r isomers, m_x/r_x, remaining during the $(n\text{-Bu})_3SnH$ reduction of DCP (see Figure 2) are plotted in Figure 4. Substituting this data into Equation (2) yields a ratio of $k_{D_m}/k_{D_r} = 1.3$. Apparently m-VV diads are 30% more reactive toward $(n\text{-Bu})_3SnH$ than are r-VV diads.

Our ^{13}C NMR analysis (**2**) of the E-V copolymers obtained via the $(n\text{-Bu})_3SnH$ reduction of PVC led to $k_m/k_r = 1.31 \pm 0.1$ in excellent agreement with the kinetics observed for the removal of chlorines from m- and r-DCP. We also found no VV diads in those E-V copolymers made by removing more than 80% of the chlorines from PVC. This observation is confirmed in the $(n\text{-Bu})_3SnH$ reduction of DCP where the chlorines in this PVC diad model compound were found to be 4 times easier to remove than the isolated chlorines in 2-chloropentane, 2-, and 4-chlorooctane.

The ^{13}C NMR spectra of TCH before and after 43% reduction with $(n\text{-Bu})_3SnH$ are shown in Figure 5. The shift assignments given in the figure and those listed in Table II were obtaind by comparison to the chemical shift data of TCH, DCP, 2-, and 4-chlorooctanes. From the relative concentrations of 2,6- and 2,4-dichloroheptane (2,6-D and 2,4-D) observed in the early stages of TCH reduction with $(n\text{-Bu})_3SnH$ we determine according to Equation (3) that the reactivity of the central chlorine in TCH is 50% greater than the terminal chlorines, ie. $k_C/k_E = 1.5$. We also find the reactivity of the central chlorine in TCH to depend on its stereoisomeric environment as follows: mm>mr or rm>rr.

In Figure 6 we have plotted and compare the triad sequences observed in the reduction of TCH and PVC with $(n\text{-Bu})_3SnH$. The curves numbered 0,1,2, and 3 correspond to triads containing 0 (EEE), 1 (VEE + EEV + EVE), 2 (VVE + EVV + VEV), and 3 (VVV) chlorine atoms. There is agreement between the curves describing the products of reduction for TCH and PVC providing strong support for considering TCH an appropriate model compound for the $(n\text{-Bu})_3SnH$ reduction of PVC. This clearly implies that the $(n\text{-Bu})_3SnH$ reduction of PVC is independent of comonomer sequences longer than triads.

The first column of Table III lists all possible ^{13}C NMR distinguishable E-V triads whose central units are V. In the next column we present the same triad structures in binary notation (0=E, 1=V) with the central unit labeled as the site of $(n\text{-Bu})_3SnH$ attack and the terminal units as either − (preceeding site) or + (following site). The final column presents the relative reactivities of the central V(1) unit in each triad toward $(n\text{-Bu})_3SnH$ based on the kinetics of reduction determined for DCP and TCH.

For the EVV (011) triads, removal of the central chlorine atoms is expected to be 3.5 (r) and 4.6 (m) times faster than for the isolated chlorine atom in the EVE (010) triad, because $k_D/k_M = 4.0$ and $k_{D_m}/k_{D_r} = 1.3$ for DCP. The central chlorines in VVV (111) triads are 6.0 (mr or rm), 4.6 (rr), and 7.8 (mm) times more reactive toward $(n\text{-Bu})_3SnH$ than the chlorine in the EVE triad based on $k_D/k_M = 4.0$ and $k_{D_m}/k_{D_r} = 1.3$ for DCP and $k_C/k_E = 1.5$ for TCH.

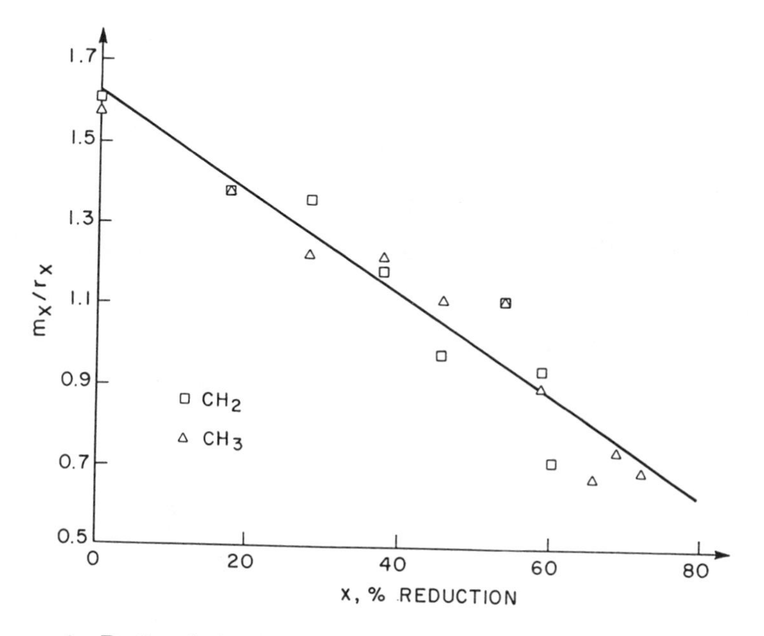

Figure 4. Ratio of the relative amounts of m and r isomers of DCP remaining after reduction by $(n\text{-}Bu)_3SnH$, as measured by the carbon-13 methylene (see Figure 2) and methyl resonances.

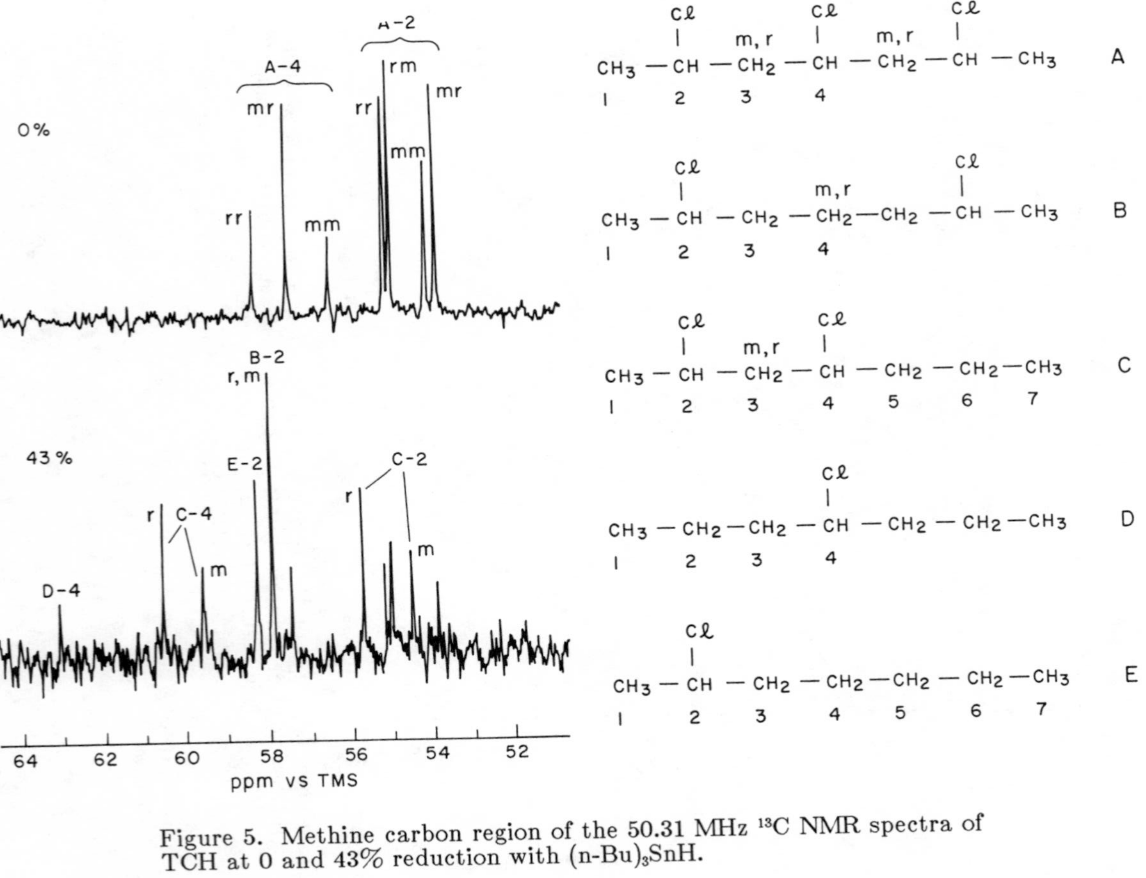

Figure 5. Methine carbon region of the 50.31 MHz ^{13}C NMR spectra of TCH at 0 and 43% reduction with $(n\text{-}Bu)_3SnH$.

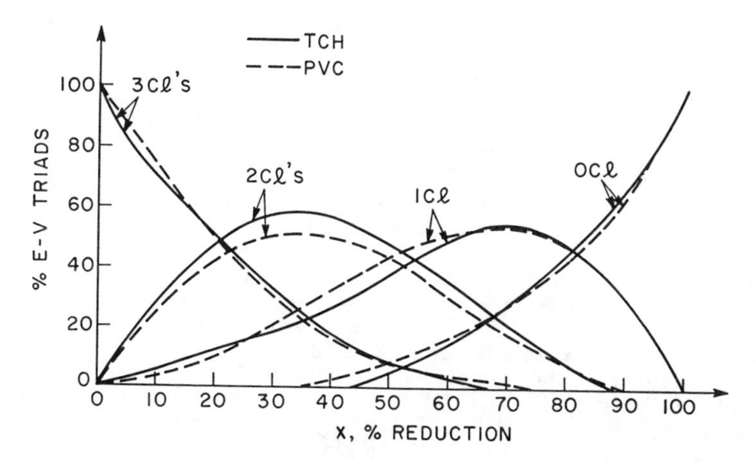

Figure 6. Comonomer triad distributions observed by ^{13}C NMR analysis during the $(n\text{-}Bu)_3SnH$ reductions of TCH (——) and PVC (- - -).

Table III

Relative Reactivities of the Central Chlorines in E-V Triads

E-V Triad	–		reduction site		+	k (relative)
EEE	0		0		0	0.0
EVE	0		1		0	1.0
EVV	0		1	r	1	3.5
EVV	0		1	m	1	4.6
VVV	1	m	1	r	1	4.0 × 1.5 = 6.0
VVV	1	r	1	r	1	6.0 ÷ 1.3 = 4.6
VVV	1	m	1	m	1	6.0 × 1.3 = 7.8

COMPUTER SIMULATION OF TCH AND PVC REDUCTION.

We begin with 100 TCH molecules reflecting the stereochemical composition of our unreduced TCH sample, ie. 52 (mr or rm), 28 (rr), and 20(mm) stereoisomers. A TCH molecule is selected by generating a random integer, I_r, where $I_r < 101$. If $I_r < 53$, then the TCH molecule chosen is a mr or rm isomer. If $52 < I_r < 81$, then the TCH is rr, and if $I_r > 80$ the TCH selected is a mm isomer.

Next we randomly choose either one or the other terminal units or the central unit of our selected TCH isomer and check to see if it is a V(1) unit or an E(0) unit. If a terminal V unit is chosen we check to see if the neighboring central unit is V or E. If the central unit is also V, then we determine whether this VV diad is m or r. For r and m VV diads the relative reactivities of the terminal V unit chlorine are 3.5 and 4.6, respectively (see Table III). If the central unit is E, then we assume the relative reactivity of the isolated terminal units in the VEE or VEV triads to be identical to the isolated central unit in the EVE triads. This assumption is supported by the identical rates of $(n\text{-}Bu)_3SnH$ reduction observed here for the 2- and 4-chlorooctanes. Finally, we select a random number between 0.0 and 1.0. If it is smaller than the relative reactivity divided by the sum of the relative reactivities of all chlorines in the VVV, EVV or VVE, VEV, VEE or EEV, and EVE isomers of TCH and partially reduced TCH (see Table III), then we remove the terminal chlorine ($1 \rightarrow 0$) and modify the relative reactivity of the central V unit in the selected TCH isomer, because its terminal neighbor has been changed from V to E.

This procedure is repeated until the desired per cent reduction, x, is reached, where
$x = 100 \times (\#\text{ of chlorines removed} \div 300)$. Each of the 100 TCH molecules is then tested for the number and sequence of V units remaining at this current value of x. In Figure 7 we plot the per cent of TCH molecules containing 3, 2, 1, and 0 chlorines, or V units, determined from our simulation and compare them to the values observed for TCH at various degrees of $(n\text{-}Bu)_3SnH$ reduction. Agreement between the simulated and observed reduction products of TCH based on the kinetics observed for both DCP and TCH is good.

Simulation of the $(n\text{-}Bu)_3SnH$ reduction of PVC is carried out in a manner similar to that described for TCH. Instead of beginning with 100 TCH molecules we take a 1000 repeat unit PVC chain that has been Monte Carlo generated to reproduce the stereosequence composition of the experimental sample of PVC used in the reduction to E-V copolymers (**2**), ie. a Bernoullian PVC with P_m=0.45. At this point we have generated a PVC chain with a chain length and a stereochemical structure that matches our experimental starting sample of PVC.

We select repeat units at random, and if they are unreduced V units a check of whether or not the units adjacent to the selected unit are E or V is made. Having determined the triad structure (both comonomer and stereosequence) of the repeat unit selected for reduction, we divide the relative reactivity of this E-V triad by the sum of relative reactivities for all V centered E-V triads as listed in Table III to obtain the probability of reduction. A random number between 0.0 and 1.0 is generated, and if it is smaller than the probability of reduction of the selected E-V triad, we remove the chlorine from the central V unit which becomes an E unit.

If either of the terminal units of the E-V triad selected are V units, then we modify their relative reactivities to reflect changing the central unit from V to E. The degree of reduction x is calculated from $100 \times (\#$

of chlorines removed ÷ 1000), and if it corresponds to the desired level of reduction we print out the numbers of each type of triad remaining in the E-V copolymer. This whole procedure is repeated for several PVC chains until the fraction of each E-V triad type at each degree of reduction remains constant when averaged over the generated set of chains.

Figure 8 presents a comparison of observed (see Table I) and simulated E-V triad composition plotted against the degree of overall reduction by $(n\text{-Bu})_3SnH$. The agreement is excellent being much improved over that found for TCH reduction. This is at least partially a consequence of the relative accuracy of the ^{13}C NMR data used to obtain the E-V triad compositions resulting from the reduction of PVC, because the TCH data is gathered during reduction and is an average over the time required to accumulate ^{13}C NMR spectra (~ 10 min.), while E-V data is obtained on static samples removed from the reduction flask.

In Figure 9 we have plotted the ratios of r/m VV diads observed by ^{13}C NMR in E-V copolymers obtained by the $(n\text{-Bu})_3SnH$ reduction of PVC (**2**). They are compared to the r/m ratios resulting from our computer simulation of PVC reduction made possible by the observation of the kinetics of $(n\text{-Bu})_3SnH$ reduction of DCP and TCH. The agreement is good, and provides us with a knowledge of E-V stereosequence as a function of comonomer composition.

The excellent agreement between the simulated and observed reduction of PVC with $(n\text{-Bu})_3SnH$ means that both DCP and TCH are appropriate model compounds for the study of PVC reduction. DCP is useful to obtain kinetic information on the relative reactivities of m- and r-diads and VV and EV diads. Reduction of TCH yields the relative reactivities of the central and terminal chlorines in the VVV triads.

The physical properties (**7-10**) of our E-V copolymers are sensitive to their microstructures. Both solution (Kerr effect or electrical birefringence) and solid-state (crystallinity, glass-transitions, blend compatibility, etc.) properties depend on the detailed microstructures of E-V copolymers, such as comonomer and stereosequence distribution. ^{13}C NMR analysis (**2**) of E-V copolymers yields microstructural information up to and including the comonomer triad level. However, properties such as crystallinity depend on E-V microstructure on a scale larger than comonomer triads.

For example, the amount and stability of the crystals formed in E-V copolymers depend on the number and length of uninterrupted, all E unit runs. Our ability to computer-simulate the $(n\text{-Bu})_3SnH$ reduction of PVC permits us to obtain this information concerning the longer comonomer sequences in the resultant E-V copolymers.

In Figure 10 we present the percentage of E units that are found in uninterrupted, all E unit runs as a function of the length of each run and the over all degree of reduction. These data were obtained in two ways: i.) simulation of the $(n\text{-Bu})_3SnH$ reduction of PVC and ii.) assuming random removal of Cl during the reduction.

The simulated data in Figure 10 make clear that the numbers and lengths of all E unit runs in E-V copolymers obtained from the reduction of PVC with $(n\text{-bu})_3SnH$ are significantly reduced compared to those resulting from random Cl removal. Though not shown in Figure 10, the percentage of E units in all E unit runs ...VE_xV... with $x \geq 29$ is 27 % for random Cl removal and just 14 % for the $(n\text{-Bu})_3SnH$ reduced PVC. This observation has to be considered when discussing the crystalline

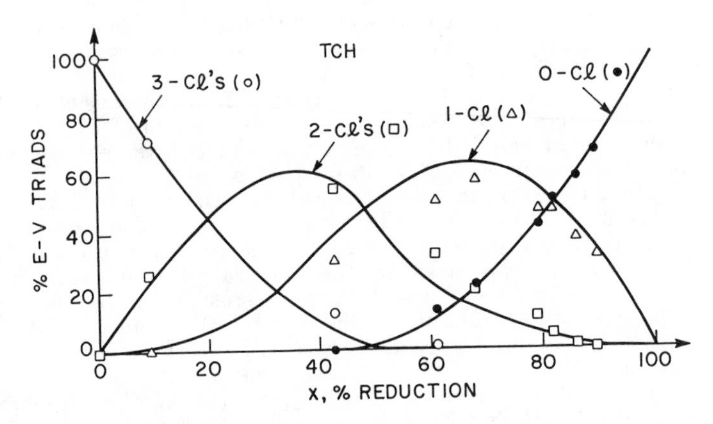

Figure 7. A comparison of the observed (symbols) and simulated (solid lines) comonomer triad distributions in $(n\text{-}Bu)_3SnH$ reduced TCH.

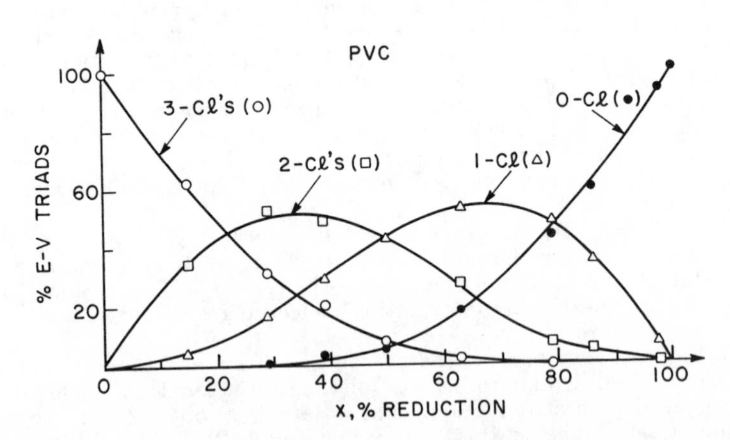

Figure 8. A comparison of the observed (symbols) and simulated (solid lines) comonomer triad distributions in $(n\text{-}Bu)_3SnH$ reduced PVC.

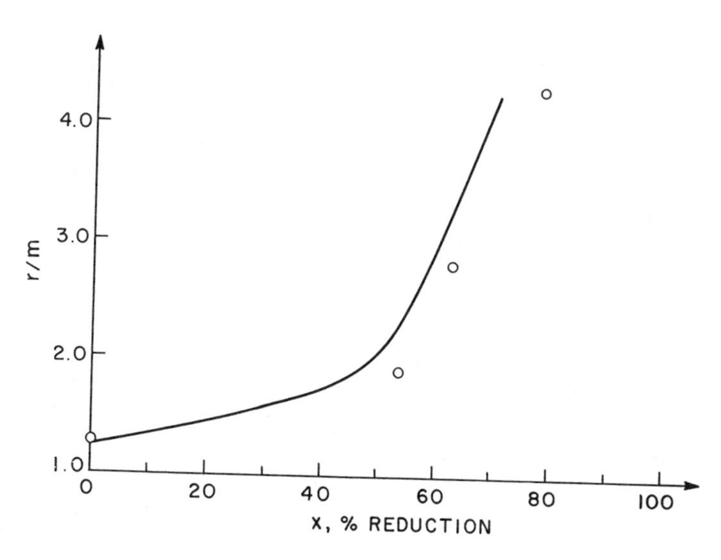

Figure 9. A comparison of the observed (O) and simulated ratios of r/m VV diads during the $(n\text{-}Bu)_3SnH$ reduction of PVC.

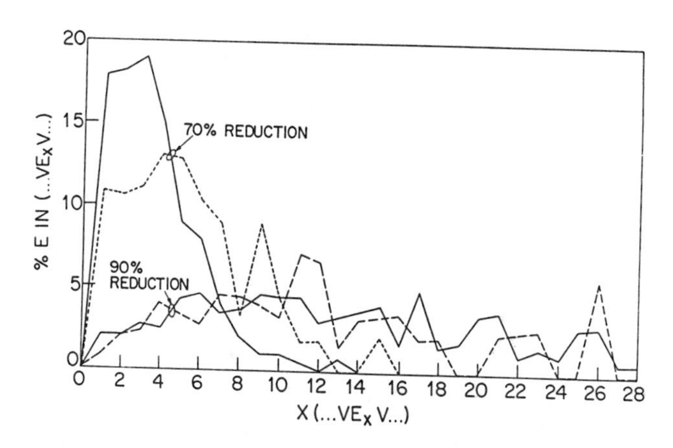

Figure 10. Percentage of E units in uninterrupted, all E unit runs as a function of run length and degree of reduction. Solid lines correspond to the computer simulation of PVC reduction with $(n\text{-}Bu)_3SnH$ and the dotted lines to the simulated results assuming random Cl removal.

morphology of E-V copolymers obtained by reduction of PVC with $(n\text{-}Bu)_3SnH$.

With the $(n\text{-}Bu)_3SnH$ reduction of PVC successfully simulated via the kinetic studies of DCP and TCH reduction, it remains to explain the mechanisms (**14**) of this reductive dechlorination. We need only consider the mechanisms of the reduction of DCP and TCH, because we have demonstrated that the kinetics of their reduction are the same as those observed for PVC when also reduced by $(n\text{-}Bu)_3SnH$.

POSSIBLE MECHANISM FOR THE REDUCTION OF PVC WITH $(n\text{-}BU)_3SnH$. The reductions of alkyl halides with $(n\text{-}Bu)_3SnH$ are known (**15**) to be free-radical chain reactions, where the $(n\text{-}Bu)_3Sn\bullet$ radical (R•) abstracts the halogen (X) from the alkyl halide (R′X) creating an alkyl radical (R′•) and $(n\text{-}Bu)_3SnX$. In Figure 11 (a,b) the rr and mm isomers of TCH are depicted in their most probable conformations (*tttt* and *gtgt*) (**16**), with their central Cl's about to be abstracted by the R•. Attack at the central chlorine by R• is hindered (**14**) by both adjacent methine protons in the *tttt* conformer of rr-TCH, while only a single methine proton obstructs the radical attack of the central chlorine in the *gtgt* conformer of mm-TCH. We would expect, as is observed, the mm isomer to be more readily reduced than the rr isomer based solely on considerations of steric interactions.

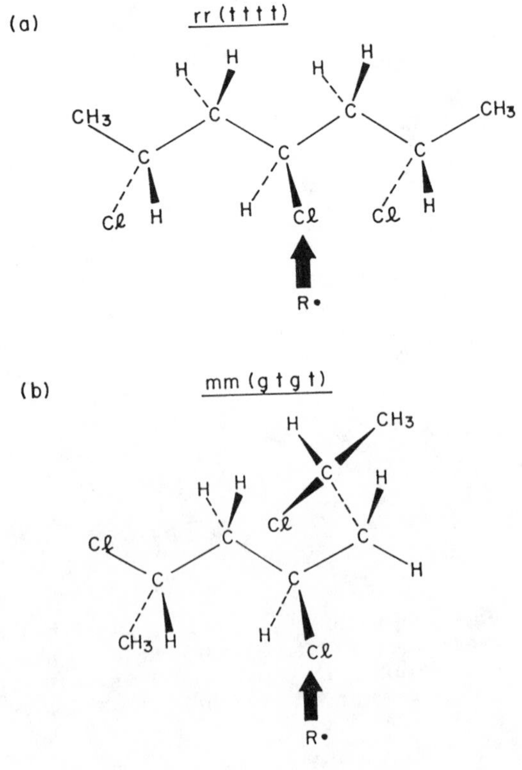

Figure 11. (a,b)-The rr and mm isomers of TCH in the *tttt* and *gtgt* conformations.

Since the $(n\text{-}Bu)_3Sn\bullet$ radical ($R\bullet$) is nucleophilic (**17**), a partial negative charge must be produced at the methine carbon whose chlorine is being abstracted. The rate of this abstraction should clearly be enhanced by electron-withdrawing groups on $R'\bullet$ due to their stabilization of this charge by inductive effects. As observed, the removal of Cl from EVV (or VV diad) is expected to be more facile than from EVE (or VE diad) as a result of a γ-halogen effect in the former structure.

Thus, the enhancements in chlorine removal from VV diads compared to EV diads and from m-VV diads compared to r-VV diads observed in the $(n\text{-}Bu)_3SnH$ reduction of DCP, TCH, and PVC are consistent with the free-radical chain reaction mechanism. Inductive effects produced by neighboring γ-Cl's tend to favor the reduction of VV diads relative to EV diads and steric interactions resulting from different preferred conformations in each isomer favor the removal of Cl from m-VV diads relative to r-VV diads.

LITERATURE CITED

1. Starnes Jr., W. H.; Schilling, F. C.; Plitz, I. M.; Cais, R. E.; Freed, D. J.; Hartless, R. L.; Bovey, F. A. *Macromolecules* **1983**, *16*, 790, and references cited therein.
2. Schilling, F. C.; Tonelli, A. E.; Valenciano, M. *Macromolecules* **1985** , *18*, 356.
3. Keller, F.; Mugge, C. *Faserforsch. Textiltech.* **1976**, *27*, 347.
4. Misono, A.; Vehida, Y.; Yamada, K. *J. Polym. Sci., Part B* **1967**, *5*, 401, and *Bull. Chem. Soc. Jpn.* **1967**, *40*, 2366.
5. Misono, A.; Uchida, Y.; Yamada, K.; Saeki, T. *Bull. Chem. Soc. Jpn.* **1968**, *41*, 2995.
6. Hagiwara, M.; Miura, T.; Kagiya, T. *J. Polym. Sci., Part A-1* **1969**, *7*, 513.
7. Tonelli, A. E.; Schilling, F. C.; Bowmer, T. N.; Valenciano, M. *Polym. Prepr., Am. Chem. Soc., Div. Polym. Chem.* **1983**, *24* (2), 211; and Tonelli, A. E.; Valenciano, M.; *Macromolecules* **1986**, *19*, 2643.
8. Bowmer, T. N.; Tonelli, A. E. *Polymer* **1985**, *26*, 1195.
9. Bowmer, T. N.; Tonelli, A. E. *Macromolecules* **1986**, *19*, 498.
10. Bowmer, T. N.; Tonelli, A. E. *J. Polym. Sci., Polym. Phys. Ed.* **1986**, *24*, 1631; and *ibid* **1987**, *25*, 1153.
11. Schilling, F. C.; Schilling, M. L. *Macromolecules* Submitted.
12. Tonelli, A. E.; Schillling, F. C.; Starnes, Jr., W. H.; Shepherd, L.; Plitz, I. M. *Macromolecules* **1979**, *12*, 78.
13. Benson, S. W. "The Foundations of Chemical Kinetics"; McGraw-Hill: New York, 1960.
14. Starnes Jr., W. H.; Schilling, F. C.; Abbas, K. B.; Plitz, I. M.; Hartless, R. L.; Bovey, F. A. *Macromolecules* **1979**, *12*, 13.
15. Carlsson, D. J.; Ingold, K. U. *J. Am. Chem. Soc.*, **1968**, *90*, 7047.
16. Flory, P. J.; Pickles Jr., C. J. *J. Chem. Soc., Faraday Trans.* **1973**, *69*, 2.
17. Grady, G. L.; Danyliw, T. J.; Rabideux, P. *J. Organomet. Chem.* **1977**, *142*, 67.

RECEIVED September 23, 1987

Chapter 27

Identification of Products from Polyolefin Oxidation by Derivatization Reactions

D. J. Carlsson, R. Brousseau, Can Zhang [1], and D. M. Wiles

Division of Chemistry, National Research Council of Canada, Ottawa K1A 0R9, Canada

A series of reactions with gases have been selected for the rapid quantification of many of the major products from the oxidation of polyolefins. Infrared spectroscopy is used to measure absorptions after gas treatments. The gases used and the groups quantified include phosgene to convert alcohols and hydroperoxides to chloroformates, diazomethane to convert acids and peracids to their respective methyl esters, sulfur tetrafluoride to convert acids to acid fluorides and nitric oxide to convert alcohols and hydroperoxides to nitrites and nitrates respectively. In some cases it is possible to differentiate between the various alkyl substituents. Primary, secondary and tertiary nitrates and nitrites all show clearly different infrared absorptions. The spectra of acid fluorides can be used to differentiate chain-end groups from pendant acid groups. Furthermore, the loss of all -OH species upon sulfur tetrafluoride exposure allows the reliable estimation of ketones, esters and lactones without the complication of hydrogen-bonding induced shifts in the spectra. Preliminary results from the use of these reactions to characterize γ-ray oxidized polyethylene and polypropylene are used to illustrate the scope of the methods.

Hydrocarbons oxidize to give a complex mixture of products which include hydroperoxides, alcohols, ketones, acids, esters, etc. (1). Polyolefins similarly can be oxidized by heat, radiation or mechano-initiated processes. The precise identification and quantification of these oxidation products are essential for the complete understanding and control of these destructive reactions. Conventional methods for the identification of oxidation products include iodome-

NOTE: This chapter was issued as NRCC No. 27914.

[1]Guest research scientist from Academia Sinica, Institute of Chemistry, Beijing, China

0097-6156/88/0364-0376$06.00/0

try and infrared (IR) spectroscopy (2-4). Iodometry sums all hydroperoxides and peracids, but also may include some dialkyl peroxides. IR cannot differentiate between hydrogen-bonded alcohol and hydroperoxide -OH groups and is only paritally successful in the resolution of the mix of carbonyl species formed. In contrast, liquid phase, nuclear magnetic resonance (n.m.r.) has the resolution to potentially differentiate all of these products (5,6). This method is limited in three areas; low sensitivity (as compared to IR for example), long data acquisition times and the need to dissolve the samples. The necessity for prolonged high temperatures to achieve solution makes questionable the final analyses because of the thermal instability of several of the key polymer oxidation products.

Derivitization reactions have previously been employed to extend the sensitivity and resolution of IR, ultraviolet and X-ray photo-electron spectroscopy (7-13). Yet no proposed method has the range to accommodate the major oxidation products from polyolefins. As part of an ongoing study of polymer oxidation and stabilization, we discuss here a series of reactions with small, reactive gas molecules. The products from these reactions can be rapidly identified and quantified by IR. Some of these reactions are new, others have already been described in the literature, although their products have not always been fully identified.

Experimental

Additive-free film samples of isotactic polypropylene (iPP, 30μm Himont Profax resin) and polyethylenes (LLDPE, 120μm, linear low density DuPont Sclair resin, and UHMW-PE, 120μm, ultra high molecular weight, high density Himont LSR 5641-1B resin) were oxidized by exposure in air to γ-radiation (AECL Gamma Cell 220, 1.0 Mrad/h). Films were stored at -20° until analysis could be carried out. Oxidized films and derivatized, oxidized films were characterized by iodometry (reflux with NaI in isopropanol/acetic acid) and by transmission Fourier Transform (FT) IR (Perkin Elmer 1500), using the spectral subtraction technique (3, 14). Free radicals were measured by the electron spin resonance technique (e.s.r., Varian E4 spectrometer).

Films were exposed to each reactive gas at room temperature in a simple flow system which could be sealed off by valves to allow reaction to proceed. After reaction, the various gases were swept out with N_2 before film analysis. The gases used included SF_4, SO_2, $COCl_2$ and NO (Matheson, used as supplied). The exception was CH_2N_2 which was generated as required in small amounts adjacent to the films by the reaction of ethanolic KOH on Diazald (Aldrich) in the reactor described by Fales et al (15). The CH_2N_2 reactions were performed with films at 22°C and also at -78°C, but the the CH_2N_2 reagents at ~ 20°C in both cases. To prevent NO_2 formation from NO-O_2 reaction, films were swept with N_2 for about 5 minutes prior to NO introduction. Because SF_4 attacks glass, these reactions were carried out in an all-polyethylene flow system with Monel valves. From the FTIR changes, all gas reactions were virtually complete in ~ 15 h for the thin iPP films and ~ 24 h for the LLDPE.

For comparison purposes, some model compounds and model polymers were reacted with the gases. Liquid models were used as dilute solutions in hexane or hexadecane; model polymers were used as solid films. n-Peroctanoic acid was synthesised from n-octanoic acid with 30% hydrogen peroxide ($-C\begin{smallmatrix}\nearrow O\\ \searrow OOH\end{smallmatrix}$ IR absorptions at 1755 cm^{-1} in hexane solution)(16). Tert.-butyl hydroperoxide (99%) was purified by the azeotropic distillation of the commercial 70% hydroperoxide/water mixture (Aldrich). n-Octanoic acid, 2-ethylhexanoic acid, stearic acid, γ-decalactone, 1,1,3,3-tetramethylbutane 1-hydroperoxide (Lucidol) were used as supplied. Polymers used to identify gas reactions included poly(methyl methacrylate), a propylene/acrylic acid copolymer, poly(vinyl alcohol) (Polysciences), Phenoxy (Union Carbide, polymer of the 2-hydroxypropylether of bisphenol A) and an ethylene/carbon monoxide copolymer.

Results and Discussion

The γ-initiated oxidation of polyolefins produces a product mix which is less complex than that resulting from photo-or thermally initiated degradation. This results from the mild conditions in the γ-cell, where the major initial oxidation product, the -OOH group, is stable. Although the derivatization methods are applicable to all types of oxidation, for simplicity only the γ-irradiated systems will be considered here.

The γ-initiated oxidation of iPP and LLDPE produces IR spectral changes in the ~ 3400 cm^{-1}, ~ 1715 cm^{-1} and ~ 1170 cm^{-1} regions (Figs. 1 and 2). These regions are broadly attributed to hydrogen-bonded alcohol and/or hydroperoxide, carbonyl and -C-O-? absorptions, respectively.

Initially, various liquid phase reagents were explored for the identification of the differing -OH species and carbonyl species in the solid films. These reagents included ethanolic sodium hydroxide [to generate IR-detectable carboxylate groups from acids (10)], and benzidine [for the colorimetric determination of peracids (17)]. However, results were disappointing with little reaction observed. This may have resulted either from lack of penetration of the reagents into the polyolefins or failure of the liquid reagents to swell the polymers. Similar problems have been reported previously (10, 12). On the other hand, successful complete reactions have been reported between gases or vapours and oxidation products (7, 8, 12, 13, 18). Apparently, small gas molecules can penetrate the polymer structure and reach most sites previously accessable to O_2. In the following sections, some of the previously proposed gas-polymer reactions are re-examined and compared with some newly developed methods.

SO_2 Reactions. Exposure to SO_2 has been proposed to lead to a quantitative reaction with -OOH groups to give a product with a marked increase in IR absorption over that of the original -OOH groups (7, 18). However, we have found the reactions with oxidized

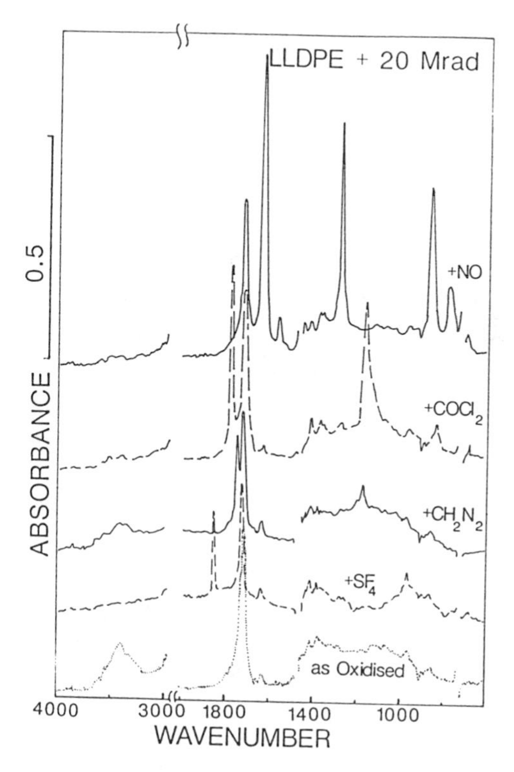

Figure 1 IR spectra of products from gas reactions with pre-oxidized LLDPE

Film oxidized in air by γ-irradiation (20 Mrad.). FTIR spectra result from the subtraction of the spectrum of non-oxidized LLDPE.

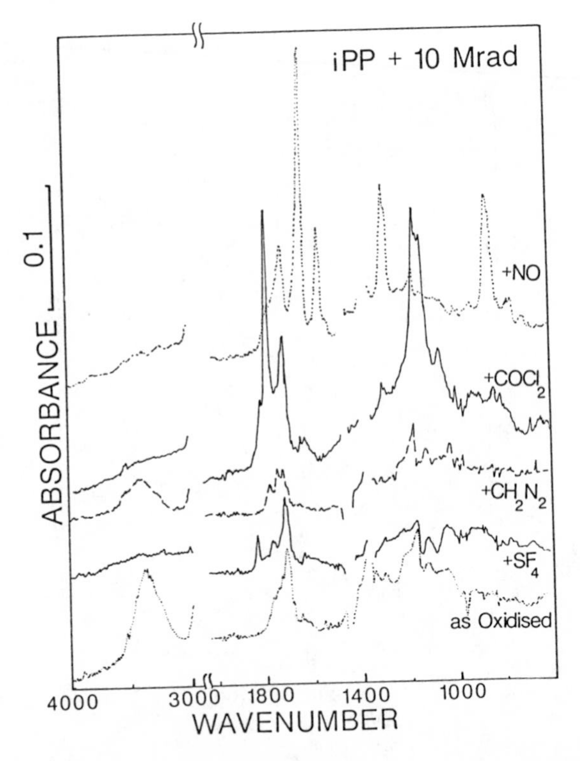

Figure 2 IR spectra of products from gas reactions with pre-oxidized iPP

Film oxidized in air by γ-irradiation (10 Mrad.). FTIR spectra result from the subtraction of the spectrum of non-oxidized iPP.

iPP and LLDPE to be complex, non-quantitative and to give only marginal enhancement in the IR absorption over the basic 3400 cm^{-1} absorption. Our results of studies of SO_2 reactions on oxidized polymers and model compounds have been reported previously (19). The SO_2 method seems to be distinctly inferior to other methods (especially NO treatment discussed below) and will not be considered further.

SF_4 Reactions. In contrast to SO_2 reactions, the SF_4 reaction as proposed over 20 years ago by Heacock is a clean, quantitative reaction although with potential beyond that originally proposed (8). From Figures 1 and 2, SF_4 exposure causes complete loss of all -OH absorptions with the generation of -C(=O)F absorptions at 1842-1848 cm^{-1} and possibly weak -C-F absorptions at ~ 1000 cm^{-1}. Oxidized LLDPE and iPP give rise to distinctly different -C(=O)F absorbances. The absorbance in LLDPE lies at 1848 cm^{-1} whereas that from iPP is at 1842 cm^{-1}, with a weaker shoulder at ~ 1848 cm^{-1}. Comparison studies on extremely linear polyethylene (UHMW-PE) also gave the 1848 cm^{-1} absorption after oxidation and SF_4 exposure. This band can reasonably be attributed to chain-end acid fluorides (reaction 1). In iPP, carboxylic acid groups can be expected from the free-radical oxidation of the methyl side groups, and to a lesser

$$-CH_2-CH_2-C(=O)OH \xrightarrow{SF_4} -CH_2-CH_2-C(=O)F \qquad 1)$$

extent from the β-scission of alkoxyl radicals (I), possibly via the reaction sequence 2.

$$\underset{\text{I}}{\sim C(CH_3)(O\cdot)\sim} \longrightarrow \sim C(CH_3)=O \;+\; \cdot CH_2-CH(CH_3)\sim \xrightarrow{\cdot O_2} \underset{\text{II}}{\cdot OO-CH_2}-CH(CH_3)\sim$$

$$\xrightarrow{RO_2\cdot} ROH + H-C(=O)-CH(CH_3)\sim \xrightarrow[RH]{RO_2\cdot} HOO-C(=O)-CH(CH_3)\sim \xrightarrow[RH]{\Delta} HO-C(=O)-CH(CH_3)\sim \qquad 2)$$

($RO_2\cdot$ represents any peroxyl radical formed from iPP, which will take part in a termination reaction with the primary peroxyl radical II). From a comparison with acid fluorides prepared from model acids (Table II) the dominant 1842 cm^{-1} absorption in PP is from the

$$\sim CH_2 - \underset{\underset{H}{|}}{\overset{\overset{O \diagdown \quad \diagup F}{C}}{\underset{|}{\overset{|}{C}}}} - CH_2 \sim$$

group whereas the 1848 cm^{-1} shoulder probably comes from the chain-end acid produced via reaction 2.

In addition to the clear generation of the acid fluoride absorptions, the SF_4 fluorination of all types of -OH groups has other advantages. Carboxylic acids may exist as very stable dimers, which absorb at ~ 1710 cm^{-1} (the normally accepted acid IR absorptions), or as free, non-hydrogen-bonded acids at 1755 cm^{-1} (20). In addition, carboxylic acids take part in hydrogen bonded complexes with other adjacent oxidation products possibly producing absorptions from 1755 through to ~ 1710 cm^{-1}. All of these extraneous absorptions are eliminated by SF_4 treatment, as shown by the narrowing of the carbonyl maximum and loss of features at ~ 1714 and ~ 1750 cm^{-1}. A further simplification of the carbonyl absorption comes from the loss of the -OOH and -OH groups, both of which cause pronounced shifts in ketone, ester and lactone absorptions (10 - 20 cm^{-1}) as compared to alkane environments. The fluoroalkyl products from SF_4 reactions with -OH and -OOH groups do cause shifts in the carbonyl absorptions, but now of only 2 - 3 cm^{-1} (21). This means that after SF_4 treatment, the residual, simplified carbonyl envelope can be interpreted by reference to the spectra of model ketones in alkanes with reasonable certainty. Attempts at interpretation of the overall carbonyl envelope prior to SF_4 treatment, as we and other have previously attempted, are now obviously quite unreliable (3, 4).

SF_4 did not react with esters, γ-lactones or ketones.

CH_2N_2 Reactions. Diazamethane (CH_2N_2) is well known to methylate acidic -OH groups (13). Consequently CH_2N_2 reactions with oxidized polyolefins can give information on carboxylic acid groups complimentary to the SF_4 reactions, with additional information on the yields of peracids (produced for example in reaction 2). From reactions with model peracids , only at -78°C is the methyl perester absorption at 1784 cm^{-1} observed; at room temperature the IR absorption of the conventional methyl ester at 1747 cm^{-1} was found from both the peracid and the normal acid. For the oxidized polyolefin, only γ-irradiated iPP showed a small, but reproducible increase in IR absorption at ~ 1785 cm^{-1} after CH_2N_2 reaction, consistent with a low level of peracid (≤ 1% of the total oxidation products). The ester absorption at 1747 cm^{-1} from CH_2N_2 reaction with oxidized iPP or LLDPE is difficult to quantify with it being superimposed upon the complex carbonyl envelope of the other oxidation products (Figs. 1 and 2); the SF_4 reaction gives a much more reliable measure. Hydroperoxide -OH groups are appreciably more acidic than alcohol -OH groups (hydroperoxides readily form sodium salts for example with sodium hydroxide). Some conversion of hydroperoxide groups to methyl peroxides was shown by n.m.r. studies of CH_2N_2-treated tert.-butyl hydroperoxide, as well as by the 1025 cm^{-1} absorption visible in the spectrum of CH_2N_2-treated, oxidized LLDPE (Fig. 1).

CH_2N_2 did not react with esters, γ-lactones, ketones or alcohols.

$COCl_2$ Reactions. The FTIR changes which result from $COCl_2$ exposure of the oxidized polyolefins (Figs. 1 and 2) are consistent with the reaction of both alcohol and hydroperoxide groups to give chloroformate groups (1785 cm^{-1}, carbonyl stretch and 1165 cm^{-1} skeletal -C(=O)-O stretch in exposed polyethylenes (22). LLDPE gives sharp peaks consistent with a single product, the secondary chloroformate (reaction 3). Although oxidized iPP also shows the 1785 and 1165 cm^{-1} peaks after $COCl_2$ exposure, other strong peaks at ~ 1140 cm^{-1} are visible. We attribute these to the sensitivity of the

$$\sim CH(-OH)\sim / \sim CH(-OOH)\sim \xrightarrow{COCl_2} \sim \underset{\substack{| \\ C(=O)Cl}}{CH} \sim + (HOCl) \qquad 3)$$

-C(C=O)Cl group to the precise structure (primary, secondary or tertiary) of the alkyl groups to which it is linked. However, our subsequent work with NO showed that its products are also sensitive to the alkyl structure yet in addition NO reacts with oxidized polymers to give distinctly different products from alcohol and hydroperoxide groups (see below). Consequently the $COCl_2$ products were not explored further.

NO Reactions. The most informative derivitization reaction of oxidized polyolefins that we have found for product identification is that with NO. The details of NO reactions with alcohols and hydroperoxides to give nitrites and nitrates respectively have been reported previously, and only the salient features are discussed here (23). The IR absorption bands of primary, secondary and tertiary nitrites and nitrates are shown in Table I. After NO treatment, γ-oxidized LLDPE shows a sharp sym.-nitrate stretch at 1276 cm^{-1} and an antisym. stretch at 1631 cm^{-1} (Fig. 1), consistent with the IR spectra of model secondary nitrates. Only a small secondary or primary nitrite peak was formed at 778 cm^{-1}. NO treatment of γ-oxidized LLDPE which had been treated by iodometry (all -OOH converted to -OH) showed strong secondary nitrite absorptions, but only traces of primary nitrite, from primary alcohol groups (distinctive 1657 cm^{-1} absorption). However, primary products were more prominent in LLDPE after photo-oxidation.

The NO treatment of γ-oxidized iPP yields a much more complex series of IR changes than does LLDPE (Fig. 2). A secondary nitrate absorption at 1278 and 1634 cm^{-1} was again found, but tert. nitrate absorption at 1302, 1290 and 1629 cm^{-1} were dominant. Small absorptions of tert.-nitrite (760 cm^{-1}) and sec.-nitrite (778 cm^{-1}) were also found. These assignments were consistent with the thermal and

TABLE I IR Bands of Nitrates and Nitrites from Polyolefin Hydroperoxides and Alcohols

Group	Absorbance Maximum (cm^{-1}) [Extinction coefficient $l.mol^{-1}cm^{-1}$][a]			
$-\overset{H}{\underset{H}{C}}-ONO_2$	1642 [2200]	1279 [1210]	860 [421]	
$-\overset{H}{\underset{\vert}{C}}-ONO_2$	1631 [1933]	1276-1278 [660]	867 [544]	
$-\overset{\vert}{\underset{\vert}{C}}-ONO_2$	1629	1302 or 1290 [260] [408]	860 [140]	
$-\overset{H}{\underset{H}{C}}-ONO$	1657-1653 [470]	----	----	778 [245]
$-\overset{H}{\underset{\vert}{C}}-ONO$	1646 [798]	----	----	778 [639]
$-\overset{\vert}{\underset{\vert}{C}}-ONO$	1638 [780]	----	----	760 [652]

a) Extinction coefficients from model compounds in hexane.

photo-instability of tert.-nitrate and tert.-nitrite as compared to the secondary species. For example, heating at 100°C for 18 h drastically reduced the 1302, 1290 and 760 cm^{-1} absorptions in oxidized NO-treated iPP whereas the absorption at 1278 and 778 cm^{-1} in all polyethylene and iPP samples were unaffected. The detection of two types of tert.-nitrate absorption (1302 and 1290 cm^{-1}) is intriguing, but difficult to rationalize. From a comparison with the NO products from oxidized model alkanes, this difference does not stem from the presence of isolated and adjacent -OOH groups after oxidation. The 1302 cm^{-1} band was found in tert.-butyl nitrate whereas the 1290 cm^{-1} absorption was found in models with more branched substituents. This implies that the 1290 cm^{-1} absorption results from the NO reaction of -OOH groups along the backbone, whereas the 1302 cm^{-1} absorption is from the reaction of NO with -OOH groups at chain ends or perhaps adjacent to other oxidation products such as ketones.

No reaction was observed between NO and ketones, γ-lactones, or carboxylic acids.

Hydroperoxide groups react with NO to give only nitrates as the dominant products, with only traces (≤ 5%) of nitrite in both oxidized polyolefins and in concentrated solutions of model hydroperoxides (-OOH levels from iodometry; -ONO and $-ONO_2$ levels by IR). As reported by Shelton and Kopczewski we have confirmed that both nitrate and nitrite result from NO reaction with dilute hydroperoxide solutions (24). Rather than the NO-induced O-O scission proposed by these authors, our evidence points to hydrogen abstraction by NO (reaction 4). (A similar scheme may explain nitrite formation from alcohols.) Both e.s.r. and FTIR evidence is

$$-\overset{|}{\underset{|}{C}}-OOH + NO \longrightarrow -\overset{|}{\underset{|}{C}}-OO\cdot + HNO \qquad (4a$$

$$-\overset{|}{\underset{|}{C}}-O-O\cdot + NO \longrightarrow \left[-\overset{|}{\underset{|}{C}}-OONO\right] \longrightarrow -\overset{|}{\underset{|}{C}}-ONO_2 \qquad (4b$$

consistent with reaction 4b. By e.s.r. spectroscopy, a population of $PPO_2\cdot$ radicals produced by γ-irradiation at -78°C was rapidly destroyed when NO was introduced even at -50°C. We have previously shown that $PPO_2\cdot$ radicals neither terminate nor propagate at ≤ -50°C (25). In addition, FTIR showed the formation of tert.-nitrate.

Other Gas Reactions. Several other reactive gases or vapours were examined but found to be unsatisfactory. No ester formation (~ 1745 cm^{-1}) was found when oxidatized films were exposed to acetic acid or formic acid vapour. Alcohol/carboxylic acid reactions in the solid state have often been suggested as the source of ester products, but not substantiated (4,5). Gaseous ammonia reacted with carboxylic acid groups to give absorptions at ~ 1550 cm^{-1} [-C(=O)-O⁻] and ~ 1300 cm^{-1} (NH_4^+). However, these absorptions were very broad and the method inferior to acid measurement by SF_4. Although N_2O did not react with oxidized polyolefins, the reaction of NO_2 with oxidized polyolefins was found to be rapid to give a complex mix of products, with nitrates and nitrites evident. However, large increases in carbonyl absorption also occurred, in part from hydroperoxide decomposition, but also from the direct reaction of NO_2 with unoxidized regions of the polyolefins. Several researchers have employed trifluoroacetic anhydride expsure to convert alcohols to fluoro-esters with a distinctive absorption at ~ 1790 cm^{-1} (12,13). However, we have found that hydroperoxide groups also react to give this same product. Gaseous hydrazine caused complete loss of all carbonyl absorptions, but no useful new IR absorptions were identified. Phosphine has been used successfully to convert hydroperoxides to alcohols (26). However, this reagent is too dangerous (pyrophoric) for routine use.

Product Analysis Scheme

From the above discussion, the following reactions are most useful for the quantification of oxidation products.

a) Iodometry to measure -OOH, but may also include peracids (separately quantified by CH_2N_2) and some of the more reactive dialkyl peroxides.

b) SF_4 to measure both carboxylic and peracids as the acid fluorides, with the ability to discriminate between secondary and primary acids.

c) CH_2N_2 to measure peracids as the peresters (1784 cm^{-1} increase). A cross check on carboxylic acids is also possible but imprecise because of overlap of the new ester absorption (~ 1748 cm^{-1}) with the residual complex carbonyl envelope.

d) NO to measure primary, secondary and tertiary alcohols and hydroperoxides.

e) Ketone, ester and γ-lactone oxidation products may be quantified directly from the residual, simplified carbonyl envelope which results from SF_4 exposure. The simplification results not only from the removal of all peracid and carboxylic acid absorptions (stretching from ~ 1755 to ~ 1710 cm^{-1}) but also from the loss of hydrogen bonding species (-OH, -OOH) which greatly perturb ketone and ester IR absorptions by up to 10 - 15 cm^{-1}).

Extinction coefficients for the various nitrates and nitrites are collected in Table I. Our experimentally determined IR extinction coefficients for other species are collected in Table II. It must be emphasized that the methods discussed above do not quantify all possible oxidation products. An obvious omission is the absence of a reliable method for dialkyl peroxides. An HI based method has been suggested in the literature, but proves to be extremely imprecise on film samples in our laboratories because of large reagent blanks (27).

TABLE II <u>IR Absorptions and Extinction Coefficients (ε) for Oxidation Product Identification</u>

Oxidation Product			Derivative			
Group	Absorbance	ε[a]	Reagent	Derivative Group	IR Maximum (cm^{-1})	ε[a] $l.mole^{-1} cm^{-1}$
$-CH_2-C(=O)OH$	1755 - 1713	----[b]	SF_4	$CH_2-C(=O)F$	1848	640
$-C-C(=O)OH$	1760 - 1710	----[b]	SF_4	$H-C-C(=O)F$	1841	690
$-C(=O)OH$	1760 - 1710	----[b]	CH_2N_2	$-CH_2-C(=O)OCH_3$	1747	510
$-C(=O)OOH$	1750	----[b]	CH_2N_2	$-CH_2-C(=O)OOCH_3$	1787	510
$-CH_2C(=O)-CH_3$	1724	320[a]				
$-CH_2-C(=O)-CH_2$	1718	350[a]				
-C - C- (ring: O=C–O–C, γ-lactone)	1794	720[a]				
$-CH_2-C(=0)-OCH_2$	1744	590[a]				

a) Extinction coefficient, $l.mol^{-1}cm^{-1}$, from model compounds in hexane.
b) Variable because of association effects.

Adhesion and Grafting

Peroxidic groups in oxidized polyolefins have frequently been employed as sources of free radicals to allow grafting of vinyl monomers to polyolefins (28). Some of the products from the gas reactions also have interesting potential as reactive sites. For example, chloroformate groups are well known to react with alcohols, and amines (29). Thus chloroformate groups could be useful for example in coupling highly oriented polyolefin fibres to resins such as epoxy based systems.

Literature Cited

1. Jensen, R.K., Korcek, S., Mahoney, L.R. and Zinbo, M., J. Amer. Chem. Soc. 1981 103, 1742.
2. Domke, W.D. and Steinke, H., J. Polym. Sci. Polym. Chem., 1986 24, 2701.
3. Carlsson, D.J. and Wiles, D.M., Macromolecules, 1969 2, 587-597.
4. Adams, J.H., J. Polym. Sci., 1970 A1 8, 1077.
5. Cheng, H.N., Schilling, F.C. and Bovey, F.A., Macromolecules 1976 9, 363.
6. Jelinski, L.W., Dumais, J.J., Luongo, J.P. and Cholli, A.L., Macromolecules 1984 17, 1650.
7. Mitchell, J. and Perkins, L.R., Appl. Polym. Symp. 1967 4, 167.
8. Heacock, J.F. J. Polym. Sci., 1963 7, 2319.
9. Kato, K., J. Appl. Polym. Sci., 1974 18 3087.
10. Holmstrom, A and Sorvic, E.M., J. Appl. Polym. Sci., 1974 18, 3153.
11. Everhart, D.S. and Reilley, C.N., Anal. Chem. 1981 53, 665.
12. Gerenser, L.J., Elman, J.F., Mason, M.G. and Pochan, J.M., Polymer 1985 26, 1162.
13. Rasmussen, J.R., Stedronsky, E.R. and Whitesides, G.M., J. Amer. Chem. Soc., 1977 99 4736.
14. Tabb, D.L., Sevcik, J.J. and Koenig, J.L., J. Polym. Sci., Polym. Phys. Ed., 1975 13, 815.
15. Fales, H.M., Jaouni, T.M. and Babashak, J.F., Anal. Chem. 1973 45 2302.
16. Parker, W.E., Ricciuti, C., Ogg, C.L. and Swern, D., J. Amer. Chem. Soc., 1955 77, 4037.
17. Feigl, F and Anger, V. "Spot Tests in Organic Analysis", Elsevier, Amsterdam, 1966.
18. Henman, T.J., Dev. Polym. Stab., 1985 6 107.
19. Carlsson, D.J., Brousseau, R. and Wiles, D.M., Polym. Deg. Stab., 1986 15, 67.
20. Swern, D., Witnauer, L.P., Eddy, C.R. and Parker, W.E., J. Amer. Chem. Soc., 1955 77, 5537.
21. Leonard, C., Halary, J.L. and Monnerie, L., Polymer 1985 26, 1507.
22. Nyquist, R.A. and Potts, W.J., Spectrochim. Acta., 1961 17 679.
23. Carlsson, D.J., Brousseau, R., Zhang, C. and Wiles, D.M., Polym. Deg. Stab., 1987 17, 303.
24. Shelton, J.R. and Kopczewski, R.F., J. Org. Chem. 1967 32, 2908

25. Carlsson, D.J., Dobbin, C.J.B. and Wiles, D.M., Macromolecules, 1985 18, 2092.
26. Clough, R.L. and Gillen, K.T., J. Polym. Sci., Polym. Chem. Ed., 1981 19, 2041.
27. Mair, R.D. and Graupner, A.J. Anal. Chem., 1964 36, 194.
28. Chapiro, A. "Radiation Induced Reactions", in Encyclopedia of Polymer Science and Technology, Vol. 11, Interscience, New York, 1969.
29. Shaefgen, J.R., J. Polym. Sci., 1968 Part C 24, 75.

RECEIVED October 7, 1987

CHEMICAL MODIFICATION FOR FUNCTIONALIZATION AND CURING

Introduction to Chemical Modification for Functionalization and Curing

Many chemical reactions on polymers are possible, as reflected in the breadth and depth of the associated literature. The sequence of preparation, modification, characterization, and utilization can be approached from many directions for a variety of reasons. Philosophies and methodologies vary widely, as do potential applications. The works are highly interdisciplinary.

The studies presented in this group stress functionalization and curing reactions on polymers. Even though they fit into this classification scheme, they still illustrate the above-mentioned range in approaches and in potential uses of chemical reactions on polymers. The functionalizations involve reduction, hydroxylation, hydroformylation, oxidation (including formation of double bonds), and various other chemical reactions on synthetic and natural polymers. They illustrate a number of opportunities in new chemistries, new materials, and new applications. The curing of polymers is stressed in studies on crosslinking of EPR, on thermal cure of specialized polyisoimides, and on other polymers _via_ use of reactive modifiers. Novel mechanisms and final network compositions/structures are discussed in detail. These papers nicely round out the coverage of chemical reactions on polymers.

Chapter 28

Catalytic Hydrogenation, Hydroformylation, and Hydroxymethylation of Polybutadiene: Synthesis and Characterization

N. A. Mohammadi and G. L. Rempel

Department of Chemical Engineering, University of Waterloo, Waterloo, Ontario N2L 3G1, Canada

The chemical modification of unsaturated polymers via homogeneous catalytic means offers a potentially useful method for introduction of desirable functional groups on the polymer chains.

A study involving hydrogenation of polybutadiene (PBD) (90% 1,2 addition) has been carried out and a number of partially hydrogenated PBD's with different degrees of unsaturation have been synthesized by monitoring hydrogen consumption. Polymer characterization has been carried out using I.R. and ^{1}H N.M.R. spectroscopy. The dilute solution viscosity measurements in toluene showed that the intrinsic viscosity of partially hydrogenated polybutadiene (HPBD) decreases linearly with the % hydrogenation.

Hydroformylation and subsequent hydrogenation of C=C and -CHO groups of PBD appear to be an appropriate means whereby a pendent hydroxy group can be introduced onto the polymer backbone. A variety of partially hydroformylated (2-20%) and hydroxymethylated polymers have been synthesized by a two-step catalytic reaction and characterized by I.R. and ^{1}H N.M.R. spectroscopy. As expected, the hydrophilic group, OH, in the polymer resulted in a greater decrease in the intrinsic viscosity as compared to the HPBD.

The chemical modification of polymers is a post polymerization process which is used in certain situations: i) to improve and optimize the chemical and mechanical properties of existing polymers or; ii) to introduce desirable functional groups in a polymer.

One of the ways to synthesize specialty polymers with certain desirable functional groups is by the polymerization of monomers with the desirable functional groups. However there are often a number of difficulties associated with polymerization of these

0097-6156/88/0364-0393$06.00/0

monomers such as: (i) difficulty in polymerization; (ii) side reactions during polymerization; (iii) difficulty in the preparation of desirable monomers; (iv) and in the case of copolymerization, unfavourable reactivity ratios. Alternatively, catalytic chemical reactions on unsaturated polymers, in which the desirable functional groups are introduced into available unsaturation sites in polymers such as polybutadiene, can be employed to synthesize certain specialty polymers. Since very stereospecific polybutadienes are available with desirable molecular weights and molecular weight distributions, the products obtained by the introduction of desirable functional groups have the same stereoregularity and chain length properties, provided that side reactions such as crosslinking, isomerization etc. are avoided.

The hydrogenation of unsaturated polymers and copolymers in the presence of a catalyst offers a potentially useful method for improving and optimizing the mechanical and chemical resistance properties of diene type polymers and copolymers. Several studies have been published describing results of physical and chemical testing of saturated diene polymers such as polybutadiene and nitrile-butadiene rubber (1-5). These reports indicate that one of the ways to overcome the weaknesses of diene polymers, especially nitrile-butadiene rubber vulcanizate, is by the hydrogenation of carbon-carbon double bonds without the transformation of other functional unsaturation such as nitrile or styrene.

Hydroformylation and subsequent hydrogenation of polybutadiene appears to be an appropriate means whereby pendent hydrophilic or reactive groups can be introduced onto the polymer hydrophobic chain. Previously there have been a number of publications on the hydroformylation and hydroxymethylation of polydienes. Catalytic methods for the preparation of polydienes containing a formyl group was reported by Lenz and coworkers (6). Hydroformylation and subsequent hydrogenation at relatively higher temperature and pressure in the presence of a cobalt catalyst has also been disclosed (7). Azuma et. al. (8) has reported a two step hydroxymethylation of polydienes *via* catalytic hydroformylation and subsequent hydrogenation using reducing agents such as sodium borohydride. So far there appears to have been no report on the catalytic hydroxymethylation of a polydiene at relatively lower temperature and pressure conditions.

The purpose of this paper is to report catalytic synthetic methods for the preparation of hydrogenated (partially and completely), hydroformylated and hydroxymethylated polybutadiene. The results from the characterization of some of the product polymers is also presented. For the chemical and microstructural characterization, I.R. and N.M.R. spectroscopic techniques were used. For the chain length properties characterization, vapor pressure osmometry, dilute solution viscometry and gel permeation chromatography were employed.

Although only the hydrogenation of PBD is presented in this paper, the reaction conditions were also found to be effective for the selective and quantitative hydrogenation of C=C (of butadiene repeating units) in NBR and SBR type copolymers.

Experimental

Apparatus. Since all the polymer modification reactions presented in this paper involved gas consumption, an automated gas consumption measuring system was designed, fabricated and used to keep constant pressure and record continuously the consumption of gas in a batch type laboratory scale reactor. Process control, data acquisition, and analysis was carried out using a personal computer (IBM) and an interface device (Lab-master, Tecmar Inc.).

The overall system essentially consists of two separate systems (i) a constant pressure reactor system and (ii) a gas monitoring system. A schematic flow diagram for the overall system is provided in Figure 1.

(i) Constant pressure system - The constant pressure system is easily identifiable in Figure 1. A drop in pressure in the Reactor (due to gas consumption) is detected by the differential pressure transducer (PT-1) and is compared with the pressure in tube AB. The pressure in the tube AB serves as a reference pressure, which can be set using the on/off valve V-1 prior to the start of reaction. A signal is generated by the PT-1 in the form of an analog voltage $\pm$ 10 volts DC which will be referred to hereafter as the measured signal. The measured signal is sent to the direct digital controller consisting of an IBM PC and a Lab-master (Tecmar Inc.) interface. The measured signals are converted to digital signals, processed by a control algorithm in the PC and converted back to analog signals. The digital/analog convertor (D/A-0) is coupled with a zero-order hold element, having high impedance output. The manipulated signals from the hold element are conditioned by a signal conditioner V/V-0 in order to make them suitable for use with low impedance elements such as control valves or relays. The manipulated signals are then fed back to an orifice control valve (CV-1) which allows the necessary amount of gas into the reactor and readjusts the pressure accordingly. As a safety precaution the magnitude of the manipulated signals are limited to a certain range outside of which a warning alarm is triggered.

(ii) Gas monitoring system - A simple principle of measurement of a drop in pressure in a gas holder (shown as HOLDER in Figure 1) due to flow of gas from it to the constant pressure system is employed in monitoring the gas consumption. The consumption of gas in the constant pressure reactor system results in the opening of CV-1 by the feed-back loop mentioned above. This results in a decrease in pressure in the gas holder which is detected by a differential pressure transducer PT-2 and compared with pressure in the tube CD. A signal is generated with a full scale of 0 - 10 V DC. The signal is sent to the direct digital controller with subsequent translation into moles of gas consumed. This on-line data is then stored on a floppy disk pending further analysis. An on line representative hydrogen consumption plot is shown in Figure 2.

A detailed description of hardware, software and mathematical analysis of the computer controlled batch reactor system is presented in reference (9).

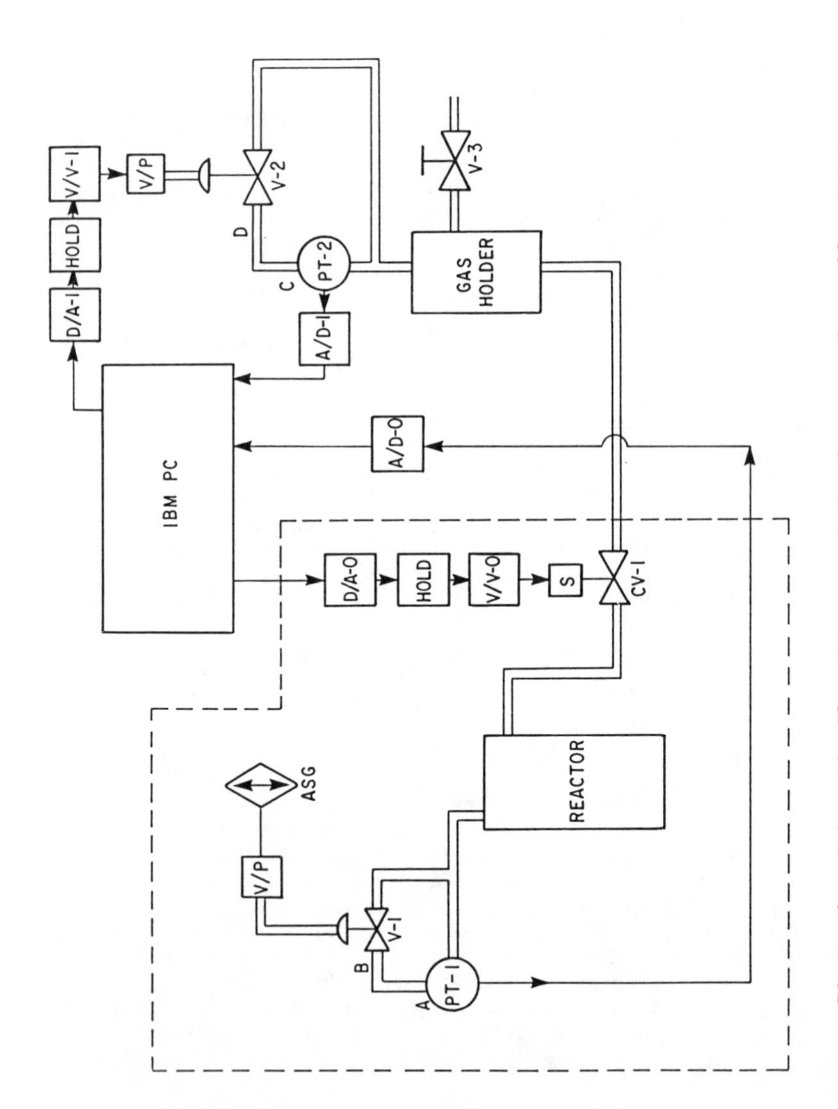

Figure 1. Schematic Flow Diagram of the Computer Controlled Batch Reactor System. (Reproduced with permission from Ref. 9. Copyright 1987 Pergamon Journals Ltd.)

Materials. Solvents - Toluene, the solvent used in all the modification reactions, was obtained from Aldrich and distilled over sodium under nitrogen. 2-propanol was used as obtained from Aldrich.

Catalysts - $RhCl(P(C_6H_5)_3)_3$ and $RhH(CO)(P(C_6H_5)_3)_3$ were prepared by the methods described in references (10) and (11) respectively. For the synthesis of $RuClH(CO)(P(C_6H_5)_3)_3$, the technique described by Ahmad et. al. (12) was employed.

Polybutadiene - Several polybutadienes with different molecular weights (Mn 9,000 to 50,000) and 90% 1,2 addition units were used for the modification reactions.

Characterization. Infrared Spectroscopic Analysis - The infrared spectra of all the chemically modified polybutadienes were obtained using a P.E. 1330 Infrared Spectrophotometer. The samples were prepared by casting polymer films on NaCl plates.

N.M.R. Spectroscopic Analysis - The ^1H N.M.R. spectra of all the polymers were recorded using a Bruker 250 MHz instrument. All the analyses were carried out in solution using benzene-d_6 as solvent.

Dilute Solution Viscometry - The hydrogenated and hydroformylated (10%) PBD were completely soluble in toluene. Intrinsic viscosity measurements were carried out in toluene at 30°C using a Cannon-Ubbelohde viscometer.

Vapor Pressure Osmometry - The number average molecular weights of polymers with Mn < 20,000 were determined using "Model 232A Molecular Weight Apparatus, Wescan Instruments, Inc., 3018 Scott Blvd., Santa Clara, CA 95050". Toluene was used as solvent and the instrument was calibrated using polystyrene of Mn 9,000 and 20,400.

Gel Permeation Chromatography - The gel permeation chromatograms of hydrogenated and hydroxymethylated polybutadienes were obtained using a Waters GPC ALC/301 instrument. The instrument conditions employed were; solvent (tetrahydrofuran), Column (Styragel, Permeability Range 10^3, 10^4, 10^5, 10^6 Å), Temperature (30°C), and detector (DRI).

Synthetic Methods. Hydrogenation of Polybutadiene - The hydrogenation of Polybutadiene was carried out using the computer controlled batch reactor described above at the following reaction conditions: PBD = 850 mol/m^3, $RhCl(P(C_6H_5)_3)_3$ = 3.14 mol/m^3, H_2 pressure (total) = 30 psig, Toluene = 75 ml, Temperature = 60°C.

For the partially hydrogenated polymers, the reaction was quenched by adding 2-propanol.

Hydroformylation of Polybutadiene - A number of hydroformylation experiments were performed using the automated batch reactor system in order to synthesize the hydroformylated polybutadiene with varying degree of reaction completion (2 to 20% of the total C=C present). The following reaction conditions were employed: PBD = 1.55×10^3 mol/m^3, $RhH(CO)(P(C_6H_5)_3)_3$ = 0.58 mol/m^3, CO : H_2 (1 : 1) = 30 psig, Toluene = 75 ml, Temperature = 60°C.

The extent of the reaction was estimated by monitoring the amount of CO + H_2 consumption. The hydroformylated PBD solution, in the form of reaction mixture in toluene, was stable. However, when the hydroformylated PBD was precipitated out using 2-propanol, the coagulated polymer was not soluble in most of the solvents including toluene, benzene, hexane and chloroform.

Hydroxymethylation of Polybutadiene - The hydroformylated PBD reaction solution obtained above was subsequently hydrogenated directly using the following reaction conditions: Hydroformylated PBD solution = 75 ml, $RuClH(CO)(P(C_6H_5)_3)_3$ = 1.05 mol/m^3, Hydrogen pressure = 600 psig, Temperature = 120°C.

During the course of reaction, samples were withdrawn at suitable time intervals from the reactor and analyzed by infrared spectroscopy. After the completion of reaction, the hydroxymethylated PBD was separated out by adding 2-propanol to the final reaction solution. The product polymer (2-20% -OH functional groups) obtained was dried under vacuum and was soluble in a variety of solvents including toluene, benzene, hexane and chloroform. Since one mole of -CHO forms one mole of -OH functional group, the -OH content in the polymer chains was readily estimated from the known -CHO functional group content.

Results and Discussion

Chemical and Microstructure of Product Polymers. The I.R. spectrum of polybutadiene is shown in Figure 3. The bands at 1640, 910, 970 and 992 cm^{-1} are due to C=C unsaturation. The I.R. spectrum of completely hydrogenated PBD indicated that all the characteristic peaks for unsaturation disappeared as shown in Figure 4. Figure 5 shows the I.R. spectrum of 5% hydroformylated PBD. Beside the peaks for the unsaturation, the characteristic peaks for the formyl group i.e.; C=O stretching at 1727 cm^{-1} and C-H stretching (of -CHO) at 2700 cm^{-1} confirm the presence of formyl groups. The I.R. spectrum of hydroxymethylated PBD is given in Figure 6. The broad peak at 3310 cm^{-1} is characteristic of the OH stretching vibrations. All the peaks for C=C and -CHO groups disappeared, thus confirming a quantitative hydrogenation.

The ^{1}H N.M.R. spectrum of polybutadiene is given in Figure 7. The peaks in the range of 5.0 to 5.5 ppm are characteristic of olefinic protons and in the range of 1 to 2.3 ppm are due to paraffinic protons. The ^{1}H N.M.R. spectrum of completely hydrogenated PBD is shown in Figure 8. The absence of peaks in the range of 5.0 to 5.5 ppm confirms a quantitative hydrogenation. A ^{1}H N.M.R. spectrum of partially hydroformylated PBD is shown in Figure 9. Beside the peaks due to unsaturation, a new peak at 9.45 ppm can be seen here. This peak is identified as due to the ^{1}H resonance of the formyl group, which confirms the presence of -CHO group. The ^{1}H N.M.R. spectrum of hydroxymethylated PBD is shown in Figure 10. A triplet at 3.62 ppm, which is due to the methylene protons adjacent to the hydroxyl group, can be seen in this spectrum.

The overall catalytic reactions involved in the two step synthesis of hydroxymethylated PBD are given by Equations 1-6. It can be seen from reactions (1-6) that the addition of -CHO functional group can occur in three ways resulting in the formation of three types of hydroformylated PBD repeating units as shown by species A, B or C. Subsequent hydrogenation of units A, B and C result in the formation of hydroxymethylated PBD repeating units A',

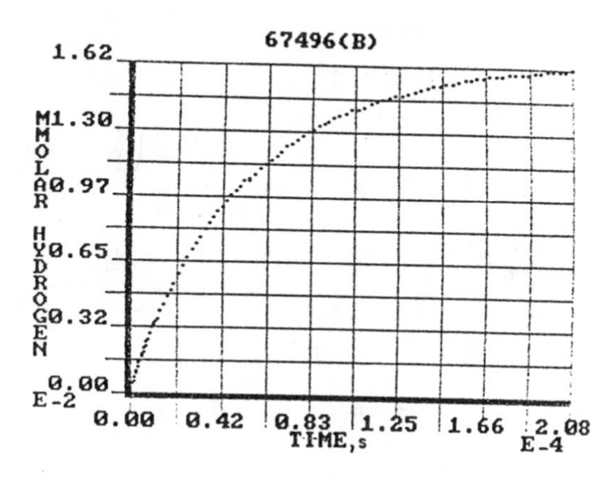

Figure 2. Representative Hydrogen Consumption Plot. (Reproduced with permission from Ref. 9. Copyright 1987 Pergamon Journals Ltd.)

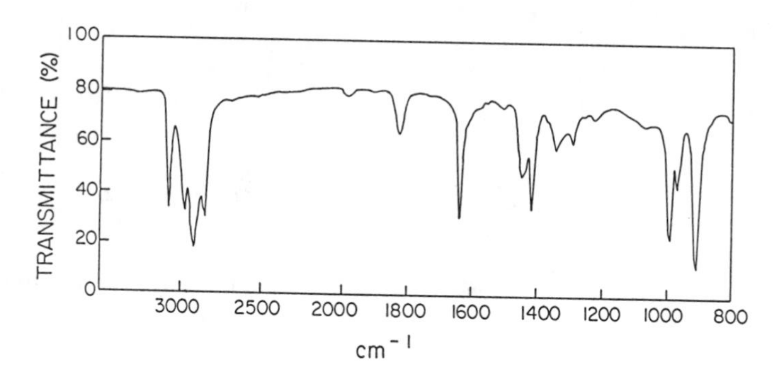

Figure 3. Infrared Spectrum of Polybutadiene

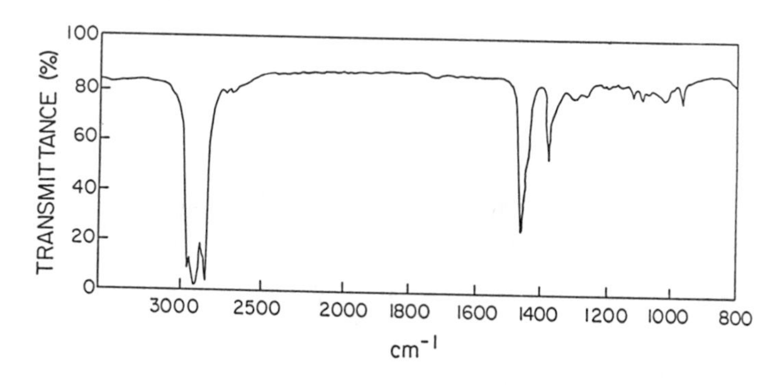

Figure 4. Infrared Spectrum of Hydrogenated Polybutadiene

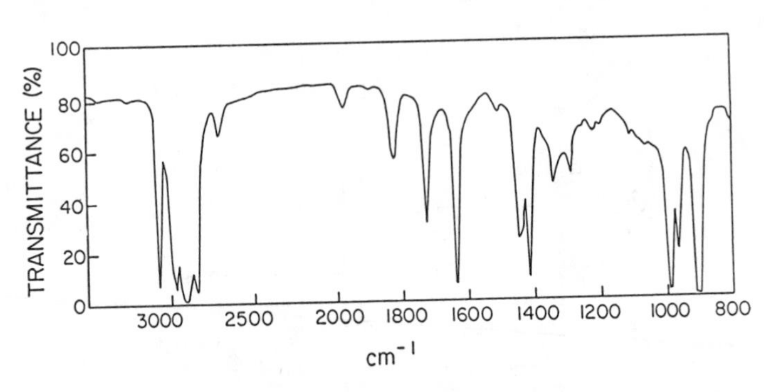

Figure 5. Infrared Spectrum of Hydroformylated Polybutadiene

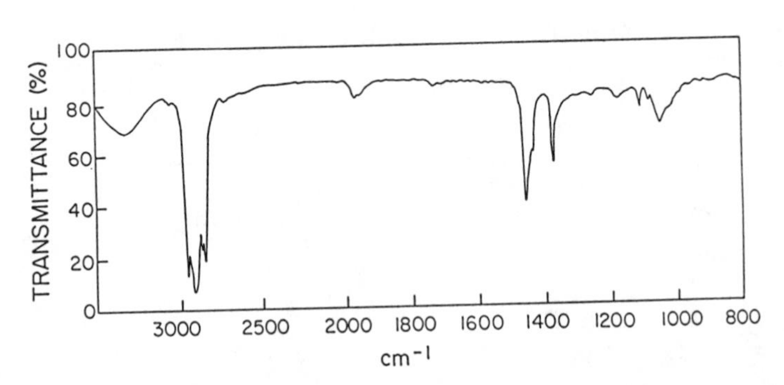

Figure 6. Infrared Spectrum of Hydroxymethylated Polybutadiene

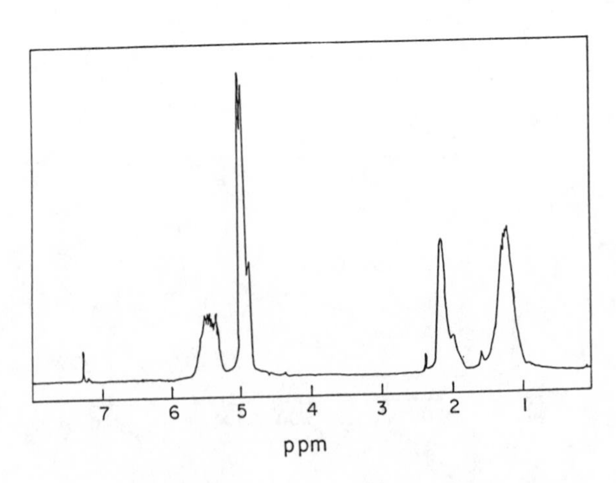

Figure 7. ^{1}H N.M.R. Spectrum of Polybutadiene

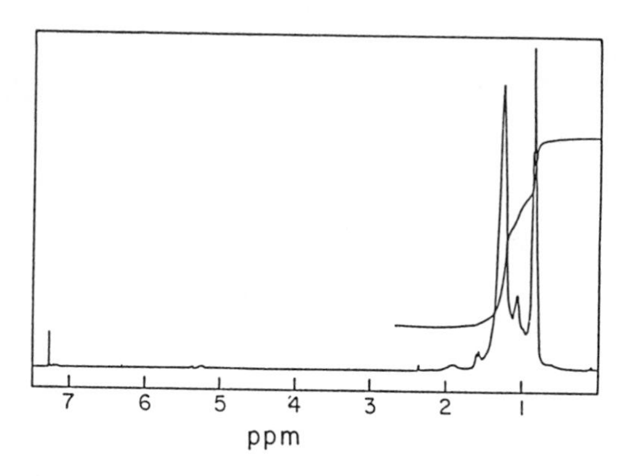

Figure 8. ^{1}H N.M.R. Spectrum of Hydrogenated Polybutadiene

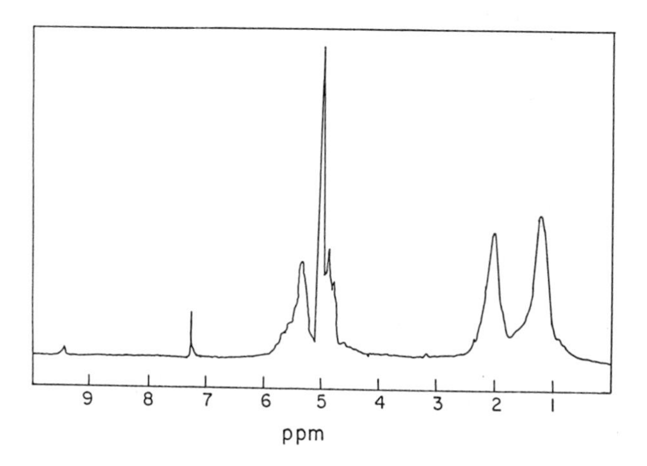

Figure 9. ^{1}H N.M.R. Spectrum of Partially Hydroformylated Polybutadiene

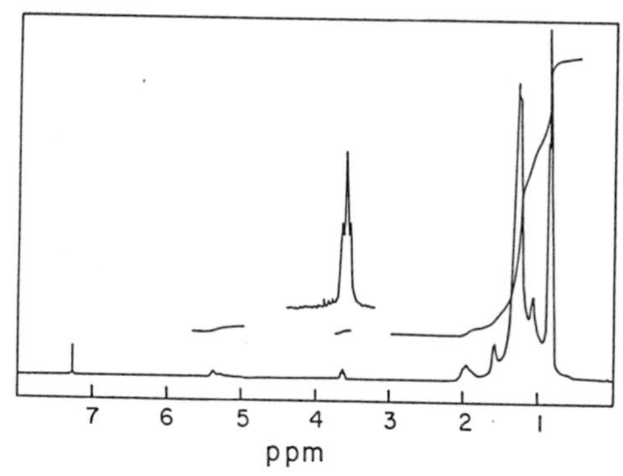

Figure 10. ^{1}H N.M.R. Spectrum of Hydroxymethylated Polybutadiene

B' and C' respectively. The contents of units A', B' or C' depend on the reaction conditions employed in the hydroformylation step. One of the most important factors for the selectivity for the formation of A, B, or C is the nature of the OXO catalyst employed to carry out the hydroformylation reaction.

$$\text{-}[CH_2\text{-}CH{=}CH\text{-}CH_2]\text{-} \xrightarrow[\text{cat. 1}]{H_2/CO} \text{-}[CH_2\text{-}C(H)(CHO)\text{-}CH_2\text{-}CH_2]\text{-}\,\text{-}[CH_2\text{-}CH{=}CH\text{-}CH_2]\text{-} \quad (1)$$

(1,4 PBD) (A)

$$\text{-}[CH_2\text{-}CH(CH{=}CH_2)]\text{-} \xrightarrow[\text{cat. 1}]{H_2/CO} \text{-}[CH_2\text{-}CH(CH(CH_3)\text{-}CH{=}O)]\text{-}\,\text{-}[CH_2\text{-}CH(CH{=}CH_2)]\text{-} \quad (2)$$

(1,2 PBD) (B)

$$\text{or} \quad \text{-}[CH_2\text{-}CH(CH_2\text{-}CH_2\text{-}HC{=}O)]\text{-}\,\text{-}[CH_2\text{-}CH(CH{=}CH_2)]\text{-} \quad (3)$$

(C)

$$\text{-}[CH_2\text{-}CH(HC{=}O)\text{-}CH_2\text{-}CH_2]\text{-}\,\text{-}[CH_2\text{-}CH{=}CH\text{-}CH_2]\text{-} \xrightarrow[\text{cat. 2}]{H_2} \text{-}[CH_2\text{-}CH(CH_2\text{-}OH)\text{-}CH_2\text{-}CH_2]\text{-} \quad (4)$$

(A')

$$\text{-}[CH_2\text{-}CH_2\text{-}CH_2\text{-}CH_2]\text{-}$$

$$\text{-}[CH_2\text{-}CH(CH(CH_3)\text{-}C(H){=}O)]\text{-}\,\text{-}[CH_2\text{-}CH(CH{=}CH_2)]\text{-} \xrightarrow[\text{cat. 2}]{H_2} \text{-}[CH_2\text{-}CH(CH(CH_3)\text{-}CH_2OH)]\text{-}\,\text{-}[CH_2\text{-}CH(CH_2\text{-}CH_3)]\text{-} \quad (5)$$

(B')

$$\left[CH_2\text{-}\underset{\underset{\underset{\underset{O=C-H}{|}}{H_2-C}}{|}}{\underset{CH_2}{|}}{CH}\right] \left[CH_2\text{-}\underset{\underset{\underset{CH_2}{\|}}{CH}}{|}}{CH}\right] \xrightarrow[\text{cat. 2}]{H_2} \left[CH_2\text{-}\underset{\underset{\underset{\underset{H_2C-OH}{|}}{CH_2}}{|}}{\underset{CH_2}{|}}{CH}\right] \left[CH_2\text{-}\underset{\underset{CH_3}{|}}{\underset{CH_2}{|}}{CH}\right] \quad (6)$$

(C')

where cat. 1 = $RhH(CO)(P(C_6H_5)_3)_3$

cat. 2 = $RuClH(CO)(P(C_6H_5)_3)_3$

It was mentioned above that the ^1H N.M.R. spectrum of 5% hydroxymethylated PBD sample (see Figure 10) shows a triplet at 3.62 ppm, which is due to the methylene protons adjacent to the hydroxyl group. From the observed line shape, splitting and ca. ^3J (HH) of 6.5 Hz, it can be concluded that the predominant structure of the hydroxymethylated PBD repeating unit is C'. The presence of only structure C' in the 5% hydroformylated PBD suggests the formation of structure C in the first i.e. hydroformylation step.

The formation of only structure C in the hydroformylation step can be explained by considering the mechanism which has previously been suggested for the $RhH(CO)(P(C_6H_5)_3)_3$ catalyzed hydroformylation of olefins as shown in Scheme (1). In the hydroformylation of simple alkenes, the catalyst $RhH(CO)(P(C_6H_5)_3)_3$ has been reported to give both straight and branched chain aldehydes (13). The ratio of straight to branched chain aldehydes is thought to be determined largely by steric factors. Observations of only straight chain aldehyde (at 5% reaction completion) from the hydroformylation of polybutadienes can be explained also in terms of steric factors. Structure 1 shows the transition state which gives rise to the alkyl-rhodium intermediate.

For the internal C=C (1,4 addition units), i.e. when R (or R') and R'' (or R''') are polymer chains, the transition state is highly unfavourable. The failure to observe species A' (i.e. A in step 1), at 5% conversion, confirms that the steric hindrance is too large to allow formation of the alkyl-rhodium intermediate.

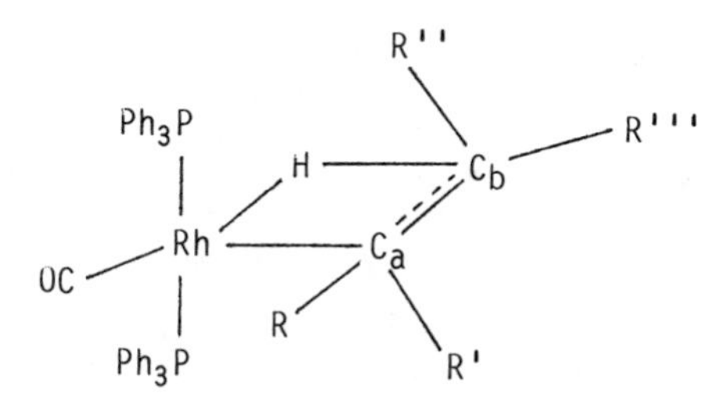

Structure 1: Transition state which leads to the formation of an alkyl-rhodium intermediate.

In the case of terminal C=C (1,2 addition units), i.e. when R=R'=H and R'' (or R''') = polymer chain, two types of hydride migration are possible, namely (i) The Markownikoff's addition which would lead to the formation of B type repeating units and (ii) The anti Markownikoff's addition which would result in the formation of the observed repeating units C. In the case of Markownikoff's type addition the hydride transfer occurs to Ca and results in the formation of branched alkyl-rhodium intermediate complex shown by Structure 2. Whereas when anti Markownikoff's addition occurs, the resulting intermediate alkyl-rhodium complex has linear alkyl ligand as shown by Structure 3.

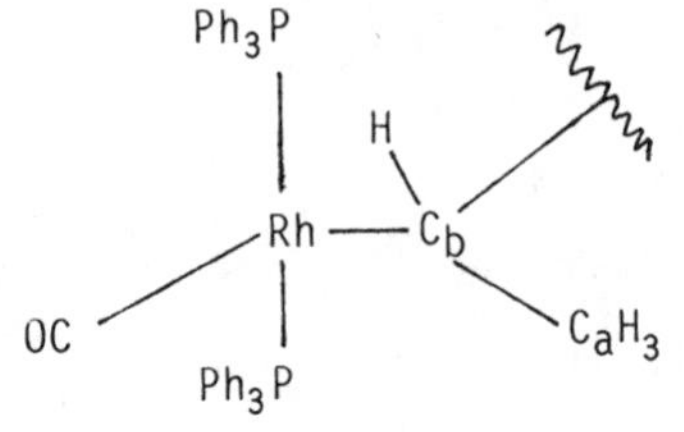

Structure 2: Alkyl-rhodium intermediate formed via Markownikoff's type addition.

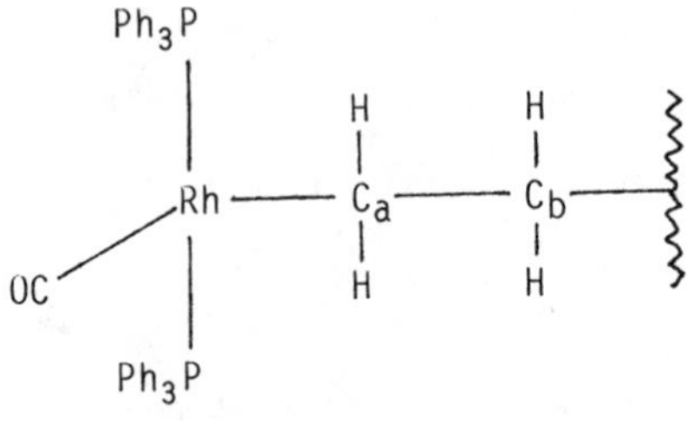

Structure 3: Alkyl-rhodium intermediate formed via anti Markownikoff's type addition.

It can be seen from Structures 2 and 3 that Structure 2 is sterically less favourable due to the two bulky triphenylphosphine ligands surrounding the alkyl ligand. In the case of Structure 3, the linear alkyl will allow the main bulk of the polymer to remain outside the immediate coordination sphere of the rhodium center. Thus the favourable linear alkyl-rhodium transition state will eventually give rise to a straight chain aldehyde. Thus it can be concluded that the observations of high selectivity in the hydroformylation of PBD to form repeating unit C (and corresponding hydroxymethylated PBD repeating unit C') can be attributed to the steric bulk of the polymer chains.

<u>Chain Length Properties of the Modified Polymers.</u> A number of partially hydrogenated and hydroxymethylated polybutadienes were analyzed using vapour pressure osmometry, dilute solution viscometry and gel permeation chromatography. The parent polybutadiene had Mn in the range of 9,000 to 50,000. In the case of vapour pressure osmometry, the data were reproducible for polymers with Mn less than 20,000. All the polymers obtained (hydrogenated and hydroxymethylated) were completely soluble in toluene and trichlorobenzene at room temperature.

The intrinsic viscosities and Mn (from vapour pressure osmometry) of a number of chemically modified polymers are listed in Table I. Figure 11 shows a plot of intrinsic viscosity versus degree of hydrogenation of polybutadiene. The GPC chromatograms of hydrogenated and hydroxymethylated (5%) polybutadienes are given in Figure 12.

Table I. Intrinsic Viscosities and Mn of Chemically Modified Polymers

	PBD (Mn 9,000)	
% hydrogenated	intrinsic viscosity dL/g	Mn by Vapour Pressure Osmometry
0	0.443	7727
50	0.384	9274
100	0.267	8139
	PBD (Mn 30,000, 90% 1,2 addition)	
% hydrogenated	% hydroxymethylated	intrinsic viscosity dL/g
0	0	0.684
25	0	0.636
50	0	0.574
75	0	0.516
100	0	0.498
	PBD (Mn 20,000, 90% 1,2 addition)	
% hydrogenated	% hydroxymethylated	intrinsic viscosity dL/g
0	0	0.569
100	0	0.346
85	15	0.248

As seen from Table I, no significant change in Mn (from VPO) occurred as a result of hydrogenation. This implies that no change in large-scale molecular structure (chain scission, crosslinking, etc.) took place during the hydrogenation process. The intrinsic viscosity however, shows a linear decrease with an increase in degree of hydrogenation (see Figure 11). If no change in chain length characteristic took place during hydrogenation, the decrease in intrinsic viscosity can be attributed to the change in solvent-polymer interactions as a result of hydrogenation. This

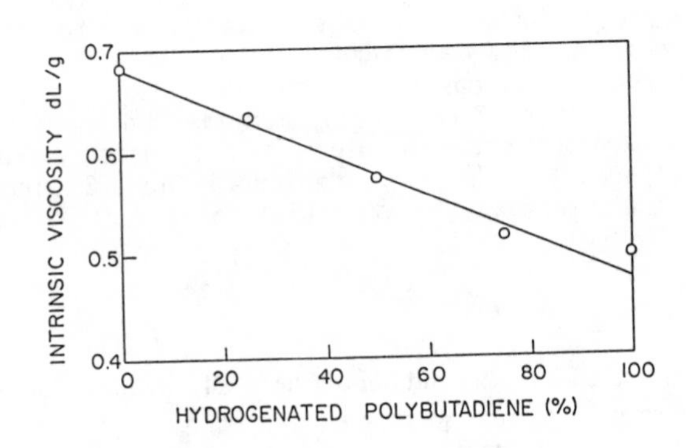

Figure 11. Intrinsic Viscosity versus % Hydrogenated PBD (Mn 30,000 90% 1,2)

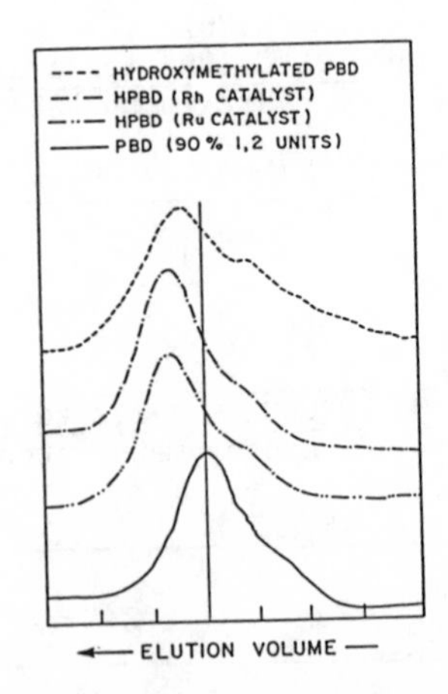

Figure 12. Gel Permeation Chromatograms of Chemically Modified PBD (90% 1,2 units)

phenomenon can be explained on the basis of decrease in hydrodynamic volume of polymer chains with a decrease in carbon-carbon double bond content. The chain segments in a polymer molecule are capable of undergoing rotations about single bonds. In the case of double bonds, such rotations are not possible. For a given chain length, a decrease in C=C content would result in an increase in chain segment rotations about the single bonds. This results in a more compact orientation of saturated polymer chains as compared to the unsaturated polybutadiene chains. Since the hydrodynamic volume is a measure of solvated polymer chain volume, a decrease in double bonds would decrease the hydrodynamic volume. This in turn, results in a decrease in the intrinsic viscosity of polymer solution.

In the case of hydroxymethylated polybutadiene, a greater decrease in the intrinsic viscosity as compared to the saturated polybutadiene was observed (see Table I). This can be attributed to the greater hydrophilic character of the -OH group on the polymer chains.

Chromatograms obtained in THF at room temperature of hydrogenated and hydroxymethylated polybutadienes, as illustrated in Figure 12, show no significant change in the molecular weight distribution. However a shift in peak maximum can be seen for the hydrogenated and hydroxymethylated polybutadienes. However, as already mentioned above, no large-scale change in chain length characteristic appears to have occurred during hydrogenation. Hence this shift in peak maximum can be due to a decrease in hydrodynamic volume of solvated polymer chains as a result of change in solvent-polymer interactions. Due to non-availability of Mark-Houwink constants (K and a) for the chemically modified polymers, a more quantitative treatment of this observation is not possible at this stage.

Conclusions

i) At the given reaction conditions for the hydrogenation of PBD, no change in the chain length properties occurred.

ii) For the chemical modification reactions involving gaseous reactants, the extent of reaction can be accurately controlled by monitoring the amount of gas consumed.

iii) $RhCl(P(C_6H_5)_3)_3$ and $RhH(CO)(P(C_6H_5)_3)_3$ are efficient catalysts for the hydrogenation and hydroformylation of C=C unsaturation in polybutadienes respectively.

iv) $RuHCl(CO)(P(C_6H_5)_3)_3$ is an efficient catalyst for quantitative hydrogenation of -CHO and C=C groups in polymers.

Acknowledgment

We wish to thank Dr. A.F. Halasa of Goodyear Tire and Rubber Company (Akron, Ohio) for supplying the polybutadiene material which made this work possible.

We would also like to thank Ms. Christine Keary for carrying out viscosity measurement experiments. Support of this work by a grant from the Natural Science and Engineering Research Council of Canada is greatly appreciated.

References

1. Thoermer, J., Mirza, J. and Shoen, N. Elastomerics, Sept. 1986.
2. Hashimoto, K., Watanabe, N., and Yoshioka, A. Rubber World, May 1984, 190(2).
3. Finch, A.M.T., Jr. U.S. Patent 3 700 637, 1970.
4. Weinstein, A.H. Rubber Chem. and Technol. 1984, 57(1), 203.
5. Krigas, T.M., Carella, J.M., Struglenski, M.J., Crist, B., and Graessley, W.W., J. Polym. Sci. 1985, Polym. Phy. Ed., 23, 509.
6. Sanui, K., KacKnight, W.J. and Lenz, R.W. Macromolecules 1974, 7, 952.
7. Esso R and E Co., Brit. Patent 1 072 796, 1967.
8. Azuma, C., Mitsuboshi, T., Sanui, K. and Ogata, N., J. Polym. Sci. 1980, Polym. Chem. Edition, 18, 781.
9. Mohammadi, N.A. and Rempel, G.L., Comput. Chem. Engng. 1987, 11, 27.
10. Osborn, J.A., Jardine, F.H., Young, J.F., and Wilkinson, G. J. Chem. Soc. 1966, (A), 1711.
11. Ahmad, N., Levison, J.J., Robinson, S.D., and Uttley, M.F. Inorg. Synth. 1974, 15, 59.
12. Ahmad, N., Levison, J.J., Robinson, S.D., and Uttley, M.F. Inorg. Synth. 1974, 15, 45.
13. Gates, B.C., Kartzer, J.R., and Schuit, G.C.A. In Chemistry of Catalytic Processes 1979, McGraw-Hill.

RECEIVED September 2, 1987

Chapter 29

Reactive Modifiers for Polymers

S. Al-Malaika

Department of Molecular Sciences, Aston University, Aston Triangle, Birmingham B4 7ET, England

Commercial processing operations of hydrocarbon polymers involve an initial mechanochemical stage resulting in the formation of unstable free radicals which are normally damaging to the polymer chain and sensitise subsequent thermal- and photo- oxidation. However, in the complete absence of oxygen and/or any other free radical trap, macroalkyl radicals initially formed can be harnessed to advantage. Thus, the processing machine is used as the "reaction vessel" and the mechanochemically formed macroalkyl radicals are made to react with properly functionalised reactive modifiers. These reactions were used to produce a polymer with improved properties, e.g., with better adhesion to reinforcing fibres, as in the case of high performance polymer composites for demanding applications. Alternately, reactions of macroalkyl radicals are exploited to enhance additive performance, e.g., by forming chemically bound antioxidant function, to render greater polymer stability under aggresive environments.

Reactive macroalkyl radicals are formed during stress-initiated scission of the polymer backbone occassioned by the application of mechanical shear during industrial processing of thermoplastic polymers. These radicals undergo further reactions with other species or reactive sites, most important of which is molecular oxygen (dissolved or trapped in the polymer feed), with deleterious consequences.

0097–6156/88/0364–0409$06.00/0

The term "mechanochemical degradation" is used here to describe chain scission of polymer backbone through the application of shearing forces on polymer melts during processing operation. Over 45 years ago Kauzman and Eyring (1) were the first workers to infer from rheological changes in polymers during shearing that mechanical scission of carbon-carbon bonds occurs in the polymer backbone. Subsequent studies by Watson and co-workers in the 1950's (2,3) showed that, in the absence of oxygen, macroalkyl radical produced by mechanochemical scission of the polymer chain could be used to initiate the polymerisation of an added vinyl monomer to form block copolymers.

Effectively, performance of additives, e.g., antioxidants and modifiers, in polymers is a function of a complex relationship involving the following factors,

(1) intrinsic activity of additive during manufacture, processing and in-service,

(2) compatibility and limit of solubility of additives in the polymer melt and in the finished product, and

(3) diffusion properties of additives under aggressive environments.

Although chemical and structural features are critical for **1** above, the physical characteristics of additives as required in **2** and **3** are more challenging. In the pursuit of competitive, more cost-effective speciality chemicals (antioxidants and modifiers) for specialised and demanding (high temperature and stress) engineering applications, **2** and **3** present the ultimate and critical test for additive substantivity. The amount of additive (initially added or its transformation products) which remains in the polymer after allowing for losses through volatilisation and extraction by solvent (loss from surface) and migration (in the presence or absence of solvent) to surface (depletion from bulk) is the effective molar mass of additive responsible for the ultimate properties of the finished article.

To circumvent the inherent problem of loss of additives, different approaches were developed but in all cases emphasis is placed on those additives which initially comply with **1** above. In the case of antioxidants, for example, the trend over the past two decades has been towards the development of oligomeric antioxidants of sufficient solubility in the host polymer. In the 1960's, Thomas and co-workers (4) have carried out pioneering and systematic approaches to the production of oligomeric antioxidants for high temperature rubbers. In the 1970's, a number of oligomeric light strabilisers which are much less readily lost from polymers under aggressive environments were developed. It was later suggested (5), however, that although such oligomeric antioxidants are more substantive at high temperatures they can still be lost by solvent leaching. During the last decade a successful approach was developed by Vogl (6) where polymeric and polymerisable uv stabilisers of several types were synthesised and, subsequently, homopolymerised and copolymerised with monomers such as styrene and methyl methacrylate.

Another approach was developed by Scott in the 1970's (7,8) which utilises the same mechanochemistry used previously by Watson to initiate the "Kharacsh-type" addition of substituted alkyl mercaptans and disulphides to olefinic double bonds in unsaturated polymers. More recently, this approach was used to react a variety of additives (both antioxidants and modifiers) other than sulphur-containing compounds with saturated hydrocarbon polymers in the melt. In this method, mechanochemically formed alkyl radicals during the processing operation are utilised to produce polymer-bound functions which can either improve the additive performance and/or modify polymer properties (Al-Malaika, S., Quinn, N., and Scott, G; Al-Malaika, S., Ibrahim, A., and Scott, G., Aston University, Birmingham, unpublished work). This has provided a potential solution to the problem of loss of antioxidants by volatilisation or extraction since such antioxidants can only be removed by breaking chemical bonds. It can also provide substantial improvement to polymer properties, for example, in composites, under aggresive environments.

Properly functionalised additives can react with polymer substrates to produce polymer-bound functions which are capable of effecting the desired modification in polymer properties, hence the use of the term "reactive modifiers". As an integral part of the polymer backbone, reactive modifiers are useful vehicles for incorporating the desired chemical functions to suit the specialised application. Being molecularly dispersed, the problem of solubility expressed under 2 above is avoided. Implicitly, the bound-nature of the function is not subjected to the normal problems of the loss of additives from the surface which are common with both high and low molecular mass additives. The bound nature of the function must be fully defined for the conditions of service.

SYNTHETIC APPROACHES

Current interest in reactive modifiers is in the areas of polymer property enhancement and improvement in additive performance. Reactive modifiers can be incorporated into commercial polymers by,

(1) copolymerisation of the reactive function of the modifier into polymers during their manufacture (synthesis).

(2) "tieing" the reactive modifier into polymer chain after their manufacture.

Researchers at Goodyear (9) have successfully exploited route 1 above to produce antioxidant-modified SBR and NBR; for example, Chemigum HR 665, is a commercial grade of nitrile rubber containing a small proportion (less than 2%) of a copolymerised monomer (**1**). It is far more resistant to high temperature and hot lubricating oils than conventionally stabilised nitrile rubber. Cost is, however, a deterrent in this approach especially when a new speciality polymer is produced in relatively small quantities for each end use.

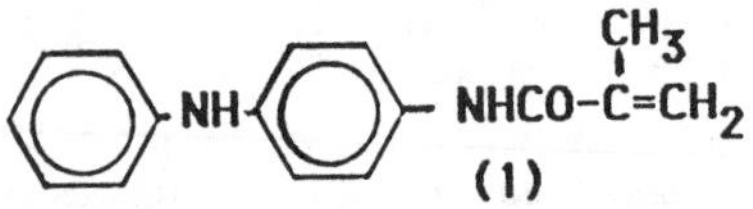

CH_3

NH NHCO-C=CH_2

(1)

Tieing of reactive modifiers into polymers after their manufacture (via route **2**), on the other hand, can be achieved (5,10) by either,

(i) carrying-out a separate chemical reaction, e.g., in rubber latices, using traditional redox initiator technology. This is limited to polymers which are manufactured as latices, e.g., emulsion polymers, or,

(ii) melt processing, where the high shearing action during the processing operation is utilised to promote the reaction between the reactive function of the modifier and the polymer often in the presence of an initiator.

Approach **ii** is applicable to all thermoplastic polymers and offers considerable practical advantage since reactively modified polymers can be readily produced by normal commercial high-shear mixing procedures at no additional cost.

REACTIVE MODIFIERS FOR IMPROVING ADDITIVE PERFORMANCE-REACTIONS DURING MELT PROCESSING

Mechanochemical chain scission of the polymer backbone occurs during the conversion of the polymer to a finished product by any of the industrial procedures which may be used for the conversion such as the use of screw extruders.

The chemistry of the mechano-oxidative breakdown of polypropylene during processing (11) is outlined in reaction scheme **1**. Shearing the polymer during the processing operation leads to mechanical scission of carbon-carbon bonds in the polymer backbone (**1a**). In the presence of free radical initiator, e.g. peroxides, reaction of the relatively long-lived mechanochemically formed tertiary macroalkyl radical and the relative contribution of pathway (**1e**) is a function of oxygen concentration (see Figure **1**, inset). Moreover, macroalkyl hydroperoxides formed (by route **1f**) are themeselves important initiators of oxidation (see reactions **1g** & **1h**) and lead to decreasing the molecular weight of the polymer which is inferred from the rapid

$\sim C(CH_3)(H) - CH_2 - CH(CH_3) - CH_2 \sim$ ⇌ (a) / (b) N_2 ⇌ $\sim \dot{C}H(CH_3) + CH(CH_3) - \dot{C}H_2$

(c) ROOR ↓

$\sim \dot{C}(CH_3) - CH_2 \sim$ (I)

O_2 (e) ↙ ↘ (d) Reactive Modifier (RM)

$\sim C(CH_3)(OO^{\bullet}) - CH_2 \sim$ $\sim C(CH_3)(RM) - CH_2 \sim$ (II)

Mechanochemically Synthesised Adduct

(f) ↓ $\sim CH(H) \sim$

$\sim C(CH_3)(OOH) - CH_2 \sim + \sim \dot{C} \sim$

(g) ↓

$\sim C(CH_3)(O^{\bullet}) - CH_2 \sim + \dot{O}H$ —(h)→ $\sim C(CH_3)=O + \dot{C}H_2 - CH(CH_3) \sim$

scheme 1

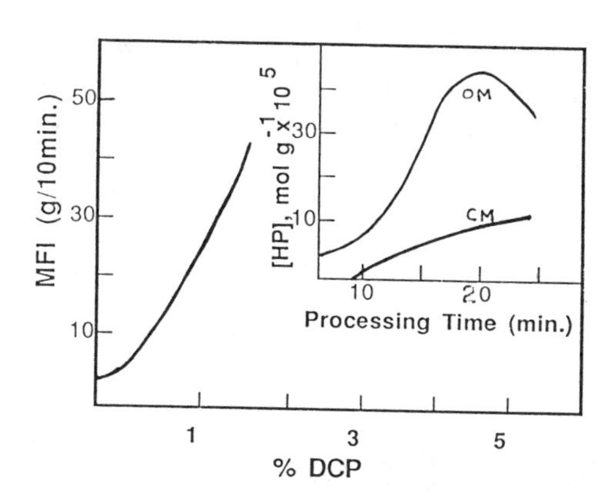

Figure 1. Effect of dicumyl peroxide (DCP) and processing conditions (excess, OM, and restricted, CM, oxygen amount) on melt stability of PP.

decrease in melt viscosity (or increase in melt flow index, MFI) observed during processing (see Figure 1).

These oxidative degradation processes are shown schematically in reaction scheme 2. The first oxidative cycle involves the formation of macroradicals. Reaction of these macroradicals with oxygen gives macroperoxyl radicals which, in turn, give rise to macrohydroperoxides. Hydroperoxides are the most important product of the cyclical auto-oxidation process because they are the main source of further radicals in the system. They give rise to reactive hydroxyl radicals which react with the substrate to give macroalkyl radicals which, in turn, feed back into the main auto-oxidation cycle, thus maintaining the chain mechanism. It is clear, therefore, that oxygen greatly affects mechanodegradation since it does not only interfere with the recombination step of macroalkyl radicals (scheme 1b) but it also gives rise to hydroperoxides which are themeselve powerful initiators of oxidation.

In the absence of oxygen or any other radical trap, however, mechanochemically formed macroalkyl radicals (scheme 1, **I**) can be made to react with chemically reactive modifiers, RM, (see scheme 1d); this forms the basis of an in-situ synthesis of polymer adducts i.e., the functionalised additive/modifier becomes chemically bound onto the polymer backbone.

This approach is not only very cost-effective, since it adopts the normal commercial processing operation as the reaction step at no additional cost, but has the added advantage also of preparing modified concentrates which are fully compatible with the unmodified polymer. Furthermore, concentrates may also be used as conventional additives in the same polymer or in different polymers. In this way a substantive antioxidant (or modifier) system can be produced with very high effectiveness especially under aggressive environments. In comparison, under such demanding conditions, conventional antioxidants will not only lead to premature failure of the polymer but also to dangerous situations such as in the case in food contact application (packaging), medical uses of polymers (artificial joints), and failure of aircraft tyres where human lifes are at risk.

REACTIVE MODIFIERS FOR DEMANDING APPLICATIONS

In demanding applications, where polymers are subjected to high temperatures and continuous exposure to aggressive extractive environments, conventional antioxidants are rendered useless: they are readily removed from the polymer and oxidative durability rather than physical stability is the major limitation. Chemically reacted antioxidants on polymers do not suffer from this problem.

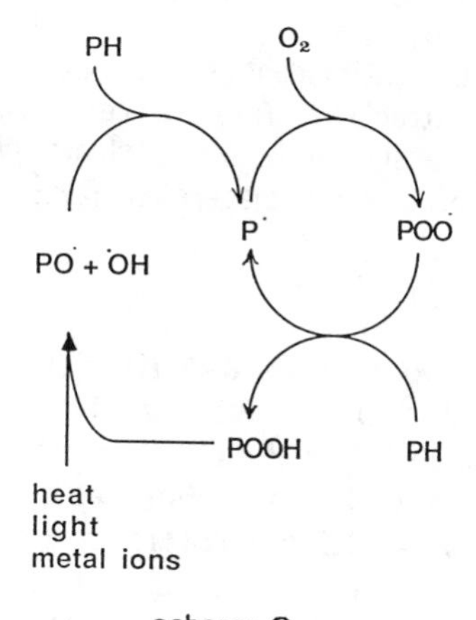

scheme 2

A recent example which highlights the above involves the incorporation of a reactive modifier with high activity antioxidant moiety during melt processing to achieve both the desired enhancement in polymer properties and in additive performance. Using this approach, a "reactive" acrylamide-PP adduct containing glass fibre was prepared (Al-Malaika, S., Quinn, N., and Scott, G., Aston University, Birmingham, unpublished work) on a Buss-Ko kneader (which offers high shearing forces due to the rotating and reciprocating action of the screw). Tables **I and II** show clearly that such system offers both high mechanical performance and exceptional oxidative durability at 150°C under exhaustive solvent extractive conditions.

The fact that the antioxidant adduct is more effective than conventional antioxidants (e.g., the hindered phenols Irganox 1076 and 1010, Ciba Geigy) after extraction (table **I**) is not surprising since the former becomes chemically attached (bound) to the polymer during the in-situ modification (during melt processing). However, it may be surprising that the bound antioxidant is much better than conventional antioxidants before extraction. This may result partly from the fact that arylamines in general have higher intrinsic molar activity than hindered phenols, although better dispersion may also be responsible in the former case.

Table I : Thermo-oxidative stability (at 150° C) of PP (ICI,HF22) without and with 30% glass fibre (GF) in the absence and presence of antioxidants. Soxhlet extraction was with chloroform and acetone

Antioxidants (AO)	Induction Period to Carbonyl Formation (h)		Number of days to craze formation	
	PP Films (250 μm) with no GF		PP Plaques (1mm) with 30% GF	
	Before Extraction	After Extraction	Before Extraction	After Extraction
Bound acrylamide	2250	2400	55	60
Irganox 1010	350	5	42	5
Irganox 1076	1200	5	22	3
Control (No AO)	1	1	2	2

Table II : Mechanical properties of unmodified and modified PP (ICI,HF22) containing 30% GF. Numbers in paranthesis are standard deviations

PP samples	Yield Strength (MPa)	Tensile Modulus (GPa)	Flexural Strength (MPa)	Impact Strength (Charpy) (KJm^{-2})
modified with acrylamide adduct	84 (1.02)	4.0 (0.14)	103 (1.07)	168 (2.10)
Control (no modification)	39 (1.30)	2.3 (0.12)	45 (1.10)	110(1.40)

Another example where antioxidant performance can be improved dramatically lies in the mechanochemically initiated addition of reactive antioxidants on rubbers (5,10) or unsaturated thermoplastics such as ABS (12). For example, using thiol antioxidants **2** and **3** as the reactive antioxidants, Kharasch-type addition of the thiol function to the polymer double bond takes place during melt processing to give bound antioxidant adduct (see Equation **1**); the polymer becomes much more substantive under aggressive environments.

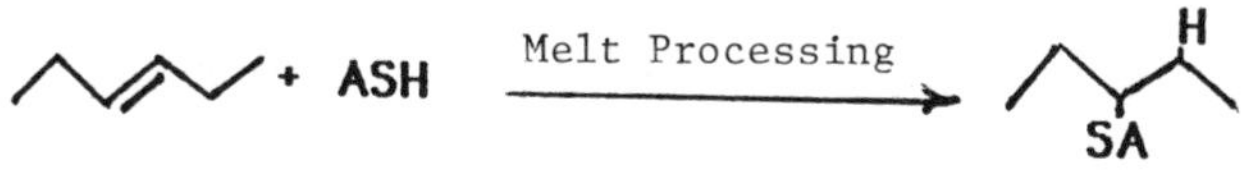

$$\text{polymer (C=C)} + \text{ASH} \xrightarrow{\text{Melt Processing}} \text{polymer(H)(SA)} \quad (1)$$

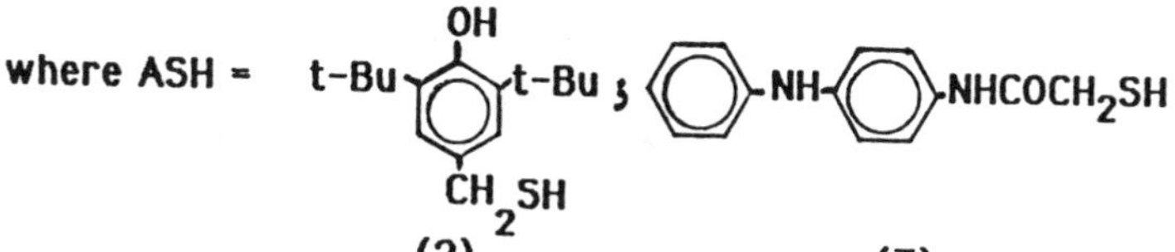

where ASH = t-Bu–[phenol, OH, t-Bu]–CH_2SH **(2)** ; C_6H_5–NH–C_6H_4–$NHCOCH_2SH$ **(3)**

However, it may be more surprising for a similar reaction to take place with saturated hydrocarbon polymers such as polypropylene (PP) and polyethylene (PE) since the level of unsaturation is insufficient. It was shown, however, that in such cases binding takes place in two stages (13). Most of the adduct is formed during the first minute of processing when the applied shear is maximum (high polymer viscosity). The first stage of binding, therefore, involves the formation of macroalkyl radicals by mechanical scission of the polymer chains, (scheme **3**). The thiol antioxidant (ASH) is extremely labile to radical attack; it attacks the mechanoradical to produce a thiyl radical which then reacts with another mechano chemically-generated radical to form the bound antioxidant.

The mechanoradical produced will react with the small amount of oxygen to form hydroperoxides; these are subsequently utilised as radical generators in the second stage. The resulting hydroxyl radical (from hydroperoxide decomposition) abstracts a hydrogen from the substrate to form macroradical which, in turn, will react with more of the thiyl radical to form more bound antioxidant. The polymer bound antioxidant made in this way is very much more resistant to solvent leaching and volatilisation when compared to commercial additives (13), see Figure **2**.

REACTIVE MODIFIERS FOR POLYMER MODIFICATION

Enhancing the properties of the relatively cheap commodity plastics through the use of small amounts of reactive modifiers during melt processing (as in method **2(ii)** above) is both attractive and rewarding. A good example of a reactive modifier which has been used (14) to enhance properties of polyolefins is maleic anhydride (MA). The formation of maleic adduct in polypropylene (PP), for example, can be used to effect several modifications; e.g. to improving hydrophilicity, adhesion and dyeability. Moreover, the polymer-maleic adduct has an availabla additional functionality to effect other chemical modifications for achieving the desired material design objectives. Reactions of MA with polymers in solution are described in the patent literature (15).

We have successfully achieved (Al-Malaika,S., Quinn, N., and Scott, G., Aston University, Birmingham, unpublished work) in-situ synthesis of maleic-polypropylene adduct during melt processing of the polymer in the presence of MA and free radical initiator. The functionalised polymer adduct was used to introduce further chemical modifications for improving the polymer properties. For example, using the above method, we have achieved improved adhesion to the surface of reinforcing glass and mineral fibres. Chemical attachment of antioxidants was also effected for the purpose of of improving additive performance.

Addition reaction of peroxide-generated macroalkyl radicals with the reactive unsaturation in MA is shown in reaction scheme **4**. The functionalised maleic-polymer adduct (**II**, scheme **4**) is the product of hydrogen abstraction reaction of the adduct radical (**I**, scheme **4**) with another PP chain. Concomitantly, a new macroalkyl radical is regenerated which feeds back into the cycle. The frequency of this feedback determines the efficiency of the cyclical mechanism, hence the degree of binding. Cross-linking reaction of **I** occurs by route **c** (scheme **4**).

The presence of an active site in the modifier is essential for the formation of polymer-modifier adduct by the in-situ method (method **2**) above. Figure **3** compares the degree of binding which can be achieved

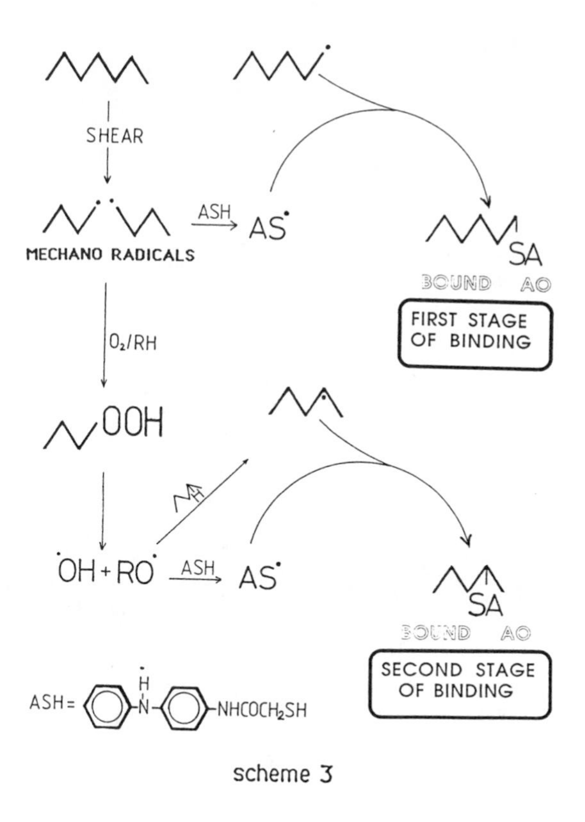

scheme 3

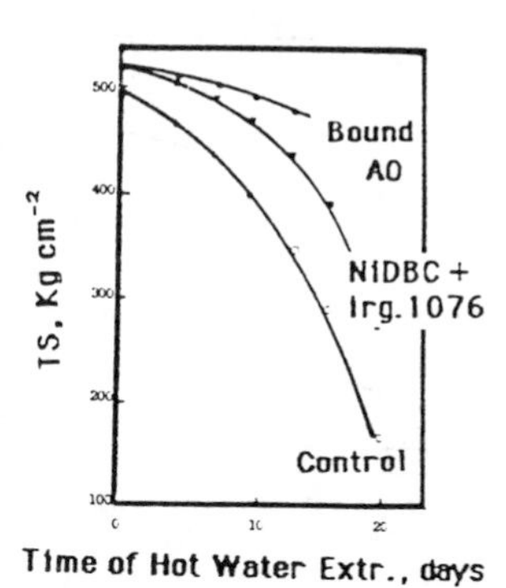

Figure 2. Effect of hot (100°C) water extraction on impact strength of PP in the absence and presence of a combination of conventional antioxidants (Nickel dibutyl dithiocarbamate and Irganox 1076, NiDBC+Irg.1076) and a thiol bound antioxidant (3), Bound AO. (Reproduced with permission from Ref. 13. Copyright 1983 App. Sci. Pub.)

(no O_2)

ROOR

(I)

Cross-Linking

(II)

Functionlised
Maleic/PP Adduct

scheme 4

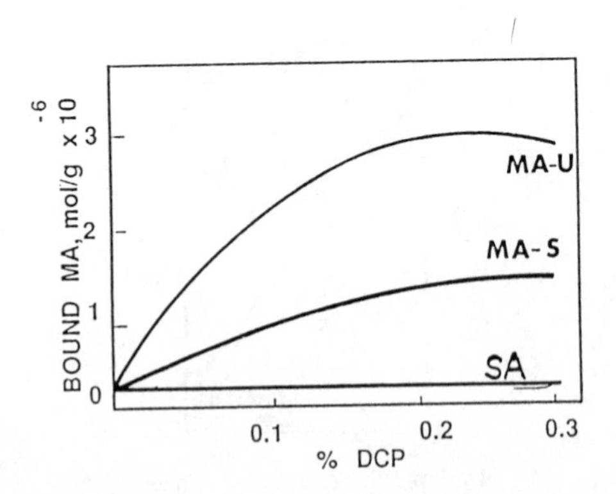

Figure 3. Effect of reactive site in modifiers (MA is maleic anhydride and SA is succinic anhydride) on binding in unstabilised (U), ICI, HF22, and stabilised (S), ICI HWM25, polypropylene.

when using analogous modifiers except for the presence (with MA) and absence (with succinic anhydride, SA) of reactive sites (C = C). Up to 55% binding of MA on PP was obtained (based on exhaustive Soxhlet extraction with chloroform and acetone) wheras no binding was achieved when SA was used. The importance of PP-MA adduct formation is further demonstrated when the above method was used in composites based on glass fibre and PP-MA and PP-SA samples; PP-SA sample exhibited very inferior mechanical properties when compared to the composite system based on PP-MA adduct (see Fig. 4).

To achieve optimum conditions for the attachment of reactive modifiers to the polymer backbone, oxygen and other radical traps should be avoided since these compete for the same macroalkyl radical sites (scheme 1e and 1d, see also figure 3, curves MA-U and MA-S). Mechanochemical attachment of MA to both unstabilised (ICI, HF22) and stabilised (ICI, HWM25) grades of PP shows that, under exactly similar processing conditions, the degree of binding of the modifier (MA) to PP is twice as much for the unstabilised grade. The mechanism of the antioxidant action of thermal stabilisers (11) can account for this apparent discrepency. Hindered phenols function by a cyclical regenerative process involving alternating chain breaking-acceptor (CB-A) and chain breaking-donor (CB-D) antioxidant steps (see reaction scheme 5). Hindered phenols form stable phenoxyl radicals in the presence of alkyl peroxyl radicals. The phenoxyl radicals react with macroalkyl radicals in a CB-A step to reform the hindered phenol which in turn reacts with more alkylperoxyl radicals in a CB-D step to regenerate the phenoxyl radical which then reacts with more alkyl radicals.

The central feature of this mechanism is, therefore, that the phenoxyl radical is reversibly reduced and re-oxidised; this leads to the continuous consumption of macroalkyl radicals. The phenoxyl radical can, therefore, react with polypropylene radicals and compete with PP-MA adduct formation in the stabilised polymer (Figure 3, curve MA-S).

REACTIVE MODIFIERS AND ADHESION

The localised reactivity which has been introduced by tieing the reactive modifier mechanochemically into the otherwise inert PP substrate is exploited here to promote adhesion to different surfaces, e.g. glass. Homogenisation of glass fibre (in a Buss-Ko kneader at 230° C), which has been treated with an amino-containing silane coupling agent (Union Carbide, A1100), with mechanochemically synthesised PP-MA adduct (using PP grade HF22) results in considerable improvement in mechanical properties. Figure 4 shows that the modified unstabilised PP composite does not only offer much improved mechanical performance over the hindered phenol stabilised and modified PP composite analogue but is also considerably better than a commercial control of 30% glass reinforced PP (ICI, HW60/GR30). The amide linkage resulting from the condensation reaction between the functionalities on the modifier adduct and the silane coupling agent (reaction scheme **6**), which takes place during the homogenisation process, was confirmed by IR spectroscopy. The improved mechanical properties (see Figure **4**) are attributed to amide linkages which promote better glass-polymer adhesion.

Examination of scanning electron micrographs of tensile fractured surfaces of the modified reinforced unstabilised PP and the hindered phenol stabilised analogue (Figures 5 and 6) shows clearly that fractured surfaces of the former exhibit short fibre pull-out (plate **1**) compared to the much longer fibre pull-out for the latter (plate **2**). The short fibre pull-out is a clear indication of effective bonding (and higher mechanical pereformance, see Figure **4**) as a result of increased adhesion between the fibre and the modified polymer interfaces. The extent of this adhesion must be a function of the level of maleic adduct present in the reactively-modified PP surface which consequently provides the neccessary functionality at the interface to enable condensation reactions to take place with the amino groups (giving rise to the amide linkages observed) of the coupling agent on the glass surface (compare curves MA-S and MA-U, Figure **3**).

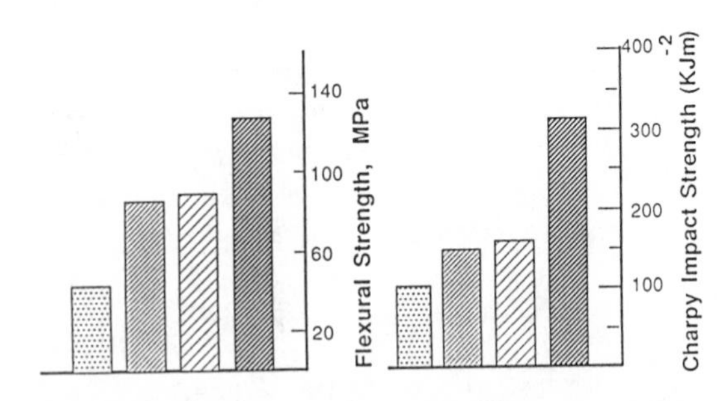

Figure 4. Comparison of mechanical properties of,
- (dark hatched) 30% glass reinforced MA-modifiedPP (unstabilised, ICI-HF22),
- (wide hatched) commercial control of 30% glass reinforced PP (ICI-HW60/GR30),
- (fine hatched) 30% glass reinforced MA-modified PP (stabilised, ICI-HWM25),
- (dotted) 30% glass reinforced SA-modified PP (unstabilised, ICI-HF22).

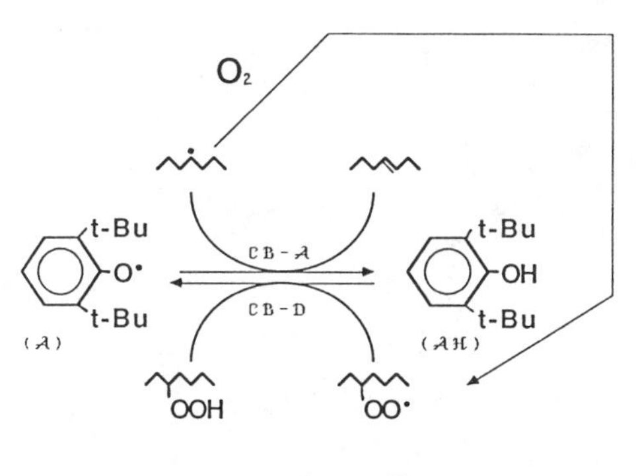

scheme 5

O-Si-(CH2)3-NH2 +

Glass Surface

Polymer Surface

EXTRUSION

O-Si-(CH2)3-NH-C(=O)-CH-
HO-C(=O)-CH-

Glass Surface

Polymer Surface

scheme 6

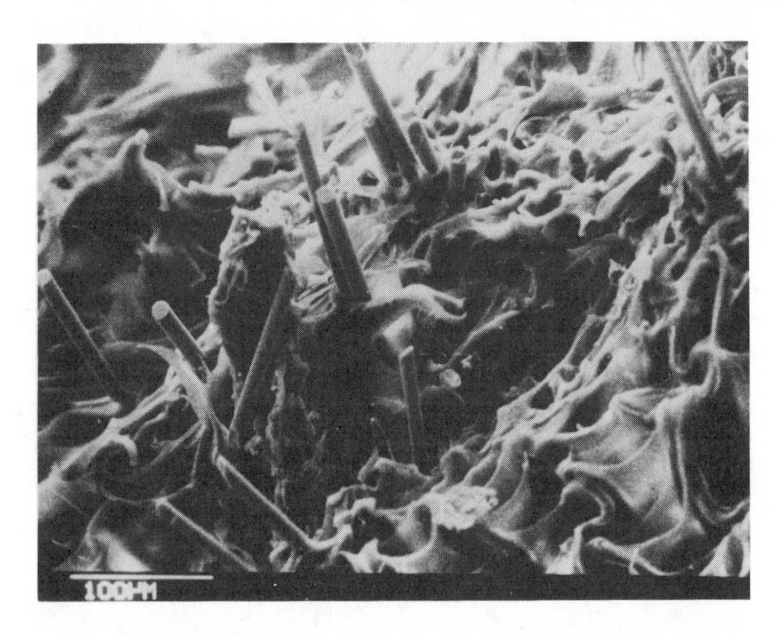

Figure 5. Scanning electron micrograph of tensile fractured surface of 30% glass reinforced MA-modified PP (unstabilised, ICl-HF22).

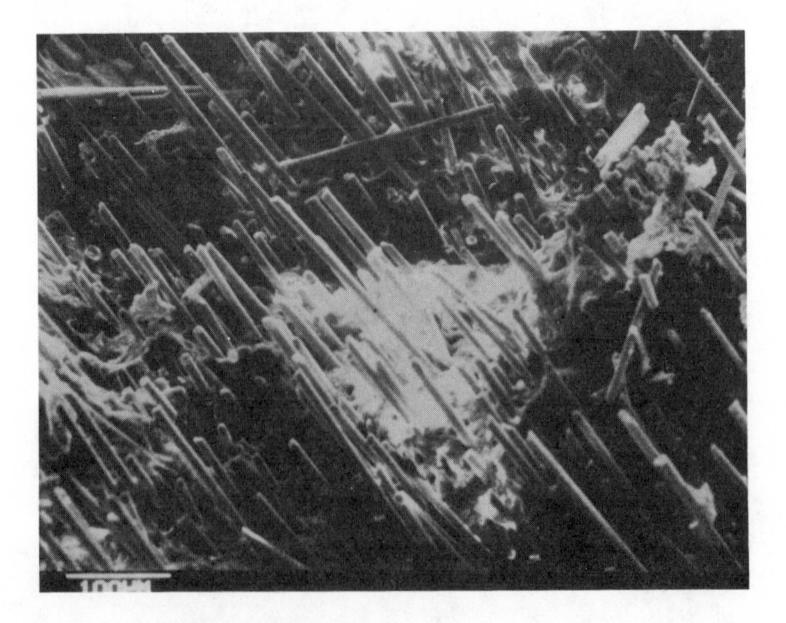

Figure 6. Scanning electron micrograph of tensile fractured surface of 30% glass reinforced MA-modified PP (stabilised with hindered phenol, ICl-HWM25).

ACKNOWLEDGMENTS

The author is grateful to Professor G. Scott and Dr N. Quinn for preliminary publication of results which will be published in full elsewhere.

LITERATURE CITED

1. Kauzman, W.; Eyring, H., J. Am. Chem. Soc. 1940, 62, 3113.
2. Ayrey, G.; Moore, C.G.; Watson, W.F., J. App. Polym. Sci. 1956, 19, 1.
3. Ceresa, R.J.; Watson, W.F., J. App. Polym. Sci. 1959, 1, 101.
4. Thomas, D.K., in Developments in Polymer Stabilisation-1, Scott, G., Ed.; App. Sci. : London, 1979; ch.4.
5. Scott, G., in Developments in Polymer Stabilisation-4, Scott, G., Ed.; App. Sci. : London, 1981; ch.6.
6. Gupta, Fu.s.; Albertson, A.C.; Vogl, O., in New Trends in the Photochemistry of Polymers, Allen, N.S., and Rabek, J.F., Ed. ; Elsevier App. Sci. : London, 1985; Ch. 15.
7. Scott, G., Pure and App. Chem., 1983, 55, 128 and 1615.
8. Scott, G., Polym. Eng. Sci., 1984, 24, 1007.
9. Meyer, G.E.; Kavchok, R.W.; Naples, T.F., Rubb. Chem. Tech., 1973, 46, 106.
10. Al-Malaika,S.; Honggokosomo, S.; Scott, G., Polym. Deg. Stab., 1986, 16, 25.
11. Al-Malaika, S.; Scott, G., in Degradation and Stabilisation of Polyolefins, Allen, N.S., Ed. ; App. Sci. : London, 1983, Ch.6.
12. Ghaemy, M.; Scott, G., Polym. Deg. Stab., 1980-81, 3, 405.
13. Scott, G.; Setoudeh,E., Polym. Deg. Stab., 1983, 5, 1.
14. Culbertson, B.M.; Trivede, B.C., in Maleic Anhydride; Plenum: NY, 1982, Ch.11.
15. Hercules Inc., US Pat., 3,437,556; 1969.

RECEIVED October 30, 1987

Chapter 30

Phase-Transfer-Catalyzed Modification of Dextran Employing Dibutyltin Dichloride and Bis(cyclopentadienyl)titanium Dichloride

Yoshinobu Naoshima [1] and Charles E. Carraher, Jr.[2]

[1]Department of Chemistry, Okayama University of Science, Ridai-cho, Okayama 700, Japan
[2]Department of Chemistry, Florida Atlantic University, Boca Raton, FL 33431

The modification of dextran was achieved employing dibutyltin dichloride and bis(cyclopentadienyl)titanium dichloride using the aqueous interfacial condensation system. The condensation was studied employing various phase transfer agent and as a function of reactant molar ratio. The parameters of percentage product yield and location of the maximum as the molar ratio of reactants is varied were employed to discern if the added phase transfer agents were functioning as a phase transfer agent. For condensations involving the titanocene dichloride, all of the employed phase transfer agents are believed to act as phase transfer agents when employing either sodium hydroxide or triethylamine as the added base. Further, triethylamine itself appears to act as a phase transfer agent. When the organostannane is employed similar results are obtained with the exception that the product percentage yields are generally less for systems employing triethylamine compared with systems employing sodium hydroxide. The roles of triethylamine for such systems is presently unknown.

Carbohydrates are the most abundant, weight-wise, organic material available. Photosynthesis produces about 400 billion tons annually. The polysaccharides are generally composed of mono- and disaccharide units.

Dextran was chosen to study for the following reasons. First, it is water soluble allowing three dimensional modification employing aqueous solution and classical interfacial condensation routes. Second, it is readily available in industrial quantities. Third, it is available in a range of molecular weight allowing product modification to be studied as a function of dextran chain size. Fourth, it is generally considered to be an under-utilized natural feedstock.

Dextran is the collective name of extracellular bacterial poly-alpha-D-glucopyranoses linked largely by 1,6 bonds, with branching occurring at the 1,2, 1,3 or 1,4 bonds. Physical properties vary

0097-6156/88/0364-0426$06.00/0

according to the amount and manner of branching, nature of endgroup, molecular weight and molecular weight distribution, processing, etc.

Dextran is produced from sucrose by a number of bacteria the major ones being the nonpathogenic bacteria Leuconostoc mesenterodes and Leuconostoc dextranicum. As expected, the structure (and consequently the properties) of the dextran is determined by the particular strain that produces it.

Dextran is the first microbial polysaccharide produced and utilized on an industrial scale. The potential importance of dextran as a structually (and property) controlled feedstock is clearly seen in light of the recent emphasis of molecular biologists and molecular engineers in the generation of microbes for feedstock production. Dextran is employed as pharmaceuticals (additives and coatings of medications), within cosmetics, as food extenders, as water-loss inhibitors in oilwell drilling muds and as the basis for a number of synthetic resins.

Svenska Sockerfabriks Aktiebolaget for Aktiebolaget Pharmacia in Sweden began large scale production of dextran about 1942 (1). By 1947, Dextran Ltd. (East Anglia Chemical Co., England) began production of dextran and in 1949 Commercial Solvents Corp. (USA) began production (2,3). In 1952, the R. K. Laros Co. (3) began the enzymic production of dextran in the presence of living cells. In an effort to standardize the dextran produced, by 1952, the majority of companies employed the L. mesenteroides NRRL B-512-(f) in the production of dextran. The production of dextran involves mixing the appropriate quantities of sucrose and enzyme under prescribed conditions.

Solubility generally decreases with increase in chain size and extent of branching. The solubility of dextran can be divided into four groups — those that are readily soluble at room temperature in water, DMF, DMSO and dilute base; those that have difficulty dissolving in water; those that are soluble in aqueous solution only in the presence of base; and, those that are soluble only under pressure, at high temperatures (> 100°C) and in the presence of base. Dextran B-512 readily dissolves in water and 6M, 2M glycine and 50% glucose aqueous solutions.

Dextran B-512, dissolved in water, approaches a form of compact spherical-like helical coils (4-6). Streaming, birefringence measurements indicate that the dextran has some flexibility. The B-512 dextran shows a refractive index increment, dn/dc, of 0.154 ml/g at 436 mu in water and an apparent radius of gyration in water on the order of 2×10^3 A. Branching occurs through about 5% of the units through the 1,3 linkage with about 80% of these branches being only one unit in length. There exists a few, less than 1%, long branches in B-512 dextran (6).

Chemically dextrans are similar to one another. The activation energy for acid hydrolysis is about 30-35 Kcal/mol (5). The C-2 hydroxyls appear to be the most reactive in most Lewis base and acid-type reactions. A wide variety of esters and ethers have been described as well as carbonates and xanthates (7,8). In alkaline solution, dextran forms a varying complex with a number of metal ions (9).

The biological properties of dextran again vary with strain. The B-512 is digestible in mammalian tissues as the liver, spleen, and kidneys, but not in the blood (10). It appears not to stimulate formation of antibodies in man upon intravenous infusion (1). On the other hand, subcutaneous injection in humans led to skin sensitivity and to the formation of precipitating antibodies (11). Antibody formation and precipitation with antiscrums decreases as chain length increases (11,12).

Linear, water-soluble dextrans have many uses. One dextran is employed in viscous water-flooding for secondary recovery of petroleum with a potential market of about 2×10^5 tons per year. This dextran is superior to carboxymethyl cellulose when employed in high calcium drilling muds (13,14). These dextran-muds show superior stability and performance at high pH and in saturated brines (15-17). Other dextrans show good resistance to deterioration in soil and the ability to stabilize aggregates (18) in soils. They can also be used in binding collagen fibers into surgical sutures (19).

An underlying assumption is that dextran is a representative polysaccharide and that results derived from its study can be applied to other polysaccharides. Effected modifications are intended to occur throughout the dextran material rather than only at the surface. This is achieved by employing solutions containing dissolved dextran.

Recently Carraher, Naoshima and coworkers effected the modification of polysaccharides employing organostannanes and bis(cyclopentadienyl)titanium dichloride, BCTD (20-25). Here we report the modification of dextran employing the interfacial condensation technique using various phase transfer agents utilizing BCTD and dibutyltin dichloride, DBTD.

Experimental

Reactions were carried out employing a one quart Kimex emulsifying jar placed on a Waring Blendor (Model 7011G) with a no load stirring rate of about 20,000 rpm. The following organic chemicals were used as received (from Aldrich unless otherwise noted): bis(cyclopentadienyl)titanium dichloride, dibutyltin dichloride, 18-crown-6, dibenzo-18-crown-6, tetra-*n*-butylammonium hydrogensulfate (Tokyo Kasei Kogyo Co., Ltd., Japan), triethylamine (Wako Pure Chemical Industries, Japan), and dextran (Wako Pure Chemical Industries; molecular weight = 2 to 3×10^5). In a typical procedure, an aqueous solution of dextran containing a base and a phase transfer catalyst (PTC) was added rapidly stirred solutions of the organometallic dichloride in chloroform. Repeated washings with organic solvent and water assisted in the product purification. Elemental analyses for titanium and tin were conducted employing the usual wet analysis procedure with $HClO_4$. Infrared spectra were obtained using Hitachi 260-10 and 270-30 spectrometers and a Digilab FTS-IMX FT-IR. EI mass spectral analysis was carried out employing a JEOL JMS-D300 GC mass spectrometer connected with a JAI JHP-2 Curie Point Pyrolyzer. DT and TG analyses were performed employing a SINKU-RIKO ULVAC TGD-500M or a DuPont 990 TGA and 900 DSC.

Results and Discussion

Structural characterization was based on solubility, thermal and elemental analyses, and infrared and mass spectroscopies. Characterization results were in agreement with the product being composed of units, including unreacted units, as depicted in Figures 1 and 2 and reported in references 21, 24 and 25.

The investigation of the chemical modification of dextran to determine the importance of various reaction parameters that may eventually allow the controlled synthesis of dextran-modified materials has began. The initial parameter chosen was reactant molar ratio, since this reaction variable has previously been found to greatly influence other interfacial condensations. Phase transfer catalysts, PTC's, have been successfully employed in the synthesis of various metal-containing polyethers and polyamines (for instance 26). Thus, the effect of various PTC's was also studied as a function of reactant molar ratio.

Two parameters were employed in evaluating if an added PTC is functioning as a PTC. These parameters were percentage yield and position of the maximum as the molar ratio of reactant is varied. If PTC activity is occurring, these parameters might vary as follows : a. differing, normally enhanced yield for systems employing an added PTC compared with the analogous systems omitting the PTC, and b. change in the position of the maximum in percentage yield. The maximizing of yield as the molar ratio is varied is common for ieterfacial systems (for instance 27-29). The molar ratio corresponding to the maximum yield corresponds to a favorable balancing of the entry of the reactants into the reaction zone. For a given set of reactants this maximin will vary dependent on the specific reaction parameters. Thus factors that affect the relative transport factors will affect the position of the maximum whereas factors that do not affect these transport factors will result in maximums occurring in the same general area as the molar ratio is varied. Increased yield is also consistent with an increase in the reactants reaching the reaction zone during the reaction. For the present study, yield differences can be considered as relative measures of transport factors.

A third, less dependable parameter that may be a measure of PTC activity is percentage organometallin incorporation. Percentage incorporation is believed to be less dependable for the following reasons. First, steric factors appear to be critical for many analogous polymer modification of poly(vinyl alcohol) and polyethyleneimine (for instance 30-33). Second, for the current systems, only moderate differences in percentage incorporation are found and these differences are believed to be mainly due directly to the difference in molar ratio of reactant. As with most analogous cases involving polymer modification, generally high proportions of incorporation are found regardless of percentage yield (for instance 30-33).

Tables I-IV contain results as a function of initial base, presence or absence of an added PTC, organometallin reactant and molar ratio of reactants. On a strictly molar ratio of reactant sites basis, the balancing of net quantity of reactants should occur at an organometallic/dextran molar ratio of 3:2. The present studies are consistent with earlier studies that showed the maximums occurring away from this 3:2 ratio. The values cited in Table V for reactant muximums should be considered only as general. Results contained in Table I will be utilized to illustrate how the parameters of percentage yield and position of the maximum of precentage yield as reactant molar ratio is varied might be employed to indicate if the added PTC

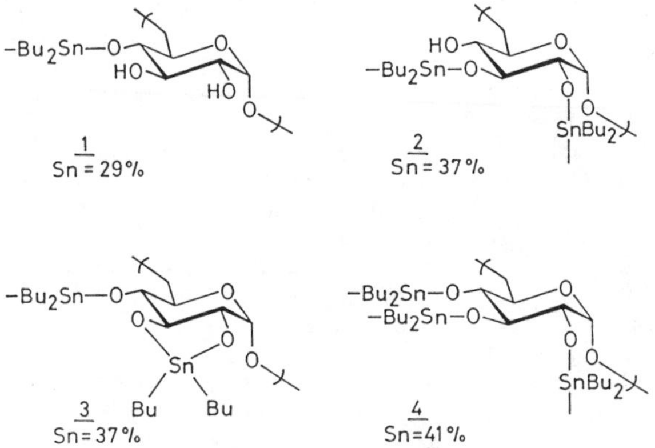

−Bu$_2$Sn—O
HO
HO
O
O

1
Sn = 29%

HO
−Bu$_2$Sn−O
O
O
O
SnBu$_2$

2
Sn = 37%

−Bu$_2$Sn—O
O
O
O
Sn
O
Bu
Bu

3
Sn = 37%

−Bu$_2$Sn−O
−Bu$_2$Sn−O
O
O
O
SnBu$_2$

4
Sn = 41%

Figure 1. Possible structures of the tin-containing units.

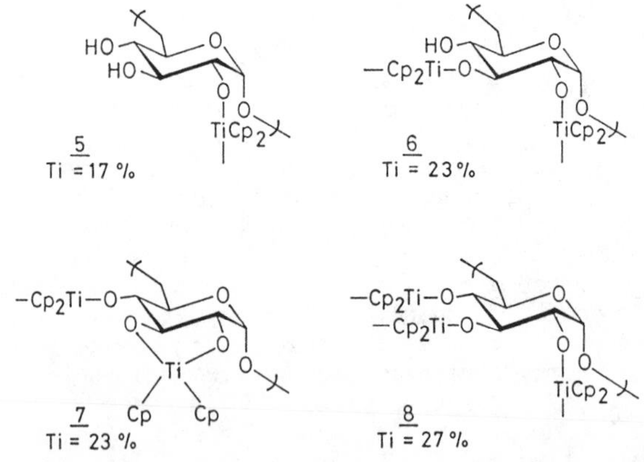

HO
HO
O
O
O
TiCp$_2$

5
Ti = 17%

HO
−Cp$_2$Ti−O
O
O
O
TiCp$_2$

6
Ti = 23%

−Cp$_2$Ti−O
O
O
O
Ti
O
Cp
Cp

7
Ti = 23%

−Cp$_2$Ti−O
−Cp$_2$Ti−O
O
O
O
TiCp$_2$

8
Ti = 27%

Figure 2. Possible structures of the titanium-containing units.

Table I. Results as a function of base and amount of DBTD for systems containing TBAHS or no phase transfer catalyst, PTC

Bu_2SnCl_2 (mmol)	Yield[a] (%) PTC	Yield[a] (%) None	Yield (g) PTC	Yield (g) None	Sn (%) PTC	Sn (%) None
A. Triethylamine						
0.50	12	7	0.015	0.008	29	29
1.00	29	19	0.07	0.05	29	29
2.00	28	22	0.14	0.11	28	37
3.00	19	38	0.14	0.28	41	37
4.00	7	4	0.07	0.035	28	37
(Dextran (3.00mmol), TEA (9.00mmol) and TBAHS (0.90mmol) in 50ml water; Bu_2SnCl_2 in 50ml $CHCl_3$; 30 sec. stirring time.)						
B. NaOH						
0.50	37	8	0.046	0.01	33	30
1.00	57	21	0.14	0.05	27	28
2.00	83	90	0.41	0.44	23	37
3.00	4	94	0.025	0.70	39	25
4.00	0	0	0	0	-	-
(Ibid. above except employing NaOH (9.00mmol) in place of TEA and 90 sec. stirring time.)						

a. Yields based on the presence of three Bu_2Sn units per saccharide unit for Tables I, II and VI.

Table II. Results as a function of base and amount of DBTD for systems containing 18-crown-6 or no phase transfer catalyst, PTC

Bu_2SnCl_2 (mmol)	Yield (%) PTC	Yield (%) None	Yield (g) PTC	Yield (g) None	Sn (%) PTC	Sn (%) None
A. Triethylamine						
0.50	16	7	0.02	0.008	37	29
1.00	37	19	0.09	0.05	29	29
2.00	61	22	0.30	0.11	28	37
3.00	24	38	0.17	0.28	29	37
4.00	15	4	0.15	0.035	28	37
(Dextran (3.00mmol), TEA (9.00mmol) and 18-crown-6 (0.90mmol) in 50ml water; Bu_2SnCl_2 in 50ml $CHCl_3$; 30 sec. stirring time.)						
B. NaOH						
0.50	25	8	0.03	0.01	23	30
1.00	82	21	0.20	0.05	29	28
2.00	97	90	0.48	0.44	29	37
3.00	98	94	0.73	0.70	27	25
4.00	0	0	0	0	-	-
(Ibid. above except employing NaOH (6.00mmol) in place of TEA and 90 sec. stirring time.)						

Table III. Results as a function of base and amount of BCTD for systems containing TBAHS or no phase transfer catalyst, PTC

Cp_2TiCl_2 (mmol)	Yield[b] (%)		Yield (g)		Ti (%)	
	PTC	None	PTC	None	PTC	None
A. Triethylamine						
0.50	15	24	0.02	0.03	12	7
1.00	78	56	0.19	0.14	12	9
2.00	77	67	0.38	0.33	18	12
3.00	69	46	0.51	0.34	17	21
4.00	28	1	0.28	0.01	21	18

(Dextran (3.00mmol), TEA (9.00mmol) and TBAHS (0.90mmol) in 50ml water; Cp_2TiCl_2 in 50ml $CHCl_3$; 30 sec. stirring time.)

Cp_2TiCl_2 (mmol)	Yield[b] (%) PTC	Yield[b] (%) None	Yield (g) PTC	Yield (g) None	Ti (%) PTC	Ti (%) None
B. NaOH						
0.50	0	0	0	0	-	-
1.00	0	0	0	0	-	-
2.00	79	0	0.39	0	13	-
3.00	69	58	0.52	0.43	15	14
4.00	77	60	0.77	0.60	14	13

(Ibid. above except employing NaOH (9.00mmol) in place of TEA and 90 sec. stirring time.)

b. Yields based on the presence of three Cp_2Ti units per saccharide unit for Tables III and IV.

Table IV. Results as a function of base and amount of BCTD for systems containing 18-crown-6 or no phase transfer catalyst, PTC

Cp_2TiCl_2 (mmol)	Yield (%)		Yield (g)		Ti (%)	
	PTC	None	PTC	None	PTC	None
A. Triethylamine						
0.50	0	24	0	0.03	-	7
1.00	4	56	0.01	0.14	12	9
2.00	16	67	0.08	0.33	15	12
3.00	25	46	0.18	0.34	17	21
4.00	39	1	0.38	0.01	21	18
(Dextran (3.00mmol), TEA (9.00mmol) and 18-crown-6 (0.90mmol) in 50ml water;Cp_2TiCl_2 in 50ml $CHCl_3$;30 sec. stirring time.)						
B. NaOH						
0.50	0	0	0	0	-	-
1.00	0	0	0	0	-	-
2.00	0	0	0	0	-	-
3.00	80	58	0.59	0.43	9	14
4.00	59	60	0.59	0.60	15	13
(Ibid. above except employing NaOH (6.00mmol) in place of TEA and 90 sec. stirring time.)						
C. NaOH						
0.50	0	0	0	0	-	-
1.00	0	0	0	0	-	-
2.00	21	0	0.11	0	17	-
3.00	86	48	0.64	0.36	17	17
4.00	86	53	0.86	0.53	21	21
(Ibid. above except employing NaOH (9.00mmol) in place of TEA and 90 sec. stirring time.)						

Source: Reproduced with permission from Ref. 34. Copyright 1987 Longman.

Table V. Summary of PTC results

OM	Initial Base	Added PTC	Yield Variation	Maximum Variation	Maximum OM/Dextran*
Bu_2SnCl_2	TEA	TBAHS	+	+	1/3
	NaOH	TBAHS	+	+	2/1
	TEA	18-C-6**	+	+	2/1
	NaOH	18-C-6	+	0	3/1
	TEA	DB-18-C-6***	+	+	2/1
	NaOH	DB-18-C-6	+	0	3/1
Cp_2TiCl_2	TEA	TBAHS	+	+	>4/1
	NaOH	TBAHS	+	+	3/1
	TEA	18-C-6	+	+	1/3
	NaOH	18-C-6	+	+	2/3

"+" indicates that a variation exists between results obtained employing a PTC with results obtained not employing a PTC.
"0" indicates that the trends are similar.
"+" indicates that there appears to be a variation, but it is not pronounced.
OM is organometallic reactant.
*for PTC containing systems.
**18-crown-6. ***dibenzo-18-crown-6.

is functioning as a PTC. First, percentage yields vary between systems employing an added PTC and those not containing an added PTC consistent with the addition of the PTC acting to influence the transport of the reactants. Second, the maximums vary with respect to the presence or absence of the added PTC. For triethylamine (TEA) systems with tetra-n-butylammonium hydrogensulfate (TBAHS), the percentage yield maximum occurs near a DBTD/dextran ratio of 1/3; for the analogous systems except omitting the PTC, the maximum occurs around 3/3. For systems employing sodium hydroxide as the added base, the maximum occurs around 2/1 when the TBAHS is present and around 3/1 when no PTC is present. The results are consistent with TBAHS acting to influence the transport of reactants. Results for other PTC's appear in Table V and are consistent with PTC's functioning to influence the transport of one or both of the reactants for the vast majority of the cases. Triethylamine is believed to act as a phase transfer agents (PTA) some interfacial systems (for instance 26). The maximum percentage yields for the modification effected employing dibutyltin dichloride both occur around a 3/3 ratio (DBTD/dextran; Table I) but with quite varying yields for systems employing 2 mmols of organostannane and greater. The approximate maximums (Table V) and yield for PTC-containing systems vary with respect to the nature of initially added base (sodium hydroxide or TEA). With the exception of coincidence of the maximums for non PTC-containing systems, the evidence is consistent with the presence of the TEA affecting the general outcome of the reaction. With most of the systems (Tables I, II) the percentage yields are lower when TEA is employed. For these systems, it is not presently known if TEA acts as a PTA or in some other manner. The product of dibutyltin dichloride and TEA is brown colored and not water soluble. The modified product is white, eliminating the possibility that the product contains significant portions of the simple organostannane-TEA product.

For systems employing BCTD the situation is different. The maximums occur at about 2/3 when employing TEA but no PTC and 4/3 when employing sodium hydroxide but no PTC. Again the maximums are dependent on the nature of the added base and added PTC. Yields vary but are not consistently lower when employing TEA compared with sodium hydroxide. It is possible that TEA may act to influence the transport of one or both of the reactants for systems employing the organotitanium reactant.

Addition of base is intended to serve two primary functions. First, to act as a scavenger, neutralizing hydrogen chloride eliminated through condensation between the organometallic dichloride and dextran. Second, to further activate, polarize the Lewis base for more ready attack at electrophilic sites. Sodium hydroxide, a strong base, is believed to further polarize Lewis bases such as amines and hydroxyls whereas TEA does not provide this added assistance to as great of an extent (see 9). It is possible that this further polarization of the dextran-hydroxyl groups by the sodium hydroxide is responsible for the relatively larger yields found compared to systems employing TEA as the added base when DBTD is used. It is possible that BCTD is sufficiently more reactive than DBTD such that additional polarization of the hydroxyls by the hydroxide ion is not significant to outcome of the reaction.

While the terms phase transfer catalyst and phase transfer agent are often employed interchangeably, the term PTA is broader and includes additives that become part of the overall reaction product. In the present study, the TEA is a PTA but not a PTC since it has been found to react with both of the metal-containing reactants. For the present products, TEA-containing moieties have been indicated employing both infrared and mass spectroscopies (24).

An additional question concerns whether the location of the PTC has an effect on the reaction. Table VI contains results where 18-crown-6 was added to the either phase. It appears that the initial location of the 18-crown-6 has little or no effect on the reaction.

Table VI. Results as a function of DBTD and original location for the PTC 18-crown-6

Bu_2SnCl_2 (mmol)	Yield (%)			Yield (g)			Sn (%)		
	None	OR*	Water	None	OR	Water	None	OR	Water
0.50	8	20	25	0.01	0.025	0.03	30	37	23
1.00	21	75	82	0.05	0.19	0.20	28	29	29
2.00	90	91	97	0.44	0.45	0.48	37	37	29
3.00	94	97	98	0.70	0.72	0.73	25	28	27
4.00	0	0	0	0	0	0	-	-	-

Dextran (3.00mmol) and sodium hydroxide (6.00mmol) in 50ml water added to stirred solutions of Bu_2SnCl_2 in 50ml $CHCl_3$ with 90 sec. stirring time.

*Organic.

O–H····$OH^{\ominus}$ O–H

9

Literature Cited

1. Gronwall, A. Dextran and Its Use in Colloidal Solutions; Academic Press, New York, **1957**.
2. Mfg. Chemist **1952**, 23(2), 49.
3. Chem. Eng. **1952**, 59, 215(Sept.) and 240(Dec.).
4. Oene, H. V.; Cragg, L. H. J. Polymer Sci. **1962**, 57, 175.
5. Antonini, E.; Bellelli, L.; Bruzzeni, A.; Caputo, A.; Chiancone, E.; Rossi-Fanelli, A. Biopolymers **1964**, 2, 35.
6. Bovey, F. J. Polymer Sci. **1959**, 35, 167 and 183.
7. Baker, P. J. Dextrans; Academic Press, New York, **1959**.

8. Flodin, P. Dextran Gels and Their Applications in Gel Filtration; Halmatd, Uppsala, **1962**.
9. Rowe, C. E. Ph. D. Thesis, Univ. Birmingham, **1956**.
10. Fischer, E.; Stein, E. Dextranases; Academic Press, New York, **1960**.
11. Kabat, E. Bull. Soc. Chim. Biol. **1960**, 42, 1549.
12. Hehre, E.; Sugg, J.; Neill, J. Ann. N. Y. Acad. Sci. **1952**, 55, 467.
13. Monaghan, P; Gidley, J. Oil Gas J. **1959**, 57, 100.
14. Dumbauld, G.; Monaghan, P. U. S. Patent 3,065,170, **1962**.
15. Mueller, E. Z. Angew. Geol. **1963**, 9, 213.
16. Owen, W. Sugar **1952**, 47(7), 50.
17. Owen, W. U. S. Patent 2,602,082, **1952**.
18. Novak, L.; Witt, E.; Hiler, M. Agr. Food Chem. **1955**, 3, 1028.
19. Novak, L. U. S. Patent 2,748,774, **1956**.
20. Carraher, C. E.; Giron, D. J.; Schroeder, J. A.; Mcneely, C. U. S. Patent 4,312,981, **1982**.
21. Naoshima, Y.; Hirono, S.; Carraher, C. E. J. Polymer Materials **1985**, 2, 43.
22. Carraher, C. E.; Gehrke, T. G.; Giron, D. J.; Cerutis, D.; Molloy, H. M. J. Macromol. Sci.-Chem. **1983**, A19, 1121.
23. Carraher, C. E.; Burt, W. R.; Giron, D. J.; Schroeder, J. A.; Taylor, M. L.; Molloy, H. M.; Tiernan, T. O. J. Appl. Polym. Sci. **1983**, 28, 1919.
24. Naoshima, Y.; Carraher, C. E.; Hess, G. G. Polym. Mat. Sci. Eng. **1983**, 49, 215.
25. Naoshima, Y.; Hirono, S.; Carraher, C. E. Polym. Mat. Sci. Eng. **1985**, 52, 29.
26. Mathias, L. J.; Carraher, C. E., Eds. Crown Ethers and Phase Transfer Catalysis in Polymer Science; Plenum Press, New York, **1984**.
27. Morgan, P. W. Condensation Polymers:By Interfacial and Solution Methods; Wiley, New York, **1965**.
28. Millich, F.; Carraher, C. E., Eds. Interfacial Synthesis, Vols. I and II; Dekker, New York, **1977**.
29. Carraher, C. E.; Preston, J., Eds. Interfacial Synthesis, Vol. III; Dekker, New York, **1982**.
30. Carraher, C. E.; Tsuda, M., Eds. Modification of Polymers; American Chemical Society, Washington, D. C., **1980**.
31. Carraher, C. E.; Moore, J. A., Eds. Modification of Polymers; Plenum Press, New York, **1983**.
32. Carraher, C. E.; Feddersen, M. F. Angew. Makromolekulare Chemie **1976**, 54, 119.
33. Carraher, C. E.; Ademu-John, C.; Fortman, J. J.; Giron, D. J.; Turner, C. J. Polymer materials **1984**, 1, 116.
34. Naoshima, Y., Applied Organometallic Chemistry 1987, 1, 245-249.

RECEIVED October 30, 1987

Chapter 31

Cross-Linking of Ethylene–Propylene Copolymer Rubber with Dicumyl Peroxide–Maleic Anhydride

Norman G. Gaylord, Mahendra Mehta, and Rajendra Mehta

Gaylord Research Institute, Inc., New Providence, NJ 07974

The reaction of EPR with dicumyl peroxide (DCP) at 180°C yielded a fraction insoluble in cyclohexane at 22°C. The presence of maleic anhydride (MAH) in the EPR-DCP reaction mixture increased the amount of cyclohexane-insoluble gel. However, the gel concentration decreased as the DCP concentration increased. The MAH content of the soluble polymer increased when either the MAH or the DCP concentration increased. The molecular weight of the soluble polymer increased with increasing MAH concentration and decreased with increasing DCP concentration in the reaction mixture. The products from the EPR-DCP and EPR-MAH-DCP reactions were soluble in refluxing xylene and were fractionated by precipitation with acetone. The presence of stearamide in the EPR-MAH-DCP reaction increased the amount and the molecular weight of the cyclohexane-soluble polymer.

The crosslinking of ethylene-propylene copolymer rubber (EPR) in the presence of organic peroxides has been investigated by Natta and/or his coworkers (1-3) and others (4,5). Co-agents such as sulfur (3,4) and unsaturated monomers (6), including maleic anhydride (MAH)(3,7) have been utilized in an effort to increase the crosslinking efficiency in the EPR-peroxide system.

As part of our continuing study of the peroxide-catalyzed reactions of MAH with saturated polyolefins, the present investigation was undertaken to determine the extent of crosslinking and/or degradation which accompanies the EPR-MAH reaction. The gel content, presumably indicative of crosslinking, was determined by extraction with cyclohexane at room temperature (22°C) for 60 hr.

EXPERIMENTAL

EPR REACTIONS. Vistalon 404 EPR (Exxon Chemical Co.)(40/60 wt ratio E/P, M_w/M_n 37) was fluxed for 2 min in the mixing chamber of a Brabender Plasticorder at 60 rpm and 180°C. A mixture of DCP, MAH and stearamide (SA), when present, was added in four portions at 2 min intervals to the 40g fluxing EPR. After the last addition, mixing

0097–6156/88/0364–0438$06.00/0

was continued for 2 min (total 10 min) and then the reaction mixture was quickly removed from the chamber.

CYCLOHEXANE EXTRACTION. A 5-6g portion of the product was cut into small pieces and stirred in 250 ml cyclohexane at room temperature for 60 hr. The insoluble fraction was separated by filtering the solution through cheesecloth. The cyclohexane-soluble fraction was recovered by distilling the solvent in vacuo and the polymer was dried in vacuo at 40°C for 24 hr.

XYLENE FRACTIONATION. A 5-6g portion of the product was cut into small pieces and heated in 200 ml refluxing xylene for 5 hr. The hot suspension was filtered through cheesecloth into 800 ml acetone. The polymer adhering to the sides of the flask and the small amount filtered from the suspension were refluxed with an additional 150 ml xylene until complete solution was noted and then precipitated in acetone. The precipitated polymers were combined as Fraction I. The combined filtrates were concentrated on a rotavaporator and the residual polymer, as well as the polymer recovered by precipitation, were dried in vacuo at 140°C for 4 hr. The polymer recovered from the filtrates was identified as Fraction II.

CHARACTERIZATION. The intrinsic viscosity of the soluble fractions was determined in toluene at 30°C. The MAH content of the soluble fractions was determined by heating a 0.5-1.0g portion in refluxing water-saturated xylene for 1 hr and titrating the hot solution with 0.05N ethanolic KOH using 1% thymol blue in DMF as indicator.

RESULTS

The [η] of EPR decreased when mixed for 10 min in a Brabender Plasticorder at 180°C. The complete solubility of the EPR in cyclohexane at 22°C was unchanged.

The presence of 0.25-0.5 wt-% DCP at 180°C resulted in the formation of about 20% of a cyclohexane-insoluble fraction. The presence of 5 wt-% MAH (based on EPR) increased the amount of cyclohexane--insoluble gel, whose concentration decreased from 65% to 27% as the DCP content increased from 0.25 to 1.0 wt-% (based on EPR), respectively. The cyclohexane-soluble polymer contained about 1 wt-% MAH which apparently increased while the [η] decreased as the DCP content increased (Table I).

When the DCP concentration was kept constant at 1 wt-% while the MAH concentration in the charge increased from 5 to 20 wt-%, the amount of cyclohexane-insoluble polymer decreased from 29 to 23%. The [η] of the cyclohexane-soluble polymer increased and its MAH content decreased, as the MAH concentration in the charge increased (Table II).

When stearamide (SA) was present in the DCP-MAH mixture added to EPR at 180°C, the amount of cyclohexane-insoluble EPR-g-MAH decreased, analogous to the effect of SA in reducing crosslinking in the PE-MAH-peroxide reaction (8,9). The [η] of the cyclohexane-soluble EPR-g-MAH increased when SA was present in the reaction mixture, analogous to the effect of SA in reducing degradation in the PP-MAH-peroxide reaction (Table III).

Table I. Effect of DCP at Constant MAH Concentration at 180°C

MAH wt-% on EP	DCP wt-% on EP	DCP wt-% on MAH	Cyclohexane Extraction[a] Soluble[b] %	Soluble[b] $[\eta]^c$ dl/g	Soluble[b] MAH wt-%	Insoluble %
0	0	0	100	1.20[d]	0	0
0	0.25	0	77	1.42	0	20
0	0.5	0	80	1.23	0	16
5	0.25	5	33	1.10	0.9	65
5	0.5	10	56	0.56	1.0	43
5	1.0	20	73	0.89	1.3	27

[a] Extraction with cyclohexane at 22°C for 60 hr
[b] Recovered by solvent removal from filtrate
[c] Toluene, 30°C
[d] Untreated EPR $[\eta]$ 1.42 dl/g

Table II. Effect of MAH at Constant DCP Concentration at 180°C

MAH wt-% on EP	DCP wt-% on EP	DCP wt-% on MAH	Cyclohexane Extraction[a] Soluble[b] %	Soluble[b] $[\eta]^c$ dl/g	Soluble[b] MAH wt-%	Insoluble %
5	1.0	20	70	1.31	0.7	29
10	1.0	10	72	1.34	0.5	26
20	1.0	5	75	1.64	0.3	23

a,b,c Footnotes to Table I

Table III. Effect of Stearamide in EPR-MAH-DCP Reaction at 180°C

MAH wt-% on EP	DCP wt-% on EP	DCP wt-% on MAH	SA[d] mole-% on MAH	Cyclohexane Extraction[a] Soluble[b] %	Soluble[b] $[\eta]^c$ dl/g	Soluble[b] MAH wt-%	Insoluble %
5	0.25	5	0	33	1.10	0.9	65
5	0.25	5	20	43	1.18	0.5	55
5	0.5	10	0	56	0.56	1.0	43
5	0.5	10	20	68	1.02	1.6	29

a,b,c Footnotes to Table I
[d] SA-MAH-DCP mixture added in 4 equal portions to molten EP

The polymer formed in the reaction of EPR with 0.5 wt-% DCP at 180°C, in the presence or absence of 5 wt-% MAH, was completely soluble in refluxing xylene, although it contained a fraction insoluble in cyclohexane at 22°C. The EPR and EPR-g-MAH were fractionated by addition of the xylene solution to acetone.

Both the acetone-precipitated polymer (I) and the polymer recovered by removal of the solvent from the acetone-xylene filtrate (II) had a higher $[\eta]$ when MAH was present in the charge, although the amount of acetone-precipitated polymer (I) was decreased when MAH had been present. The presence of stearamide in the EPR-MAH-DCP reaction mixture increased the amount and [η] of the acetone precipitated polymer (Table IV).

Table IV. Fractionation of EPR-MAH-DCP Product in Xylene[a]

MAH	DCP		SA	CH	Xylene Fractionation[c]					
					Fraction I[d]			Fraction II[e]		
wt-%	wt-%		mole-%	Insol[b]		[η][f]	MAH		[η][f]	MAH
on EP	on EP	on MAH	on MAH	%	%	dl/g	wt-%	%	dl/g	wt-%
0	0.0	0	0	0	82	2.44	0	15	0.60	0
0	0.5	0	0	16	85	1.17	0	15	0.37	0
5	0.5	10	0	43	48	1.38	1.9	46	1.23	2.9
5	0.5	10	20	29	68	1.71	1.2	28	0.94	2.7

[a] EPR reactions conducted in Plasticorder at 180°C
[b] Extraction with cyclohexane at 22°C for 60 hr
[c] Extraction with refluxing xylene for 5 hr
[d] Fraction I recovered by precipitating xylene solution in acetone
[e] Fraction II recovered by removal of solvent from filtrate
[f] Toluene, 30°C

DISCUSSION

The complete solubility in refluxing xylene of the EPR and EPR-g-MAH produced by reaction with DCP at 180°C suggests that treatment with cyclohexane at 22°C does not separate the crosslinked polymer from the uncrosslinked polymer but rather fractionates the uncrosslinked polymer based on molecular weight. Alternatively, the lightly crosslinked polymer is swollen and the crosslinks are broken in refluxing xylene.

The crosslinking of EPR in the presence of radicals from peroxide decomposition is attributable to attack on the secondary CH_2 moieties and the generation of polymer radicals which couple either with similar secondary polymer radicals or polymer radicals generated by attack on the tertiary CH moieties on the propylene units in the chain. The generation of tertiary radicals by the latter route is the predominant reaction. The tertiary radicals preferentially undergo disproportionation, resulting in degradation rather than crosslinking. Increasing the DCP concentration results in an increase in the amount of soluble polymer and a decrease in the molecular weight of the polymer due to increased tertiary CH attack, followed by disproportionation.

The rapid decomposition of a peroxide in the presence of MAH results in the excitation and homopolymerization of MAH. When the latter is conducted in the presence of polypropylene, the latter undergoes degradation (10) while polyethylene is crosslinked under the same conditions (11). This has been attributed to the presence of excited MAH which increases the radical generation on the polymer beyond that due to the radicals from the peroxide.

At a fixed MAH concentration, increasing the peroxide content results in a higher concentration of excited MAH and a greater generation of polymer radicals. The latter undergo disproportionation and the amount of soluble polymer increases while its molecular weight decreases. Notwithstanding, at a given peroxide concentration, the presence of MAH increases the amount of cyclohexane-insoluble polymer approximately threefold. At a fixed DCP concentration, increasing the MAH content has little effect on the concentration of excited MAH and therefore a negligible effect on the amount of soluble polymer. Further, the MAH content of the polymer decreases with an increase in the amount of MAH in the charge due to the quenching of excited MAH by excess ground state MAH.

Although it has been suggested that crosslinking in the presence of MAH involves coupling of appended MAH radicals with other appended MAH radicals or with polymer radicals (7), the former is improbable due to the tendency for disproportionation rather than coupling between radicals derived from strong electron acceptor monomers such as MAH.

The increased crosslinking may be attributed to the preferential generation of secondary polymer radicals due to the presence of excited MAH and therefore an increase in the amount of coupling of secondary polymer radicals rather than of pendant MAH radicals.

Stearamide is one of many electron donors which donate an electron to the cationic moiety in excited MAH or in propagating -MAH chains. This results in the inhibition of the homopolymerization of MAH and decreases the crosslinking of polyethylene and the degradation of polypropylene which accompany the peroxide-catalyzed reaction of MAH with these polyolefins (8,9).

The presence of stearamide in the EPR-MAH-DCP reaction increases the yield of soluble polymer, i.e. decreases the crosslinking, which polymer has a higher molecular weight, i.e. decreases the degradation, than noted in the absence of stearamide.

Literature Cited

1. Crespi, G.; Bruzzone, M. Chim. e Ind. (Milan) 1959, 41, 741.
2. Natta, G.; Crespi, G.; Bruzzone, M. Kautschuk u. Gummi 1961, 14 (3), WT54.
3. Natta, G.; Crespi, G.; Valvassori, A.; Sartori, G. Rubber Chem. Technol. 1963, 36, 1583.
4. Loan, L. D. J. Polym. Sci. 1964, A2, 3053.
5. Harpell, G. A.; Walrod, D. H. Rubber Chem. Technol. 1973, 46, 1007.
6. Lenas, L. P. Ind. Eng. Chem., Prod. Res. Dev. 1963, 2, 202.
7. Natta, G.; Crespi, G.; Borsini, G. (to Montecatini Soc. Gen.) U.S. Patent 3 236 917, 1966.
8. Gaylord, N. G. (to Gaylord Research Institute, Inc.) U.S. Patent 4 506 056, 1985.
9. Gaylord, N. G.; Mehta, R. Polymer Preprints 1985, 26 (1), 61.
10. Gaylord, N. G.; Mishra, M. K. J. Polym. Sci., Polym. Lett. Ed. 1983, 21, 23.
11. Gaylord, N. G.; Mehta, M. J. Polym. Sci., Polym. Lett. Ed. 1982, 20, 481.

RECEIVED October 13, 1987

Chapter 32

Synthesis of Conjugated Polymers via Polymer Elimination Reactions

Samson A. Jenekhe

Physical Sciences Center, Honeywell, Inc., 1071 Lyndale Avenue, South Bloomington, MN 55420

The transformation of nonconjugated polymers into conjugated polymers using elimination reactions is described. Heterocyclic conjugated polymers containing alternating aromatic and quinonoid sections in the main chain are synthesized by chemical or electrochemical redox elimination reaction on soluble precursor polymers containing *sp*^{3}-carbon atom bridges between the aromatic heterocyclic units. Progress of the redox elimination process is followed by infrared and electronic spectra as well as by cyclic voltammetry. A reaction mechanism in which the precursor polymer undergoes a redox reaction followed by loss of the bridge hydrogens is proposed. The resulting conjugated aromatic/quinonoid polymers generally have very small semiconductor band gaps in accord with predictions of recent theoretical calculations. A brief review of related syntheses of conjugated polymers from nonconjugated precursor polymers is also given.

Conjugated polymers are currently of wide interest because of their electronic (1), electrochemical (2), and nonlinear optical (3) properties which originate from their delocalized π-electron systems. Unfortunately, the high density and geometrical disposition of π-bonds in conjugated polymers which confine the desirable electronic and optical properties also make them more insoluble and infusible relative to nonconjugated polymers. The many synthetic routes to conjugated polymers may be classified into two broad categories: (1) those in which the target conjugated polymer is obtained directly from conventional addition and condensation polymerization processes (4), and (2) those involving transformation of an existing nonconjugated precursor polymer into the target conjugated polymer (5). The latter approach is especially attractive since the nonconjugated precursor polymer can be more readily processed into films and other forms prior to conversion to the

0097–6156/88/0364–0443$06.00/0

conjugated derivative. Also, the nonconjugated precursor polymer route will allow the tuning of physical properties of the conjugated derivative, including density, morphology, crystallinity, and electronic and optical properties. Among the possible polymer reactions which could be used to convert nonconjugated polymers into conjugated ones are elimination, addition, and isomerization (5), but our primary interest here is polymer elimination reactions.

One of the earliest known synthesis of a conjugated polymer by elimination on a nonconjugated precursor was achieved by Marvel et al (6) who demonstrated that polyacetylene was obtained by elimination of HCl from polyvinyl chloride as illustrated in Figure 1. Feast and his co-workers have recently described an elegant synthesis of polyacetylene (PA) films by thermal elimination of aromatic hydrocarbons such as 1,2-bis(trifluoromethyl)benzene, naphthalene and anthracene from films of soluble nonconjugated precursor polymers (7), Figure 2. The resulting polyacetylene films, now known in the literature as Durham polyacetylene, have physical properties that are significantly different from the Shirakawa polyacetylene: Durham PA is largely amorphous and has a non-fibrous morphology compared to Shirakawa PA which is highly crystalline and fibrous (8-9). Also recently, Lenz, Karasz and co-workers (10) have reported the synthesis of poly(p-phenylenevinylene) (PPV) films by thermal elimination of $(CH_3)_2S$ and HCl from poly(p-xylene-α-dimethylsulfonium chloride), a soluble polyelectrolyte (see Figure 3). Highly oriented Durham PA and PPV films with stretch ratios up to 20 have been obtained by stretching the precursor polymer films during the transformation to conjugated derivatives; such oriented films when doped give rise to conductors with large anisotropy in electrical conductivity. The Durham route to polyacetylene and the above PPV synthesis demonstrate the rich potentials of synthesis of conjugated polymers via elimination reactions on nonconjugated precursor polymers.

Recently, we discovered a novel type of polymer elimination reaction for producing conjugated polymers from the class of nonconjugated polymers containing alternating sp^3-carbon atom (-C(R)H-) and conjugated sections in the main chain (11-13). Under chemical or electrochemical oxidative or reductive conditions such polymers undergo irreversible reactions leading to loss of the bridge hydrogens, converting the sp^3-carbon to sp^2-carbon as illustrated in Figure 4, where D or A denotes π-electron donor or acceptor conjugated sections. Initial observations on this redox elimination reaction were made on poly(3,6-N-methylcarbazolediyl methylene) (PMCZM) (11) and poly(3,6-N-methylcarbazolediyl benzylidene) (PMCZB) (12). For example, upon oxidation of PMCZM and PMCZB with bromine or iodine it was shown that a bridge hydrogen was eliminated as HBr or HI resulting in fully conjugated doped derivatives which had dc conductivities (0.1-1 S/cm) as high as the parent poly(3,6-N-methylcarbazolediyl) (11-12). Figure 5 shows the optical absorption spectra of a film of PMCZM before and after a prolonged exposure to iodine vapor. The large red shift, corresponding to a narrowing of the optical band gap or increase in the degree of conjugation is to be noted. Also, it is significant that elimination of the bridge hydrogens (-C(R)H-) in PMCZM and PMZCB produced conjugated polymers containing quinonoid sections in the main chain. These initial

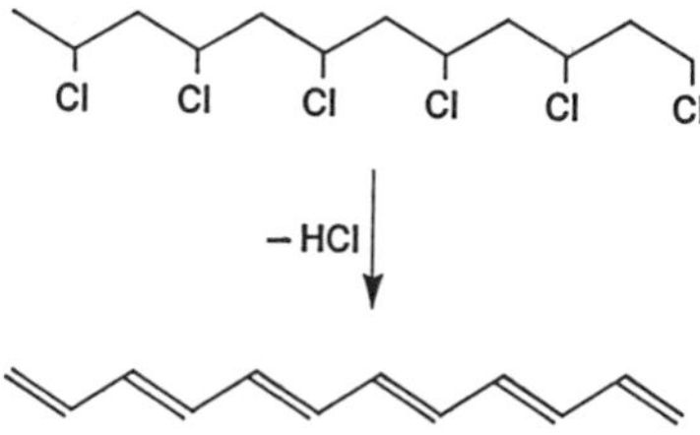

Cl Cl Cl Cl Cl Cl

− HCl

Figure 1. Synthesis of polyacetylene from polyvinyl chloride by elimination of HC1.

CF_3 CF_3 CF_3 CF_3 CF_3 CF_3

I II III

IV V VI

VII

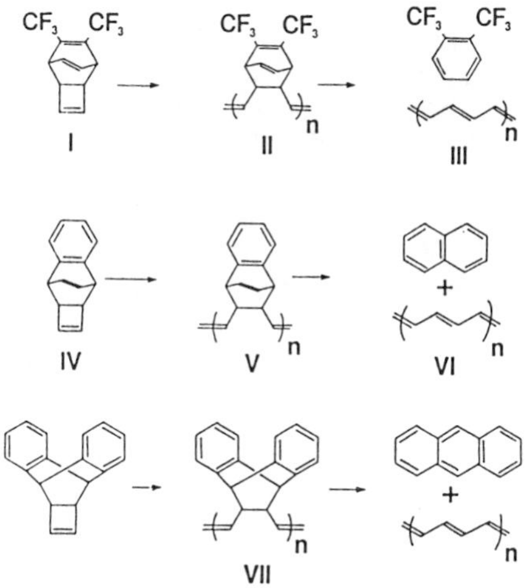

Figure 2. The soluble nonconjugated precursor polymer route to Durham polyacetylene via thermal elimination.

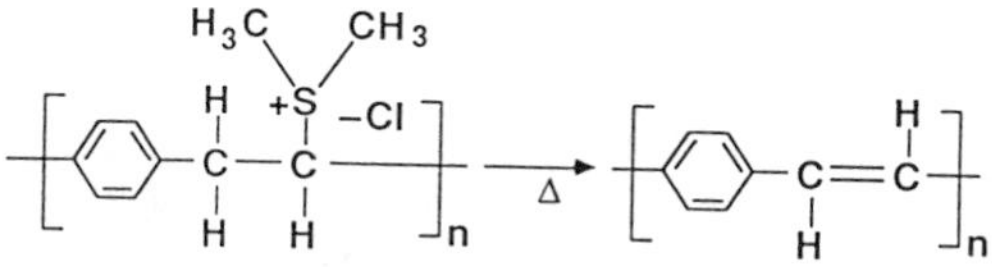

H_3C CH_3 $+S$ $-Cl$ Δ

Figure 3. Synthesis of poly(p-phenylene vinylene) films by thermal elimination on a soluble polyelectrolyte.

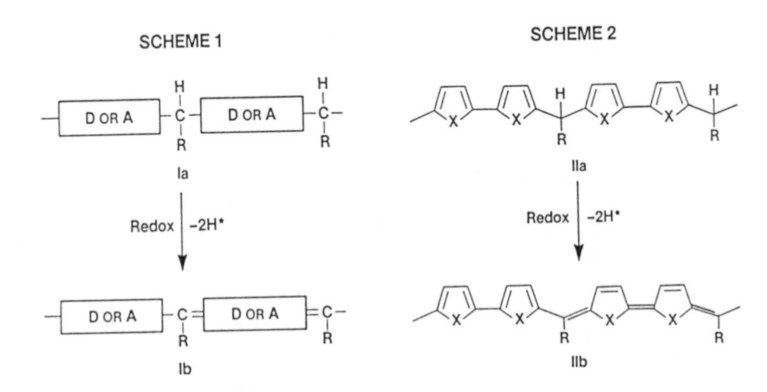

Figure 4. General scheme of irreversible redox elimination reaction on the class of nonconjugated polymers containing alternating -C(R)H- and conjugated sections in the main chain yielding conjugated polymers.

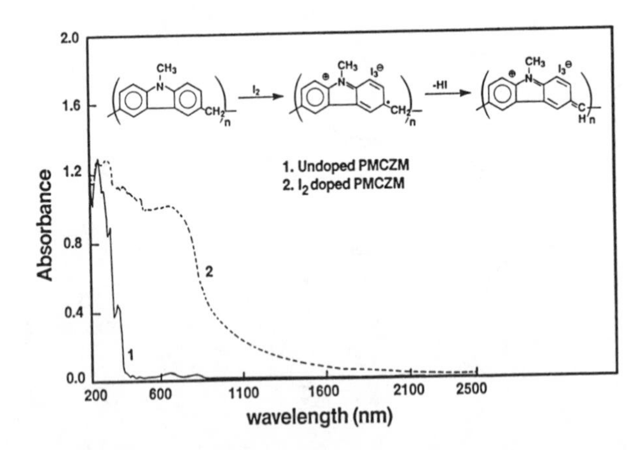

Figure 5. Optical absorption spectra of a thin film of precursor PMCZM (1) and after exposure to iodine vapor (2). Also shown is the proposed scheme of redox elimination yielding a doped conjugated conducting polymer.

observations lead us to pursue the synthesis of conjugated polymers containing alternating aromatic and quinonoid sequences in the main chain via redox elimination on precursors, such as the heterocyclic conjugated polymers of Figure 6. Extensive recent theoretical calculations predict that very narrow band gaps would be obtained in conjugated polymers containing quinonoid sections (14-18). We recently reported the smallest bandgap in organic polymers for such conjugated polymers (19).

In this paper we present results on the polymer redox elimination reaction used in the synthesis of the polymers in Figure 6. Preliminary results on electrochemical redox elimination on precursor polymers are also presented. A mechanism of the polymer elimination reaction is proposed. Related recent experimental observations at other laboratories that can be described within the framework of the scheme of Figure 4 are discussed.

Experimental

Precursor Polymers. The synthesis and characterization of the nonconjugated precursor polymers containing alternating heteroaromatic units and -C(R)H- in the main chain are described elsewhere (20-21). These nonconjugated precursor polymers include polythiophenes, polypyrroles and polyfurans; however, only the studies done with some of the polythiophenes and their related conjugated derivatives are described in detail here: poly(2,5-thiophenediyl benzylidene) (PTB), poly[α-(5,5'-bithiophenediyl) benzylidene] (PBTB), poly[α-(5-5'-bithiophenediyl)p-acetoxybenzylidene] (PBTAB), and poly[α-(5-5"-terthiophenediyl) benzylidene] (PTTB). These precursor polymers are soluble in organic solvents including tetrahydrofuran (THF), methylene chloride and N,N-dimethylformamide (DMF). Thin films used in spectroscopy and for cyclic voltammetry were cast from THF, methylene chloride or DMF.

Elimination Procedures. Chemical redox-induced elimination was performed on precursor thin films by exposure to bromine or iodine vapor or by immersion of films in hexane solutions of these halides.

In order to follow progress of elimination, reactions were also performed on thin films in a special sealed glass cell which permitted *in situ* monitoring of the electronic or infrared spectra at room temperature (23°C). Typically, the infrared or electronic spectrum of the pristine precursor polymer film was obtained and then bromide vapor was introduced into the reaction vessel. *In situ* FTIR spectra in the 250-4000 cm^{-1} region were recorded every 90 sec with a Digilab Model FTS-14 spectrometer and optical absorption spectra in the 185-3200 nm (0.39-6.70 eV) range were recorded every 15 min with a Perkin-Elmer Model Lambda 9 UV-vis-NIR spectrophotometer. The reactions were continued until no visible changes were detected in the spectra.

Cyclic voltammetry was performed on precursor polymer thin films cast on platinum electrodes in order to assess the possibility of electrochemical redox elimination and consequently as an alternative means of monitoring the process. All electrochemical experiments were performed in a three-electrode, single-compartment cell using a double junction $Ag/Ag^{+}(AgNO_3)$ reference electrode in 0.1M tetraethylammonium perchlorate (TEAP)/acetonitrile. The voltage regulator was a Princeton Applied Research Model 175 Universal programmer. Current measurement was done using a Princeton Applied Research Model 173 Potentiostat/Galvanostat in conjunction with Princeton Applied Research Model 376 logarithmic current converter. The cyclic voltammograms were recorded on a HP Model 7040A X-Y recorder.

Results

Figure 7 shows the FTIR spectra of initial precursor PBTB film (A) and at two subsequent times (B and C) during in situ elimination reaction with bromine vapor at 23°C. Several main changes in the infrared spectrum of the precursor film are observed. The aliphatic C-H stretching vibration bands in the 2800-3000 cm^{-1} region attributed to the bridge sp^3 C-H are significantly modified, generally decreasing with elimination reaction time and eventually disappearing. In contrast, the absorption bands in the 3060-3250 cm^{-1} region attributed to the side chain phenyl and thiophene C-Hs remain relatively the same. A new gas phase absorption band at 2400-2800 cm^{-1} assigned to evolved HBr gas appears as seen in Figure 7 (B and C). The gas phase absorption band attributed to HBr gas appeared in the first spectrum after bromine was introduced into the reaction cell and intensified with reaction time. New absorption bands appeared and intensity of existing ones increased in the carbon-carbon double bond absorption region, 1000-1600 cm^{-1}. There was no evidence of a new band in the 500-650 cm^{-1} region that would be attributable to a C-Br bond due to substitution or bromination reactions on the polymer chain. These IR spectra results clearly show that the bridge hydrogens (-C(R)H-) of PBTB are eliminated as HBr by reaction with bromine, yielding the anticipated aromatic/ quinonoid conjugated polymers of Figure 6.

The electronic absorption spectra at different times of elimination reaction on PBTB are shown in Figure 8. As-synthesized PBTB is a green polymer (Curve 1, Figure 8) due to partial conversion of the -C(R)H- to -C(R)= bridges in the main chain (19-21). After 15 min. reaction time, the electronic spectrum is greatly red-shifted with the absorption edge now located at about 1200 nm (1.03 eV). Progress in elimination further moves the absorption edge to longer wavelength as curves 2 to 13 in Figure 8 show. No observable changes in the electronic absorption spectra were seen after 24 hours (Curve 13). Visually, the sample changes from green (Curve 1) to metallic gray in color. The absorption edge determined from Curve 13 is about 1500 nm (0.83 eV). This represents an extremely large increase in the degree of π-electron delocalization. The relatively sharp absorption edges are also noteworthy.

Figure 9 shows the electronic absorption spectrum of a PTTB film which has undergone extensive but incomplete reaction with bromine in a non-in-situ experiment. The absorption spectrum is that expected for a one-dimensional conjugated polymer. The sharpest absorption edge is at about 1490 nm (o.83 eV) and the absorption maximum is located at 1240 nm (1.0 eV). Thus, this material has a bandgap of about 0.83 eV. Note that two small

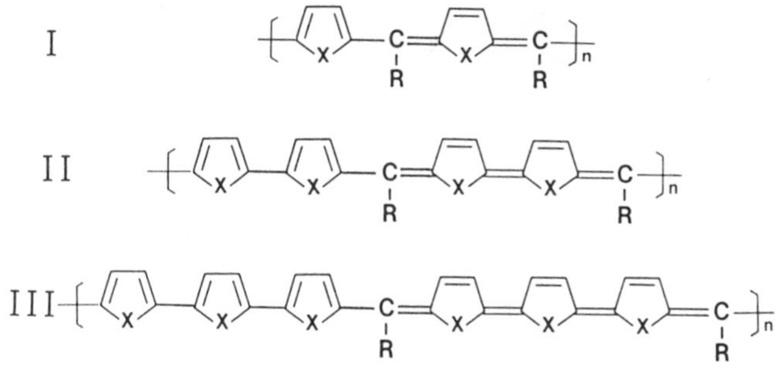

I

II

III

Figure 6. Novel conjugated heteroaromatic polymers synthesized by redox elimination.

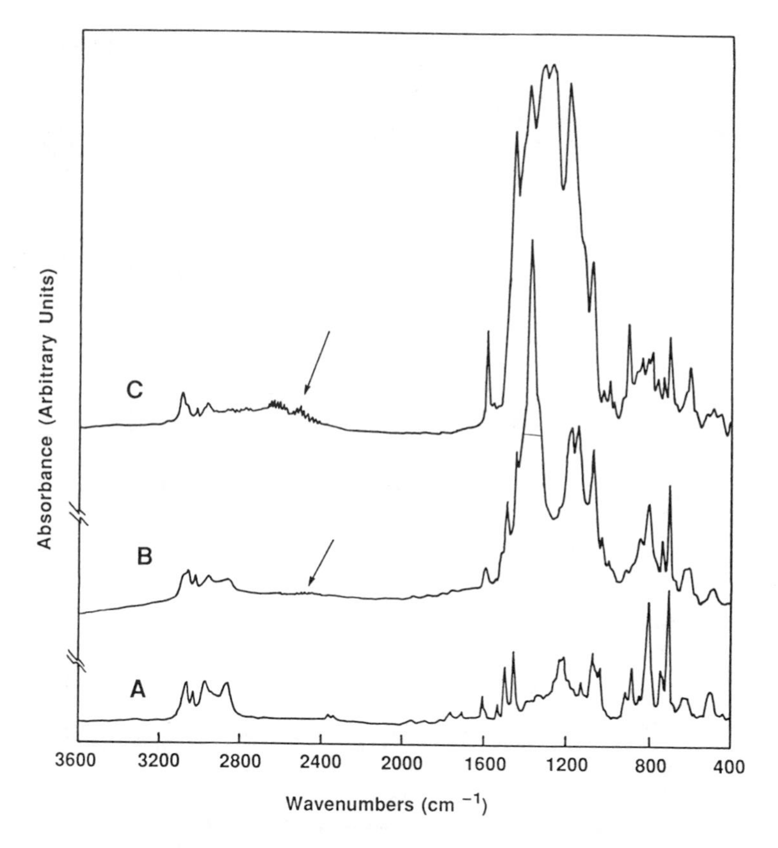

Figure 7. FTIR absorption spectra of precursor PBTB thin film (A) and during in-situ elimination (B and C) in bromine vapor at 23°C. Arrow indicates a superposed gas phase absorption band due to evolved HBr gas.

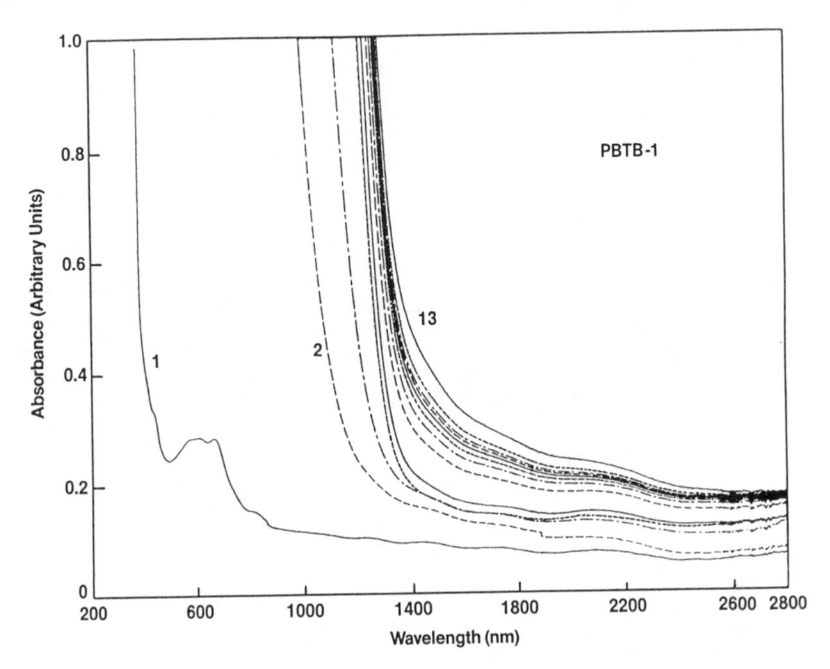

Figure 8. Electronic absorption spectra of precursor PBTB thin film (1) and conjugated derivatives (2-13) at different times during <u>in-situ</u> elimination reaction at 23°C.

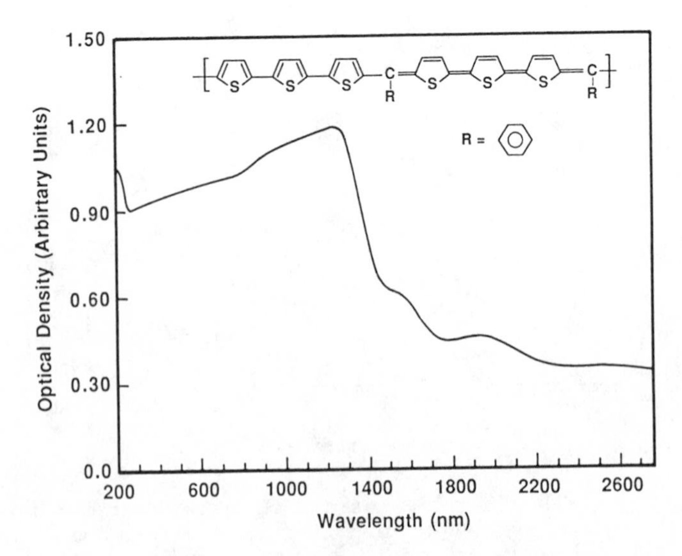

Figure 9. Electronic absorption spectra of a conjugated derivative of PTTB.

absorptions can be seen below the gap at 1560 nm (0.79 eV) and 1950 nm (0.64 eV). The two absorptions below the bandgap may be interpreted as evidence of charged bipolarons (dications) (22). However, no such below bandgap absorptions are observed in the completely reacted material.

Some of the results of cyclic voltammetric studies of a film of PBTAB at various stages of redox elimination are shown in Figures 10A and 10B. The cyclic voltammogram of Figure 10A shows the behavior of a precursor PBTAB film on platinum electrode under repeated cycling at 200mV/s. Initially, an oxidation peak at about 0 V (vs. Ag/Ag^+), a main reduction peak at -0.56 V and a small broad reduction peak centered at -0.15 V are observed. This initial votammogram is very unstable under cycling: (i) the redox peaks at -0.56 V and 0.0 V are decreasing; and (ii) evolving very broad peaks that form a redox couple are observed in the range +0.14-0.21 and -0.09 to -0.23 V. After numerous cycling (~100) over a period of hours the resulting polymer film was washed and its cyclic voltammogram was again obtained in a fresh electrolyte. A reversible redox couple with oxidation peak at +0.21 V and reduction peak at -0.24 V was obtained as shown in Figure 10B. This later cyclic voltammogram is quite stable under repeated cycling and is what one might expect for an electroactive conjugated polymer. These results suggest that the electrochemical redox reaction induces elimination of the -C(R)H- bridge hydrogens, yielding -C(R)= bridges.

Preliminary measurements of electrical conductivity of the conjugated derivatives of PBTAB, PBTB and PTTB obtained by the above treatment with bromine vapor are poor semiconductors with a conductivity of the order 10^{-6}S/cm which apparently is not due to doping. Subsequent electrochemical or chemical doping of these polymers lead to 4-6 orders of magnitude increase in conductivity. Ongoing studies of the electrical properties of these conjugated polymers with alternating aromatic/quinonoid units will be reported elsewhere.

Similar redox elimination reactions have been performed on several heteroaromatic precursor polymers within the class shown in Figure 4 and whose conjugated derivatives are shown in Figure 6. In general, the results and observations are similar to those described here. The major differences observed are due to differences in the heteroatom (X) while minor differences are observed with different R groups. For example, the smallest bandgaps were obtained when X=S in the polymers of Figure 6. On the other hand, faster elimination kinetics were observed for aromatic R compared to aliphatic R side groups.

Discussion

The electronic absorption spectra of Figures 8 and 9, and those not shown, reveal that the conjugated polymers of Figure 6 generally exhibit intrinsic semiconductor bandgaps that are very small. This is in accord with extensive recent theoretical calculations which predict that introduction of quinoidal geometry into the main chain of aromatic conjugated polymers (e.g. polythiophene, polypyrrole, and poly-p-phenylene) reduces the bandgap. Especially relevant are the

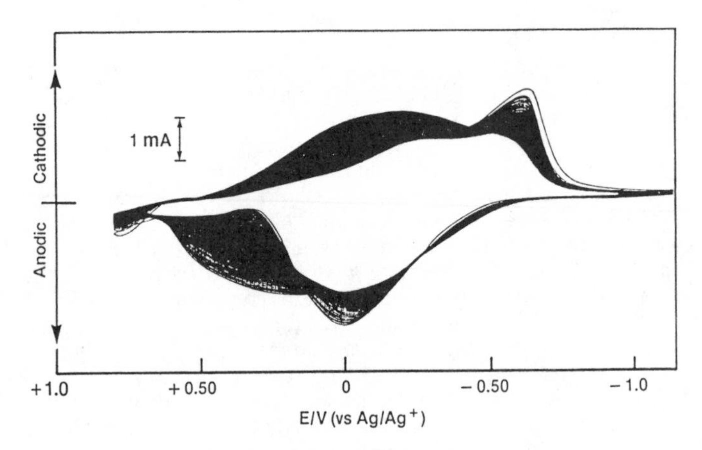

Figure 10A. Cyclic voltammogram of a precursor PBTAB thin film on Pt electrode in TEAP/acetonitrile at a scan rate of 200 mV/s.

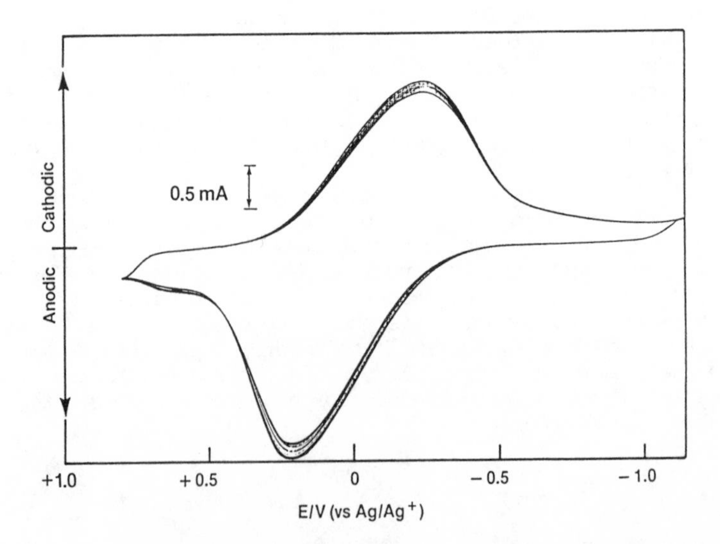

Figure 10B. Cyclic voltammogram of the same film in Figure 10A obtained after washing and running in a fresh TEAP/ acetonitrile at a scan rate of 200 mV/s.

valence effective hamiltonian (VEH) calculations of Bredas and co-workers (14-16) which are known to be very accurate in reproducing experimental data for conjugated polymers. In particular, Bredas has shown that as quinonoid structure is introduced into poly(2,5-thiophene) geometry the bandgap decreases linearly with increasing quinonoid character (14). More recently, the electronic band structures of the aromatic/quinonoid polymers of Figure 6 have been obtained using MNDO polymer geometry optimized total energy calculations (23). The theoretical energy bandgaps (23) were found to be in agreement with the experimental results (19).

Mechanism of Redox Elimination. The proposed elimination reaction mechanism is illustrated with a methylene-bridged polybithiophene in Figure 11. If one considers attempts to oxidize the nonconjugated precursor polymer by electron withdrawal or reduce it by electron addition, one finds that the added charge cannot be accommodated by delocalization because of the sp^3 hybridized carbon ($-CH_2-$) bridges. Consequently, highly unstable charged polymeric chains are generated. Depending on the sign of charge added to the precursor chain, such a high energy unstable intermediate would be best stabilized by loss of H^+ or H^- from adjacent bridges, yielding the neutral conjugated polymer chain containing alternating aromatic and quinoidal units. This mechanism can be used to explain the electrochemical results. For example, the observed decreasing redox peaks in the cyclic voltammogram of Figure 10A is attributed to the consumption of the charged intermediates. The optical absorption data of Figure 9 also appear to evidence bipolarons (dications) as intermediate and hence, consistent with this mechanism.

The general scheme of Figure 11 is for both electrochemical and chemical redox elimination. If we consider the case of oxidative elimination with bromine, generation of the necessary radical cation or dication intermediate is due to: $Br_2 + 2e^- \longrightarrow 2Br^-$. The bromide ions presumably then abstract protons from the ionized precursor.

Related Polymer Systems and Synthetic Methods. Figure 12A shows a hypothetical synthesis of poly (p-phenylene methide) (PPM) from polybenzyl by redox-induced elimination. In principle, it should be possible to accomplish this experimentally under similar chemical and electrochemical redox conditions as those used here for the related polythiophenes. The electronic properties of PPM have recently been theoretically calculated by Boudreaux et al (16), including: bandgap (1.17 eV); bandwidth (0.44 eV); ionization potential (4.2 eV); electron affinity (3.03 eV); oxidation potential (-0.20 vs SCE); reduction potential (-1.37 eV vs SCE). PPM has recently been synthesized and doped to a semiconductor (24).

The two limiting structural forms of polyaniline are shown in Figure 12B: (i) the nonconjugated leuco base polymer in which the imine nitrogens are completely protonated (PAN-1); and (ii) the neutral conjugated polymer containing alternating aromatic and quinoidal units (PAN-2). The highly conducting (10 S/cm) polyaniline chemically or electrochemically synthesized from aniline is believed to be a partially oxidized form with a structure intermediated between PAN-1 and PAN-2 (25-28). The complicated electrochemical behavior of polyaniline, especially in aqueous media, is a result of redox reactions coupled with reversible deprotonation/

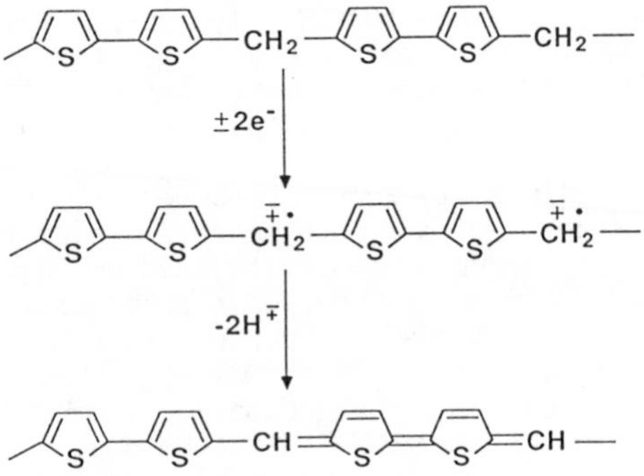

Figure 11. Illustration of the proposed mechanism of redox-induced elimination.

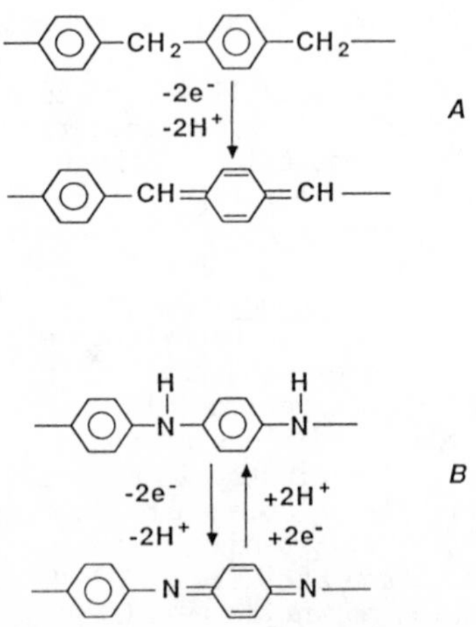

Figure 12. A possible synthesis of poly(p-phenylene methine) from polybenzyl via redox elimination (A). The two limiting structures of polyaniline (B).

protonation reaction. This polymer system with reversible conversion between PAN-1 and PAN-2 is to be contrasted with the irreversible polymer systems of Figures 4 and 12A. Theoretical calculations (28) of the electronic properties of both PAN-1 and PAN-2 show the best properties when the dihedral angle between adjacent rings on the chain is zero; PAN-1: bandgap (3.6 eV), bandwidth (3.0 eV), ionization potential (4.2 eV), oxidation potential (-0.2 V *vs* SCE), reduction potential (-3.8 V *vs* SCE); PAN-2: bandgap (0.7 eV), bandwidth (1.5 eV), ionization potential (5.3 eV), oxidation potential (+0.9 *vs* SCE), reduction potential (-0.2 V *vs* SCE). It is noteworthy that PAN-2 has a dramatically smaller bandgap than PAN-1.

Figure 13 shows the irreversible conversion of a nonconjugated poly (p-phenylene pentadienylene) to a lithium-doped conjugated derivative which has a semiconducting level of conductivity (0.1 to 1.0 S/cm) (29). Obviously, the neutral conjugated derivative of poly (p-phenylene pentadienylene) can then be reversibly generated from the n-type doped material by electrochemical undoping or by p-type compensation. A very similar synthetic method for the conversion of poly(acetylene-co-1,3-butadiene) to polyacetylene has been reported (30), Figure 14. This synthesis of polyacetylene from a nonconjugated precursor polymer containing isolated CH_2 units in an otherwise conjugated chain is to be contrasted with the early approach of Marvel *et al* (6) in which an all-sp^3 carbon chain was employed.

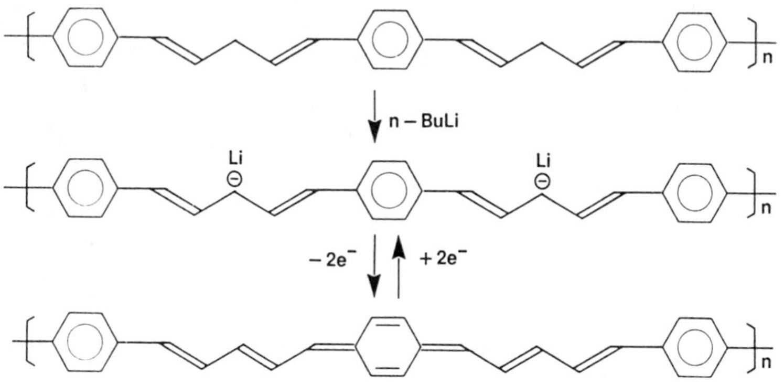

n – BuLi

Li ⊖ Li ⊖

– 2e⁻ + 2e⁻

Figure 13. Synthesis of a conjugated derivative from poly(p-phenylene pentadienylene).

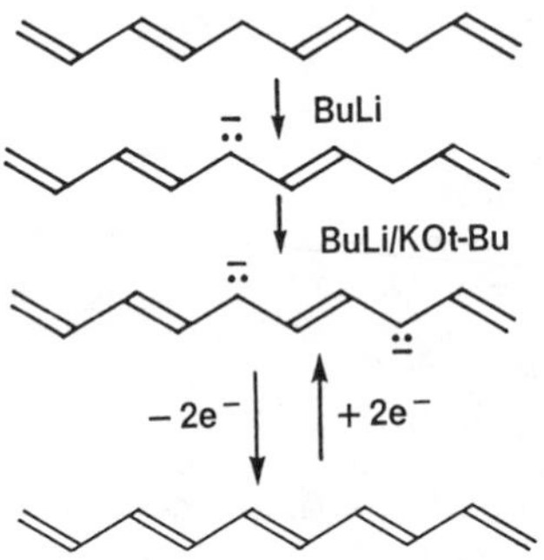

Figure 14. Synthesis of polyacetylene from poly(acetylene-co-1,3-butadiene).

Conclusions

Heterocyclic conjugated polymers containing alternating aromatic and quinonoid sections in the main chain have been synthesized by chemical or electrochemical redox elimination reaction on soluble precursor polymers containing -C(R)H- bridges betwen aromatic heterocyclic units. Progress of the redox elimination reaction converting the nonconjugated precursors to the conjugated polymers was monitored by infrared spectra, electronic absorption spectra and cyclic voltammetry. Some of the resultant conjugated polymers have the smallest known bandgaps among organic polymers (19) and the generally small semiconductor bandgaps are in accord with predictions of recent theoretical calculations (14-18, 23).

A reaction mechanism for the elimination reaction in which the precursor polymer undergoes redox reaction followed by loss of the bridge hydrogens is proposed. A brief review of related syntheses of conjugated polymers via elimination reactions on nonconjugated precursor polymers is given. It is suggested that this synthetic approach to conjugated polymers holds promise not only for achieving processing advantages but also for tuning electronic, optical and other physical properties of the target conjugated polymer.

Acknowledgments

This research was supported in part by the Office of Naval Research. The technical contributions of Marcia K. Hansen, James R. Peterson, and Lee Hallgren to the work described here is appreciated.

Literature Cited

1. Skotheim, T.A., (Ed.) Handbook of Conducting Polymers, Vols. 1 and 2, Marcel Dekker, New York, 1986.
2. (a) Chidsey, C.E.D.; Murray, R.W. Science 1986, 231, 25-31. (b) MacDiamid, A.G.; Kamer, R.B. in:ref 1, 689-727.
3. (a) Williams, D.J., (Ed.) Nonlinear Optical Properties of Organic and Polymeric Materials, Am. Chem. Soc., Washington, D.C., 1983. (b) Agrawal, G.P.; Cojan, C.; Flytzanis, C. Phys. Rev. 1978, B17,, 778. (c) Rustang, K.C.; Ducuing, J. Opt. Commun. 1974, 10, 258-261. (d) Sauteret, C.; Hermann, J.P.; Frey, R.; Pradere, F.; Ducuing, J.; Baughman, R.H.; Change, R.R. Phys. Rev. Lett. 1976, 38, 956-959.
4. Odian, G. Principles of Polymerization, 2nd Ed., Wiley, New York, 1981.
5. (a) Davydov, B.E.; Krentsel, F.A. Adv. Polym. Sci. 1977, 25, 1. (b) Feast, W.J., in:ref. 1, 1-43.
6. Marvel, C.S.; Sample, J.H.; Roy, M.F. J. Am. Chem. Soc. 1939, 61, 3241-3244.
7. (a) Edwards, J.H.; Feast, W.J. Polymer 1980, 21, 595. (b) Bott, D.C.; Chai, C.K.: Edwards, J.H.; Feast, W.J.; Friend, R.H.; Horton, M.E. J. de Phys. (Paris) 1983, 44, C3-143.
8. Friend, R.H.; Bott, D.C.; Bradley, D.D.C.; Chai, C.K.; Feast, W.J.; Foot, P.J.S.; Giles, J.R.M.; Horton, M.E.; Pereira, C.M.; Townsend, P.D. Phil. Trans. R. Soc. Lond. 1985, A314, 37-49.
9. Brandley, D.D.C.; Friend, R.H.; Hartmann, T.; Marseglia, E.A.; Sokolowski, M.M.; Townsend, P.D. Synthetic Metals, in press.
10. (a) Capistran, J.D.; Gagnon, D.R.; Antoun, S.; Lenz, R.W.; Karasz, F.E. Polym. Preprints 1984, 25(2), 282. (b) Gagnon, D.R.; Capistran, J.D.; Karasz, F.E.; Lenz, R.W. Polym. Preprints 1984, 25(2), 284. (c) Gagnon, D.R.; Capistran, J.D.; Karasz, F.E.; Lenz, R.W. Polym. Bulletin 1984, 12, 293. (d) Karasz, F.E.; Capistran, J.D.; Gagnon, D.R.; Lenz, R.W. Mol. Cryst. Liq. Cryst. 1985, 118, 327.
11. (a) Jenekhe, S.A.; Wellinghoff, S.T.; Deng, Z. Synthetic Metals 1985, 10, 281. (b) Jenekhe, S.A.; Wellinghoff, S.T.; Deng, Z. Polym. Preprints 1984, 25(2), 240.
12. Jenekhe, S.A., to be submitted.
13. Jenekhe, S.A. Macromolecules 1986, 19, 2663-2664.
14. Bredas, J.L. Mol. Cryst. Liq. Cryst. 1985, 118, 49-56; J. Chem. Phys. 1986, 82, 3808.
15. Bredas, J.L.; Themas, B.; Fripiat, J.G.; Andre, J.M.; Chance, R.R. Phys. Rev. B. 1984, 29, 6761.
16. Boudreaux, D.S.; Chance, R.R.; Elsenbaumer, R.L.; Frommer, J.E.; Bredas, J.L.; Silbey, R. Phys. Rev. 1985, B31, 652-655.
17. Whangbo, M.H.; Hoffmann, R.; Woodward, R.B. Proc. Royal Soc. Lond. 1979, A366, 23-46.
18. Wennerstrom, O. Macromolecules 1985, 18, 1977.
19. Jenekhe, S.A. Nature (London) 1986, 332, 345.
20. Jenekhe, S.A., to be published.
21. Jenekhe, S.A., submitted for publication.

22. Bredas, J.L.; Street, G.B. Acc. Chem. Res. 1985, 18, 309-315.
23. Kertesz, M.; Lee, Y.S. Am. Chem. Soc. National Mtg., Denver, CO, Div. Phys. Chem., Abstract #204, April 5-10, 1987. Also, J. Phys. Chem. 1987, in press.
24. Al-Jumah, K.; Fernandez, J.E.; Peramunage, D.; Garcia-Rubio, L.H. Abstracts, Am. Chem. Soc., 193rd National Mtg., Denver, CO, April 5-10, 1987.
25. MacDiarmid, A.G.; Chiang, J.C.; Halpern, M; Huang, W.S.; Mu, S.L.; Somasiri, N.L.D.; Wu, W.; Yaniger, S.I. Mol. Cryst. Liq. Cryst. 1985, 121, 173.
26 (a) MacDiarmid, A.G.; Chiang, J.C.; Huang, W.S.; Humphrey, B.D.; Somasiri, N.L.D. Mol. Cryst. Liq. Cryst. 1985 125, 309.
(b) Salaneck, W.R.; Lundstrom, I.; Huang, W.S.; MacDiarmid, A.G. Synth. Metals 1986, 13, 291.
27. Wnek, G.E. Synth. Metals 1986, 15, 213-218.
28. Chance, R.R.; Boudreaux, D.S.; Wolf, J.F.; Shacklette, L.W.; Sibley, R.; Themas, B.; Andre, J.M.; Bredas, T.L. Synth. Metals 1986, 15, 105-114.
29. Gordon, III, B.; Hancock, L.F. Abstract and Preprints of paper presented at Speciality Polymers '86, Baltimore, MD, August 6-8, 1986.
30. Tolbert, L.M.; Schomaker, J.A.; Holler, F.J. Synth. Metals 1986, 15, 195-199.

RECEIVED August 27, 1987

Chapter 33

Cross-Linking and Isomerization Reactions of an Acetylene-Terminated Polyisoimide Prepolymer

R. H. Bott, L. T. Taylor [1], and T. C. Ward

Department of Chemistry and Polymer Materials and Interfaces Laboratory, Virginia Polytechnic Institute and State University, Blacksburg, VA 24061-0699

Acetylene terminated polyimide prepolymers have many advantages over conventional polyimides in the areas of processing and solvent resistance. In addition, the presence of the isoimide structure further extends the the utility of these systems by modification of the solubility properties and glass transition temperature. This work discusses the thermal crosslinking and isomerization reactions occurring in the acetylene terminated isoimide prepolymer: Thermid IP600. The techniques of Fourier Transform Infrared Spectrometry and Differential Scanning Calorimetry are used to determine the contribution of these two reactions during the thermal cure including their kinetics at 183°C.

The increasing need for high service temperature adhesives and structural matrix resins has led to the development of many new polymeric systems in recent years. One of the most interesting and potentially useful of these new polymers is polyimides. Polyimides are noted for their excellent thermal and mechanical properties but their utility has been severely limited due to problems with fabrication and processing of these polymers (1-3). Nevertheless, the careful design of polyimides can lead to enhanced processability. In this respect, several approaches have been investigated and found to be useful. One design method which has improved the processability of linear aromatic polyimides is the introduction of meta-substituted aromatic diamines for para substituted analogs (4,5). This procedure, while improving the processability, also has the possible detrimental effect of lowering the glass transition temperature. Another method which has been successfully utilized in improving polyimide processing and solubility characteristics is the incorporation of bulky side groups such as phenylated diamine monomers (6). Although these materials maintain a high glass transition temperature their resistance to solvents may be sacrificed. Processability can also

[1]Correspondence should be addressed to this author.

0097-6156/88/0364-0459$06.00/0

be improved by diluting the rigid imide functionality in the polymer chain through the use of block copolymerization with a flexible segment such as a siloxane (7-9). These approaches all rely on enhancing the thermoplasticity of the polyimides through incorporation of flexibilizing linkages. Finally, processability and fabrication aspects of polyimides have been improved through the use of low molecular weight imide oligomers terminated with acetylenic groups (10-11). The material (MC-600) shown in Figure 1 is an example of a commercially available product (National Starch and Chemical Co.). These materials have improved solubility and processing characteristics while maintaining both a high glass transition temperature and good solvent resistance due to their highly crosslinked nature following thermal cure of the acetylene groups. This approach also has problems in terms of processing parameters. Preliminary reactions of the terminal acetylene groups during thermal cure lead to a restriction of flow and wetting properties before good contact is achieved (12). In addition, since the glass transition temperature of these imide oligomers is quite high (~200°C) the crosslinking reaction proceeds very rapidly resulting in an infusible, rigid network. Once the glass transition temperature has been exceeded, enough mobility is available in the system for rapid crosslinking of the terminal acetylene groups. In this case, above 200°C (the Tg of the uncured polymer) this crosslinking reaction proceeds very rapidly. The gel time for MC-600 has been estimated at less than three minutes at a temperature of 250°C (13). This short gel time severely restricts the uses of the material in applications such as matrix resins and adhesives where good flow is necessary prior to the onset of gellation.

In order to circumvent this problem of rapid gellation an isomeric imide structure, termed isoimide, has been introduced into these systems (13,14). Materials with this functionality (i.e., IP-600) (Figure 2) exhibit improved solubility as well as longer gel times and lower glass transition temperatures (~160°C vs ~200°C for the corresponding imide oligomer). Initially, it was thought that the presence of the isoimide structure, as an unfavorable side reaction product in polyimides, led to premature thermal decomposition of polyimides through loss of CO_2 from the imino-lactone heterocyclic ring (15). However, later work (16) showed that the isoimide functionality thermally isomerized to the imide functionality (Figure 3) prior to any significant degradation of the polymer backbone.

Since the utility of these materials is improved by the incorporation of these reactive functionalities without severely decreasing other favorable properties such as thermooxidative stability and solvent resistance the chemistry of the isoimide isomerization and acetylene crosslinking reactions is of considerable interest. Previous work in our laboratory has shown that these materials, when loaded with metal powders, provide a convenient and effective method of optimizing the electrical conductance and thermal stability of aluminum conductor joints. The goal of this work is to elucidate the relationship between the thermal isomerization and crosslinking reactions occurring in this acetylene terminated polyisoimide oligomer: Thermid IP600. The

MC-600

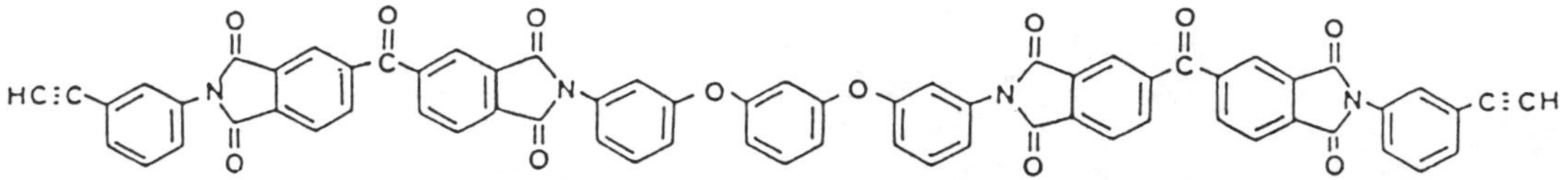

Figure 1. Structure of Thermid MC-600 Acetylene Terminated Imide Oligomer.

IP-600

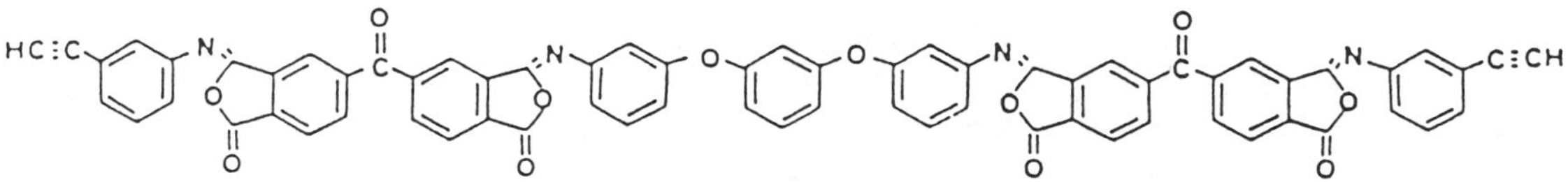

Figure 2. Structure of Thermid IP-600 Acetylene Terminated Isoimide Oligomer.

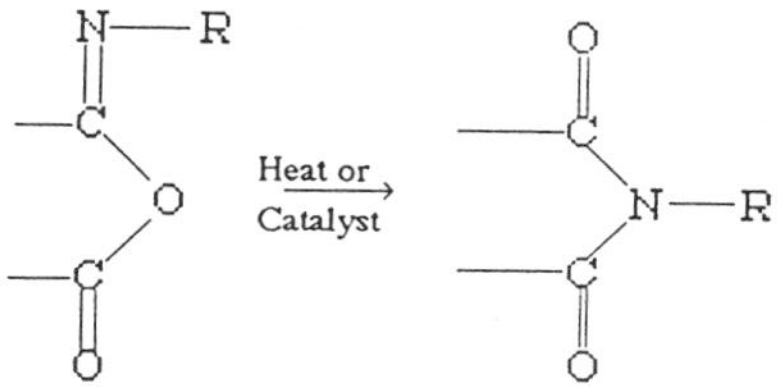

Heat or Catalyst

Figure 3. Isomerization Reaction of Isoimide to Imide.

techniques of Fourier Transform Infrared spectrometry and differential scanning calorimetry have been shown to be useful in determining the cure states of acetylene-terminated resins such as imides and sulfones (17-20). These techniques will be applied to the cure reactions occurring in the IP600 acetylene terminated isoimide system.

Experimental

Fourier Transform Infrared Spectrometry (FTIR) was performed on a Nicolet 6000 Infrared Spectrometer. Spectra were obtained at 2 cm^{-1} resolution and were an average of 10 scans. The samples were prepared by casting from a 5% (w/w) solution in THF a thin film onto either a KBr or KCl plate. The solvent was removed by vacuum drying at 60-80°C. The FTIR cure experiments were performed using a Barnes Model C019-020 heated cell. In order to more accurately monitor the temperature of the curing reactions at the KBr plate, holes were drilled into the side of the sample plates and a thermocouple was inserted. In this way the temperature of the crystal could be independently monitored. Differential Scanning Calorimetry (DSC) was performed on a Perkin Elmer DSC-4 at heating rates of 10, 5 and 2.5°C/min with a nitrogen purge. Thermogravimetry (TG) was performed in a Perkin Elmer TGS-2 thermogravimetric analyzer at 10°C/min in a dynamic air atmosphere at a purge rate of 50 cc/min. Thermid IP-600 was obtained from National Starch and Chemical Corporation, P.O. Box 6500, Bridgewater, NJ, 08807.

Results and Discussion

The incorporation of the isoimide functionality into acetylene terminated oligomers has been shown to result in enhanced solubility, a lower glass transition temperature and improved processability compared to the analogous imide oligomers (19). Chemical changes occurring in curing systems can be monitored in a variety of ways, two of the most common of which are DSC and FTIR. DSC provides an overall energetic profile of the reactions occurring in a given temperature range; while FTIR provides detailed structural information concerning the relative concentrations of functionalities present in the system. By establishing a reaction energy profile using DSC one can effectively choose the proper temperatures at which to conduct isothermal FTIR experiments. These isothermal FTIR experiments can then be used to establish the nature and kinetics of the chemical changes occurring in the system as a function of time at the given isothermal temperature.

The DSC thermograms of MC-600 and IP-600 are shown in Figure 4. It is apparent from these scans that the isoimide oligomer (IP-600) has not only a lower glass transition temperature but also a larger exotherm of reaction than the corresponding imide oligomer (~ -80 cal/g for IP-600 vs ~ -40 cal/g for MC-600). This observation along with the fact that these oligomers are of the same nominal molecular weight (~1100 amu) leads to the conclusion that the isomerization of the isoimide to the imide functionality is sufficiently exothermic to contribute significantly to the exotherm of the net reaction measured by DSC. Activation energies for these two samples were

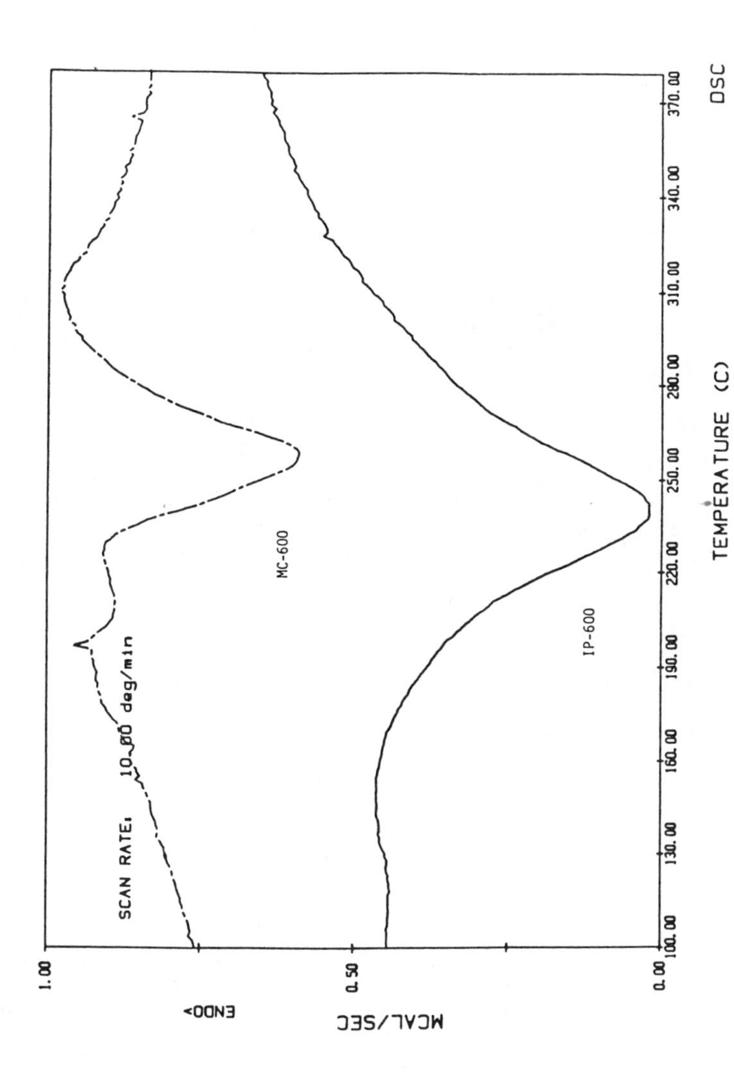

Figure 4. DSC Thermograms of MC-600 and IP-600 Oligomers.

calculated from the variation of peak reaction temperature with heating rate. The activation energies so calculated along with the equation used in this calculation are shown in Table I. The non-agreement of activation energies suggest a substantial difference

Table I. Calculated Activation Energies From DSC DATA*

Heating Rate:	10-5°/min	5-2.5°/min	10-2.5°/min
IP-600	14.1	13.1	13.6
MC-600	20.4	19.1	19.5

Method of Calculation:

$$E_a \quad \frac{-R}{1.052} \quad \frac{\Delta \ln \phi}{\Delta (1/T_p)}$$

*Kcal/mole

ϕ = Heating Rate (°K/min)

T_p = Peak Reaction Temperature (°K)

in reaction pathways for these two samples. The method chosen for extracting activation energies is expected to be more accurate than analysis of a single scan for situations where two overlapping peaks occur in the DSC scan (21). The activation energy calculated for the imide oligomer, MC-600, is 19.7 ± 0.7 Kcal/mole which agrees with values obtained for similar model systems such as sulfones (24.2 ± 0.7 Kcal/mole) (22) and phenoxy phenyl acetylene (23.2 ± 1 Kcal/mole) (23). The activation energy for the isoimide oligomer is substantially lower (13.6 ± 0.5 Kcal/mole). This result is consistent with the lower initial reaction temperature of the isoimide oligomer and also suggests that the isomerization reaction may precede the crosslinking reaction to some extent.

In order to better characterize the kinetics of these two processes (i.e., isomerization and crosslinking) in the isoimide oligomer, heated cell FTIR experiments were conducted. The infrared spectrum of uncured IP-600 isoimide oligomer is shown in Figure 5. This spectrum shows characteristic absorbances for the acetylene functionality at 3295 cm^{-1} and 940 cm^{-1} (20) and for the isoimide functionality at 1805 cm^{-1} and 930 cm^{-1} (24-25). A small amount of imide is also evident by the presence of a shoulder at 1725 cm^{-1}. To minimize the amount of interference due to adjacent bands we chose the 3295 cm^{-1} band for monitoring the concentration of acetylene functionality and the 1805 cm^{-1} band for monitoring the concentration of isoimide functionality. Some interference was still observed in the case of the isoimide band due to a 1775 cm^{-1} band assignable to the imide structure. A sample was placed in the cell and the cell was heated to a temperature of 183°C. Infrared spectra of the sample were then recorded as a function of time at this temperature. Figure 6 shows the isoimide-imide region of the spectrum as a function of time. It can be seen that the isoimide peak decreases rapidly during the first six minutes of the

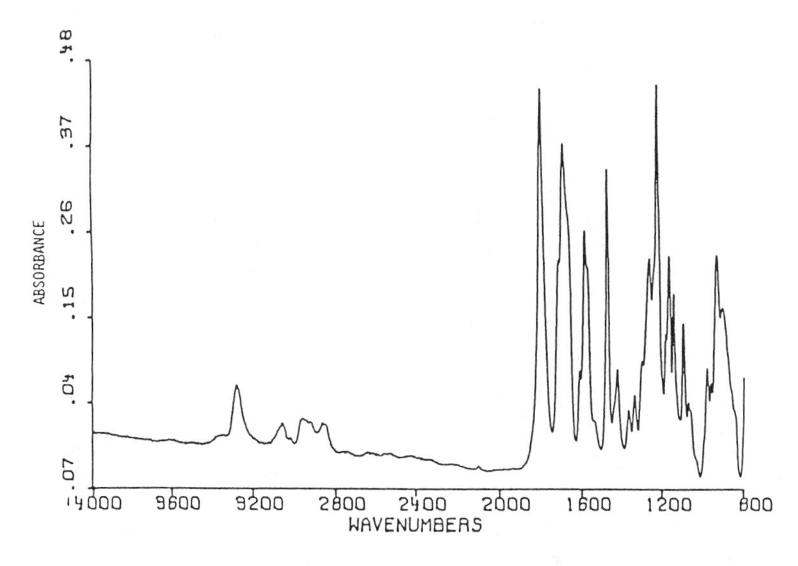

Figure 5. Infrared Spectrum of Uncured IP-600 Isoimide Oligomer.

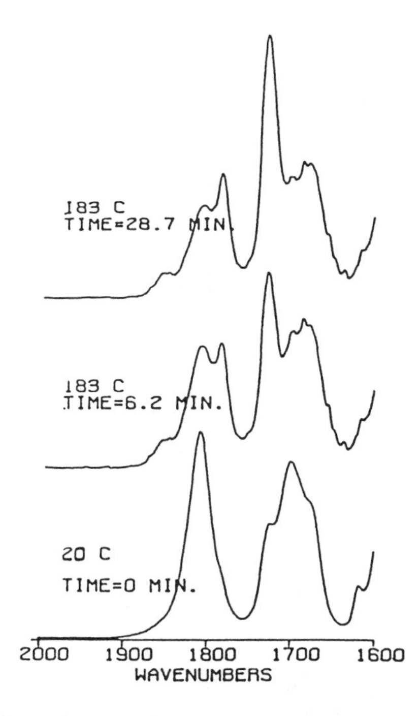

Figure 6. Isoimide-Imide Region of Infrared Spectrum of IP-600 as a Function of Time at 183°C.

experiment (the amount of time necessary for the 183°C isothermal temperature to be reached) and then continues to decrease at a steady rate. Figure 7 shows the acetylene region of the spectrum at the same times. Here it appears that the decrease in absorbance is not as rapid as in the case of the isoimide band.

Figures 8 and 9 show the first order kinetic plots for the isomerization and crosslinking reactions, respectively. In the data analysis the area of the isoimide peak was measured between consistent limits chosen to exclude any contribution from the 1775 cm^{-1} imide band. These data were generated by measuring the area of the appropriate peak in a baseline corrected spectrum and ratioing this area to that of a reference peak (which was invarient during the experiment) in the same spectrum. This concentration indicative number was then ratioed to the concentration ratio observed on the initial scan. Plots of the log of the ratio of the concentration of the functionality at time "t" to the concentration of the functionality at t = 0 were then constructed. In order to insure that the trends in the data were not artifacts of this procedure or of the baseline correction routine, we also plotted the data in terms of peak intensity in absorbance units and observed the same trends but with more scatter in the data.

The first order plot for the isomerization reaction shows a good linear fit (correlation coefficient = 0.998); while there is

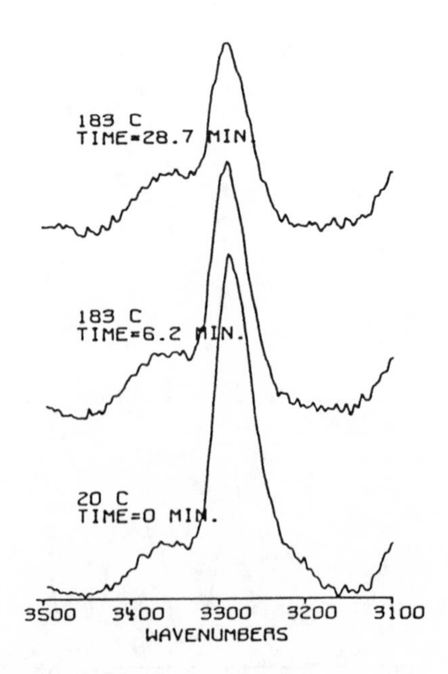

Figure 7. Acetylene Region of Infrared Spectrum of IP-600 as a Function of Time at 183°C.

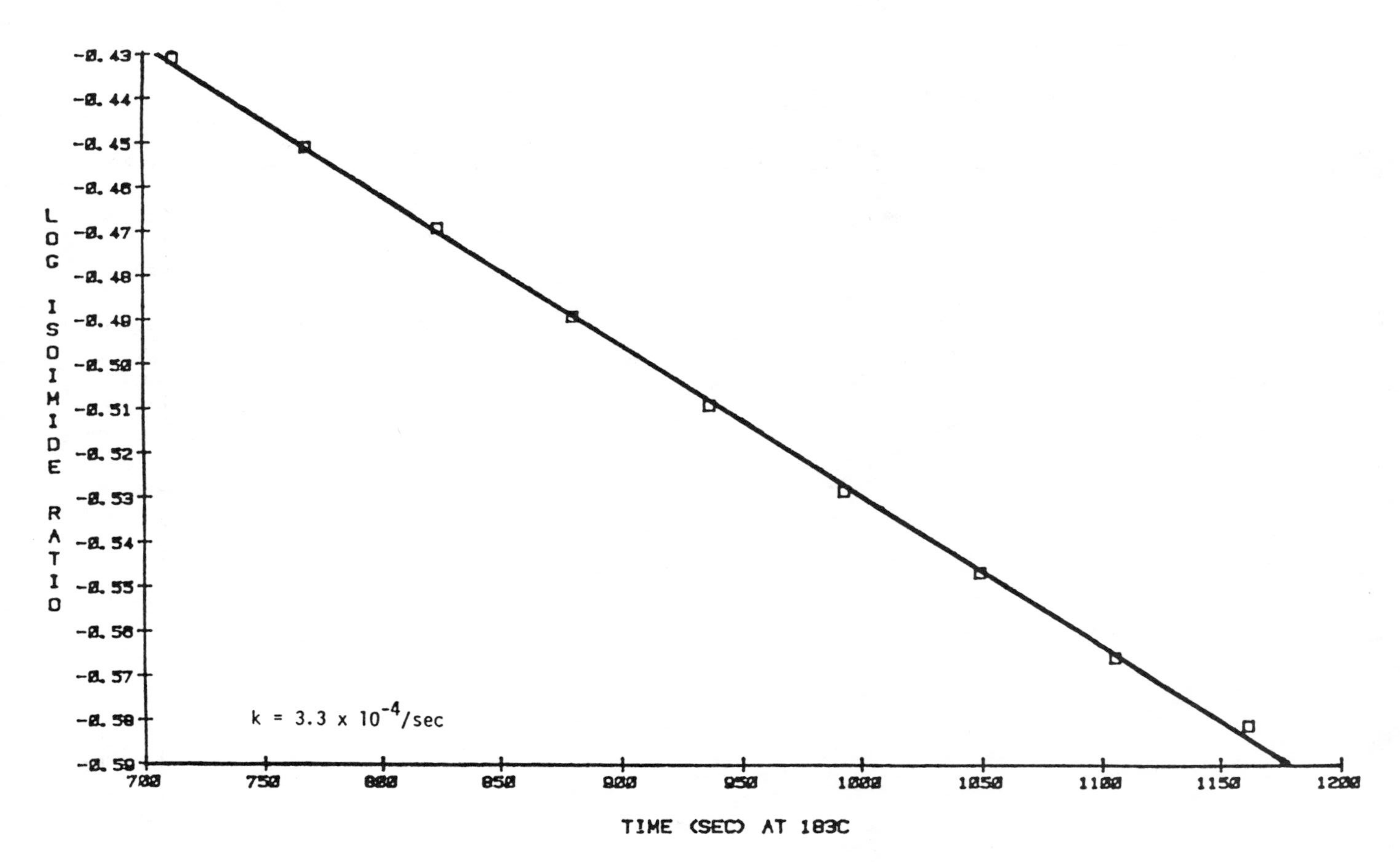

Figure 8. First Order Plot of Isomerization Reaction at 183°C.

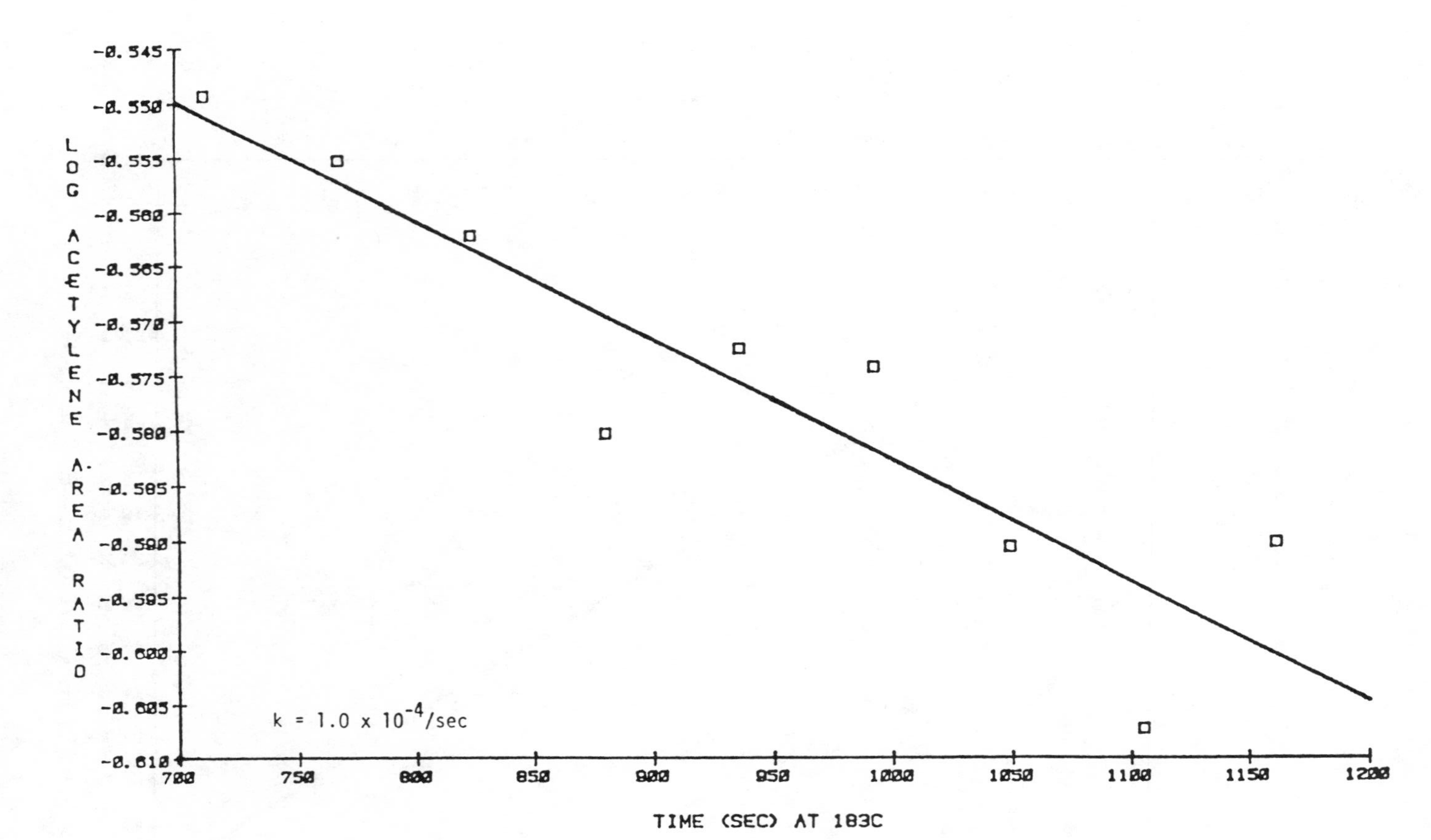

Figure 9. First Order Plot of Acetylene Reaction at 183°C.

more scatter in the acetylene first order plot. The apparent rate constants calculated from these data are 3.3×10^{-4}/sec for the isomerization of the isoimide to the imide and 1.0×10^{-4}/sec for the loss of the acetylene group. These are apparent rate constants since this first order fit does not take into account changes in mobility which undoubtedly occur during these processes nor is it established that either of these processes is uniquely first order in nature. We can, however, interpret the trend in the apparent rate constants in terms of the isomerization reaction proceeding more rapidly than the crosslinking reaction of the acetylene group.

Future work in this area will involve the extension of these techniques to other temperatures in an effort to better characterize the overall reaction kinetics of these two processes. In addition, degree of cure obtained through isothermal DSC measurements will be compared with the fraction of acetylene consumed as measured by isothermal FTIR experiments for the same temperature and time. Also, the effect of the incorporation of metal fillers on the isomerization and crosslinking reactions will be addressed.

Acknowledgments

The authors gratefully acknowledge ALCOA Corporation and the Virginia Center for Innovative Technology for sponsoring this research. Also, the assistance of John Hellgeth in conducting the FTIR experiments is greatly appreciated.

Literature Cited

1. Burks, H. D.; St. Clair, T. L. "Polyimides: Synthesis, Characterization, and Applications", Vol. 1, K. L. Mittal, ed., Plenum, NY, 1984, pp. 117-135.
2. Young, P. R.; Wakely, N. T. Proceedings from the Second International Conference on Polyimides, Ellenville, NY, 1985, pp. 414-425.
3. Critchley, J. P.; White, M. J. Polym. Sci., Polym. Chem. Ed., 10, 1809 (1972).
4. Bell, V. L., U. S. Patent 4,094,862, June 13, 1978.
5. St. Clair, A. K.; St. Clair, T. L.; Smith, E. N. Polymer Preprints, 17, 359 (1976).
6. Harris, F. W.; Norris, S. O; Lanier, L. H.; Reinhardt, B. A.; Case, R. D. Varaprath, S.; lPakaki, S. M.; Torres, M.; Feld, W. A. "Polyimides: Synthesis, Characterization and Applications", Vol. 1., K. L. Mittal, ed., Plenum, NY, 1984, pp. 3-14.
7. Maudgal, S.; St. Clair, T. L. Int. J. Adhesion and Adhesives, 4, 87 (1984).
8. Maudgal, S.; St. Clair, T. L. Proceedings from the Second International Conference on Polyimides, Ellenville, NY, 1985, pp. 47-73.
9. Summers, J. D.; Arnold, C. A.; Bott, R. H.; Taylor, L. T.; Ward, T. C.; McGrath, J. E. Polymer Preprints, 27, 403 (1986).
10. Landis, A. L.; Miller L. J. U. S. Pat. 3,845,018 (Oct. 29, 1974) N. Bilow, (Hughes Aircraft Co.)

11. Landis, A. L.; Bilow, M.; Boscham, R. H.; Lawrence, B. E.; Aponyi, T. J. Polymer Preprints, 19, 23 (1978).
12. Hergenrother, P. M. Encyclopedia of Polymer Sci. and Eng., Vol. 1, 61(1985), John Wiley and Sons.
13. Landis, A. L.; Naselow, A. B. "Polyimides: Synthesis, Characterization and Applicatons", Vol. 1, Mittal, K. L., Ed., Plenum, NY, 1984, pp. 39-49.
14. Landis, A. L.; Naselow, A. B. Natl. SAMPE Tech. Conf. Ser. 14, 236 (1982).
15. Gay, F. P.; Berr, C. E. J. Polym. Sci., A-1, 6, 1935 (1968).
16. Zurakowska-Orszagh, J.; Chreptowlez, T.; Orteszko, A.; Kaminski, J. European Polymer Journal, 15, 409 (1978).
17. Eddy, L. T. R.; Lucarelli, A. M.; Helminiak, T.; Jones, W.; Picklesimer, L. F. An Evaluation of An Acetylene Terminated Sulfone Oligomer, Internal Report, A FWAL/MLBC, January 1983.
18. Landis, A. L. Chemistry of Procesible Acetylene-Terminated Imides, Final Report, AFWAL/ML Contract F33615-82-C-5016, August 1983.
19. Koenig, J. L. Spectroscopic Characterization of the Cured State of An Acetylene-Terminated Resin, Final Report, AFWAL/ML Contract F33615-82-K-5038, August 1983.
20. Lind, A. C.; Saundreczki, T. C.; Levy, R. L. Characterization of Acetylene Terminated Resin Cures States, Interim Technical Report, AFWAL/ML Contract F33615-80-C-5170, August 1984.
21. Prime, R. B. Thermal Characterization of Polymeric Materials, Turi, E. A., Ed., Academic Press, NY, 1981, pp. 435-562.
22. Pickard, J. M.; Jones, E. G.; Goldfarb, I. J. Polymer Preprints, 20 (2), 370 (1979).
23. Pickard, J. M.; Jones, E. G.; Goldfarb, I. J. Macromolecules, 12 (5), 895 (1979).
24. Cotter, R. J.; Sauers, C. K.; Whelan, J. M. J. Org. Chem., 26, 10 (1961).
25. Roderick, W. R.; Bhatia, P. L. J. Org. Chem., 28, 2018 (1963).

RECEIVED September 18, 1987

INDEXES

Author Index

Affiliation Index

Subject Index

Production and Indexing by Keith B. Belton and Linda R. Ross
Jacket design by Carla L. Clemens

Elements typeset by Hot Type Ltd., Washington, DC
Printed and bound by Maple Press, York, PA

Recent Books

Personal Computers for Scientists: A Byte at a Time
By Glenn I. Ouchi
276 pp; clothbound; ISBN 0–8412–1000–4

The ACS Style Guide: A Manual for Authors and Editors
Edited by Janet S. Dodd
264 pp; clothbound; ISBN 0–8412–0917–0

Silent Spring Revisited
Edited by Gino J. Marco, Robert M. Hollingworth, and William Durham
214 pp; clothbound; ISBN 0–8412–0980–4

Chemical Demonstrations: A Sourcebook for Teachers
By Lee R. Summerlin and James L. Ealy, Jr.
192 pp; spiral bound; ISBN 0–8412–0923–5

Phosphorus Chemistry in Everyday Living, Second Edition
By Arthur D. F. Toy and Edward N. Walsh
362 pp; clothbound; ISBN 0–8412–1002–0

Pharmacokinetics: Processes and Mathematics
By Peter G. Welling
ACS Monograph 185; 290 pp; ISBN 0–8412–0967–7

Synthesis and Chemistry of Agrochemicals
Edited by Don R. Baker, Joseph G. Fenyes, William K. Moberg, and Barrington Cross
ACS Symposium Series 355; 474 pp; 0–8412–1434–4

Nutritional Bioavailability of Manganese
Edited by Constance Kies
ACS Symposium Series 354; 155 pp; 0–8412–1433–6

Supercomputer Research in Chemistry and Chemical Engineering
Edited by Klavs F. Jensen and Donald G. Truhlar
ACS Symposium Series 353; 436 pp; 0–8412–1430–1

Sources and Fates of Aquatic Pollutants
Edited by Ronald A. Hites and S. J. Eisenreich
Advances in Chemistry Series 216; 558 pp; ISBN 0–8412–0983–9

Nucleophilicity
Edited by J. Milton Harris and Samuel P. McManus
Advances in Chemistry Series 215; 494 pp; ISBN 0–8412–0952–9

For further information and a free catalog of ACS books, contact:
American Chemical Society
Distribution Office, Department 225
1155 16th Street, NW, Washington, DC 20036
Telephone 800-227-5558